Chemical Processes for Environmental Engineering

Chemical Processes for Environmental Engineering

Teh Fu Yen

University of Southern California, USA

ICP Imperial College Press

Published by

Imperial College Press
57 Shelton Street
Covent Garden
London WC2H 9HE

Distributed by

World Scientific Publishing Co. Pte. Ltd.
5 Toh Tuck Link, Singapore 596224
USA office: 27 Warren Street, Suite 401-402, Hackensack, NJ 07601
UK office: 57 Shelton Street, Covent Garden, London WC2H 9HE

British Library Cataloguing-in-Publication Data
A catalogue record for this book is available from the British Library.

CHEMICAL PROCESSES FOR ENVIRONMENTAL ENGINEERING

ISBN-13 978-1-86094-759-9
ISBN-10 1-86094-759-X

Printed in Singapore by B & JO Enterprise

This book is dedicated to my wife Shiao Ping Siao Yen, Ph. D.,

of the Jet Propulsion Laboratory,

California Institute of Technology,

for her support, understanding, and sacrifice.

FOREWORD

Although industrialization does bring about an improvement in some physical comforts for mankind, it also brings with it rapid population growth and resulting environmental degradation. Human beings are thus faced with ten serious worldwide environmental problems: air pollution, destruction of the ozone layer, global warming, marine pollution, groundwater contamination, desertification, disappearance of forests, extinction of species, toxic chemicals/hazardous wastes, and the worsening of public health. Most of these global problems are directly related to chemistry.

Environmental engineers have been working hard to fix the aforementioned problems; however, sometimes what seems like a feasible solution can result in negative side affects that lead to more harm than good. This has led to a trend in recent years toward green chemistry or green technology and the development of sustainable technology to curtail pollution.

As we move deeper into the new millennium, the aim of environmental engineering is constantly evolving. Treatment processes are no longer limited to wastewater and polluted air. Pollutants appear not only in a single sphere such as the hydrosphere, but also in the biosphere and pedosphere. To cope with these new conditions, environmental engineering must now take on a broader multidisciplinary approach, one which is not merely regional, but international in scope. Scientists must thus have a sufficient background in multiple scientific fields in order to function in today's multifaceted network.

Of all the processes for treatment and technology involved in environmental practice, the chemical processes are the most important because they are involved in all different steps of chemical treatment. Thus, if we can understand the processes discussed in this book, we will be better qualified to design a more efficient system to solve environmental problems. As our analytical techniques improve, new species will emerge as pollutants that will require the development of new techniques. Just as current techniques rely on knowledge of unit process and unit operations developed in the old fields of sanitary and chemical engineering, future solutions will rely on a basic foundation of processes described in this book. Thus, more sophisticated tools and procedures will be combined with present knowledge to design solutions of the future.

PREFACE

Chemists and chemical engineers both in training and at work will find this text quite useful. It provides comprehensive information about how chemistry applies to environmental practice, so that it will be easy for chemists and chemical engineers to become environmental engineers. Professional environmental and civil engineers can also use this volume as a constant companion for easy reference.

In addition to environmental sustainability, policy-making and the political infrastructure make up another crucial element in creating a better environment. As well as providing instruction for professional environmental engineers, this book also provides ample information for non-science, non-engineering professionals, such as political officials, legal professionals, and journalists. They will be able to use information contained in these pages to develop better, more informed decision-making.

In order to tackle environmentally related chemical problems, scientists must have a grasp of both chemistry and environmental engineering. This volume includes introductions to fundamental engineering concepts such as fluid mechanics, hydraulics, mass transfer, and diffusion phenomena are blended into the scientific segments. We have purposely omitted the design and implementation of processes, since these can be found in other excellent sources.

In every chapter, mathematical expressions are used to rationalize certain thoughts and ideas. The level of mathematics needed for this volume extends to calculus and differential equations, which are included in the Bachelor of Science requirements of most colleges and universities. This volume can also serve as a refresher before a modeling course is taught.

Both Metric and English units are used throughout the text. Often, the choice depends upon the preference of the user. For conversion of a given system, both the popular and expert usages are listed; those that are defined by the International Union of Pure and Applied Chemistry (IUPAC) are recommended.

To economize time and space, the secondary lists of authors' references in many tables are eliminated. However, the names of the authors are maintained for the user's convenience, in order to locate the references through the available computer database. Some portions of this book have been used in Environmental Chemistry published by Prentice Hall in two volumes. Permission has been given to use that material for the present version with the appropriate corrections and additions. These volumes have been used to teach the basics of chemical processes and environmental chemistry to students in universities for close to twenty years.

The author would like to take this opportunity to thank all of the colleagues and previous students, too many to mention, who have taken the course and who have contributed to shaping this book. He would also like to acknowledge the scientists who have summarized work done in the past, without which the present volume would not be possible. Lastly, the author would like to name Ngo Yeung Chan, Zixuan Chen, Vivek Gupta, Kinjal Modi, Abhay Nadkarni, Arpitha Shetty, Karan Singh and Rachel Tucker for their assistance.

Teh Fu Yen

CONTENTS

CHAPTER 1

CHEMICAL EQUILIBRIUM

Impurities in natural water or wastewater include chemical species consisting of both organic and inorganic compounds — compounds that are either dissolved or suspended. Thermodynamic concepts are applied to these systems when either the entire or parts of the system are in thermodynamic equilibrium*. In this manner, we can characterize the composition and predict the change of direction under certain conditions.

In this chapter we will first introduce the free energy of a real system. We will discuss the concept of activity and the activity coefficient with organic compounds in an aqueous system. The consideration of ionic strength in an aqueous system and its computation will also be explored.

* Review thermodynamics in any physical chemistry textbook, for e.g. T. F. Yen, ***Chemistry for Engineers*** to be published by Imperial College Press.

Deviation from the ideal system will be assessed in light of the Debye-Hückel theory. In particular, the deviation of the Debye-Hückel limiting law for most salts will be shown step by step. The approximation of activity coefficients by different methods will be compiled. Finally, the calculation of equilibrium constants from thermodynamic quantities will be demonstrated.

1.1 IDEAL VERSUS REAL SYSTEM

In a typical chemical system, we encounter both the flow of mass and the flow of energy. In thermodynamics we consider a closed chemical system, one in which we are limited to the exchange of energy. There are three postulates for free energy. First, the free energy of a system is the sum of free energies of its components, so that:

$$G = \sum n_i u_i \qquad [1\text{-}1]$$

where μ_i is chemical potential of the *i*th component, which is related to the free energy of the system.

$$\mu_i = \left[\frac{\partial G}{\partial n_i}\right]_{T,P,n_1,\cdots,n_i} \qquad [1\text{-}2]$$

The second postulate is that the equilibrium of a system is given at the lowest possible energy state. The last postulate maintains that the free energy of a system is a state property, namely the energy given to a reference state. For example,

$$\mu_i = \mu_i^{ref} + RT \ln \frac{x_i}{x_i^{ref}} \qquad [1\text{-}3a]$$

or

$$\mu_i - \mu_i^{ref} = RT \ln x_i - RT \ln x_i^{ref}$$

which yields:

$$\mu_i = \mu_i^0 + RT \ln x_i \qquad [1\text{-}3]$$

where

$$\mu_i^0 = \mu_i^{ref} - RT \ln x_i^{ref}$$

in which x_i is the mole fraction concentration of the ith component.

For an ideal gas, the free energy[†] in the expansion is:

$$dG = VdP - SdT$$

For the isothermal process, the last term is dropped.

$$dG_T = VdP$$

Substituting the ideal gas law, $PV = nRT$, we find that:

$$dG = nRT\, d \ln P \quad (T = \text{constant}) \qquad [1\text{-}4a]$$

After integrating from P' to P,

$$\Delta G = G - G' = nRT \ln\left(\frac{P}{P'}\right) \qquad [1\text{-}4]$$

From the first law of thermodynamics, the reversible work done for $dE = 0$ is:

$$dQ = dW = PdV$$

Thus,

$$W_{rev} = Q = nRT \int_{V'}^{V} V^{-1} dV = nRT \ln\left(\frac{V}{V'}\right) = nRT \ln\left(\frac{P'}{P}\right) \qquad [1\text{-}5]$$

In this manner,

[†] For thermodynamic functions, see Appendix A.

$$\Delta G = -W_{rev} \qquad [1\text{-}6]$$

The significance of both ΔG and W_{rev} can be seen in Figure 1-1, where the regular hyperbola can be represented by an ideal gas through Boyle's law:

$$W_{rev} = \int_{V_1}^{V_2} PdV = \text{Area } ABCD \qquad [1\text{-}6a]$$

$$\Delta G = \int_{P_1}^{P_2} VdP = \text{Area } ABEF \qquad [1\text{-}6b]$$

However, the above equation for free energy (Eq. [1-6b]) does not hold true for real gas. The indicated integration may be carried out by substituting for V in terms of P or for dP in terms of V and dV, provided the equation of its state is known. In general, for real gas, an idealized pressure termed **fugacity**, f, is introduced (compare with Eq. [1-4a]).

$$dG = nRT\, d(\ln f) \qquad [1\text{-}7]$$

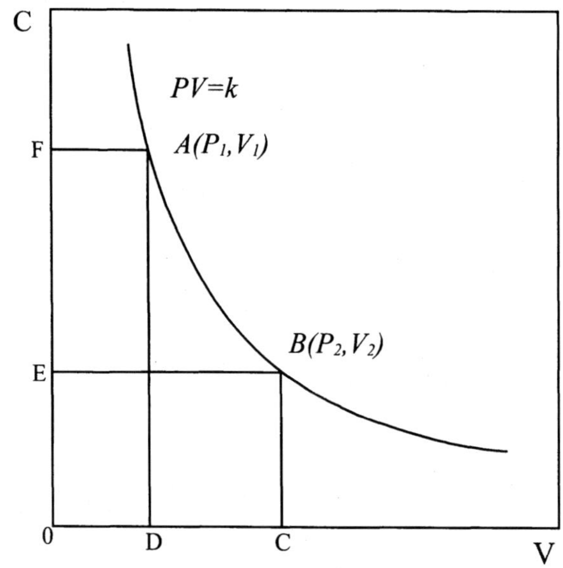

Figure 1-1. Illustration of ΔG and W_{rev} for isothermal expansion of an ideal gas.

or

$$\Delta G\,T = G - G' = nRT \ln \frac{f}{f'} \qquad [1\text{-}8]$$

and

$$\lim_{P \to 0} \frac{f}{P} = \frac{f^*}{P^*} = 1 \qquad [1\text{-}9]$$

For the gaseous state, fugacity is the pressure of the gaseous state at zero pressure or $f \to P$ when $P = 0$.

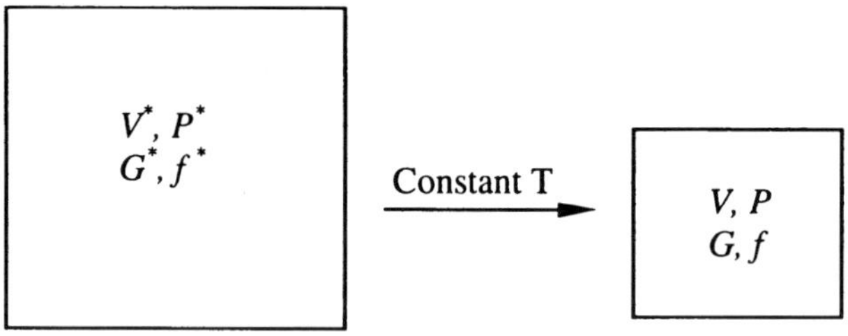

Figure 1-2. Calculation of activity coefficients for gases.

Consider an isothermal system in which one mole of gas, originally at an extremely low pressure in a large volume, changes to a higher pressure and correspondingly smaller volume, as illustrated in Figure 1-2. In the first state, the pressure, volume, free energy, and fugacity are designated as P^*, V^*, and f^*, respectively. For the real gas,

$$G - G^* = RT \ln\left(\frac{f}{f^*}\right) = \int_{P^*}^{P} V\,dP \qquad [1\text{-}10]$$

and for the ideal gas,

$$G_i - G_i^* = RT\ln\left(\frac{P}{P^*}\right) = \int_{P^*}^{P} V_i dP \qquad [1\text{-}11]$$

Subtraction of Equation [1-11] from Equation [1-10] gives us:

$$RT\ln\frac{f}{P} = RT\ln\left(\frac{f^*}{P^*}\right) - \int_{P^*}^{P}(V_i - V)dP \qquad [1\text{-}12]$$

Since $f^*/P^* = 1$ and for the first state, $P^* \to 0$,

$$RT\ln\left(\frac{f}{P}\right) = \int_{P^*}^{P}(V_i - V)dP \qquad [1\text{-}12a]$$

Here, $V_i - V$ is the difference between the molar volume of the ideal gas and that of the real gas, also referred to as the volume residue or α. If we define the ratio f/P as the **activity coefficient** γ and

$$\gamma = \frac{f}{P} \qquad [1\text{-}13]$$

then

$$\ln\gamma = \ln f/P = -\frac{1}{RT}\int_{P^*}^{P}\alpha dP \qquad [1\text{-}14]$$

Experimentally, as P approaches zero, α approaches a finite value. Here we can evaluate fugacity or the activity coefficient by plotting α versus P, extrapolating α to zero pressure, and graphically integrating the area under the curve as shown in Figure 1-3.

For liquids and solids, a **relative fugacity** or **activity**, a, is introduced where

$$a = \frac{f}{f^0} \qquad [1\text{-}15]$$

The f used here denotes a set of given conditions, while f^0 is at a standard state of an arbitrary condition. If $f = f^0$, then $a = 1$; if $f^0 = 1$, then $a = f$, as in a gaseous state

$$G - G^0 = RT \ln \frac{f}{f^0} = RT \ln a \qquad [1\text{-}15a]$$

This relative fugacity or activity, a, is similar to fugacity, f, which has an **activity coefficient**, γ:

$$\gamma = a/x \qquad [1\text{-}16]$$

where x is the mole fraction concentration. Usually this can be expressed as molar concentration:

$$a/c = \gamma \qquad [1\text{-}16a]$$

or molar concentration:

$$a/m = \gamma \qquad [1\text{-}16b]$$

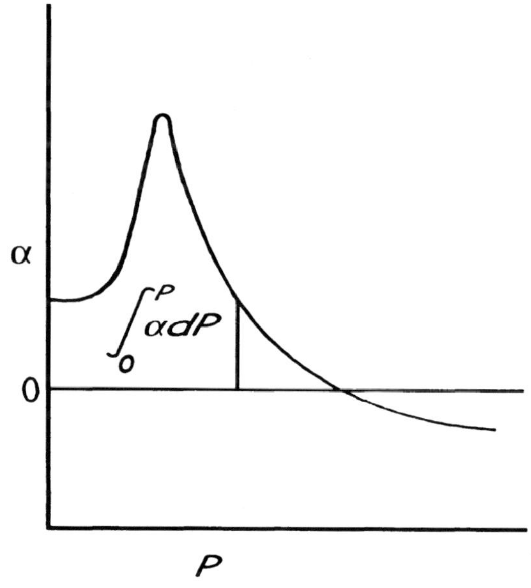

Figure 1-3. Determination of γ for gases from volume residue α.

In the ideal case of $\gamma = 1$ and $a = x$, it can also be expressed in γ_c and γ_m molar activity coefficient or molal activity coefficient.

In general, for an ideal system, the ideal gas law is valid, $PV = nRT$. A real system deviates from the ideal system, hence an activity coefficient, □, is developed so that the concentration in a real system can be converted to the "ideal" activity (in a thermodynamic sense) as:

$$a_i = \text{activity}$$

$$= □_i\, C_i$$

1.1.1 Activity Coefficient for a Binary System in Solution

When applied in solution the activity coefficient may be taken as a correction factor for nonideal behavior. Assuming that for two liquids at a given temperature, where the number of moles for the first component is n_1 and that for the second is n_2, the mole fractions of the two are, respectively,

$$x_1 = \frac{n_1}{(n_1 + n_2)} \qquad [1\text{-}17]$$

$$x_2 = \frac{n_2}{(n_1 + n_2)} \qquad [1\text{-}18]$$

If the vapor pressure of the pure liquids are P_1^0 and P_2^0, respectively, then the partial pressure for any mixture of component 1 is:

$$P_1 = x_1 P_1^0$$

Similarly,

$$P_2 = x_2 P_2^0 \qquad [1\text{-}19]$$

The above equations are for an ideal solution. In a real system, the two component liquids form a nonideal mixture. Thus,

$$P_1 = \gamma_1 x_1 P_1^0$$

$$P_2 = \gamma_2 x_2 P_2^0 \qquad [1\text{-}20]$$

Therefore, for a pure ideal system,

$$\gamma_1 = 1$$

$$\gamma_2 = 1$$

The real system will reduce to the ideal system. Also,

$$\gamma_1 x_1 = a_1$$

$$\gamma_2 x_2 = a_2 \qquad [1\text{-}21]$$

To dilute solution x_1 into x_2, as x_1 decreases γ_1 will increase, and $\gamma_1 > 1$ in a given solution. The value of γ_1 will increase in proportion to x_1's decrease, until reaching a certain limiting value, γ_1^∞. Therefore, the limiting value of x_1 (in this case it may be referred to as solubility of x_1 in x_2) is S_1.

$$(\gamma_1^\infty)^{-1} = x_1^0 = S_1$$

This concept is applicable to x_2 in x_1

$$(\gamma_2^\infty)^{-1} = x_2^0 = S_2 \qquad [1\text{-}22]$$

Therefore, we are able to approximate mutual solubility limits, Henry's law constants, and octanolwater partition coefficients for any pair of organic liquids, etc. In the case of Henry's law constants,

$$H = \frac{P_{vp1}}{S_1} \qquad [1\text{-}23]$$

where P_{vp1} is the vapor pressure in atmosphere of component 1 at system temperature 25°C and S_1 is solubility expressed in moles/m^3. Usually the mole fraction is multiplied by 55.5×10^3 if component 2 is water. Generally, the octanol-water partition coefficient K_{ow} is proportional to the ratio of activity coefficients of component A in water and in octanol. Thus,

$$K_{ow} = 0.151 \frac{\left(\gamma_{w}^{A}\right)^{\infty}}{\left(\gamma_{0}^{A}\right)^{\infty}} \qquad [1\text{-}24]$$

[Example 1-1] Estimate the solubility of benzene in water and water in benzene at 25°C.

Usually the infinite dilution activity coefficient, γ^{∞}, can be obtained by relating the molecular structures of the solute and solvent with an equation containing the number of carbon atoms for both species.

$$\log \gamma_1^{\infty} = A_{1,2} + B_2 \frac{N_1}{N_2} + \frac{C_1}{N_1} + D(N_1 - N_2)^2 + \frac{F_2}{N_2} \qquad [1\text{-}25]$$

where subscript 1 refers to solute, 2 to solvent, and

$A_{1,2}$ = coefficient that depends on the nature of solute and solvent functional groups

B_2 = coefficient that depends only on the nature of solvent functional group

C_1 = coefficient that depends only on the solute functional group

D = coefficient independent of the solute and solvent functional groups

F_2 = coefficient that essentially depends only on the nature of the solvent functional group

N_1, N_2 = number of carbon atoms in the solute and solvent, respectively

using Table 1-1 for an approximation and consulting the notes in the last column and at the end of the table for usage.

From the structure of benzene, $N_1 = 6$; for water, $N_2 = 0$. However, this does not lead to an infinite B_2 term because footnote f of Table 1-1 dictates the use of a modified B_2 term.

From Table 1-1 the correlation coefficients for alkylbenzenes in water are $A_{1,2} = 3.554$, $B_2 = 0.622$, and $C_1 = -0.466$. There are no D and F_2 terms. Under this modification, the equation for $\log \gamma_1^\infty$ becomes:

$$\log \gamma_1^\infty = A_{1,2} + B_2 (N_1 - 6) + \frac{C_1}{(N_1 - 4)}$$

$$\log \gamma_1^\infty = 3.544 + 0.622\,(6 - 6) - \frac{0.466}{(6-4)} = 3.321$$

γ_1^∞ = antilog (3.321) = 2090. The experimental value is 2400 for an error of −12.8%

Repeating Step 2 with water as solute and benzene as solvent, the correlation coefficients are $A_{1,2} = 3.04$ and $F_2 = -3.14$. The equation is used as is with $B_2 = C_1 = D = 0$.

$$\log \gamma_2^\infty = 3.04 - \frac{3.14}{6} = 2.517$$

where γ_2^∞ = antilog (2.517) = 330. The experimental value is 430 for an error of −23.5%.

For benzene in water, the solubility is inversely proportional to γ_1^∞ as $\gamma_1^\infty > 1000$; hence, $x_1 = 1/\gamma_1^\infty = 4.8 \times 10^{-4}$ mole fractions. The experimental value for x_1 is 4.2×10^{-4}. For water in benzene, $50 < \gamma_2^\infty < 1000$; hence, $x_2 = 3.5 \times 10^{-3}$ mole fraction. The experimental value for x_1 is 3.1×10^{-3}.

Table 1-1. Correlating Constants for Activity Copefficients at Infinite Dilution, Homologous Series of Solutes and Solvents.

Solute (1)	Solvent (2)	Temp (°C)	$A_{1,2}$	B_2	C_1	D	F_2	See Note
n-Acids	Water	25	-1.00	0.622	0.490	0	0	a
		50	-0.80	0.590	0.290	0	0	a
		100	-0.620	0.517	0.140	0		a
n-Primary alcohols	Water	25	-0.995	0.622	0.558	0		a
		60	-0.755	0.583	0.460	0		a
		100	-0.420	0.517	0.230	0		a
Secondary alcohols	Water	25	-1.220	0.622	0.170	0		b
		60	-1.023	0.583	0.252	0		b
		100	-0.870	0.517	0.400	0		b
Tertiary alcohols	Water	25	-1.740	0.622	0.170	0		c
		60	-1.477	0.583	0.252	0		c
		100	-1.291	0.517	0.400	0		c
Alcohols, general	Water	25	.0.525	0.622	0.475	0		d
		60	-0.33	0.583	0.39	0		d
		100	-0.15	0.517	0.34	0		d
n-Allyl alcohols	Water	25	-1.180	0.622	0.558	0		a
		60	-0.929	0.583	0.460	0		a
		100	-0.650	0.517	0.230	0		a
n-Aldehydes	Water	25	-0.750	0.622	0.320	0		a
		60	-0.400	0.583	0.210	0		a
		100	-0.3	0.517	0	0		a

Table 1-1. Continued

Solute (1)	Solvent (2)	Temp (°C)	$A_{1,2}$	B_2	C_1	D	F_2	See Note
n-Alkenene Aldehydes	Water	25	-0.720	0.622	0.320	0		a
		60	-0.540	0.593	0.210	0		a
		100	-0.298	0.517	0	0		a
n-Ketones	Water	25	-1.475	0.622	0.500	0		b
		60	-1.040	0.583	0.330	0		b
		100	-0.621	0.517	0.200	0		b
n-Acetals	Water	25	-2.556	0.622	0.486	0		e
		60	-2.184	0.583	0.451	0		e
		100	-1.780	0.517	0.426	0		e
n-Ethers	Water	20	-0.770	0.640	0.195	0		b
n-Nitriles	Water	25	-0.587	0.622	0.760	0		a
		60	-0.368	0.583	0.413	0		a
		100	.0.095	0.517	0	0		a
n-Alkene nitriles	Water	25	-0.520	0.622	0.760	0		a
		60	-0.323	0.583	0.413	0		a
		100	-0.074	0.517	0	0		a
n-Esters	Water	20	-0.930	0.640	0.260	0		b
n-Formates	Water	20	-0.585	0.640	0.260	0		a
n-Monoalkyl chlorices	Water	20	-1.265	0.640	0.073	0		a
n-Paraffins	Water	16	0.688	0.642	0	0		a
n-Alkylbenzenes	Water	25	3.554	0.622	-0.466	0		f
Alcohols, general	Paraffins	25	1.960	0	0.475	-0.00049		d
		60	1.460	0	0.390	-0.00057		d
		100	1.070	0	0.340	-0.00061		d

Table 1-1. Continued

Solute (1)	Solvent (2)	Temp (°C)	$A_{1,2}$	B_2	C_1	D	F_2	See Note
n-Ketones	Paraffins	25	0.0877	0	0.757	-0.00049		
		60	0.016	0	0.680	-0.00057		
		100	-0.067	0	0605	-0.00061		
Water	*n*-Alcohols	25	0.760	0	0	0	-0.630	
		60	0.680	0	0	0	-0.440	
		100	0.617	0	0	0	-0.280	
Water	Sec-Alcohols	80	1.208	0	0	0	-0.690	
Water	*n*-Ketones	25	1.857	0	0	0	-1.019	
		60	1.493	0	0	0	-0.73	
		100	1.231	0	0	0	-0.557	
Ketones	*n*-Alcohols	25	-0.088	0.176	050	-0.00049	-0.630	
		60	-0.035	0.138	0.33	-0.00057	-0.440	
		100	-0.035	0.112	0.20	-0.00061	-0.280	
Aldehydes	*n*-Alcohols	25	-0.701	0.176	0.320	-0.00049	-0.630	
		60	-0.239	0.138	0.210	-0.00057	-0.440	
Esters	*n*-Alcohols	25	0.212	0.176	0.260	-0.00049	-0.630	
		60	0.055	0.138	0.240	-0.00057	-0.440	
		100	0	0.112	0.220	-0.00061	-0.630	
Acetals	*n*-Alcohols	60	-1.10	0.138	0.451	-0.00057	-0.440	
Paraffins	Ketones	25	0	0.1821	0	-0.00049	0.402	
		60	0	0.1145	0	-0.00057	0.402	
		90	0	0.0746	0	-0.00061	0.402	

Table 1.1. Continued

Solute (1)	Solvent (2)	Temp (°C)	$A_{1,2}$	B_2	C_1	D	F_2	See Note
Paraffins	*n*-alcohols	25	0.87	0.176	0	-0.00049	-0.630	
		60	0.80	0.138	0	-0.00057	-0.440	
		100	0.72	0.112	0	-0.00061	-0.280	
Water	Paraffins	25	2.55	0	0	0	3.38	
Water	*n*-Alkylbenzenes	25	3.04	0	0	0	-3.14	

Note: Modification of terms in Equation [1-25]

a. B_2N_1 b. B_2N_1, $C_1\left(\frac{1}{N_1'}+\frac{1}{N_1''}\right)$ c. B_2N_1, $C_1\left(\frac{1}{N_1'}+\frac{1}{N_1''}+\frac{1}{N_1'''}\right)$ $F_2\left(\frac{1}{N_2'}+\frac{1}{N_2''}\right)$

d. B_2N_1, $C_1\left(\frac{1}{N_1'}+\frac{1}{N_1''}+\frac{1}{N_1'''}-3\right)$ c. B_2N_1, $C_1\left(\frac{1}{N_1'}+\frac{1}{N_1''}+\frac{2}{N_1'''}\right)$; N_1' relates to R''' in R' (R'') C (OR'),

f. $B_2(N_1-6)$, $C_1\left(\frac{1}{N_1-4}\right)$ g. $C_1\left(\frac{1}{N_1'}+\frac{1}{N_1''}\right)$

h. $C_1\left(\frac{1}{N_1'}+\frac{1}{N_1''}+\frac{2}{N_1'''}\right)$; N_1''' relates to K'' in R' (R'') C (OR'''), i. $P_2\left(\frac{}{N_2'}+\frac{1}{N_2''}\right)$

N1, N2 = total number of carbon atoms in molecules 1 and 2, respectively.

N', N'', N''' = number of carbon atoms in respective branches of branched compounds, counting the polar grouping; thus, for t-butanol, N' = N'' = N''' = 2

Source: Pierotti, Deal, and Derr as modified by Reid, Prausnitz, and Sherwood. This table is modified from Table 11-4 of W. J. Lyman, W.F. Reehl and D. H. Rosenblatt, *Handbook of Chemical Property Estimation Methods*, McGraw-Hill, New York, 1982

[Example 1-2] Find Henry's law constant for benzene in water at 25°C.

From Equation 1-23,

$$H = \frac{P_{vp1}}{S} \tag{a}$$

where P_{vp1} is the vapor pressure of component 1 at 25°C and S is the solubility expressed in M/m^3 at 25°C. From the Antonine equation,

$$\ln P_{vp} = \frac{A_2 - B_2}{(T - C_2)} \tag{b}$$

$$A_2 = \frac{B_2}{(T_b - C_2)}$$

and

$$B_2 = \frac{\Delta H_{vb}}{\Delta Z_b R T_b^2}(T_b - C_2) \tag{c}$$

After substitution,

$$\ln P_{vp} = \frac{\Delta H_{vb}(T_b - C_2)^2}{\Delta Z_b R T_b^2}\left[\frac{1}{(T_b - C_2)} - \frac{1}{(T - C_2)}\right] \tag{d}$$

T_b for benzene is 353.1K (boiling point).

$$\Delta Z_b = 0.97$$

$$C_2 = -18 + 0.19\, T_b$$

$$\frac{\Delta H_{vb}}{T_b} = \Delta S_{vb} = K_F\left(8.75 + R\ln T_b\right) \tag{e}$$

where ΔH_{vb} is the heat of vaporization at boiling constant and K_F = 1.00. Substituting for T_b and other values,

$$\frac{\Delta H_{vb}}{T_b} = 20.4 \text{ cal/M.K}$$

$$C_2 = 49.09$$

and

$$\ln P_{vp} = -2.02$$

Thus,

$$P_{vp} = \text{antilog } (-2.24) \times 760 = 100.8 \text{ mm Hg}$$

From Example 1-1,

$$S = 4.8 \times 10^{-4} \times 55.5 \times 10^3 = 26.6 \text{ M/m}^3$$

Thus,

$$H = \frac{\frac{100.8}{760}}{26.6} = 5.0 \times 10^{-3} \text{ atm·m}^3\text{/M}$$

The experimental value is 5.5×10^{-3} atm·m^3/M.

1.2 Electrolyte and Ionic Strength

Electrovalent compounds, such as inorganic salts or bases, dissociate into relative free positive and negative ions in a solvent of high dielectric constant. For example,

$$KCl_{(s)} = K^+ + Cl^- \text{ (solution)}$$

In solvents of low dielectric constant, a large fraction of the oppositely charged solute ions may form ion pairs without any net charge. A covalent solute may react with the solvent to form positive and negative ions. For example,

$$CH_3COOH + H_2O = H_3O^+ + CH_3COO^-$$

For a dilute solution, the osmotic pressure π of one mole of nonelectrolyte in V liters is:

$$\pi V = RT$$

Since the experimental value is always greater than what is calculated,

$$\pi V = iRT \qquad [1\text{-}26]$$

where i is Van't Hoff's factor, which measures the deviation from the ideal dilute solution. For molarity used as concentration,

$$\pi = imRT \qquad [1\text{-}26a]$$

For a weak polyelectrolyte, the i factor can be related to the degree of dissociation, α, of ionic solute

$$i = 1 + (\nu - 1)\,\alpha \qquad [1\text{-}27]$$

where ν is the number of ions yielded by an electrolyte and

$$\alpha = \frac{(i-1)}{(\nu-1)}$$

α can be measured by conductance ratio.

$$\alpha = \frac{\Lambda_c}{\Lambda_0}$$

Λ_c represents equivalent conductance, and Λ_0 represents equivalent conductance at infinite dilution.

The concept of **ionic strength**, μ, is defined as half of the sum of stoichiometric ionic concentration, where each is multiplied by the square of the valence. For the electrolyte, z is commonly used for valence. For molal concentration,

$$\mu_m = \frac{1}{2}\sum m_i z_i^2$$

and for molar concentration,

$$\mu_c = \frac{1}{2}\sum c_i z_i^2 \qquad [1\text{-}28]$$

The former relation is frequently used in thermodynamic calculations, while the latter is used in solubility and reaction-rate studies.

[Example 1-3] What is the effect of the solubility of AgCl in an aqueous medium to which the same molar concentration of nitrate salts of potassium, barium, and lanthanum are added? Explain why deviation of the bivalent salt is threefold and of the trivalent salt sixfold, as shown in Figure 1-4. Also, determine its solubility product.

For the bivalent salt,

$$\mu_c = \frac{1}{2}\,(c \times 2^2 + 2c \times 1^2) = 3c$$

and

$$\frac{\mu_c}{c} = 3$$

It follows for the trivalent salt that

$$\frac{\mu_c}{c} = 6$$

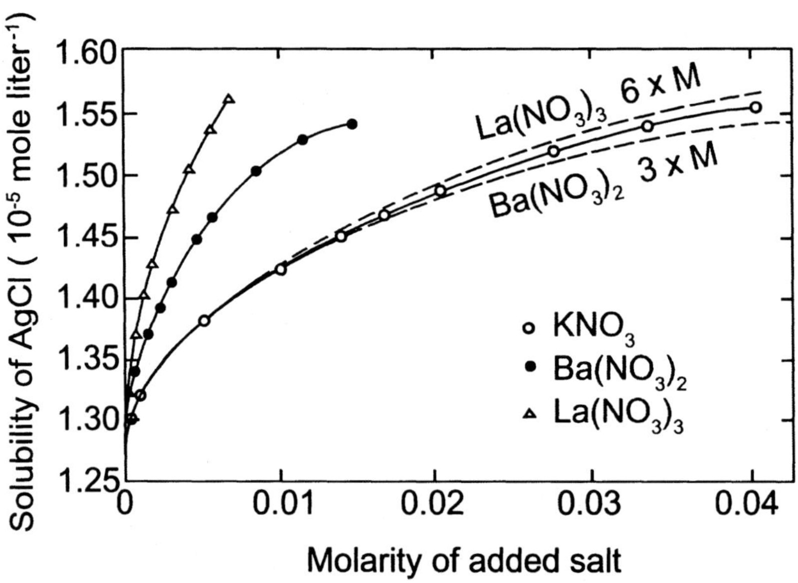

Figure 1-4. Solubility of AgCl in aqueous solutions of various salts at 25°C. Abscissas of dashed lines are 3 and 6 times the molarity of $Ba(NO_3)_2$, respectively.

Next, we consider a solubility product K_s, which is defined as:

$$K_s = \left(a_{Ag^+}\right)\left(a_{\mathrm{Cl}^-}\right) = \left(\gamma_{\mathrm{Ag}^+} \times C_{\mathrm{Ag}^+}\right)\left(\gamma_{\mathrm{Cl}^-} \times C_{\mathrm{Cl}^-}\right)$$

where $C_{Ag^+} = C_{Cl^-} = S$. S is the molar solubility and γ is designated as the molar activity coefficient. Usually the **geometric mean activity** and **mean molar activity** coefficients can be expressed as:

$$a_\pm = \sqrt{a_{Ag^+} \cdot a_{Cl^-}}$$

and

$$\gamma_\pm = \sqrt{\gamma_{\mathrm{Ag}^+} \cdot \gamma_{\mathrm{Cl}^-}}$$

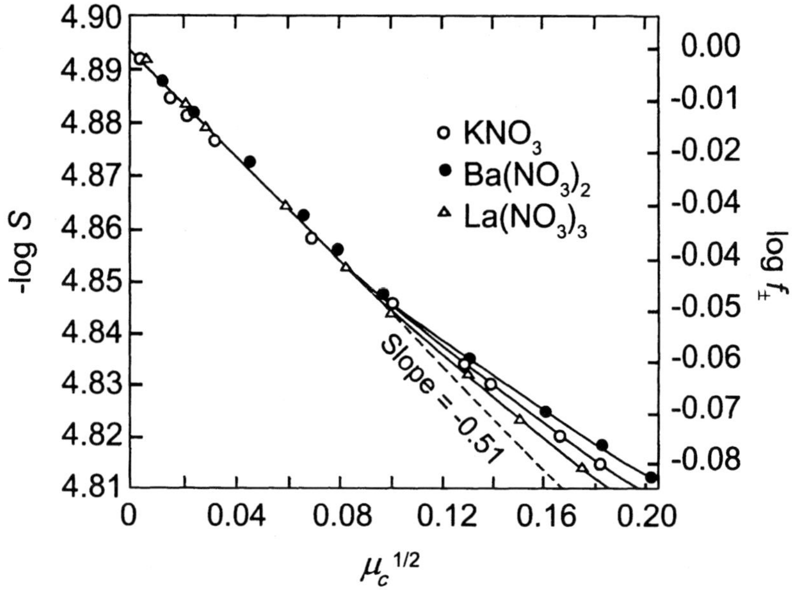

Figure 1-5. Logarithm of solubility and activity coefficient of AgCl in aqueous solutions at 25°C plotted against $\mu_c^{1/2}$.

In a saturated solution (see Fig. 1-5),

$$K_s = a_\pm^2 = \gamma_\pm^2 S^2 \qquad [1\text{-}29]$$

$$\frac{1}{2} \log K_s = \log \gamma_\pm + \log S$$

At low concentration S, $-\log S$ decreases linearly with $\mu^{1/2}$ as a power series.

$$-\log S = -\frac{1}{2} \log K_s + \log \gamma_\pm = A - B\mu^{1/2} + C\mu + \ldots$$

As μ approaches zero, $\log \gamma_\pm$ also approaches zero, and $\gamma_\pm$ approaches 1. Hence, the intercept is $A = 1/2 \log K_s$, and

$$-\log S = \frac{1}{2}\log K_s - B\mu^{1/2} + C\mu + \ldots \qquad [1\text{-}30]$$

Therefore, plotting –log S versus $\mu^{1/2}$ (see Fig. 1-5) yields –1/2 log K_s as the intercept and log $\gamma_\pm$ as the vertical distance from the ordinate of the intercept to the curve at any particular ionic strength.

$$\log \gamma_\pm = -\log S - \frac{1}{2}\log K_s$$

We can determine from the figure that K_s is 1.62×10^{-10}.

For asymmetric salts such as silver chromate (Ag_2CrO_4),

$$\begin{aligned} K_s &= \left(a^2_{Ag^+} \cdot a_{CrO_4^{2-}}\right) \\ &= \left(\gamma_{Ag^+} \cdot S_{Ag^+}\right)^2 \left(\gamma_{CrO_4^{2-}} \cdot S_{CrO_4^{2-}}\right) \\ &= \left(\gamma^2_{Ag^+} \cdot \gamma_{CrO_4^{2-}}\right)(2S)^2(S) \\ &= \left(\gamma^2_{Ag^+} \cdot \gamma_{CrO_4^{2-}}\right)\left(2^2 S^3\right) \end{aligned} \qquad [1\text{-}31]$$

The mean properties are (geometric mean):

$$a_\pm = \left(a^2_{Ag^+} \cdot a_{CrO_4^{2-}}\right)^{1/3}$$

$$\gamma_\pm = \left(\gamma^2_{Ag^+} \cdot \gamma_{CrO_4^{2-}}\right)^{1/3}$$

$$C_\pm = \left(C^2_{Ag+} \cdot C_{CrO_4^{2-}}\right)^{1/3} = 2^{2/3}C$$

Thus,

$$K_s = a^3_\pm = \gamma^3_\pm \cdot S^3_\pm = 2^{2/3}S\gamma^3_\pm \qquad [1\text{-}32]$$

In general, if a salt yields ν ions, with $\nu+$ being positive and $\nu-$ negative, then

$$a_{\pm} = \left(a_{+}^{\nu+} \cdot a_{-}^{\nu-}\right)^{1/\nu}$$

$$\gamma_{\pm} = \left(\gamma_{+}^{\nu+} \cdot \gamma_{-}^{\nu-}\right)^{1/\nu} \qquad [1\text{-}33]$$

$$C_{\pm} = \left(\nu_{+}^{\nu+} \cdot \nu_{-}^{\nu-}\right)^{1/\nu} C$$

Thus,

$$K_s = a_{\pm}^{\ \nu} = \gamma_{\pm}^{\ \nu} \cdot S_{\pm}^{\ \nu} = \gamma_{\pm}^{\ \nu}\,(\nu_{+}^{\ \nu+} \cdot \nu_{-}^{\ \nu-}) S^{\nu}$$

$$\log \gamma_{\pm} = -\log S_{\pm} + 1/\nu \log K_s \qquad [1\text{-}34]$$

where

$$S_{\pm} = (\nu_{+}^{\ \nu+} \cdot \nu_{-}^{\ \nu-})^{1/\nu} S$$

Similarly, the molal activity coefficient $\gamma_{\pm}$ can be developed.

There are also a number of empirical approximations for evaluating ionic strength, such as the **Langelier approximation**

$$\mu = 2.5 \times 10^{-5}\ \text{TDS} \qquad [1\text{-}35]$$

where TDS is the total dissolved solids in mg/L and the **Russell correlation**

$$\mu = 1.6 \times 10^{-5} \wedge \qquad [1\text{-}36]$$

where $\wedge$ is specific conductance in μmho/cm as shown in Figure 1-6.

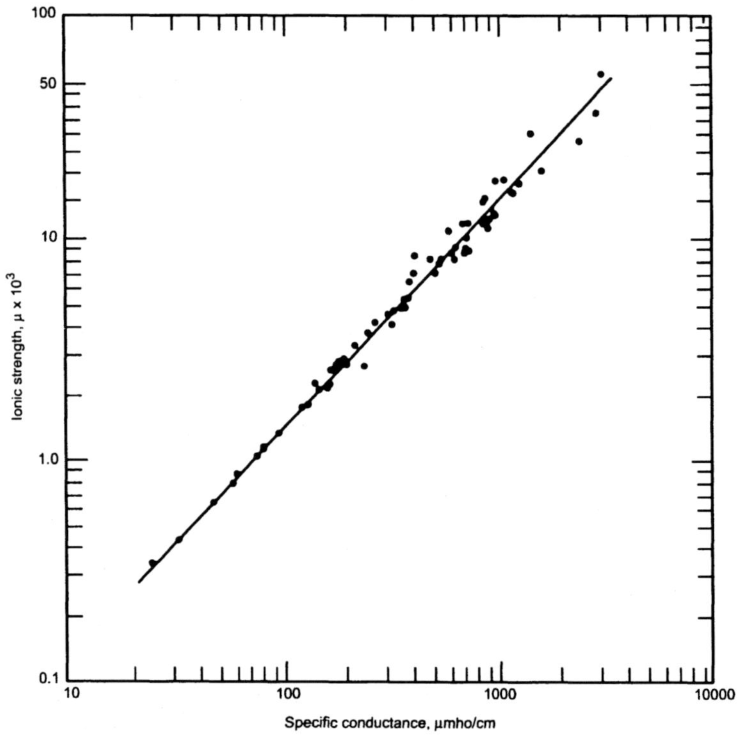

Figure 1-6. Specific conductance as a means of estimating ionic strength. Adapted from C. J. Lind, *U.S. Geological Survey Professional Paper* 700D, 1970, pp. D272-D280, and V. L. Snoekink and D. Jenkins, *Water Chemistry*, John Wiley, New York, 1980.

Ionic strength can also be approximated from **hardness and alkalinity** as

$$\mu = 4H - T \qquad [1\text{-}37]$$

where H is total hardness in moles/L and T is alkalinity in eq/L.

1.3 DEBYE-HÜCKEL THEORY

The theory of Debye and Hückel is based on the assumption that strong electrolytes are completely dissociated into ions and supposed to have ideal behaviors. Any deviation from such ideal behavior must be due to electrical interactions between the ions. It is necessary to calculate the extra free energy arising from those electrostatic interactions. Our problem is to calculate the average electric potential, U, of a given ion in the solution when all the other ions are surrounding it. Knowing U, we can calculate the work that must be expanded to charge the ions reversibly to the potential, and the work will be the free energy due to electrostatic interactions. The extra free energy is directly related to the activity coefficient, since both are a measure of the deviation from the ideal.

In order to complete U, Poisson's differential equation is employed:

$$\nabla^2 U = -\frac{4\pi\rho}{\varepsilon} \qquad [1\text{-}38]$$

where ρ is the average charge density (e.s.u./mL) and ε is the dielectric constant in solution. Expressing Poisson's equation in a spherical symmetric field using r as a radius, we have

$$\frac{1}{r^2}\frac{d}{dr}(r^2\frac{dU}{dr}) = -\frac{4\pi\rho}{\varepsilon} \qquad [1\text{-}39]$$

The charge density, ρ, can be approximated from the Boltzmann distribution equation:

$$\rho = \sum n_i q_i \exp\left(\frac{-q_i U}{kT}\right) \qquad [1\text{-}40]$$

where q_i is the charge and n_i is the average number in unit volume.

In a dilute solution where the interionic potential energy is small compared to its thermal energy and $q_i U << kT$, then the experimental function of Equation [1-40] can be expanded into

$$\exp\left[\frac{-q_iU}{kT}\right]=1-\frac{q_iU}{kT}+\frac{1}{2!}\left(\frac{q_iU}{kT}\right)^2-\cdots \qquad [1\text{-}41]$$

Terms higher than the second being negligible, we get

$$\rho\approx\sum n_iq_i-\frac{U}{kT}\sum n_iq_i^2 \qquad [1\text{-}42]$$

Due to overall electrical neutrality, the first term vanishes and $q_i = z_ie$ (z_i is valence). Thus,

$$\rho=-\frac{e^2U}{kT}\sum n_i z_i^2 \qquad [1\text{-}43]$$

Substituting into Poisson's equation (Eq. [1-39]), we derive

$$\frac{1}{r^2}\frac{d}{dr}\left(r^2\frac{dU}{dr}\right)=\frac{4\pi e^2U}{\varepsilon kT}\sum n_i z_i^2 \qquad [1\text{-}44]$$

Let

$$b^2=\frac{4\pi e^2}{\varepsilon\, kT}\sum n_iz_i^2 \qquad [1\text{-}45]$$

and b^{-1} has a dimension of a length called the Debye length. Hence,

$$\frac{d}{dr}\left(r^2\frac{dU}{dr}\right)=b^2r^2U \qquad [1\text{-}46]$$

Solving the differential equation,

$$U=\frac{A}{r}e^{-br}+\frac{B}{r}e^{br} \qquad [1\text{-}47]$$

with A and B as constants of integration. U must vanish as r approaches infinity.

$$0 = \frac{Ae^{-\infty}}{\infty} + \frac{Be^{\infty}}{\infty}$$

This implies that $B = 0$, since the limit of e_r/r as r approaches infinity is not zero. Thus,

$$U = \frac{A}{r} e^{-br} \tag{1-48}$$

The constant A is determined by the fact that when $b = 0$ the concentration is zero. Thus, a single ion in the absence of other charges,

$$U = \frac{ze}{\varepsilon r} = \frac{A}{r} \tag{1-49}$$

or $A = ze/\varepsilon$, where ze is charge and ε is the dielectric constant of medium. Hence, for a dilute solution

$$U = \frac{ze}{\varepsilon r} \exp(-br) \approx \frac{ze}{\varepsilon r}(1 - br)$$

$$U \approx \frac{ze}{\varepsilon r} - \frac{zeb}{\varepsilon} \tag{1-50}$$

The first term is the potential due to a certain ion, and the second term is the potential due to other ions.

In this manner, we only consider the extra potential due to interaction of other ions. For a single ion, the extra free energy, which is equal to the work, must be expanded to charge the ion reversibly to the required potential of $U = -zeb/\varepsilon$, or

$$\Delta G = W = \int_0^U zedU'$$

$$= \int_0^U \left(-\frac{\varepsilon U'}{b}\right) dU'$$

$$= -\frac{\varepsilon U^2}{2b} = -\frac{b}{2\varepsilon} z^2 e^2 \qquad [1\text{-}51]$$

Now the chemical potential of an ionic species, *I*, is

$$\mu_i = \mu_i^0 + RT \ln a_i = \mu_i^0 + RT \ln C_i + RT \ln \gamma_I \qquad [1\text{-}52]$$

The first two terms are results of an ideal case, whereas the last term is the result of electrical interaction. Since the above equation is expressed per mole in a per ion base, the last term is

$$kT \ln \gamma_i = W = -\frac{b z_i^2 e^2}{2\varepsilon} \qquad [1\text{-}53]$$

We may now calculate the b value in Equation [1-45]:

$$b = \left(\frac{4\pi e^2}{\varepsilon kT} \sum n_i z_i^2\right)^{\frac{1}{2}} = \left(\frac{8\pi N^2 e^2 \rho}{1000 \varepsilon RT}\right)^{\frac{1}{2}} \mu^{\frac{1}{2}} = B\mu^{\frac{1}{2}} \qquad [1\text{-}54]$$

We have changed n_i (ions/mL) into C_i (mole/L) using $n_i = C_i N/1000$ and replacing $C_i = m_i\, \rho$, where ρ is the solvent density. Now let us transform the activity relations into base-10 log form and take e as 4.8×10^{-10} esu, $R = 8.31 \times 10^7$ erg/°C-mole, producing

$$\log \gamma = -1.825 \times 10^6 z^2 (\mu/\varepsilon^3 T^3 \rho)^{1/2} = -Az^2 \mu^{1/2} \qquad [1\text{-}55]$$

This is known as the **Debye-Hückel limiting law**. For water at 25°C, $\varepsilon = 78.54$, $\rho = 0.997$, this equation with $A = 0.509$ is

$$\log \gamma = -0.509\, z^2 \mu^{1/2} \qquad [1\text{-}56]$$

Later Debye and Hückel made an improvement by considering the ionic size effect, which may be called the extended form:

$$\log \gamma = -\frac{Az^2 \mu^{\frac{1}{2}}}{1 + aB\mu^{\frac{1}{2}}} \qquad [1\text{-}57]$$

where a is the average effective diameter of an ion. A number of ions with varying a values and activity coefficients are listed in Table 1-2. The value of B is approximated as 3.28×10^7, and for most ions, the diameters of hydrated ions (except for H^+) are between 3 and 4 Å (except high-valence ions). Since we know that $(a\ B) \approx 1$, the second term in the denominator of the above equation can be simplified. This will reduce to the **Güntelberg approximation**:

$$\log \gamma = -\frac{0.5 z_i^2 \mu^{\frac{1}{2}}}{1 + \mu^{\frac{1}{2}}} \qquad [1\text{-}58]$$

See Figure 1-7 for further information. Furthermore, the **Davies equation** is another type of approximation: (A $\approx$ 0.5)

$$\log \gamma = -Az^2 \left[\frac{\mu^{\frac{1}{2}}}{1 + \mu^{\frac{1}{2}}} - 0.2\mu \right] \qquad [1\text{-}59]$$

Table 1-2. Parameter *a* and Individual Ion Activity Coefficients

Ion Size Parameter, a (Å)	Ion	Activity Coefficients Calculated for Ionic Strength				
		10^{-4}	10^{-3}	10^{-2}	0.05	10^{-1}
9	H^+	0.99	0.97	0.91	0.86	0.83
	Al^{3+}, Fe^{3+}, La^{3+}, Ce^{3+}	0.90	0.74	0.44	0.24	0.18
8	Mg^{2+}, Be^{2+}	0.96	0.87	0.69	0.52	0.45
6	Ca^{2+}, Zn^{2+}, Cu^{2+}, Sn^{2+}, Mn^{2+}, Fe^{2+}	0.96	0.87	0.68	0.48	0.40
5	Ba^{2+}, Sr^{2+}, Pb^{2+}, CO_3^{2-}	0.96	0.87	0.67	0.46	0.39
4	Na^+, HCO_3^-, $H_2PO_4^-$, CH_3COO^-	0.99	0.96	0.90	0.81	0.77
	SO_4^{2-}, HPO_4^{2-}	0.96	0.87	0.66	0.44	0.36
	PO_4^{3-}	0.90	0.72	0.40	0.16	0.10
3	K^+, Ag^+, NH_4^+, OH^-, Cl^-, ClO_4^-, NO_3^-, I^-, HS^-	0.99	0.96	0.90	0.80	0.76

Source: Adapted from J. Kielland, *J. Am. Chem. Soc.*, 59, 1675 (1937).

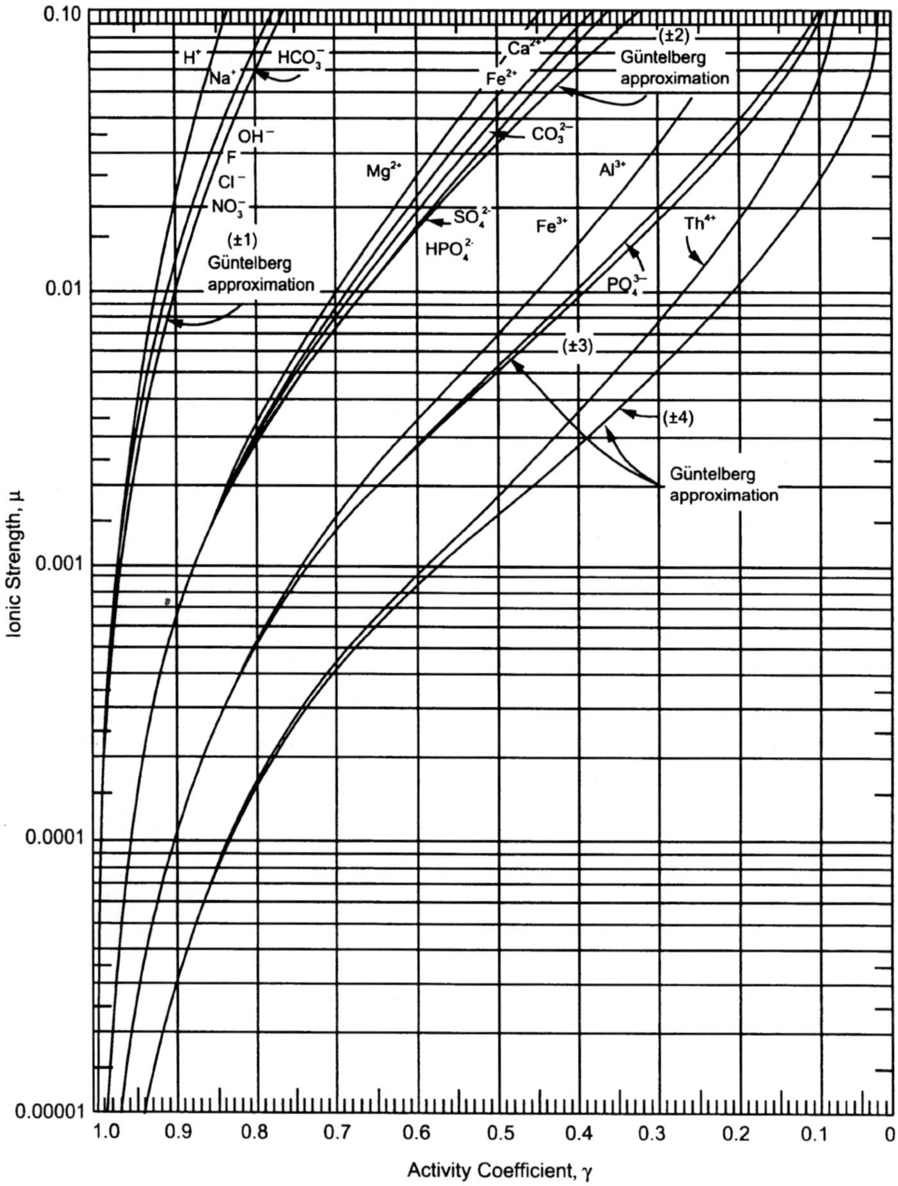

Figure 1-7. Activity coefficients of aqueous ions based on the extended Debye-Hückel equation and the Güntelberg approximation.

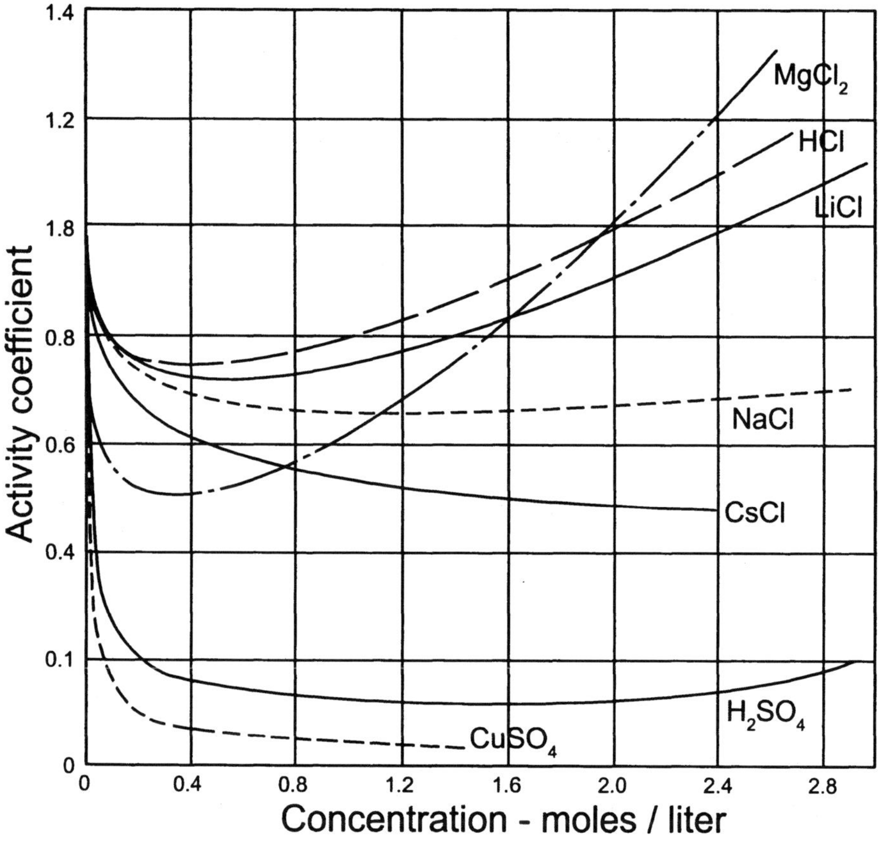

Figure 1-8. Activity coefficient of electrolytes.

The individual ion activity coefficients are summarized in Table 1-3. Deviations from these approximations can be seen in Figure 1-8.

Table 1-3. Mean Activity Coefficients of Representative Electroytes at 25°C

molality	0.00	0.01	0.05	0.10	0.30	0.50	1.00	1.50	2.00	3.00	5.00	15.00
HCl	1.000	.904	.830	.796	.756	.757	.810	.903	1.019	1.320	2.38	4204
HI	1.000	.908	.845	.818	.811	.839	.963	1.136	1.356	2.015		
LiCl	1.000	.904	.828	.790	.744	.739	.774	.838	.921	1.156	2.02	37.9
NaOH	1.000	.901	.816	.766	.708	.690	.678	.689	.709	.784	1.077	11.58
NaCl	1.000	.902	.819	.774	.710	.681	.657	.656	.668	.714	.874	
NaBr	1.000	.903	.823	.782	.719	.697	.687	.702	.731	.812		
NaI	1.000	.904	.826	.787	.735	.723	.736	.771	.820	.963		
KCl	1.000	.899	.815	.764	.683	.644	.604	.583	.573	.569		
RbCl	1.000	.901	.813	.764	.675	.634	.583	.559	.546	.536	.546	
CsCl	1.000	.900	.809	.756	.656	.606	.544	.513	.495	.478	.474	
NaNO3	1.000	.901	.812	.762	.665	.617	.548	.507	.478	.437	.386	
AgNO3	1.000	.897.	.798	.734	.606	.536	.429	.362	.316	.252	.181	
CaCl2	1.000	.731	.582	.520	.445	.448	.500	.616	.792	1.483	5.89	
SrCl2	1.000	729.	.571	.512	.442	.430	.461	.547	.670	1.126		
BaCl2	1.000	.723	.554.	.500	.419	.397	.395	.426			.621	
CuCl2	1.000	.728	.576	..508	.429	.411	.417	.439	.466	.520		
FeCl2	1.000	730	.579	. 518	.454	.450	.506	.623	.794		.354	2.18
ZnCl2	1.000	.731	.579	. 515	.432	.394	.339	.304	.289	.287	.208	1.30
H2SO4	1.000	.544	.340	266	.183	.156	.132	.126	.128	.142		
Na2SO4	1.000	.712	.529	. .445	.320	.266	.201	.170	.152	.137		
CuSO4	1.000	.407.	.206	150	.083	.062	.043	.036				
ZnSO4	1.000	.421	.216	.155	.083	.063	.043	.037	.035	.041		
LaCl2	1.000	560	.388	.320	.263	.266	.342	.513	.825			
Th(NO3)4	1.000			.279	.203	.189	.207	.257	.326	.486	.791	

Source: Adapted from R.H. Stokes, *Trans. Faraday Soc.*, 44. 295-307 (1948); R.A. Robinson and R.H. Stokes, *Trans. Faraday Soc.*, 45, 612-624 (1949); T.S. Sky, *J. Am. Chem. Soc.*, 72, 3680-3682 (1950)

[Example 1-4] Calculate the activity coefficients for 0.25 molar magnesium chloride in water with various models, and compare with the experimental value of a mean activity coefficient of 0.53.

First compute the ion strength.

$$MgCl_2 = Mg^{2+} + 2\ Cl^-$$

$$\mu = \frac{1}{2}\sum C_i z_i^2 = \frac{1}{2}\left(0.25 \times 2^2 + 0.5 \times 1^2\right) = 0.75$$

For the Debye-Hückel limiting equation,

$$\log \gamma = -Az^2\mu^{1/2}$$

$$\log \gamma_{Mg} = -0.509\ (2)^2\,(0.75)^{1/2}$$

$$\gamma_{Mg} = \gamma_+ = 0.0173$$

$$\log \gamma_{Cl} = -0.509\ (1^2)\ (0.75)^{1/2}$$

$$\gamma_{Cl} = \gamma_- = 0.362$$

$$\gamma_\pm^{\,3} = \gamma_+\,\gamma_-^{\,2} = 0.0173\ (0.362)^2$$

$$\gamma_\pm = 0.13$$

For the Güntelberg equation or the Debye-Hückel extended equation, similar values are found:

$$\gamma_+ = 0.114$$

$$\gamma_- = 0.580$$

$$\gamma_\pm = 0.34$$

and for the Davies equation,

$$\gamma_+ = 0.229$$

$$\gamma_- = 0.692$$

$$\gamma_\pm = 0.48$$

Apparently, the results of the Davies equation are closer to the experimental value.

[Example 1-5] If a given amount of seawater contains sodium ions at 0.47 M and chloride ions at 0.55 M, the ion strength is found to be 0.65 M. If this seawater is allowed to evaporate in order to precipitate the sodium chloride, what concentration will the seawater have?

Given that $NaCl_{(s)} = Na^+ + Cl^-$ and $pK_s = 1.54$, and assuming the increase of concentration in all components is q such that

$$(Na^+) = q(Na^+)^0 = 0.47q$$

$$(Cl^-) = q(Cl^-)^0 = 0.55q$$

$$\mu = q\mu^0 = 0.65q$$

then the solubility product is:

$$\gamma_{Na}\, q(Na^+)^0 \cdot \gamma_{Cl} q(Cl^-)^0 = K_s$$

or

$$2 \log q + \log \gamma_{Na} + \log \gamma_{Cl} = \log K_s - \log 0.47 - \log 0.55 = 2.13$$

Using the Davies equation,

$$2 \log q - \frac{(0.65q)^{\frac{1}{2}}}{1+(0.65q)^{\frac{1}{2}}} + 0.13q = 2.13$$

when solving by the graphical method where $q = 7.9$. This value is lower than the actual value of $q = 10.5$.

1.4 EQUILIBRIUM OF CHEMICAL REACTION

Considering a reaction in a mixture of reactants and products,

$$bB + cC = rR + sS$$

As T and P are constants, the change of free energy is:

$$\Delta G = rG_R + sG_S - bG_B - cG_C \quad [1\text{-}60]$$

For component I,

$$\mu_i = \mu_i^0 + RT \ln a_i$$

and

$$G_i = G_i^0 + RT \ln a_i$$

Then the free energy of the reaction is:

$$\Delta G = r(G^0{}_R + RT \ln a_R) + s(G^0{}_S + RT \ln a_S) - b(G^0{}_B + \mathrm{RT} \ln a_B) - c(G^0{}_C + RT \ln a_C)$$

but

$$\Delta G^0 = rG^0{}_R + sG^0{}_S - bG^0{}_B - cG^0{}_C$$

or

$$\Delta G = \Delta G^0 + RT \ln \frac{a_R^r a_S^s}{a_B^b a_C^c}$$

At equilibrium, $\Delta G = 0$, and

$$\Delta G^0 = -RT \ln \frac{a_R^r a_S^s}{a_B^b a_C^c} \tag{1-61}$$

Often the equilibrium constant is defined as:

$$K = \frac{a_R^r a_S^s}{a_B^b a_C^c}$$

Therefore,

$$\Delta G^0 = -RT \ln K \tag{1-62}$$

In an actual case, we can measure a concentration C_i readily, and

$$Q = \frac{C_R^r C_S^s}{C_B^b C_C^c} \tag{1-63}$$

If $a = \gamma C$ is used, then

$$\Delta G = \Delta G^0 + RT \ln Q + RT \ln \gamma$$

Thus, when $\Delta G = 0$,

$$K = Q\gamma \tag{1-63a}$$

where γ can be expressed as follows:

$$\gamma = \frac{\gamma_R \gamma_S}{\gamma_B \gamma_C} \tag{1-64}$$

In general, for the ith species of reactants and products in a solution, individual concentration i for calculation may be

$$K = \prod_i i^{\nu_i} \prod_i \gamma_i \tag{1-65}$$

If many (let us say n) reactions are involved, then

$$K_n = \prod_{j=1}^{n} K_j \qquad [1\text{-}66]$$

and

$$pK_n = \sum_{j=1}^{n} \left(-\log K_j\right) = \sum_{j=1}^{n} pK_j \qquad [1\text{-}67]$$

[Example 1-6] Determine the equilibrium constant for the reaction of liquid water dissociated to H^+ and OH^- at 25°C.

From the thermodynamic constants shown in Table 1-4, the free energy can be calculated. Assuming that $H_2O_{(l)}$ = –56.69, H^+ = 0, and OH^- = –37.60, then

$$\Delta G^0 = (1)(0) + (1)(-37.60) - (1)(-56.69) = 19.09 \text{ kcal} = -RT \ln K$$

$$\ln K = \frac{(-19.09)}{\left(1.987 \times 10^{-3}\right)(298)} = -32.24$$

$$K = 9.96 \times 10^{-15} \approx 1 \times 10^{-14}$$

Table 1-4. Thermodynamic Constants for Species of Chemical Importance for Water

Species	$\Delta \overline{H}_f^0$ (kcal/mole)	$\Delta \overline{G}_f^0$ (kcal/mole)
$Ca^{2+}_{(aq)}$	−129.77	−132.18
$CaCO_{3(s)}$, calcite	−288.45	−269.78
$CaO_{(s)}$	−151.9	−144.4
$C_{(s)}$, graphite	0	0
$CO_{2(g)}$	−94.05	−94.26
$CO_{2(aq)}$	−98.69	−92.31
$CH_{4(g)}$	−17.889	−12.140
$H_2CO_3{}^*{}_{(aq)}$	−167.0	−149.00
$HCO_3{}^-{}_{(aq)}$	−165.18	−140.31
$CO_3{}^{2-}{}_{(aq)}$	−161.63	−126.22
CH_3COO^-, acetate	−116.84	−89.0
$H^+_{(aq)}$	0	0
$H_{2(g)}$	0	0
$Fe^{2+}_{(aq)}$	−21.0	−20.30
$Fe^{3+}_{(aq)}$	−11.4	−2.52
$Fe(OH)_{3(s)}$	−197.0	−166.0
$Mn^{2+}_{(aq)}$	−53.3	−54.4
$MnO_{2(s)}$	−124.2	−111.1
$Mg^{2+}_{(aq)}$	−110.41	−108.99
$Mg(OH)_{2(s)}$	−221.00	−199.27
$NO^{3-}_{(aq)}$	−49.372	−26.43
$NH_{3(g)}$	−11.04	−3.976
$NH_{3(aq)}$	−19.32	−6.37
$NH_4{}^+{}_{(aq)}$	−31.74	−19.00
$HNO_{3(aq)}$	−49.372	−26.41
$O_{2(aq)}$	−3.9	3.93
$O_{2(g)}$	0	0
$OH^-_{(aq)}$	−54.957	−37.595
$H_2O_{(g)}$	−57.7979	−54.6357
$H_2O_{(l)}$	−68.3174	−56.690
$SO_4{}^{2-}{}_{(aq)}$	−216.90	−177.34
$HS^-_{(aq)}$	−4.22	3.01
$H_2S_{(g)}$	−4.815	−7.892
$H_2S_{(aq)}$	−9.4	−6.54

a. For a hypothetical ideal state of unit molarity, which is approximately equal to that of unit molarity. Source: Adapted from R.M. Garrels and C.L. Christ, *Solutions, Minerals, and Equilibria*, Harper & Row, New York, 1965; and *Handbook of Chemistry and Physics*, Chemical Rubber Publishing Company, Cleveland, Ohio.

REFERENCES

1-1. Amis, E. S., *Solvent Effects on Reaction Rates and Mechanisms*, Academic Press, New York, 1966.

1-2. Bard, A. J., *Chemical Equilibrium*, Harper and Row, New York, 1966.

1-3. Benefield, L. D., J. F. Judkins, and B. L. Weand, *Process Chemistry for Water and Wastewater Treatment*, Prentice-Hall, Englewood Cliffs, NJ, 1982.

1-4. K. G. Denbigh, *The Principles of Chemical Equilibrium with Applications in Chemistry and Chemical Engineering*, 4th ed., Cambridge University Press, Cambridge, United Kingdom, 1981.

1-5. Fair, G. M., J. C. Geyer, and D.A. Okun, *Water and Wastewater Engineering*, Vol. 2, *Water Purification and Wastewater Treatment and Disposal*, Wiley, New York, 1968.

1-6. Freiser, H. and Q. Fernando, *Ionic Equilibrium in Analytical Chemistry*, Wiley, New York, 1963.

1-7. Lyman, W. J., W. F. Reehl, and D. H. Rosenblatt, *Handbook of Chemical Property Estimation Methods and Environmental Behavior of Organic Compounds*, McGraw-Hill, New York, 1982.

1-8. Moore, W. J., *Physical Chemistry*, 2nd ed., Prentice-Hall, Englewood Cliffs, New Jersey, 1955.

1-9. Prausnitz, J. M., *Molecular Thermodynamics of Fluid Phase Equilibrium*, Prentice-Hall, 1969.

1-10. Stumm, W. and J. J. Morgan, *Aquatic Chemistry, Chemical Equilibrium and Rates in Natural Waters*, 3rd ed., Wiley, New York, 1995.

PROBLEM SET

1. From enthalpy and entropy data, compute the equilibrium constant for the solution of gaseous ammonia in water at 25°C. The standard molar enthalpies of fraction and the standard molar entropies are:

Species	$\overline{H}_f$ (kcal · mol^{-1})	$\overline{S}$ (cal · dg^{-1} · mol^{-1})
NH_3 (g)	−11.04	46.01
NH_3 (aq)	−19.32	26.3

From the equilibrium constant, estimate the partial vapor pressure of ammonia over an acid rain sample containing 5×10^{-9} M NH_3 (aq).

2. Estimate the solubility (mg/L) of gasoline in water, assuming that gasoline is *n*-octane.

CHAPTER 2

WATER SOFTENING

Carbon is widely distributed in our ecosystem through five major spheres, namely the lithosphere, hydrosphere, atmosphere, pedosphere, and biosphere (see Table 2-1). Sediments and ocean water contain the greatest amounts of inorganic carbon. The carbonate in natural waters may have originated from mineral dissolution processes such as that of calcite or albite. For example, albite breaks down to form kaolinite, the most important alkaline species of the carbonate system.

$$NaAlSi_3O_6(s) + CO_2 + 11/2\ H_2O = Na^+ + HCO_3^- + 2H_4SiO_4 + 1/2Al_2Si_2O_5(OH)_4(s)$$

Due to the dissolution process, certain dications such as Ca^{2+}, Mg^{2+}, Si^{2+}, Mn^{2+}, etc., often cause hardness in water. The major anions associated with them are HCO_3^-, SO_4^{2-}, SiO_3^{2-}, Cl^-, NO_3^-, etc. Boiler scale is a good example, resulting from the decomposition of bicarbonate in steam lines:

$$Ca^{2+} + 2HCO_3^- = CaCO_3 + H_2O + CO_2$$

Table 2-1. Carbon in Major Global Cycles

	Total on Earth (10^{18} mol C)	Total on Earth (units of atmospheric CO_2, A_0)
Sediments		
Carbonate	5,100	94,000
Organic carbon	1,000	18,500
Land		
Organic carbon	0.1	1.8
Ocean		
$CO_2 + H_2CO_3$	0.019	0.35
HCO_3^-	2.9	54
CO_3^{2-}	0.36	7
Dead organic	0.4	7
Living organic	0.0007	0.01
Atmospheric		
CO_2 (A_0)	0.0540	1.0
Representative Concentrations	C_r $(M)^a$	Alkalinity (eq lilter^{-1})[b]
Seawater	2.3×10^{-3}	2.5×10^{-3}
River waters, average	~ 10^{-3}	~ 10^{-3}
River waters, typical range	$10^{-4} - 5 \times 10^{-3}$	$10^{-4} - 5 \times 10^{-3}$
Groundwaters, typical range, the United States	$5 \times 10^{-4} - 8 \times 10^{-3}$	$10^{-4} - 5 \times 10^{-3}$
Rainwater, typical range	$10^{-5} - 5 \times 10^{-5}$	$0 - 4 \times 10^{-5}$[c]

Atmospheric CO_2 = 0.033% by volume in dry air: preindustrial $p_{CO_2} = 2.9 \times 10^{-4}$ atm; $p_{CO_2} = 3.3 \times 10^{-4}$ atm

[a] $C_r = [CO_2 (aq)] + [H_2CO_3] + [HCO_3^-] + [CO_3^{2-}]$ (Oceanographers often use the symbol ΣCO_2)

[b] Alkalinity, the acid-neutralizing capacity, is expressed as mol protons per liter, or equivalents per liter.

[c] Some rainwaters contain mineral acidity of up to 10^{-3} eq liter^{-1}.

Source: Stumm and Morgan

Atmospheric CO_2 will interact and dissolve in almost any kind of water to establish equilibrium. In this chapter we will first review the carbonate system and explain its common equilibrium constants. Then we will proceed to discuss examples of carbonate species in open and closed systems. Next, we will discuss the system points and equivalent points in relation to the different endpoints of acidic or alkaline titrations. One of the most economic processes used for water softening is the addition of lime and soda ash, which we have illustrated with typical examples.

It is important to understand the carbonate system in gas-solution-solid equilibrium situations. For example, in the acid titration of carbonates, equilibrium can be either closed or open to the atmosphere, solution or gas-solution type. In a natural water system, the hypolimnion of a lake is considered to be a solution, a solid equilibrium, since the top layer, the epilimnion, can act as a barrier to the atmosphere. In a manmade system, such as the oxidation pond of an activated sludge plant, the pond is considered open to the atmosphere but closed to the solid, because aeration occurs so vigorously within the pond by air or oxygen that any CO_2 from carbohydrate degradation and the accompanying nitrification process will escape from the surface.

On the other hand, the lime-soda ash softening process is treated as closed to the atmosphere but open to the solid precipitation. Because the ratio of the surface area to the volume of the treatment basin is small, the solid — both added and formed — is abundant and the contact time is short. In the lime-soda ash process, if excess soda ash is used, recarbonation will be necessary after the first stage; split treatment will be required if the excess lime is consumed by CO_2 or the remaining portion of raw water from the split stream. Recarbonation is also often used for the removal of H_2S from stagnant water. The primary function of CO_2 is to lower the pH, and consequently the dissolved sulfur species will be greatly decreased (e.g., at pH 5 the total undissolved species of sulfide is only 2%).

2.1 Carbonate System

Carbonate occurs in natural waters due to the dissolution of atmospheric carbon dioxide as well as lithospheric mineral deposits of carbonate salts. In aquatic chemistry, [ ] is used to indicate concentration, and { } is used to indicate chemical activity. In this book, we use ( ) to indicate concentration, except in a few cases. In general, the governing equilibrium reactions and their constants are as follows:

1. Henry's Law Constant

$$\mathrm{p}K_H = 1.5 \qquad CO_{2(g)} = CO_{2(aq)} \tag{2-1}$$

Equation [2-1] can have an equilibrium constant given by

$$K_D = \frac{(CO_2)_{aq}}{(CO_2)_g} \tag{2-2}$$

Using Dalton's law,

$$(CO_2)_g = \frac{P_{CO_2}}{RT} \tag{2-3}$$

Thus, combining the above two expressions (Eqs. [2-2] and [2-3]),

$$(CO_2)_{aq} = \left(\frac{K_D}{RT}\right) P_{CO_2} = K_H P_{CO_2} \tag{2-4}$$

where K_H, the Henry's law constant, has a dimension measured as mole/L-atm.

$$K_H = \frac{K_D}{RT} = \frac{(CO_2)_{aq}}{P_{CO_2}} \tag{2-5}$$

K_H is also related to the Bunsen absorption coefficient, α_B, by

$$\alpha_B = K_H(22.414) \tag{2-6}$$

which has a dimension of atm^{-1}. The variable P_{CO2} is the partial pressure of CO_2, which is related to the volume fraction X_{CO2}, the total pressure P_T, and the water vapor pressure W by

$$P_{CO_2} = X_{CO_2}(P_T - W) \tag{2-7}$$

Usually the value is:

$$P_{CO_2} = 10^{-3.5}\ \text{atm}$$

and usually

$$(CO_2)_{aq} = K_H P_{CO_2} = 10^{-1.5}10^{-3.5} = 10^{-5}\ \text{M} \tag{2-8}$$

More discussion of Henry's law can be found in Chapter of Aeration and Stripping.

2. Hydration Constant

$$pK_h = 2.8 \quad CO_{2(aq)} + H_2O = H_2CO_3 \qquad [2\text{-}9]$$

Thus,

$$K_h = \frac{(H_2CO_3)}{(CO_2)_{aq}} = 1.6 \times 10^{-3}$$

which means that (H_2CO_3) is only 0.16% of $(CO_2)_{aq}$. When $(CO_2)_{aq}$ is in water, then

$$(CO_2)_{aq} >> (H_2CO_3)$$

and the majority of the species is the $CO_{2(aq)}$ form. It is customary to combine the two forms into a new form:

$$(H_2CO_3) + (CO_2)_{aq} = (H_2CO_3^*) \qquad [2\text{-}10]$$

Since the (H_2CO_3) content is very low,

$$(H_2CO_3^*) \cong (CO_2)_{aq} \qquad [2\text{-}10a]$$

3. Pseudo–First Ionization Constant

$$pK_1' = 3.5 \qquad H_2CO_3 = H^+ + HCO_3^-$$

$$\frac{(H^+)(HCO_3^-)}{(H_2CO_3)} = K_1' \qquad [2\text{-}11]$$

4. First Ionization Constant

$$pK_1 - 6.3 \qquad H_2CO_3^* = H^+ + HCO_3^-$$

$$\frac{(H^+)(HCO_3^-)}{(H_2CO_3^*)} = K_1 \qquad [2\text{-}12]$$

It can be easily seen that

$$K_1 \cong K_1' K_h = 10^{-3.5}\, 10^{-2.8} = 10^{-6.3} \qquad [2\text{-}13]$$

To verify this, K_1 can be written as

$$K_I = \frac{(H^+)(HCO_3^-)}{(H_2CO_3) + (CO_2)_{aq}} = \frac{\dfrac{(H^+)(HCO_3^-)}{(H_2CO_3)}}{1 + \left(\dfrac{1}{K_h}\right)} = \frac{K_1'}{1 + K_h^{-1}} \qquad [2\text{-}13a]$$

Since $K_h^{-1} = 1/10^{-2.8} = 631$, $K_h^{-1} >> 1$, then $K_1 \approx K_1' K_h$, and Equation [2-3] is valid.

5. Second Ionization Constant

$pK_2 = 10.3 \qquad HCO_3^- = H^+ + CO_3^{2-}$

$$\frac{(H^+)(CO_3^{2-})}{(HCO_3^-)} = K_2 \qquad [2\text{-}14]$$

6. Solubility Product

$pK_s = 8.3 \qquad CaCO_{3(s)} = Ca^{2+} + CO_3^{2-}$

$$K_s = (Ca^{2+})(CO_3^{2-}) \qquad [2\text{-}15]$$

Let us define the total concentration of dissolved carbonate species as:

$$C_T = (H_2CO_3^*) + (HCO_3^-) + (CO_3^{2-}) \qquad [2\text{-}16]$$

We can also write the interrelation between any two carbonate species of the three species from Equation [2-12]:

$$(HCO_3^-) = K_1 (H^+)^{-1} (H_2CO_3^*) \qquad [2\text{-}12]$$

and from Equation [2-14]:

$$(CO_3^{2-}) = K_2 (H^+)^{-1} (HCO_3^-)= K_1K_2 (H^+)^{-2} (H_2CO_3^*) \qquad [2\text{-}17]$$

If we define a **distribution coefficient**, α, for the three carbonate species as

$$\alpha_0 = (H_2CO_3^*)/C_T$$

$$\alpha_1 = (HCO_3^-)/C_T$$

$$\alpha_2 = (CO_3^{2-})/C_T \qquad [2\text{-}18]$$

and

$$\alpha_0 + \alpha_1 + \alpha_2 = 1 \qquad [2\text{-}19]$$

then

$$\alpha_0 = [1 + K_1(H^+)^{-1} + K_1K_2(H^+)^{-2}]^{-1} \qquad [2\text{-}19a]$$

$$\alpha_1 = [K_1^{-1}(H^+) + 1 + K_2(H^+)^{-1}]^{-1} \qquad [2\text{-}19b]$$

$$\alpha_2 = [(K_1K_2)^{-1}(H^+)^2 + K_2^{-1}(H^+) + 1]^{-1} \qquad [2\text{-}19c]$$

A diagram can be constructed (Fig. 2-1) to have the same general format as the log C-pH diagram used widely in aquatic chemistry.

N.B.

But $C_T \neq (H^+)^2 + K_1(H^+) + K_1 K_2$.

Another way to express this is:

$$\alpha_0 = (H^+)^2/[(H^+)^2 + k_1(H^+) + k_1k_2] \qquad [2\text{-}19d]$$

$$\alpha_1 = K_1(H^+) \ /[(H^+)^2 + k_1(H^+) + k_1k_2] \qquad [2\text{-}19e]$$

$$\alpha_2 = K_1 K_2/[(H^+)^2 + k_1(H^+) + k_1k_2] \qquad [2\text{-}19f]$$

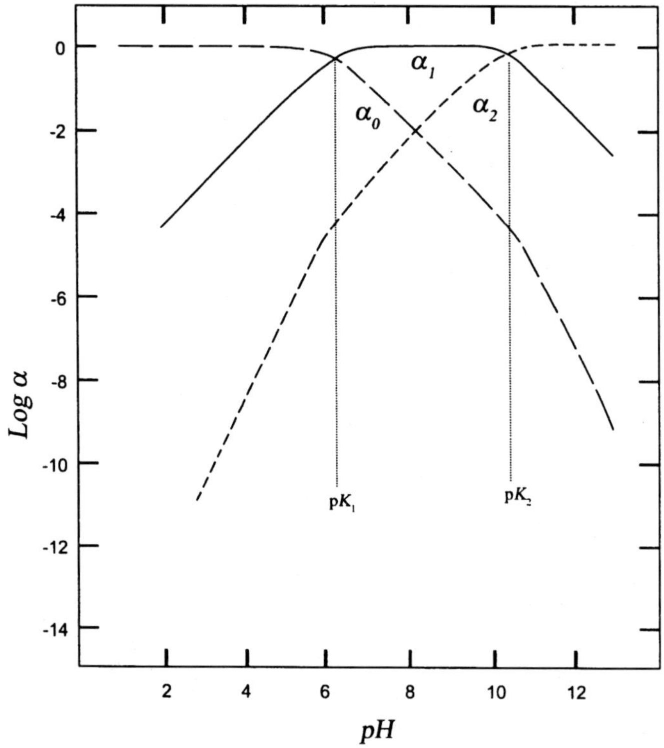

Figure 2-1. Ionization fractions of the carbonate species as a function of pH: α_0 = $(H_2CO_3^-)/CT$; $\alpha_1 = (HCO_3^-)/CT$; $\alpha_2 = (CO_3^{2-})/CT$.

[Example 2-1] Construct the diagram of log α versus pH for the carbonate species as Figure 2-1. Identify the slopes of each curve: α_0, α_1, and α_2.

Let us determine the following three regions:

Region 1 $pH < pK_1 < pK_2$
Region 2 $pK_1 < pH < pK_2$
Region 3 $pK_1 < pK_2 < pH$

for each α curve. We have nine zones to work out; assumptions and approximations can be made (see Table 2-2 for calculation details).

Table 2-2. Estimation of Ionization Fraction of Carbonate Species at Three Different Regions of pH

	Region 1 pH < pK_1 < pK_2	Region 2 pK_1 < pH < pK_2	Region 3 pK_1 < pK_2 < pH
α_0	≈ 1	$\approx 1/(K_1/(H^+)) \approx (H^+)/K_1$	$\approx (H^+)^2/K_1K_2$
Slope	0	−1	−2
α_1	$\approx 1/((H^+)/K_1) \approx K_1/(H^+)$	≈ 1	$(H^+)/K_2$
Slope	+1	0	−1
α_2	$\approx 1/((H^+)^2/K_1K_2) \approx K_1K_2/(H^+)^2$	$\approx 1/((H^+)/K_2) \approx K_2/(H^+)$	≈ 1
slope	+2	+1	0

Table 2-3 lists the essential carbonate equilibrium and solubility in terms of various constants, K, and always pK = −log K. It is useful to recall that reactions add and that constants multiply. Consequently, when reactions subtract, then constants divide. These constants usually follow Arrhenius-type temperature dependence (for equilibrium constants at various temperatures, see Physical Chemistry, Environmental Chemistry in this series). The salinity corrections for the activity coefficients are also listed in Table 2-3. Since the K constants are computed based on thermodynamic principles, the operational K constants (usually referred to as ^cK), based on nonideal cases, must correlate with the activity coefficients. For example,

$$pC = pa + \log \gamma \qquad [2\text{-}20]$$

Since $a = \gamma C$ and $-\log C = -\log a + \log\gamma$, the C is related to operational constants, and a is related to thermodynamic principles. Therefore,

$$p^cK = pK - \frac{a\mu^{\frac{1}{2}}}{1+b\mu^{\frac{1}{2}}} \qquad [2\text{-}21]$$

Table 2-3. $CaCO_3$ Solubility and Carbonate Equlibria in Terms of Dissociation Constants, p*K* = (–log*K*)

Reaction[a]	Temperature, C						
	5	10	15	20	25	40	60
$CO_{2(S)} + H_2O = H_2CO_3^*$; K	1.20	1.27	1.34	1.41	1.47	1.64	1.8
$H_2CO_3 = HCO_3^- + H^+$; K_1	6.52	6.46	6.42	6.38	6.35	6.30	6.30
$HCO_3^- = CO_3^{2-} + H^+$; K_2	10.56	10.49	10.43	10.38	10.33	10.22	10.14
$CaCO_{3(S)} = Ca^{2+} + CO_3^{2-}$; K_s	8.09	8.15	8.22	8.28	8.34	8.51	8.74
$CaCO_{3(S)} + H^+ = Ca^{++} + HCO_3^-$; (K_s/K_2)	-2.47	-2.34	-2.21	-2.10	-1.99	-1.71	-1.40
$CaCO_{3(S)} + H_2CO_3^* = Ca^{2+} + 2HCO_3^-$; (K_sK_1/K_2)	4.05	4.12	4.21	4.28	4.36	4.59	4.90
$H_2O = H^+ + OH^-$; K_w	14.73	14.54	14.35	14.17	14.00	13.54	13.02

Salinity Corrections for Equilibrium Constants[b]

$pK_1^1 = pK_1 - (\mu)^{.05}/[1 + 1.4(\mu)^{.05}]$ $\qquad pK_2^1 = pK_2 - 2(\mu)^{.05}/[1 + 1.4(\mu)^{.05}]$

$PK_s^1 = pK_s - 4(\mu)^{.05}/[1 + 3.9(\mu)^{.05}]$ $\qquad \text{Log } K_s^1/K_2^1 = \log K_s/K_2 + 2.5(\mu)^{.05}/[1 + 5.3(\mu)^{.05} + 5.5(\mu)^{.05}]$

Estimation of μ

$\mu \approx 2.5 \times 10^{-5} S_d$ where S_d is the total dissolved solids content of the water, or
$\mu \approx 4H - T$, where H = total hardness in moles/L; and T = alkalinity in eq/L (10^3 me/L)

[a]H_2CO_3 refers to the sum of dissolved CO_2 and H_2CO_3. Because $[H_2CO_3] << [CO_2]$, $[CO_2]$ is essentially equal to $[H_2CO_3^*]$.
[b]K and K^1 are thermodynamic and operational constants, respectively. The operational constants can be used with mass law expressions containing concentration terms, with the exception of H^+, which is always expressed in activities.

2.2 CARBONATE CONCENTRATION DIAGRAMS

In the following section, we will discuss reversed carbonate equilibrium as well as carbonate concentration diagrams.

2.2.1 A Closed System of Dissolved Carbonates

In an aqueous system,

$$pK_w = 14 \qquad H_2O = H^+ + OH^-$$

$$K_w = (H^+)(OH^-)$$

or

$$(OH^-) = K_w(H^+)^{-1} \qquad [2\text{-}22]$$

For material balance, again,

$$C_T = (H_2CO_3^*) + (HCO_3^-) + (CO_3^{2-}) \qquad [2\text{-}23]$$

The electroneutrality of a carbonate solution is:

$$(H^+) = (OH^-) + (HCO_3^-) + 2(CO_3^{2-}) \qquad [2\text{-}24]$$

In the presence of a base or an acid, this will become:

$$(H^+) + (C_B) = (C_A) + (OH^-) + (HCO_3^-) + 2\,(CO_3^{2-}) \qquad [2\text{-}25]$$

Usually the three equilibrium concentrations of a carbonate system can be calculated by simultaneously solving three equations involving K constants.

[Example 2-2] Determine the operational component concentration of the carbonate species of a 3×10^{-3} molar carbonate solution where the pH is 7.5, based on thermodynamic considerations.

Thermodynamically derived equilibrium constants can be correlated with the operationally concentration-based constants by an activity coefficient:

$$K = \gamma^{c}K \qquad [2\text{-}26]$$

K_1 and K_2 can be derived based on free energy. For example,

$$H_2CO_{3(aq)} = H^+_{(aq)} + HCO_3^-{}_{(aq)}$$

$$-149 \qquad 0 \qquad -140.31 \qquad \Delta G^0{}_{298}\ \text{(cal/mol)}$$

$$\Delta G^0 = -140.31 + 0 - (-149.0) = 8.69\ \text{kcal}$$

$$\ln K_1 = -\Delta G/RT = -14.65$$

$$K_1 = 4.36 \times 10^{-7} \text{ and } pK_1 = 6.36$$

Similarly, $K_2 = 5.01 \times 10^{-11}$ and $pK_2 = 10.30$.

Since the individual concentration is unknown, ionic strength is difficult to evaluate (although approximates can be made), and we can calculate the concentrations by assessing as follows:

(Case 1), $\gamma = 1$

$$K_1 = (H^+)(HCO_3^-)/(H_2CO_3^*) = 4.35 \times 10^{-7}$$

$$K_2 = (H^+)(CO_3^{2-})/(HCO_3^-) = 5.01 \times 10^{-11}$$

$$K_w = (10^{-7.5})(OH^-) = 1 \times 10^{-14}$$

$$C_T = (H_2CO_3^*) + (HCO_3^-) + (CO_3^{2-}) = 3 \times 10^{-3}$$

Solving,

$$(HCO_3^-) = 2.80 \times 10^{-3}\ M$$

$$(H_2CO_3^*) = 1.97 \times 10^{-4}\ M$$

$$(CO_3^{2-}) = 4.42 \times 10^{-6}\ M$$

$$(OH^-) = 3.16 \times 10^{-7}\ M$$

(Case 2) $\mu = 1.4 \times 10^{-3}$

Based on HCO_3^- and two other species and from the Davies equation, the activity coefficients can be determined.

$$\gamma_{HCO_3^-} = 0.96$$

$$\gamma_{CO_3^{2-}} = 0.85$$

$$\gamma_{OH^-} = 0.96$$

We can now use ^cK for the evaluation of operational constants and a more accurate component concentration:

$$K_1 = \frac{(H^+)(HCO_3^-)(0.96)}{(H_2CO_3^*)} = 4.35 \times 10^{-7}$$

$$K_2 = \frac{(H^+)(CO_3^{2-})(0.85)}{(HCO_3^-)(0.96)} = 5.01 \times 10^{-11}$$

$$K_w = (10^{-7.5})(OH^-)(0.96) = 1 \times 10^{-14}$$

The C_T equation is the same as for Case 1. The solution becomes:

$$(HCO_3^-) = 2.81 \times 10^{-3}\ M$$

$$(H_2CO_3^*) = 1.89 \times 10^{-4}\ M$$

$$(CO_3^{2-}) = 5.04 \times 10^{-6}\ M$$

The values for both Case 1 and Case 2 are very close.

2.2.2 Dissolved Carbonates in an Open System of Atmospheric Carbon Dioxide

This system is exemplified by Figure 2-2, in which six lines are formed when pC versus pH is plotted. We will show how each line is constructed. Here Henry's law becomes important:

$$(CO_2)_{aq} = K_H P_{CO_2} = 10^{-1.5}\ 10^{-3.5} = 10^{-5} \cong (H_2CO_3^*) = C_T\alpha_0 \quad \text{(from Eq. [2-18])}$$

For the above relation, the normal aqueous carbonate system open to the atmosphere contains $P_{CO_2} = 10^{-3.5}$ atm. Therefore,

$$-\log\,(H_2CO_3^*) = 5 \qquad [2\text{-}27]$$

This is **line 1** in Figure 2-2 with a slope = 0.

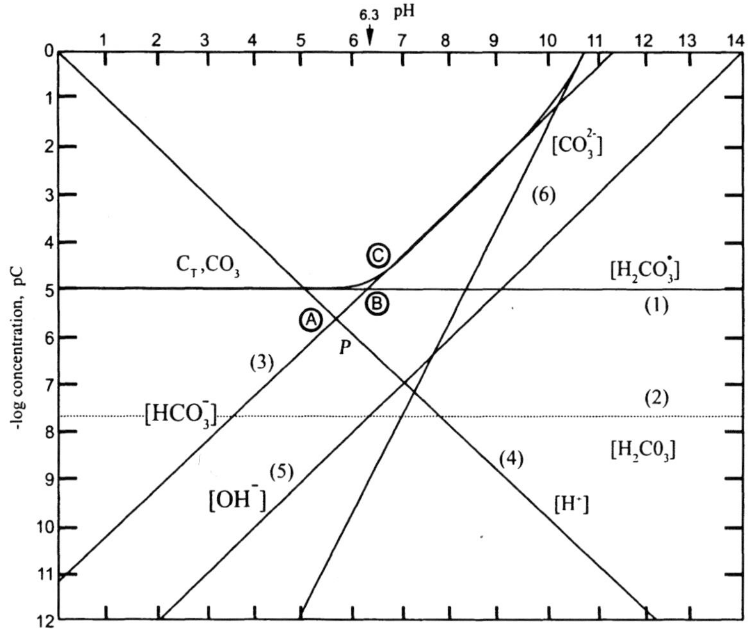

Figure 2-2. Total concentration of carbonates, C_{T,CO_3}, versus pH for an open system at 25°C. (Modified from Ref. 2-1.)

Also,

$$-\log (CO_2)_{aq} = 5$$

from the hydration constant,

$$\frac{(CO_2)_{aq}}{(H_2CO_3)} = K_h^{-1}$$

Thus,

$$\log (CO_2)_{aq} - \log (H_2CO_3) = pK_h$$

or

$$-\log (H_2CO_3) = 5 + 2.8 = 7.8 \qquad [2\text{-}28]$$

which is **line 2** in Figure 2-2.

For (HCO_3^-) species (using Eqs. [2-18], [2-10a], [2-8]),

$$\left(HCO_3^-\right)=\left(\frac{\alpha_1}{\alpha_0}\right)K_H P_{CO_2}=K_1\left(H^+\right)^{-1}K_H P_{CO_2}$$

Therefore,

$$-\log\,(HCO_3^-) = pK_1 - \log K_H - \log\, P_{CO_2} - pH = 11.3 - pH \qquad [2\text{-}29]$$

This is **line 3** for (HCO_3^-) with a slope of +1.

Now for the (H^+) line, **line 4:**

$$-\log\,(H^+) = pH$$

The intersection of line 3 and line 4 or the equivalent concentration point (point P) is:

$$-\log\,(H^+) = -\log\,(HCO_3^-) = 5.65 \qquad [2\text{-}30]$$

which can be obtained by the simultaneous solution of

$$-y = x$$

$$-y = 11.3 - x$$

where $y = \log\,(H^+)$ and $x = pH$.

The (OH^-) line, **line 5**, is then

$$\log\,(OH^-) = -pK_w + pH = -14 + pH \qquad [2\text{-}31]$$

which also has a slope of +1.

Finally, the (CO_3^{2-}) can be expressed by:

$$\left(CO_3^{2-}\right)=\left(\frac{\alpha_2}{\alpha_0}\right)K_H P_{CO_2}=K_1 K_2\left(H^+\right)^{-2}K_H P_{CO_2}$$

Thus,

$$-\log (CO_3^{2-}) = 21.6 - 2\ pH \qquad [2\text{-}32]$$

which has a slope of +2 and is represented by **line 6**.

2.2.3 Dissolved Carbonates in Contact with Solid Calcium Carbonates (Open to Solid Minerals)

The dissolution of $CaCO_3(s)$ in water assumes that the Ca^{2+} dissolved must be equal in concentration to the dissolved carbonate species. In other words,

$$(Ca^{2+}) = C_T \qquad [2\text{-}33]$$

Furthermore, the electroneutrality condition must be:

$$2(Ca^{2+}) + 2(H^+) = (HCO_3^-) + 2(CO_3^{2-}) + (OH^-) \qquad [2\text{-}34]$$

Our problem is to calculate the concentrations of all possible species in the water, which include Ca^{2+}, $H_2CO_3^*$, HCO_3^-, CO_3^{2-}, H^+, and OH^-. This will require the solution of six equations for the six unknowns. We already have two equations (Eqs. [2-33] and [2-34]). We can add the first and second ionization constants for the carbonic acid, the ion product of water, and the solubility product of calcium carbonate (K_1, K_2, K_w, and K_s).

In solving the equations, it is easy to start with the solubility product.

$$(Ca^{2+}) = K_s(CO_3^{2-})^{-1}$$

Using the relations of ionization fractions,

$$(CO_3^{2-}) = \alpha_2 C_T$$

$$(HCO_3^-) = \alpha_1 C_T$$

then

$$(Ca^{2+}) = K_s/\alpha_2 C_T = C_T = (K_s/\alpha_2)^{1/2} \qquad [2\text{-}35]$$

After substitution into the neutrality equation (Eq. [2-34]),

$$(K_s/\alpha)^{1/2}(2 - \alpha_1 - 2\alpha_2) + (H^+) - K_w(H^+)^{-1} = 0 \qquad [2\text{-}36]$$

This equation can be solved by trial and error. The results are pH = 9.95, $(Ca^{2+}) = 1.27 \times 10^{-4}$ M, $(HCO_3^-) = 8.9 \times 10^{-5}$ M, $(CO_3^{2-}) = 3.8 \times 10^{-5}$ M, $(H_2CO_3^*) = 2.5 \times 10^{-8}$ M, alkalinity = 2.54×10^{-4} eq/L, pOH = 4.05.

If the pH is adjusted with a strong acid or base, the neutrality condition becomes similar to Equation [2-24]

$$(C_B) + 2(Ca^{2+}) + 2(H^+) = (C_A) + (HCO_3^-) + 2(CO_3^{2-}) + (OH^-) \quad [2\text{-}37]$$

Here, for example, if the strong base were NaOH, $(C_B) = (Na^+)$ and if the strong acid were HCl, then $(C_A) = (Cl^-)$. A logarithmic concentration versus pH plot can be constructed as shown in Figure 2-3. Some of the pertinent elements and principles are listed below.

In order to see how these log C versus pH diagrams are constructed (e.g., the Ca^{2+} line or the C_T line), the pH again can be separated into three regions.

1st region:	$pH < pK_1 < pK_2$
2nd region:	$pK_1 < pH < pK_2$
3rd region:	$pK_1 < pK_2 < pH$

From Equation [2-35],

$$(Ca^{2+}) = (K_s/\alpha_2)^{1/2}$$

$$\log (Ca^{2+}) = 1/2 \log K_s - 1/2 \log \alpha_2 = 1/2 \log K_s + 1/2 \log [1 + K_2^{-1}(H^+) + K_1^{-1} K_2^{-1}(H^+)^2]$$

Using Table 2-2 as for Example 2-1,

$$\log (Ca^{2+}) \cong 1/2 \log K_s + 1/2 \log [K_1^{-1}K_2^{-1}(H^+)^2]$$

$$\cong -1/2\ pK_s - pH + 1/2\ pK_1 + 1/2\ pK_2$$

$$\cong 4.15 - pH \qquad [2\text{-}38]$$

and for the slope,

$$\frac{d\log(Ca^{2+})}{dpH} = -2 \times 0.5 = -1 \quad [2\text{-}39]$$

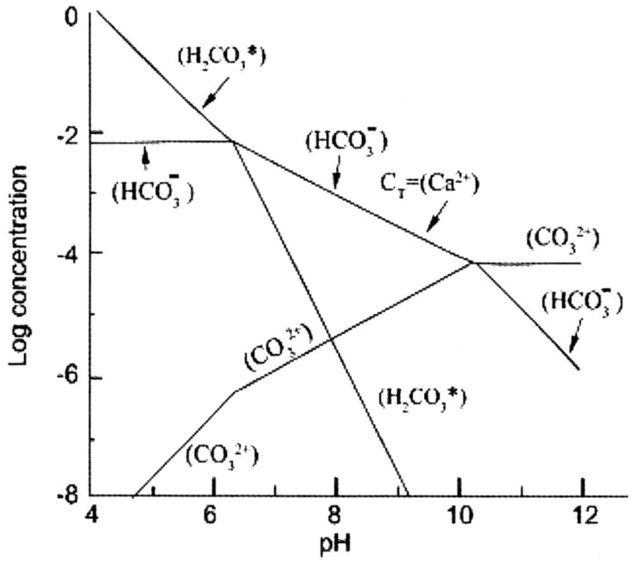

Figure 2-3. Logarithmic concentration diagram for carbonate solution in contact with solid calcium carbonate. (Adapted from Ref. 2.4.)

In the 2nd region,

$$\log(Ca^{2+}) \cong 1/2 \log K_s + 1/2 \log K_2^{-1}(H^+)$$

$$\cong -1/2\, pK_s - 1/2\, pH + 1/2\, pK_2 \cong 1 - 1/2\, pH \quad [2\text{-}40]$$

and $$\frac{d\log(Ca^{2+})}{dpH} = -0.05 \quad [2\text{-}41]$$

For the 3rd region,

$$\log\,(\mathrm{Ca}^{2+}) \cong 1/2\,\log K_s + 1/2\,\log 1 \cong -1/2\;\mathrm{p}K_s \cong 4.15 \qquad [2\text{-}42]$$

and

$$\frac{d\log\left(Ca^{2+}\right)}{dpH} \cong 0 \qquad [2\text{-}43]$$

Lines connecting the three regions are shown in Figure 2-3 as the Ca^{2+} line or C_T line.

Similarly, for the $(\mathrm{HCO_3}^-)$ line,

$$\left(\mathrm{HCO_3^-}\right) = \alpha_1 C_T = \alpha_1 K_s^{\frac{1}{2}} \alpha_2^{-\frac{1}{2}} = \frac{K_s^{\frac{1}{2}}\left[K_1^{-1}\left(\mathrm{H}^+\right) + 1 + K_2\left(\mathrm{H}^+\right)^{-1}\right]^{-1}}{\left[\left(K_2K_1\right)^{-1}\left(\mathrm{H}^+\right)^2 + K_2^{-1}\left(\mathrm{H}^+\right) + 1\right]^{-\frac{1}{2}}}$$

For the first region,

$$\log\left(\mathrm{HCO_3^-}\right) \cong \log\frac{K_s^{\frac{1}{2}}\left[\left(K_1K_2\right)^{-1}\left(\mathrm{H}^+\right)^2\right]^{\frac{1}{2}}}{K_1^{-1}\left(\mathrm{H}^+\right)} \cong \frac{1}{2}\log K_s + \frac{1}{2}\log\left(\frac{\left(\mathrm{H}^+\right)^2}{K_1K_2}\right) - \log\left(\mathrm{H}^+\right)K_1^{-1}$$

$$\cong -\log\left(\mathrm{HCO_3^-}\right) \cong \frac{1}{2}pK_s + pH - \frac{1}{2}pK_1 - \frac{1}{2}pK_2 - pH + pK_1 \cong 2.2 \qquad [2\text{-}44]$$

and the slope of this line is zero.

Others values can be similarly obtained and the diagrams completed.

2.3 System Points and Equivalence Points

In general, alkalinity is a measure of the compounds that when added to water will neutralize a strong acid. In characterizing natural water systems, alkalinity is referred to as acid-neutralizing capacity. Most natural waters contain dissolved carbonate species, and therefore, it is best to look more closely at the carbonate system.

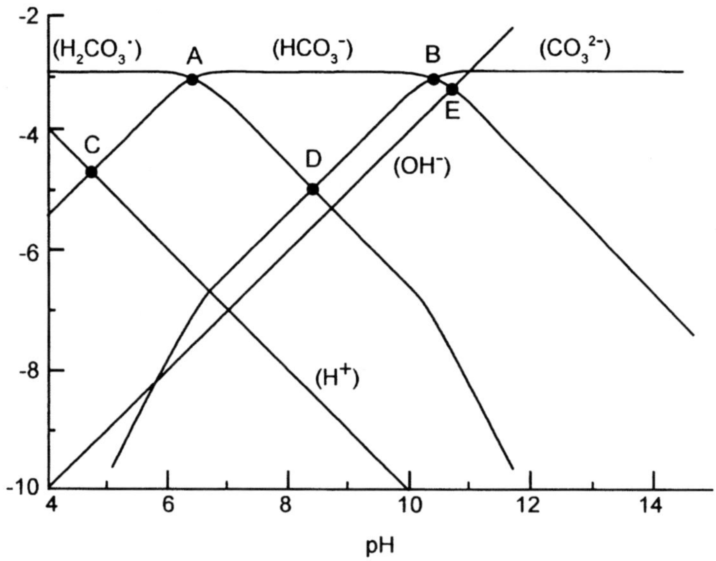

Figure 2-4. Logarithmic concentration diagram for 10^{-3} molar carbonate solution. (Adapted from Ref. 2.2.) Part A, B are system points; Part C, D and E are equivalence points.

In the logarithmic concentration of a 10^{-3} molar carbonate solution, such as in Figure 2-4, we first introduce the **system points,** *A* and *B*. At point *A*,

$$H_2CO_3^* = HCO_3^- + H^+ \quad pH \cong 6.5 \qquad [2\text{-}45]$$

At point *B*,

$$HCO_3^- = CO_3^{2-} + H^+, \quad pH \cong 10 \qquad [2\text{-}46]$$

In this instance, at point *A* ($H_2CO_3^*$) = (HCO_3^-), and pH = pK_1, C_T = ($H_2CO_3^*$) + (HCO_3^-). At point *A*, $C = C_T$, but ($H_2CO_3^*$) = (HCO_3^-), so $C = C_T/2$ or $C_T/C = 2$ Therefore, the actual concentration of point *A* is different from the p*K* value by log 2 = 0.3 or 0.3 unit less in log scale. Similarly, the system point *B* can be determined. Points *C*, *D*, and *E* are termed **equivalence points**, at which a strong acid or base is needed to achieve a reference equivalence point.

At point *C*, $H_2CO_3^*$ predominates:

$$(H^+) = (HCO_3^-) + 2\,(CO_3^{2-}) + (OH^-)$$

We consider here only the K_1 equation:

$$\frac{(H^+)(HCO_3^-)}{(H_2CO_3^*)} = K_1$$

Assuming that $(H^+) = (HCO_3^-)$ and that in this case $C_T = (H_2CO_3^*)$ as an approximation,

$$(H^+) = (C_T K_1)^{1/2}$$

and

$$pH = -1/2 \log C_T + 1/2\, pK_l = 4.5 \qquad [2\text{-}47]$$

Point *C* is often referred to as the **total alkalinity** (and mineral acidity) **equivalence point.**

Next, at point *D*, the proton condition is:

$$(H^+) + (H_2CO_3^*) = (OH^-) + (CO_3^{2-})$$

At pH >7, $(H^+) < (OH^-)$, and

$$(H_2CO_3^*) = (OH^-) + (CO_3^{2-})$$

Substituting K_1, K_w, and K_2,

$$\frac{(H^+)(HCO_3^-)}{K_1} = \frac{K_w}{(H^+)} + \frac{(HCO_3^-)K_2}{(H^+)}$$

In this case, $C_T = (HCO_3^-)$, and therefore,

$$(H^+) = \left[\left(\frac{K_w K_1}{C_T}\right) + (K_1 K_2)\right]^{\frac{1}{2}}$$

For $C_T \geq 10^{-3}$ M, the first term is dropped, and

$$pH \cong \frac{1}{2}\,pK_1 + \frac{1}{2}\,pK_2 = 8.3 \qquad [2\text{-}48]$$

Point D is usually referred to as the **carbonate alkalinity** or carbon dioxide acidity **equivalence point**. Of course, at point D, $(H_2CO_3^*) = (CO_3^{2-})$. Thus,

$$\frac{(H^+)(HCO_3^-)}{K_1} = \frac{(HCO_3^-)K_2}{(H^+)}$$

or

$$(H^+) = (K_1K_2)^{1/2}$$

Similarly, point E has a proton situation as follows:

$$2(H_2CO_3^*) + (HCO_3^-) + (H^+) = (OH^-)$$

At pH >> 7, (HCO_3^*) and (H^+) are negligible. Therefore,

$$(HCO_3^-) = (OH^-)$$

and

$$C_T = (CO_3^{2-}) \text{ and } (H^+) = (K_2K_w/C_T)^{1/2}$$

$$pH = 1/2\,pK_2 + 1/2\,pK_w - 1/2\log C_T = 10.6$$

Point E is called the **caustic alkalinity**, total acidity **equivalence point,** or endpoint.

In practice titration using phenolphthalein with an endpoint at pH 8.3 is considered for the equivalence of carbonate alkalinity. Further titration using methyl orange with an endpoint at pH 4.5 is considered for the total alkalinity. The titration of water samples containing various forms of alkalinity can be demonstrated as in Figure 2-5 (e.g., the case of hydroxide only, carbonate only, mixed hydroxide-carbonate, mixed carbonate-bicarbonate, and bicarbonate only).

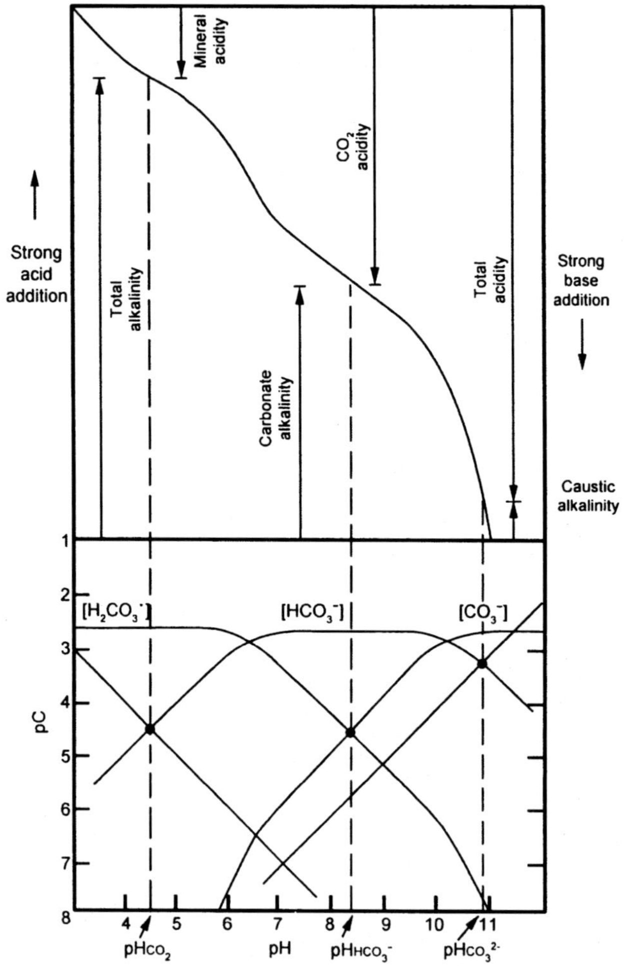

Figure 2-5. Relationship of the pC vs. pH diagram to the titration curve for the carbonate system at 25°C.

[Example 2-3] Locate all of the equivalence points (x,y) of a carbonate system if the carbonate solution is 10^{-3} M.

For point C, in the graph of pC versus pH (see Fig. 2-4),

$$C_T = (H_2CO_3^*)$$

$$(HCO_3^-) = \alpha_1 C_T$$

$$K_1 = \frac{(\alpha_1 C_T)(H^+)}{C_T} = \alpha_1 (H^+)$$

$$\log \alpha_1 = -pK_1 + pH = -6.5 + 4.5 = -2$$

$$\log C = -3 - 2 = -5$$

$$\therefore C \text{ is } (4.5, 5)$$

For point *D*,

$$C_T = (HCO_3^-)$$

$$(CO_3^{2-}) = \alpha_2 C_T = \frac{K_2 (HCO_3^-)}{(H^+)} = K_2 C_T (H^+)^{-1}$$

$$\log \alpha_2 = pH - pK_2 = 8.3 - 10.5 = -2.2$$

$$\log C = -3 - 2.2 = -5.2$$

$$\therefore D \text{ is } (8.3, 5.2)$$

For point E,

$$C_T = (CO_3^{2-}) = K_2(HCO_3^-)(H^+)^{-1} = K_2 \alpha_1 C_T (H^+)^{-1}$$

$$(H^+) = K_2\alpha_1$$

$$\log \alpha_1 = pK_2 - pH = 10.5 - 10.6 = -0.1$$

$$\log C = -3 - 0.1 = -3.1$$

$$\therefore E \text{ is } (10.6, 3.1)$$

[Example 2-4] A recarbonated water having pH 9.5 is chosen for phenolphthalein and methyl orange titrations, where in both titrations 1/50 N of H_2SO_4 is used for a 200-mL water sample. The first titration consumes 2.0 mL, and subsequently, 24.0 mL of acid is consumed. Find the total carbonate and the bicarbonate alkalinities of the water sample.

Carbonate alkalinity = 2(meq of acid/L to reach phenolphthalein endpoint)

= 2(2.0 mL × 0.02 eq/L × 1000 meq/eq × 1/200 mL sample)

= 0.40 meq/L

Total alkalinity = meq of acid/L to reach methyl orange endpoint

= (24.0 + 2.0) mL × 0.02 eq/L × 1000 meq/eq × 1/200 mL sample

= 2.6 meq/L

Bicarbonate alkalinity = total alkalinity – carbonate alkalinity = 2.6 – 0.40 = 2.2 meq/L

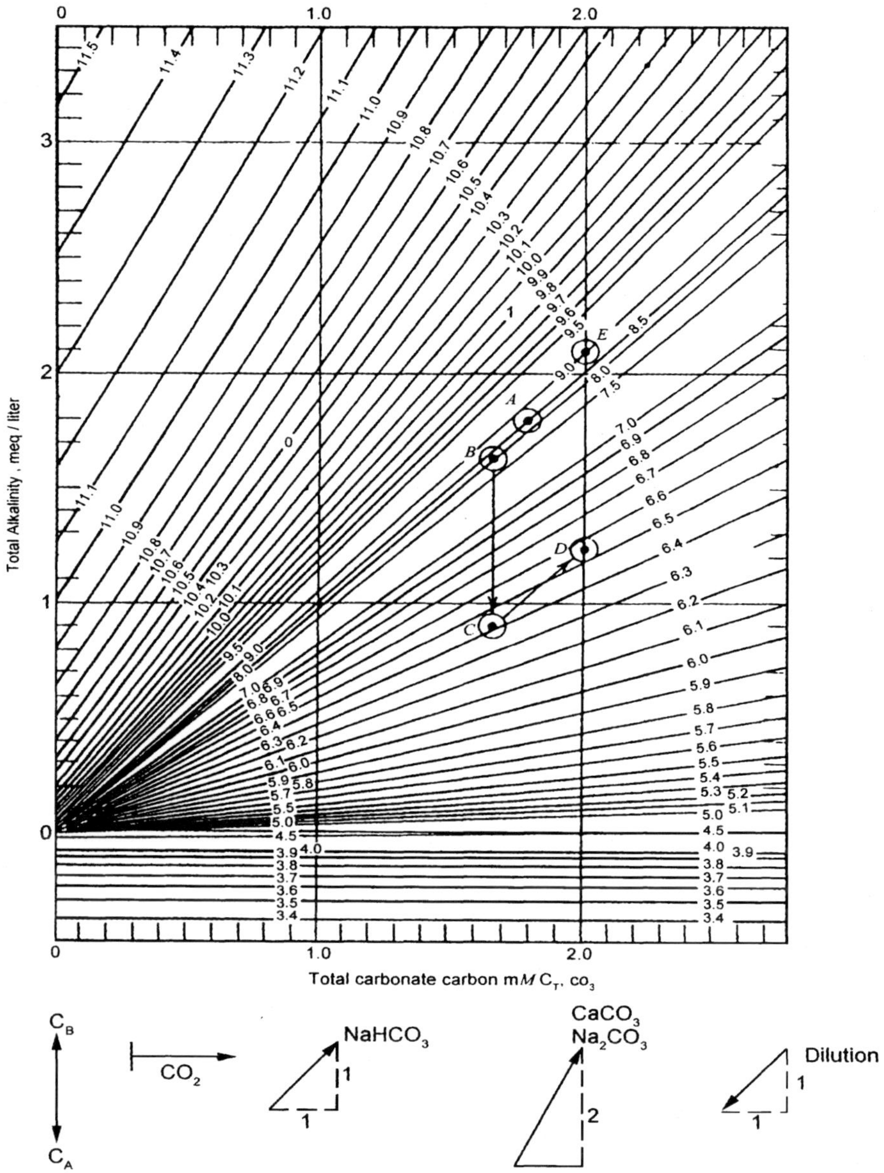

Figure 2-6. The Deffeyes diagram for 25°C. (From K. S. Deffeyes, Limnol. Ocean. 10: 412, 1965.)

A well-known Deffeyes diagram is used for the correlation among alkalinity, C_T, and pH. Using the diagram, if two of the three properties are known, then the third property can be evaluated (Fig. 2-6). Also the removal of CO_2 can cause a horizontal shift; the addition or removal of bicarbonate causes a 45° movement. Finally, the carbonate addition or removal causes a 60° movement from the horizontal. It can also be seen from the figure that dilution will cause a 45° downward movement.

2.4 Lime-Soda Ash Process

The definition of alkalinity has been discussed in Chapter 3, especially Example 3-1 (see Eq. [3-10]), which illustrates how the alkalinity of solution species is expressed as a $CaCO_3$ equivalent. Like alkalinity, hardness can be expressed in terms of mg/L of $CaCO_3$. Therefore, the **total hardness**, TH, is:

$$TH = [(Ca^{2+}) + (Mg^{2+})]\ 100 \qquad [2\text{-}49]$$

The **carbonate hardness**, CH, is the portion of the total hardness that is offset by bicarbonate and carbonate ions in the water:

$$CH = [1/2\ (HCO_3^-) + CO_3^{2-}]\ 100 \qquad [2\text{-}50]$$

The portion of hardness in excess of the original carbonate hardness is termed **noncarbonate hardness**, or NCH,

$$NCH = TH - CH \qquad [2\text{-}51]$$

The NCH actually results from the Ca^{2+} and Mg^{2+} associated, for example, with chloride, sulfate anions, or other anions.

The steps of the lime–soda ash water softening process are as follows:

(a) raise pH of water solution by adding lime to allow for precipitation for removal of CH

$$H_2CO_3^* + Ca(OH)_2 = CaCO_3\downarrow + 2H_2O \qquad (a)$$

(b) use Le Chetelier's principle to remove CH and Ca^{2+}

$$Ca^{2+} + 2HCO_3^- + Ca(OH)_2 = 2CaCO_3\downarrow + 2H_2O \qquad (b)$$

(c) remove Mg^{2+}

$$Mg^{2+} + 2HCO_3^- + 2Ca(OH)_2 = Mg(OH)_2\downarrow + 2CaCO_3 + 2H_2O \quad \text{(c)}$$

(d) use soda ash

$$Ca^{2+} + Na_2CO_3 = CaCO_3\downarrow + 2Na^+ \quad \text{(d')}$$

$$Mg^{2+} + Ca(OH)_2 = Mg(OH)_2\downarrow + Ca^{2+}$$

$$Mg^{2+} + Ca(OH)_2 + Na_2CO_3 = Mg(OH)_2\downarrow + CaCO_3\downarrow + 2Na^+ \quad \text{(d)}$$

In practice, a stoichiometric model is used where a decision is made about (HCO_3^-) and 2(Ca^{2+}). If (HCO_3^-) > 2(Ca^{2+}), the remaining (HCO_3^-) goes to Equation (c) and all the (Ca^{2+}) goes to Equation (b), then the remaining Mg^{2+} all goes to Equation (d). On the other hand, if 2(Ca^{2+}) > (HCO_3^-), then put all (HCO_3^-) in Equation (b); do not use Equation (c); enter the left-over (Ca^{2+}) to Equation (d′) and the left-over (Mg^{2+}) to Equation (d) (see Fig. 2-7).

The analytes can be expressed either in mM or mg/L of $CaCO_3$ using a bar diagram. The stoichiometric principle is based on:

$$(Ca^{2+})_{\text{final}} = (C_T)_{\text{final}} \quad [2\text{-}52]$$

That is,

$$(Ca^{2+})_0 + (Ca^{2+})_a = (C_T)_0 + (C_T)_a \quad [2\text{-}53]$$

Here, the subscript 0 is original, the subscript a is added, the $(Ca^{2+})_a$ pertains to added lime, and the $(C_T)_a$ pertains to added soda ash. Example 2-5 illustrates the principles of Equations [2-52] and [2-53].

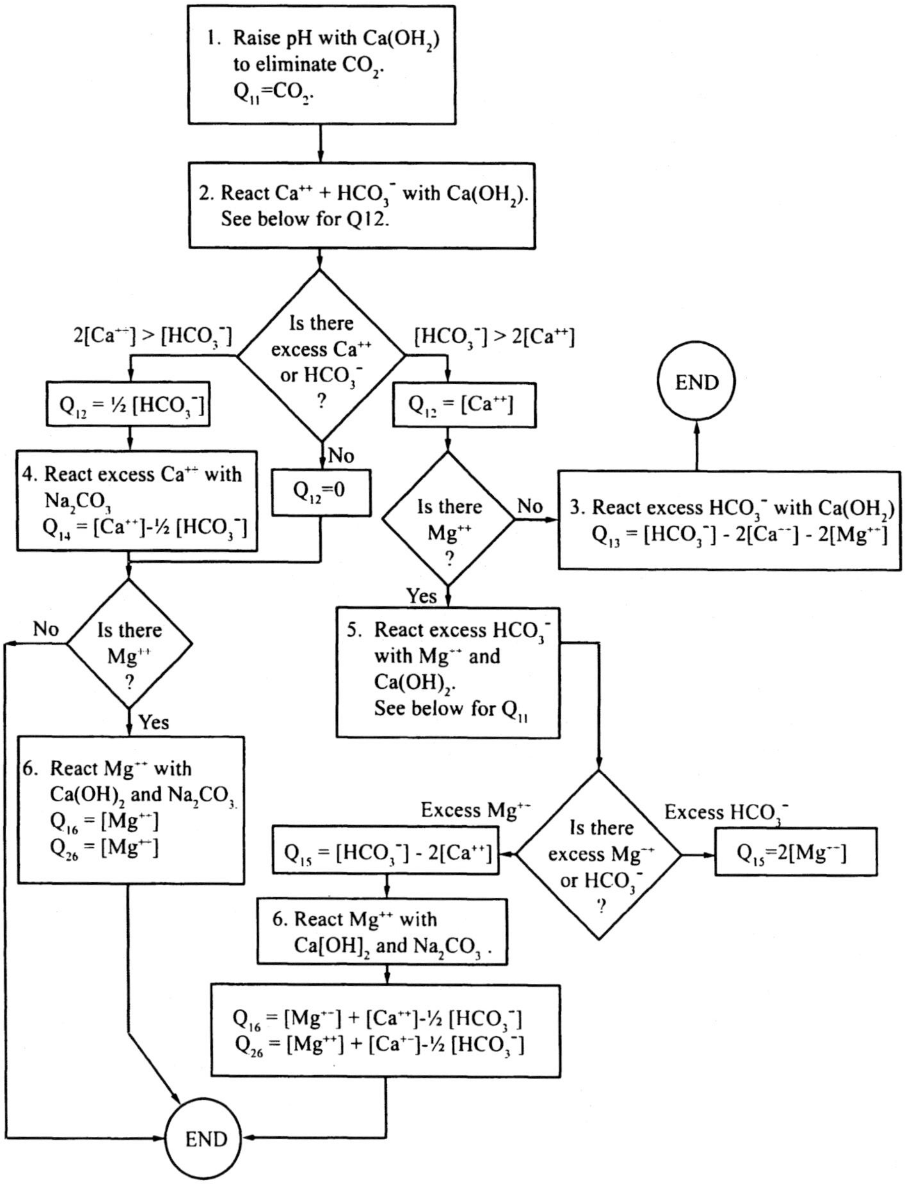

Figure 2-7. Logical decision structure of the stoichiometric model for the lime–soda water softening process. (From Ref. 2.3)

[Example 2-5] For the following two raw waters, how much lime and soda ash are required?

Water 1	Species	mg/L	mM
	Total hardness	215 (as $CaCO_3$)	2.15
	Mg^{2+}	15.8	0.65
	Na^+	8	0.35
	SO_4^{2-}	28.6	0.3
	Cl^-	10	0.285
	Alkalinity	185	3.7
	CO_2	25.8 (as CO_2)	0.59
	pH = 6.7		

Water 2			
	Ca^{2+}		5
	Mg^{2+}		2
	HCO_3^-		4
	$H_2CO_3^*$		4
	SO_4^{2-}		6
	Cl^-		0.8
	pH = 6.3		

More information can be added from the analytes. That is,

Water 1

$$Ca^{2+}\text{ (from TH)}(2.15 - 0.65 = 1.5\text{ mM}$$

$$HCO_3^-(= \text{alk}) = 3.7\text{ mM}$$

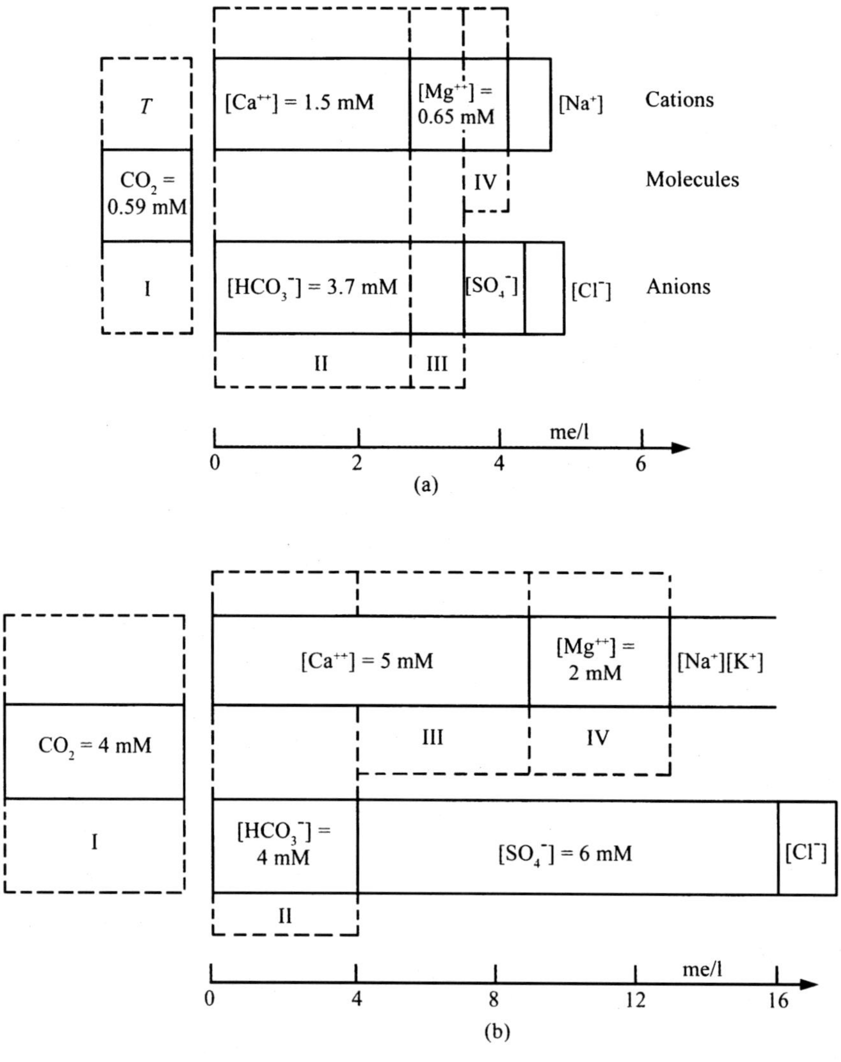

Figure 2-8. Cation-anion balance of waters to be softened by the lime–soda water process. (From Ref. 2.4.)

	CO_2 ($H_2CO_3^*$)	HCO_3^-	Ca^{2+}	Mg^{2+}	Lime	Soda
Eq(a)	1				1	
water 1	0.59				0.59	
water 2	4				4	
Eq(b)		2	1		1	
water 1		3	1.5		1.5	
water 2		4	2		2	
Eq(c)		2		1	2	
water 1		3.7-3		0.7/2	0.7	
water 2		–		–	–	
Eq(d′)			1			1
water 1						–
water 2			5-2			3
Eq(d)				1	1	1
water 1				0.65-0.7/2	0.3	0.3
water 2				2	2	2
Σ (lime) = $(Ca^{2+})_a$						
water 1					3.09	
water 2					8	
Σ (soda) = $(C_T)_a$						
water 1						0.3
water 2						5
$(Ca^{2+})_0$						
water 1			1.5			
water 2			5			
$(C_T)_0$						
water 1		0.59 +3.7				
water 2		4+4				

Underlined values indicate the stoichiometric requirement from each reaction from (a) to (d) and (d′).

Using Equations (a) to (d) for both waters (see also bar diagram in Fig. 2-8), then:

$$(\mathrm{Ca}^{2+})_0 + (\mathrm{Ca}^{2+})_a = (C_T)_0 + (C_T)_a$$

water 1 4.59

water 2 13

Thus, the requirements are in Σ (lime) and Σ (soda) as above in mM, but also can be expressed in other units.

REFERENCES

2-1. V. L. Snoeyink and D. Jenkins, *Water Chemistry*, Wiley, New York, 1980.

2-2. D. W. Sundstrom and W. E. Klee, *Wastewater Treatment*, Prentice-Hall, Englewood Cliffs, New Jersey, 1979.

2-3. J. N. Butler, *Carbon Dioxide Equilibria and Their Applications*, Addison-Wesley, Reading, Massachusetts, 1982.

2-4. G. M. Fair, J. C. Geyer, and D. A. Okun, *Water and Wastewater Engineering, Vol.2, Water Purification and Wastewater Treatment and Disposal*, Wiley, New York, 1968.

2-5. R. E. Lowenthal and G.V.R. Marais, *Carbonate Chemistry of Aquatic Systems: Theory and Application*, Ann Arbor Science, Ann Arbor, Michigan, 1976.

2-6. J. W. Clark, W. Viessman, and M. J. Hammer, *Water Supply and Pollution Control*, Harper and Row, New York, 1977.

PROBLEM SET

1. The total alkalinity of a sample of water with a pH of 11.0 is 200 mg/L. Find the distribution of the alkalinity as (a) bicarbonate, (b) carbonate, and (c) hydroxide alkalinity in mg/L as $CaCO_3$. Assume a water temperature of 25°C and that the effect of salinity can be discounted.

2. A water that is to be softened by the lime-soda process has the following carbon dioxide, calcium, magnesium, and bicarbonate concentrations:

   CO_2 = 8 meq/L Mg^{2+} = 32 meq/L

   Ca^{2+} = 65 meq/L HCO_3^- = 260 meq/L

   Determine the calcium oxide and soda ash required per million gallons if the purities are 85 and 95%, respectively. Use meq/L as the basis of computation for both alkalinity and hardness instead of mg/L as $CaCO_3$.

3. Compute the pH of a solution of limewater containing 200 mg/L of $Ca(OH)_2$ at 15°C.

4. (a) Compute the pounds of lime (CaO = 56) and soda ash (Na_2CO_3 = 106) required to soften 1 million gallons of water with analysis as shown:

   pH 7.2

   Total hardness 300 ppm as $CaCO_3$

   Total alkalinity 150 ppm as $CaCO_3$

   Calcium 200 ppm as $CaCO_3$

   Magnesium 100 ppm as $CaCO_3$

   (b) If the solubility products K_s of $CaCO_3$ and $Mg(OH)_2$ are, respectively, 0.87×10^{-8} and 5.2×10^{-11} for moles per liter, and the final effluent from the softening plant has a pH of 10.5 and a normal carbonate (CO_3^{2-}) alkalinity of 300 ppm as $CaCO_3$, calculate the theoretical residual of calcium and magnesium hardness, both as $CaCO_3$ (100).

CHAPTER 3

PRECIPITATION AND STABILIZATION

For municipal and industrial wastewater treatment, processes involving precipitation and stabilization are often of paramount importance. In this chapter, we will emphasize some of the basic principles on which both precipitation and stabilization are founded. The concept of thermodynamics involving equilibrium of reactions is extensively useful not only to explain the phenomena arising from practical water treatment, but to illustrate that similar principles can be adopted for mineral compositions of subsurface waters, geochemical modification of ore deposits, and chemostasis in oceanography. Formation and the dissolution of solid phase will be discussed.

First we will discuss the solubility product and solubility in terms of environmental practices, such as their roles in water softening and pH adjustment. Then we will look at bivalent and trivalent iron in natural waters. We will study phosphate in wastewater and its removal by a common alum coagulation material. The Langelier Index, which is discussed in this chapter, is a quantitative criterion for evaluating water stabilization. In the last section, we will look into the complexing reactions and explain masking and demasking for certain metals in water. In many cases, a certain step or unit process of a multistep treatment is separated during solid formation by precipitation, and in those situations masking by a complexing reagent is essential.

3.1 Dissolution and Solubility Product

In Chapters 1 and 2 we briefly discussed solubilities and the solubility product (K_s). The dissolution of a given mineral in water is controlled by its solubility product (see Table 3-1). This can be illustrated with A_zB_y.

$$A_z B_{y(s)} = zA^{y+} + yB^{z-} \quad [3\text{-}1]$$

and the solubility product is:

$$K_s = \frac{(A^{y+})^z (B^{z-})^y}{(A_z B_y)_S} \quad [3\text{-}2]$$

The concentration is expressed in terms of activity, a, making $(A_zB_y)_s = 1$. The concentration product, ${}^cK_s$, can be related to the equilibrium constant, K_s, by the activity coefficient, γ.

$${}^cK_s = (A^{y+})^z (B^{z-})^y = \frac{K_s}{(\gamma_{A^{y+}})^z (\gamma_{B^{z-}})^y} \quad [3\text{-}3]$$

If we delete the y^+ and z^-, then

$$K_s = (A)^z (B)^y \gamma_A{}^z \gamma_B{}^y \quad [3\text{-}4]$$

and if the Debye-Hückel extended form is used for correction,

$$\log K_s = \log {}^cK_s - (zZ_A{}^2 + yZ_B{}^2)(A\mu^{1/2}/(1 + Ba\mu^{1/2})) \quad [3\text{-}5]$$

It can also be written as:

$$pK_s = p^cK_s + \left(z Z_A^2 + y Z_B^2\right)\left(\frac{0.5\mu^{\frac{1}{2}}}{1+\mu^{\frac{1}{2}}}\right) \quad [3\text{-}6]$$

Table 3-1. Solubility Products of Various Minerals

		Log K[a]
Chlorides		
Halite	$NaCl(s) = Na^+ + Cl^-$	1.54
Sylvite	$KCl(s) = K^+ + Cl^-$	0.98
Chlorargyite	$AgCl = Ag^+ + Cl^-$	−9.74
Sulfates		
Gypsum	$CaSO_4{\cdot}2H_2O(s) = Ca^{2+} + SO_4^{2-} + 2H_2O$	−4.62
Celestite	$SrSO_4(s) = Sr^{2+} + SO_4^{2-}$	−6.50
Barite	$BaSO_4(s) = Ba^{2+} + SO_4^{2-}$	−9.96
Oxides and Hydroxides		
Calcium hydroxide	$Ca(OH)_2(s) = Ca^{2+} + 2OH^-$	−5.19
	$CaOH^+ = Ca^{2+} + OH^-$	−1.15
Brucite	$Mg(OH)_2(s) = Mg^{2+} + 2OH^-$	−11.1
	$MgOH^+ = Mg^{2+} + OH^-$	−2.6
	$Mg_4(OH)_4^{4+} = 4Mg^{2+} + 4OH^-$	−16.3
Gibbsite	$Al(OH)_3(s) = Al^{3+} + 3OH^-$	−33.5
	$AlOH_2^+ = Al^{3+} + OH^-$	−9.0
	$Al(OH)_2^+ = Al^{3+} + 2OH^-$	−18.7
	$Al(OH)_3 = Al^{3+} + 3OH^-$	−27.0
	$Al(OH)_4^- = Al^{3+} + 4OH^-$	−33.0
	$Al_2(OH)_2^{4+} = 2Al^{3+} + 2OH^-$	−20.3
	$Al_3(OH)_4^{5+} = 3Al^{3+} + 4OH^-$	−42.1
	$MnOH^+ = Mn^{2+} + OH^-$	−3.4

Table 3-1. Continued

Ferrous hydroxide	$Fe(OH)_2(s) = Fe^{2+} + 2OH^-$	–15.1
	$FeOH^+ = Fe^{2+} + OH^-$	–4.5
	$Fe(OH)_2 = Fe^{2+} + 2OH^-$	–7.4
	$Fe(OH)_3^- = Fe^{2+} + 3OH^-$	–11.0
	$Fe(OH)_4^{2-} = Fe^{2+} + 4OH^-$	–10.0
Goethite	$\alpha\text{-}FeOOH(s) + H_2O = Fe^{3+} + 2OH^-$	–41.5
Ferric hydroxide	$am\text{-}Fe(OH)_3(s) = Fe^{3+} + 3OH^-$	–38.8
Hematite	$\frac{1}{2}\ \alpha\text{-}Fe_2O_3(s) + 3/2\ H_2O = Fe^{3+} + 3OH^-$	–42.7
	$FeOH^{2+} = Fe^{3+} + OH^-$	–11.8
	$Fe(OH)_2^+ = Fe^{3+} + 2OH^-$	–22.3
	$Fe(OH)_4^- = Fe^{3+} + 4OH^-$	–34.4
	$Fe_2(OH)_2^{4+} = 2Fe^{3+} + 2OH^-$	–25.1
	$Fe_3(OH)_4^{5+} = 3Fe^{3+} + 4OH^-$	–49.7
Tenorite	$CuO(s) + H_2O = Cu^{2+} + 2OH^-$	–20.4
Cupric hydroxide	$Cu(OH)_2(s) = Cu^{2+} + 2OH^-$	–19.4
	$Cu(OH)_4^{2-} = Cu^{2+} + 4OH^-$	–16.4
	$Cu_2(OH)_2^{2+} = 2Cu^{2+} + 2OH^-$	–17.6
Litharge (red)	$PbO(s) + H_2O = Pb^{2+} + 2OH^-$	–15.3
Massicot (yellow)	$PbO(s) + H_2O = Pb^{2+} + 2OH^-$	–15.1
	$PbOH^- = Pb^{2+} + OH^-$	–6.3
	$Pb(OH)_2 = Pb^{2+} + 2OH^-$	–10.9
	$Pb(OH)_3^- = Pb^{2+} + 3OH^-$	–13.9
	$Pb_2(OH)^{3+} = 2Pb^{2+} + OH^-$	–7.6
	$Pb_3(OH)_4^{2+} = 3Pb^{2+} + 4OH^-$	–32.1
	$Pb_4(OH)_4^{4+} = 4Pb^{2+} + 4OH^-$	–35.1
	$Pb_6(OH)_8^{4+} = 6Pb^{2+} + 8OH^-$	–68.4
Carbonates		
Aragonite	$CaCO_3(s) = Ca^{2+} + CO_3^{2-}$	–8.22
Calcite	$CaCO_3(s) = Ca^{2+} + CO_3^{2-}$	–8.35

Table 3-1. Continued

Magnesite	$MgCO_3(s) = Mg^{2+} + CO_3^{2-}$	–7.46
Nesquehonite	$MgCO_3{\cdot}3H_2O(s) = Mg^{2+} + CO_3^{2-} + 3H_2O$	–4.67
Dolomite	$CaMg(CO_3)_2(s)$ = Calcite + Magnesite	–1.70
Disordered dolomite	$CaMg(CO_3)_2(s)$ = Calcite + Magnesite	–0.08
Strontianite	$SrCO_3(s) = Sr^{2+} + CO_3^{2-}$	–9.0
Rhodochrosite	$MnCO_3(s) = Mn^{2+} + CO_3^{2-}$	–10.4
Siderite	$FeCO_3(s) = Fe^{2+} + CO_3^{2-}$	–10.7
Malachite	$Cu_2CO_3(OH)_2(s) = 2Cu^{2+} + CO_3^{2-} + 2OH^-$	–33.8
Azurite	$Cu_3(CO_3)_2(OH)_2(s) = 3Cu^{2+} + 2CO_3^{2-} + 2OH^-$	–46.0
Cerussite	$PbCO_3(s) = Pb^{2+} + CO_3^{2-}$	–13.1
Phosphates		
Brushite	$CaHPO_4{\cdot}2H_2O(s) = Ca^{2+} + HPO_4^{2-} + 2H_2O$	–6.6
Hydroxylapatite	$Ca_5(PO_4)_3OH(s) = 5Ca^{2+} + 3PO_4^{3-} + OH^-$	–55.6
Newberyite	$MgHPO_4{\cdot}3H_2O(s) = Mg^{2+} + HPO_4^{2-} + 3H_2O$	–5.8
Bobierrite	$Mg_3(PO_4)_2{\cdot}8H_2O(s) = 3Mg^{2+} + 2PO_4^{2-} + 8H_2O$	–25.2
Vivianite	$Fe_3(PO_4)_2{\cdot}8H_2O(s) = 3Fe^{2+} + 2PO_4^{3-} + 8H_2O$	–36.0
Strengite	$FePO_4{\cdot}2H_2O(s) = Fe^{3+} + PO_4^{3-} + 2H_2O$	–26.4
Berlinite	$AlPO_4(s) = Al^{3+} + PO_4^{3-}$	–20.6
Sulfides		
Pyrrhotite	$FeS(s) = Fe^{2+} + S^{2-}$	–18.1
Pyrite	$FeS_2(s)$ = Pyrrhotite + S^0	–10.4
Alabandite	$MnS(s) = Mn^{2+} + S^{2-}$	–13.5
Covellite	$CuS(s) = Cu^{2+} + S^{2-}$	–36.1
Galena	$PbS(s) = Pb^{2+} + S^{2-}$	–27.5
Chalcopyrite	$CuFeS_2(s)$ = Covellite + Pyrrhotite	–6.0
Silicates		
Quartz ($\alpha + \beta$)	$SiO_2(s) + H_2O = H_2SiO_3$	–4.00
Cristobalite ($\alpha + \beta$)	$SiO_2(s) + H_2O = H_2SiO$	–3.45

Table 3-1. Continued

Amorphous silica	$SiO_2(s) + H_2O = H_2SiO_3$	–2.71
Fosterite	¼ $(2MgO \cdot SiO_2)(s) + H^+ =$ ½ Mg^{2+} + ¼ H_2SiO_3 + ¼ H_2O	7.11
Fayalite	¼ $(2FeO \cdot SiO_2)(s) + H^+ =$ ½ Fe^{2+} + ¼ H_2SiO_3 + ¼ H_2O	4.21
Enstatite	½ $(MgO \cdot SiO_2)(s) + H^+ =$ ½ Mg^{2+} + ½ H_2SiO_3	5.82
Wollastonite	½ $(CaO \cdot SiO_2)(s) + H^+ =$ ½ Ca^{2+} + ½ H_2SiO_3	6.82
Hedenbergite	¼ $(CaO \cdot FeO \cdot 2SiO_2)(s) + H^+ =$ ¼ Ca^{2+} + ¼ Fe^{2+} + ½ H_2SiO_3	4.60
Diopside	¼ $(CaO \cdot MgO \cdot 2SiO_2)(s) + H^+ =$ ¼ Ca^{2+} + ¼ Mg^{2+} + ½ H_2SiO_3	5.30
Tremolite	$1/14\ (2CaO \cdot 5MgO \cdot 8SiO_2 \cdot H_2O)(s) + H^+ = 1/7\ Ca^{2+} + 5/14\ Mg^{2+} + 4/7\ H_2SiO_3$	4.46
Alumino-Silicates		
Gibbsite	$Al(OH)_3(s) + H_2SiO_3 =$ ½ Kaolinite + $3/2\ H_2O$	4.25
Phlogopite	$1/14\ (K_2O \cdot 6MgO \cdot Al_2O_3 \cdot 6SiO_2 \cdot 2H_2O)(s) + H^+ = 1/14$ Kaolinite + $1/7\ K^+ + 3/7\ Mg^{2+} + 2/7\ H_2SiO_3 + 3/14\ H_2O$	5.01
Muscovite	½ $(K_2O \cdot 3Al_2O_3 \cdot 6SiO_2 \cdot 2H_2O)(s) + H^+ + 3/2\ H_2O = 3/2$ Kaolinite + K^+	3.51
Anorthite	½ $(CaO \cdot Al_2O_3 \cdot 2SiO_2)(s) + H^+ + 1/2\ H_2O = 1/2$ Kaolinite + ½ Ca^{2+}	9.83
Albite	½ $(Na_2O \cdot Al_2O_3 \cdot 6SiO_2)(s) + H^+ + 5/2\ H_2O = 1/2$ Kaolinite + $2H_2SiO_3 + Na^+$	–0.68
K-feldspar	½ $(K_2O \cdot Al_2O_3 \cdot 6SiO_2)(s) + H^+ + 5/2\ H_2O = 1/2$ Kaolinite + $2H_2SiO_3 + K^+$	–3.54
Na-Montmorillonite [b]	½ $(Na_2O \cdot 7Al_2O_3 \cdot 22SiO_2 \cdot 6H_2O)(s) + H^+ + 15/2\ H_2O =$ 7/2 Kaolinite + $4H_2SiO_3 + Na^+$	–9.1
Ca-Montmorillonite [b]	½ $(CaO \cdot 7Al_2O_3 \cdot 22SiO_2 \cdot 6H_2O)(s) + H^+ + 15/2\ H_2O =$ 7/2 Kaolinite + $4H_2SiO_3$ + ½ Ca^{2+}	–7.7

a Except where noted all constants are valid for 25°C and ionic strength $I = 0\ M$.

b Thermodynamic properties of minerals with continuously variable stoichiometric composition (e.g., montmorillonites and illites) are not simply defined. The values reported here allow for representative calculations of water composition, not rigorous thermodynamic geochemical analysis.

Source: Adapted from Ref. 3-1.

Since in the above, $A \cong 0.5$, $Ba \cong 1$, and if the Davies equation is used,

$$\mathrm{p}K_s = \mathrm{p}^{\mathrm{c}}K_s + \left(zZ_A^2 + yZ_B^2\right)\left(\frac{0.5\mu^{\frac{1}{2}}}{1+\mu^{\frac{1}{2}}} - 0.2\mu\right) \qquad [3\text{-}7]$$

An important conclusion from the preceding equations is that solubility increases with increasing ionic strength. The effect is pronounced for species containing multiple ions. Also, if a salt is introduced to an aqueous solution containing one of the ions, solubility will be depressed. This is called the **common ion effect**, as indicated by Example 3-1.

[Example 3-1] Evaluate the solubility of CaF_2 in pure water and in a 10^{-3} molar solution of $CaCl_2$.

$$CaF_2 = Ca^{2+} + 2F^-$$

Let $(Ca^{2+}) = S$, then $(F^-) = 2S$ and

$$K_s = \left(Ca^{2+}\right)\gamma_{Ca^{2+}}\left(F^-\right)^2\gamma_{F^-}^2 = S(2S)^2\,\gamma_{Ca^{2+}}\gamma^2_{F^-} = 4.0 \times 10^{-11}$$

For the Debye-Hückel limiting law,

$$\gamma_{Ca^{2+}} = 0.79, \gamma_{F^-} = 0.94$$

and for the extended form

$$\gamma_{Ca^{2+}} = 0.84, \gamma_{F^-} = 0.96$$

After the correction, it becomes

$$S = 2.3 \times 10^{-4}\ \mathrm{M} \qquad (a)$$

Compare this with the uncorrected calculation:

$$S = 2.1 \times 10^{-4}\ \mathrm{M} \qquad (b)$$

If the salt is added to a solution containing $CaCl_2$, the total Ca^{2+} ion should be:

$$(Ca^{2+}) = S + 10^{-3}$$

Then

$$K_s = \left(S + 10^{-3}\right)\left(2S^2\right)\gamma_{Ca^{2+}}\gamma_{F^-}^{\ 2} = 4.0 \times 10^{-11}$$

Again by trial and error,

$$\gamma_{Ca^{2+}} = 0.78, \gamma_{F^-} = 0.94$$

$$S = 1.1 \times 10^{-4}\ \text{M} \qquad \text{(c)}$$

Comparing with the corrected (a) and uncorrected (b) solubilities, the solubility value (c) is depressed.

[Example 3-2] In the gravimetric method for determination of chloride, why is potassium dichromate used as an endpoint indicator?

From Table 3-2,

$$K_{s\,(AgCl)} = 1 \times 10^{-10}$$

$$K_{s\,(Ag_2CrO_4)} = 2.5 \times 10^{-12}$$

Neglecting ionic strength,

$$(S) = (Ag^+) = (Cl^-)$$

$$(S)(S) = 10^{-10}$$

Thus,

$$S = 10^{-5}\ \text{M}$$

and for Ag_2CrO_4

$$(S')\,(2S')^2 = 2.5 \times 10^{-12}$$

$$S' = 8.6 \times 10^{-5}\ \text{M}$$

Table 3-2. Solubility Constants at 25°C

Substance	Ion Product	$pK_s = -\log K_s$	Application
$Al(OH)_3$	$(Al^{3+})\,(OH^-)^3$	32.9	Coagulation
$AlPO_4$	$(Al^{3+})\,(PO_4^{3-})$	22.0	Phosphate removal
$AgCl$	$(Ag^+)\,(Cl^-)$	9.75	Analysis
$BaSO_4$	$(Ba^{2+})\,(SO_4^{2-})$	9.96	Analysis
$CaCO_3$	$(Ca^{2+})\,(CO_3^{2-})$	8.32	Softening; corrosion control
CaF_2	$(Ca^{2+})\,(F^-)^2$	10.41	Fluoridation
$Ca(OH)_2$	$(Ca^{2+})\,(OH^-)^2$	5.26	Softening
$Ca_3(PO_4)_2$	$(Ca^{2+})^3\,(PO_4^{3-})^2$	26.0	Softening; phosphate cycle
$CaHPO_4$	$(Ca^{2+})\,(HPO_4^{2-})$	7.0	Phosphate cycle
$CaSO_4{\cdot}2H_2O$	$(Ca^{2+})\,(SO_4^{2-})$	4.62	Hardness; scale formation
$Cu(OH)_2$	$(Cu^{2+})\,(OH^-)^2$	19.25	Algae control
$Fe(OH)_2$	$(Fe^{2+})\,(OH^-)^2$	15.1	Corrosion; deferrization
$FeCO_3$	$(Fe^{2+})\,(CO_3^{2-})$	10.68	Iron cycle
FeS	$(Fe^{2+})\,(S^{2-})$	17.3	Anaerobic corrosion
$Fe(OH)_3$	$(Fe^{3+})\,(OH^-)^3$	38.04	Coagulation; deferrization
$FePO_4$	$(Fe^{3+})\,(PO_4^{3-})$	21.9	Phosphate cycle
$MgCO_3{\cdot}3H_2O$	$(Mg^{2+})\,(CO_3^{2-})$	4.25	Hardness
MgF_2	$(Mg^{2+})\,(F^-)^2$	8.18	Defluoridation
$Mg(OH)_2$	$(Mg^{2+})\,(OH^-)^2$	10.74	Softening
$MnCO_3$	$(Mn^{2+})\,(CO_3^{2-})$	10.06	Manganese cycle
$Mn(OH)_2$	$(Mn^{2+})\,(OH^-)^2$	12.8	Demanganization
$Zn(OH)_2$	$(Zn^{2+})\,(OH^-)^2$	17.15	Corrosion

When we compare S' with S, notice that Ag_2CrO_4 is eight times more soluble than AgCl. Also, (Ag^+) is about 17 times greater in the Ag_2CrO_4

solution. K_2CrO_4 (yellow) is used as an indicator when an unknown chloride solution is titrated with $AgNO_3$ solution. A white precipitate (AgCl) is formed, and when the red color (Ag_2CrO_4 precipitate) appears, an endpoint is obtained, because Ag^+ is added to lower the Cl^- concentration to less than 10^{-5} M.

The solubility product (Table 3-2) finds practical use in a lime–soda ash softening process. For example, the K_s of the following species are:

$CaCO_3$	$10^{-8.32}$
$Ca(OH)_2$	$10^{-5.26}$
$MgCO_3$	$10^{-4.25}$
$Mg(OH)_2$	$10^{-10.74}$

We can very easily precipitate Ca^{2+} as $CaCO_3$ and Mg^{2+} as $Mg(OH)_2$, but never as $MgCO_3$. The same reasoning applies to lime, $Ca(OH)_2$, which is quite soluble. Therefore, it is used to add to a "hard" water.

In general, if for A_zB_y,

$$(A^{y+})^z\,(B^{z-})^y < K_s$$

there will be no precipitation, whereas if

$$(A^{y+})^z\,(B^{z-})^y > K_s$$

the solution will be supersaturated and will finally precipitate. In analytical chemistry, for a qualitative group analysis of a mixture of inorganic ions such as Fe, Ti, Cr, Al, Be, etc., the solution is initially made basic at a high pH. If the metal concentration Me < 10^{-5} M at pH 9, the divalent ion is:

$$(Me^{2+})\,(OH^-)^2 = (Me^{2+})\,(10^{-10}) \leq 10^{-15}$$

and the trivalent ion is:

$$(Me^{3+})\,(OH^-)^3 = (Me^{3+})\,(10^{-15}) \leq 10^{-20}$$

Suppose we change the pH to 5; then for the divalent ion, the solubility product is $\leq 10^{-23}$, and for the trivalent ion, the solubility product is $\leq 10^{-32}$. Thus, more ions will tend to precipitate on the basic side. For example, the K_s of

$Ni(OH)_2$ is 10^{-17} at pH 5. Ni^{2+} ion will not precipitate, whereas if we adjust the pH to 9, it will. Obviously,

$$10^{-15}\ (\text{ppt.}) > 10^{-17} > (\text{no ppt.})\ 10^{-23}$$

All adjustments are made by adding acetic acid or ammonia.

$$HAc = H^+ + Ac^-$$

$$K_a = \frac{(H^+)(Ac^-)}{(HAc)} \cong 10^{-5} \qquad [3\text{-}8]$$

$$NH_3 + H^+ = NH_4^+$$

$$K_b = \frac{(NH_4^+)}{(H^+)(NH_3)} \cong 10^{-9} \qquad [3\text{-}9]$$

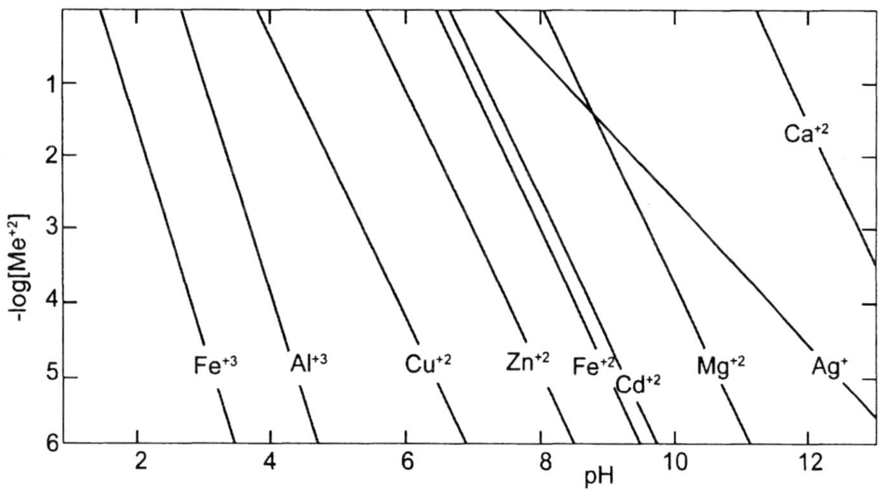

Figure 3-1. Free metal ion concentration in equilibrium with solid oxides or hydroxides. The occurrence of hydroxo metal complexes must be considered for evaluation of complete solubility.

Figure 3-1 lists examples for the oxides and hydroxides, such as

$$K_a \cdot K_b = K_w = 10^{-14}$$

$$\text{Me (OH)}_2 = \text{Me}^{2+} + 2\ \text{OH}^-$$

$$(Me^{2+})\ (OH^-)^2 = K_s$$

$$\text{Me (OH)}_2 + 2\ \text{H}^+ = \text{Me}^{2+} + 2\ \text{H}_2\text{O}$$

or

$$\text{MeO} + 2\text{H}^+ = \text{Me}^{2+} + \text{H}_2\text{O}$$

$$K = \frac{(Me^{2+})}{(H^+)^2} = \frac{K_s}{K_w^{\,2}}$$

$$\log (Me^{2+}) = \log K - 2\ \text{pH} = -\text{p}K_s + 2\ \text{p}K_w - 2\ \text{pH} \qquad [3\text{-}10]$$

The solubility of salts from a weak acid or a weak base may be more complicated. The ionization fractions may be strongly dependent on pH. For a monoprotic acid–conjugated base pair,

$$\alpha_0 = \frac{(H^+)}{(H^+) + K_a} = \frac{(HA)}{C_{T_1A}} \qquad [3\text{-}11]$$

$$\alpha_1 = \frac{K_a}{(H^+) + K_a} = \frac{(A^-)}{C_{T_1A}} \qquad [3\text{-}12]$$

In this case the ionic strength effect is negligible, and $K_a = {}^cK_a$.

$$C_{T_1A} = (HA) + (A^-) \qquad [3\text{-}12a]$$

and

$$K_a = \frac{(H^+)(A^-)}{(HA)}$$

[Example 3-3] Find the solubility of AgCN in a closed system as a function of pH.

There are six unknowns: (S), (Ag^+), (CN^-), (HCN), (H^+), and (OH^-). We only have five known relations:

$$HCN = H^+ + CN^-;\ pK_a = 9.15$$

$$AgCN = Ag^+ + CN^-;\ pK_s = 13.8$$

$$(S) = (Ag^+) = C_{T,\mathrm{Ag}}$$

$$(S) = (HCN) + (CN^-) = C_{T,\mathrm{CN}}$$

$$K_w = (H^+)(OH^-)$$

Put K_a in the $C_{T,\mathrm{CN}}$ relation.

$$(S) = (HCN) + (CN^-) = \frac{(H^+)(CN^-)}{K_a} + (CN^-) = \frac{(CN^-)[(H^+) + K_a]}{K_a} = \frac{(CN^-)}{\alpha_1} \quad \text{(a)}$$

From K_s,

$$(CN^-) = K_s/(Ag^+) = K_s/(S) \quad \text{(b)}$$

Therefore, from (a) and (b),

$$(S)\alpha_1 = K_s/(S) \quad \text{(c)}$$

or

$$(S)=\left(\frac{K_s}{\alpha_1}\right)^{\frac{1}{2}}=\left\{\left(\frac{K_s}{K_a}\right)\left[\left(H^+\right)+K_a\right]\right\}^{\frac{1}{2}} \quad \text{(d)}$$

Thus, when $(H^+) << K_s$,

$$(S) = (K_s)^{1/2}$$

or

$$\log (S) = 1/2 \log K_s = -6.9$$

When pH > 9.15, solubility is $10^{-6.9}$ M, and when $(H^+) >> K_a$, then

$$(S) = [K_s[H^+]/K_a]^{1/2}$$

$$\log (S) = 1/2 (-pK_s + pK_a - pH) = 1/2 (-4.65 - pH)$$

Thus, when pH < 9.15, solubility increases as the pH decreases.

3.2 A Heterogeneous System of Iron

Some notes should be made concerning the precipitation and dissolution before we discuss iron and manganese present in natural waters. Often in natural water systems, certain minerals are formed while others are being dissolved. Under suitable conditions, a cluster of ions or molecules selects the appropriate composition to form a solid to initiate the crystal growth. In general, if the ion activity product in an oversaturated solution is higher than the solubility product, a solid phase will result from interactions between ions or molecules for nucleus formation. These nuclei, by interaction with more ions or molecules for the growth of crystals, form the crystallites. Large crystals will eventually separate by sedimentation, which is referred as **ripening**. The formation of mass transport by the solid phase through molecular diffusion becomes important. A linear growth rate is inversely proportional to the boundary layer width. The hydrodynamic modifications, both in dissolution and crystal formation, which will affect the boundary layer thickness, have been identified.

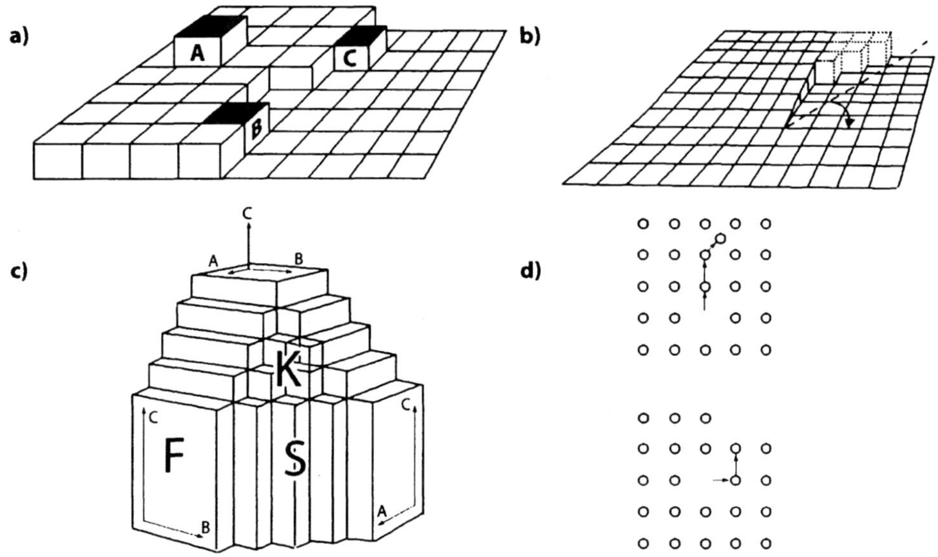

Figure 3-2. Crystal growth and defects in lattice. (a) KSV theory: sites for the attachment of absorbed atoms A, surface site; B, ledge-kink site; and C, ledge site. (b) BCF theory: nucleation site with broken lines at surface step produced by an emergent screw dislocation. (c) F (flat), S (stepped) and K (kinked) faces according to PBC theory. (d) Upper diagram expresses the Frankel pair of defects of interstitial and valong. Lower diagram shows the Schottky defect with atom migrating to the surface. (Modified and assembled from A.W. Vere, *Crystal Growth Principles and Progress*, Plenum Press, New York, 1987.)

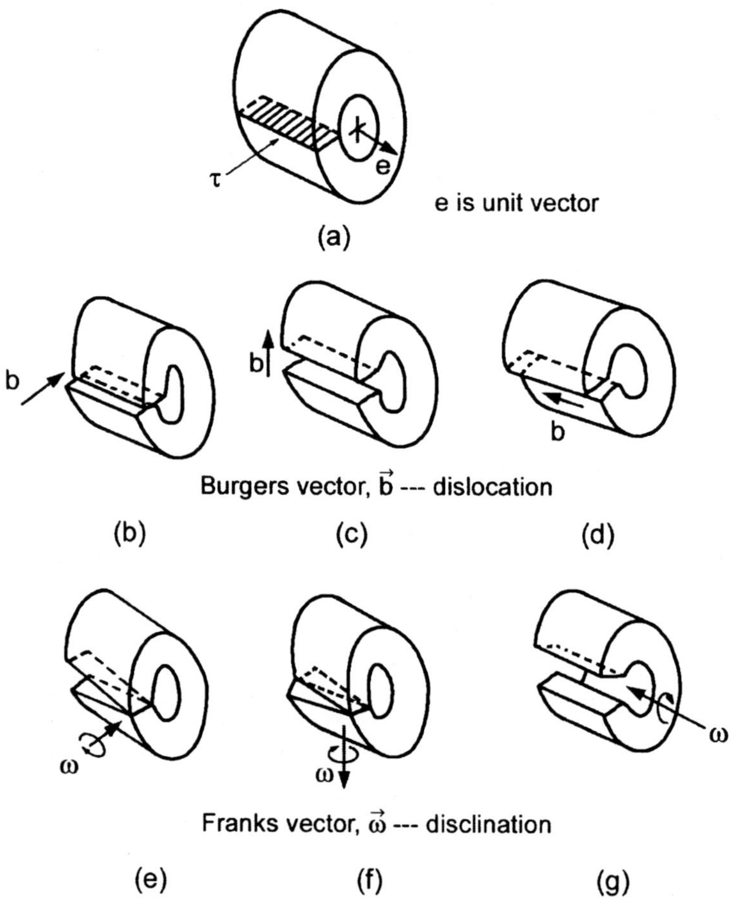

Figure 3-3. Volterra distoration in solid state for dislocations and disclinations. (Modified after T.F. Yen et al., *Advances in the Applications of Membrane-Mimetic Chemistry*, Plenum Press, New York, 1991, p. 255.)

For nucleation, the interfacial energy is important due to the thermodynamic and kinetic considerations. Nucleation of a more soluble phase is favored kinetically over that of a less soluble one because the former has a lower solid-solution interfacial tension. The **Ostwald step rule** states that the precipitate with the highest solubility (that is, the least stable solid phase) will form first in a series of consecutive precipitation reactions. The crystal growth depends upon the physical and chemical states of the growth surface. Volmer indicated that macroscopic surface roughening increases the nucleation probability. Possible lattice sites can be used for the attachment of **adsorbed atom** (or **adatoms**). **Kossel, Stranski and Volmer (KSV) theory** allows the sites at the discontinuities to be origin and features for nucleation.

The drawback of KSV theory, however, is that if the kinked ledge received adatoms that moved from the edge of the crystal as indicated in Figure 3-2a, it no longer functions as a low energy nucleation site. It was for this reason that the **Burton, Cabrera, and Frank (BCF) theory** was established. It indicates that the emergence points of dislocations with screw components at crystal surfaces are continuous generations of surface ledge, as shown in Figure 3-2b. The BCF theory emphasizes the surface roughness, which is the ledge and kink density, to account for the interatomic bonding forces at a certain crystallographic plane. This surface roughness led to the development of the **periodic bond chain (PBC) theory** as shown in Figure 3-2c. In this theory, the crystal structure is defined in terms of strongly bonded chains of atoms to form the PBC. Any face that contains more than two PBCs is a flat face (as *F*) upon which nucleation is difficult and growth is retarded. Stepped face (or S) contains only one PBC, and the rate is next. The kinked force (as *K*) contains no PBC and, therefore, has fast growth.

Solid state chemistry refers to lattice point defects such as the **Frankel pair,** in which interstitials are formed, or the **Schottky defect**, in which new atoms are formed in a free surface, as indicated in Figure 3-2d. The transport of partial charges can result in having an electron (negative) or hole (positive) movement in lattice, which results in an n-p gap of semiconductors (see Volume 4A, Section 6.4). In Volterra system, there are six types of **distortions** in linear defect propagation. Three types are due to translational shift, known as **dislocation**; the other three types are due to the angular shift or Franks vector, which are termed **disclinations** as shown in Figure 3-3. Another term, **dispiration**, is used to describe long sequences of repetitive dislocations.

Most of the surface roughness can be easily characterized by the atomic force microscopy (AFM), which uses a tip to study the nano-scale range of the surface. Furthermore, the surface can be discovered through fractal chemistry (fractal geometry applied to chemistry). For example, a rough surface may have dimensions of any value between two and three. For random growth process of disordered material, as illustrated by Figure 9-4, the resulting solids can be crystals or flocs. The growth process is controlled by the diffusion process. This so-called **diffusion limited aggregation (DLA)** makes up the basis of cluster formation. After the cluster is formed, solid and precipitates can be derived from cluster–cluster interactions through sticking probabilities. In some cases, super-aggregates can be formed.

Iron and manganese present in water are almost always in the bivalent state. The higher valency from the oxidation process will evolve from the precipitation of oxides. The bivalent iron can also remain as bivalent carbonates. Within the pH range of natural waters, Fe^{2+}, Mg^{2+}, $FeOH^{+}$, and $MnOH^{+}$ frequently occur. In the presence of carbonates, the solubility of both bivalent iron and manganese will not depend on their hydroxides, but on their carbonates (Fig. 3-5). The maximum concentration of soluble Fe^{2+} usually depends on the pH and alkalinity, T.

$$(Fe^{2+})(CO_3^{2-}) = K_s = 10^{-10.7} \qquad [3\text{-}13]$$

If it includes the K_2 of carbonate, then

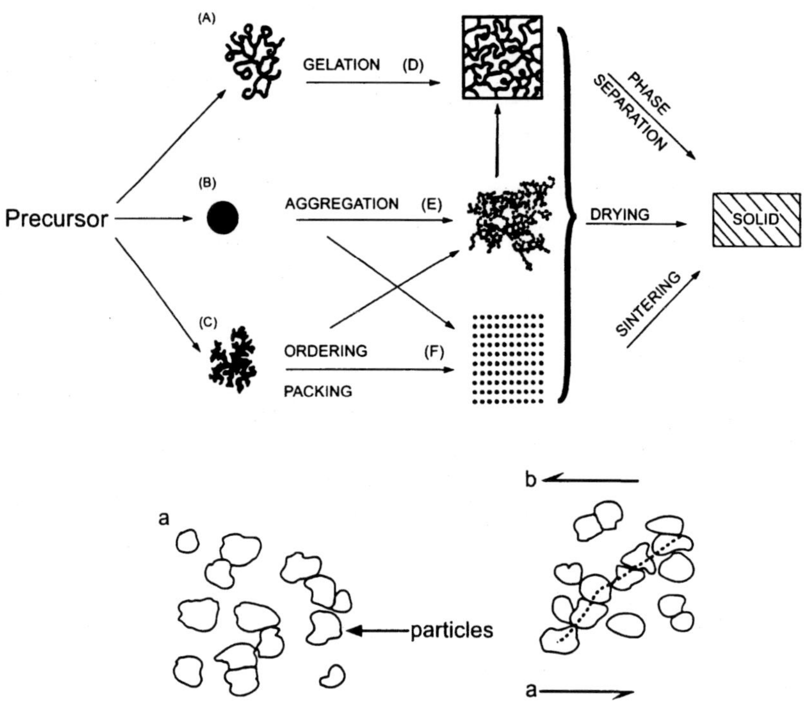

Figure 3-4. Material can become ordered lattice (crystal) or gelation and aggregation (amorphous flocs or precipitate). The lower curve indicates dilutant fluid under shear becoming agglomerates. (From T.F. Yen, Mesoscaled structure and membrane mimetic chemistry in Membrane Mimetic Chemistry, Plenum Press, New York, 1994)

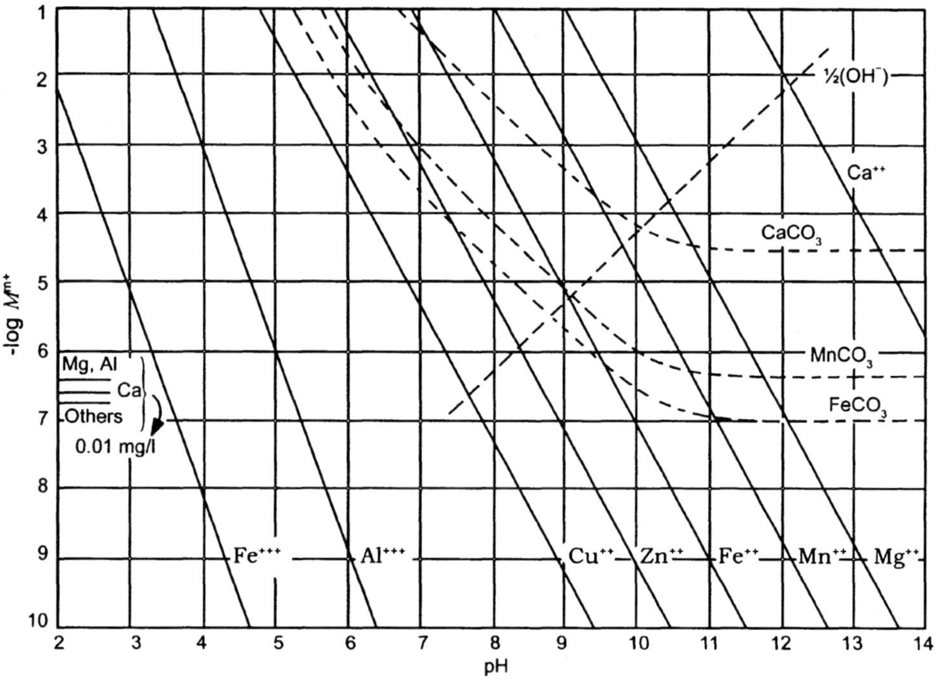

Figure 3-5. Logarithmic concentration diagrams for solubilities of hydroxides and carbonates. (Adapted from Ref. 1-5.)

$$K_s/K_2 = (Fe^{2+})(HCO_3^-)/(H^+) \qquad [3\text{-}14]$$

or

$$(Fe^{2+}) = (K_s/K_2)(H^+)/(HCO_3^-)$$

and (HCO_3^-) can be expressed in terms of alkalinity as:

$$(HCO_3^-) = (Alk - K_w/(H^+) + (H^+))/[1 + 2K_2/(H^+)] \qquad [3\text{-}14a]$$

since

$$Alk = (HCO_3^-) + 2(CO_3^{2-}) + (OH^-) - (H^+) \qquad [3\text{-}15]$$

Also,

$$C_T = (HCO_3^-) + 2CO_3^{2-} = (HCO_3^-)\,[1 + 2K_2/(H^+)] = Alk - (OH^-) + (H^+)$$

Therefore, an expression of (Fe^{2+}) can be made in terms of alkalinity and pH. For example, the solubility of (Fe^{2+}) is 2.8×10^{-4} M (15.7 mg/L) for water with an alkalinity of 10 mg/L as $CaCO_3$ and a pH of 7.5. If the pH is above 10, then the solubility of (Fe^{2+}) is governed by

$$(Fe^{2+}) = K_{s[Fe(OH)2]}/(OH^-)^2 \qquad [3\text{-}16]$$

which would yield higher amount of (Fe^{2+}). In the presence of carbonate, however, this would not be true, and the maximum amount would be controlled by carbonates.

The trivalent iron (Fe^{3+}) is controlled by the equilibria conditions of the hydroxo complexes or the hydrolysis products. The equilibria can belong to a homogeneous system as well as to a heterogeneous one. In the former case, the amount and composition of soluble species of Fe^{3+} can be very easily calculated. Among the major iron complexation species, the following three equilibria are important:

$$Fe^{3+} + H_2O = FeOH^{2+} + H^+;\ pK_{1h} = 2.16 \qquad [3\text{-}17]$$

$$Fe^{3+} + 2H_2O = Fe(OH)_2^+ + 2H^+;\ pK_{2h} = 6.74 \qquad [3\text{-}18]$$

$$2Fe^{3+} + 2H_2O = Fe_2(OH)_2^{4+} + 2H^+;\ pK_{3h} = 2.85 \qquad [3\text{-}19]$$

Two additional equilibria are written in a heterogeneous system:

$$Fe(OH)_3(s) = Fe^{3+} + 3OH^-;\ pK_{s4} = 38 \qquad [3\text{-}20]$$

$$Fe(OH)_3(s) + H_2O = Fe(OH)_4^- + H^+;\ pK_{s5} = 19 \qquad [3\text{-}21]$$

Usually the soluble total Fe^{3+} ion can be represented as:

$$C_{T,Fe} = (Fe^{3+}) + (FeOH^{2+}) + (Fe(OH)_2^+) + 2(Fe_2(OH)_2^{4+}) + (Fe(OH)_4^-) \qquad [3\text{-}22]$$

All five equilibria of Fe^{3+} can be readily expressed as log C versus pH forms:

$$\log (FeOH^{2+}) = 1.84 - 2pH, \text{ line 1} \qquad [3\text{-}17a]$$

$$\log [Fe(OH)_2^+] = -2.74 - pH, \text{ line 2} \qquad [3\text{-}18a]$$

$$\log [Fe_2(OH)_2^{4+}] = 5.2 - 4pH, \text{ line 4} \qquad [3\text{-}19a]$$

$$\log (Fe^{3+}) = 4 - 3pH, \text{ line 5} \qquad [3\text{-}20a]$$

$$\log [Fe(OH)_4^-] = pH - 19, \text{ line 3} \qquad [3\text{-}21a]$$

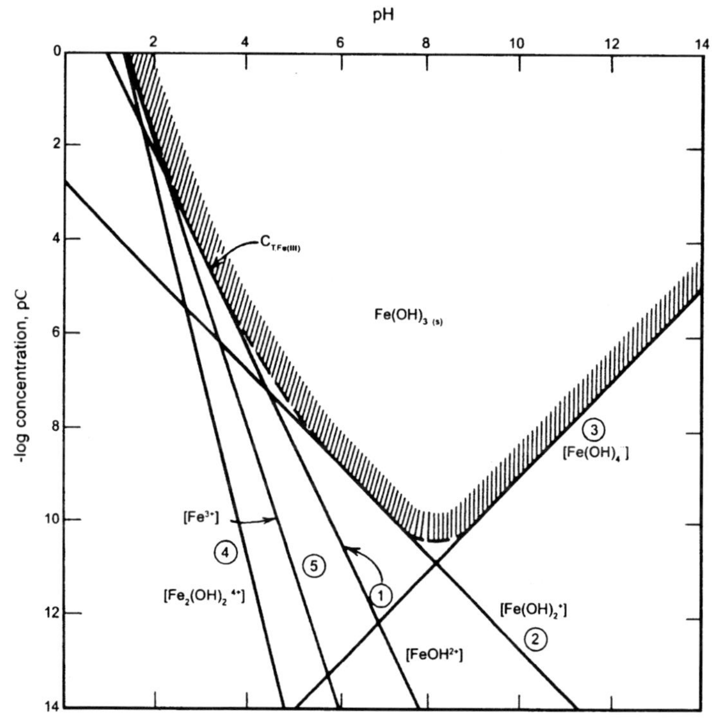

Figure 3-6. Equilibrium concentration of hydroxo iron(III) complexes in a solution in contact with freshly precipitated $Fe(OH)_3(s)$ at 25°C. (From Ref. 2.1.)

The above five lines are shown in Figure 3-6, and the surrounding $Fe(OH)_3(s)$ is indicated by the hatched area. Individual concentrations of the soluble Fe^{3+} can be obtained at any given pH level.

3.2.1 Aluminum Phosphate Precipitation

Similar to iron, aluminum forms the $Al(OH)_3(s)$ precipitate in water.

$$Al^{3+} + H_2O = AlOH^{2+} + H^+;\ pK_{1h} = 5 \qquad [3\text{-}23]$$

$$2Al^{3+} + 2H_2O = Al(OH)_2^{4+} + 2H^+;\ pK_{2h} = 6.3 \qquad [3\text{-}24]$$

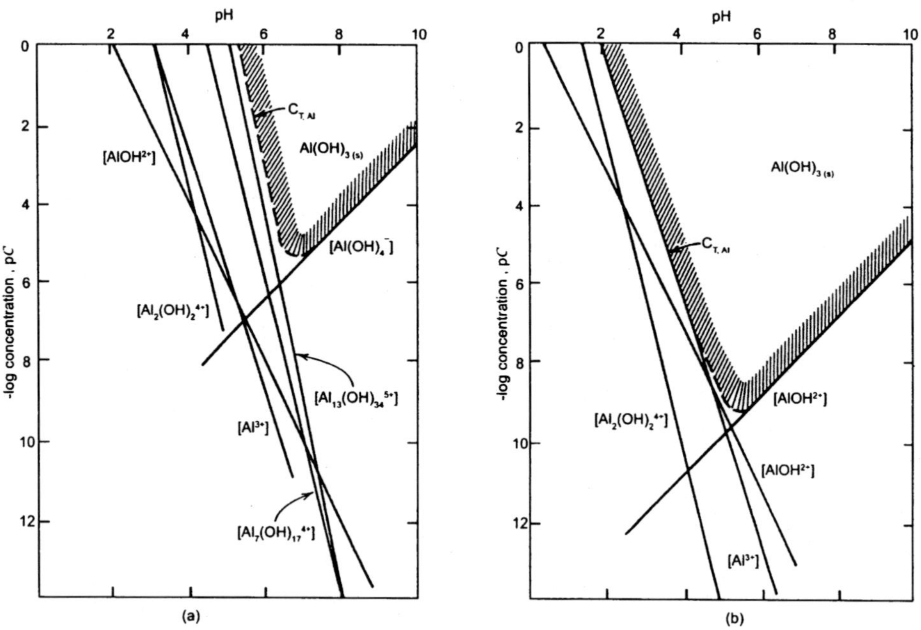

Figure 3-7. Equilibrium concentrations of hydroxo aluminum(III) complexes in a solution in contact with $Al(OH)_3(s)$ at 25°C. (a) Freshly precipitated $Al(OH)_3(s)$; (b) aged $Al(OH)_3(s)$, gibbsite ($Al_2O_3 \cdot 3H_2O$). (From Ref. 2-1.)

$$7Al^{3+} + 17H_2O = Al_7(OH)_{17}^{4+} + 17H^+;\ pK_{3h} = 48.8 \qquad [3\text{-}25]$$

$$13Al^{3+} + 34H_2O = Al_{13}(OH)_{34}^{5+} + 34H^+;\ pK_{4h} = 97.4 \qquad [3\text{-}26]$$

$$Al(OH)_3(s) = Al^{3+} + 3OH^-; \; pK_{s1} = 33 \qquad [3\text{-}27]$$

$$Al(OH)_3(s) + OH^- = Al(OH)_4^-; \; pK_{s2} = -1.3 \qquad [3\text{-}28]$$

The pC versus pH plot is found in Figure 3-7a. Note that the polymeric species such as $Al_7(OH)_{17}^{4+}$ or $Al_{13}(OH)_{34}^{5+}$ regulate the $Al(OH)_3(s)$ at about pH 6. The aged $Al(OH)_3(s)$, gibbsite, predominates in as wide an area as that of the freshly prepared $Al(OH)_3(s)$ precipitate (see Fig. 3-7b), and the polymeric species become precipitates.

Soluble phosphate species in natural waters may consist of orthophosphates (PO_4^{3-}, HPO_4^{2-}, $H_2PO_4^-$, $CaH_2PO_4^+$), pyrophosphates ($P_2O_7^{4+}$, $H_3P_2O_7^{2-}$, $H_2P_2O_7^{2}$, $HP_2O_7^{3-}$), tripolyphosphates ($H_3P_3O_{10}^{2-}$, $H_2P_3O_{10}^{3-}$, $HP_3O_{10}^{4-}$, $P_3O_{10}^{5-}$), and metaphosphates ($P_3O_9^{3-}$, $HP_3O_9^{3-}$). The miscible solids may be derived from calcium. For example,

$$Ca(PO_4)_2(s) = 3Ca^{2+} + 2PO_4^{3-}; \; pK = 26 \qquad [3\text{-}29]$$

$$CaHPO_4(s) = Ca^{2+} + HPO_4^{2-}; \; pK = 7 \qquad [3\text{-}30]$$

$$Ca_5OH(PO_4)_3(s) = 5Ca^{2+} + OH^- + 3PO_4^{3-}; \; pK = 55.6 \qquad [3\text{-}31]$$

Based on this principle, phosphorus can be removed from wastewater by adding lime either before the primary classifier or after the biological reactor. In the above species, hydroxyapatite, $Ca_5OH(PO_4)_3$, the principal constituent of teeth, is extremely stable.

Alum or aluminum sulfate is added to wastewater for the removal of phosphate. In this case aluminum phosphate is a precipitate.

$$AlPO_4(s) = Al^{3+} + PO_4^{3-}; \; pK_s = 21 \qquad [3\text{-}32]$$

Again, we assume the major phosphate species are from othophosphates:

$$C_{T,\,PO_4} = (H_3PO_4) + (H_2PO_4^-) + (HPO_4^{2-}) + (PO_4^{3-}) \qquad [3\text{-}33]$$

$$H_3PO_4 = H^+ + H_2PO_4^-; \; pK_1 = 2.1 \qquad [3\text{-}34]$$

$$H_2PO_4 = H^+ + HPO_4^{2-};\ pK_2 = 7.2 \qquad [3\text{-}35]$$

$$HPO_4^{2-} = H^+ + PO_4^{3-};\ pK_3 = 12.3 \qquad [3\text{-}36]$$

Expressing (HPO_4^{2-}), ($H_2PO_4^-$), and (H_3PO_4) in terms of (PO_4^{3-}), we obtain

$$C_{T,\,PO_4} = (PO_4^{3})[\ ((H^+)^3/(K_1K_2K_3)) + ((H^+)^2/(K_2K_3)) + ((H^+)/(K_3)) + 1\ ] \qquad [3\text{-}37]$$

or

$$(PO_4^{3}) = C_{T,\,PO_4}[\ ((H^+)^3/(K_1K_2K_3)) + ((H^+)^2/(K_2K_3)) + ((H^+)/(K_3)) + 1\ ]^{-1} \qquad [3\text{-}37a]$$

For the aluminum species we choose the more prominent ones: Al^{3+}, $AlOH^{2+}$, and $Al(OH)_4^-$.

$$C_{T,\,Al} = (Al^{3+}) + (AlOH^{2+}) + (Al(OH)_4^-) \qquad [3\text{-}38]$$

We have discussed the Al species previously.

$$Al^{3+} + H_2O = AlOH^{2+} + H^+;\ pK_{1h} = 5 \qquad [3\text{-}23]$$

$$Al^{3+} + 4H_2O = AlOH_4^- + 4H^+;\ pK_{5h} = 20.3 \qquad [3\text{-}39]$$

$$pK_{5h} = pK_{s2} - pK_{s1} + pK_w$$

Similar to phosphate, expressing ($AlOH^+$) and ($Al(OH)_4^-$) in terms of (Al^{3+}), we obtain

$$C_{T,\,Al} = (Al^{3+})(\ 1 + (K_{1h}/(H^+)) + (K_{5h}/(H^+)^4)\ ) \qquad [3\text{-}40]$$

or

$$(Al^{3+}) = C_{T,\,Al}(\ 1 + (K_{1h}/(H^+)) + (K_{5h}/(H^+)^4)\ )^{-1} \qquad [3\text{-}40a]$$

Assuming the stoichiometric ratio of the precipitate,

$$K_{s,AlPO_4} = C_{T,\,Al}\left(1 + \left(\frac{K_{1h}}{H^+}\right) + \left(\frac{K_{5h}}{(H^+)^4}\right)\right)^{-1} C_{T,PO_4}\left(\frac{(H^+)^3}{K_1K_2K_3}\right) + \frac{(H^+)^2}{K_2K_3} + \frac{(H^+)}{K_3} + 1)^{-1}$$

$$= 10^{-21} \qquad [3\text{-}41]$$

Given $C_{T,\mathrm{PO_4}}$, we can approximate the pH value for this operation (see Fig. 3-8). Evidently from the figure one should not select a pH below 3 or above 6 for the precipitation as $AlPO_4$(s). In practice, an optimum pH of 6 is used, and aluminum used is doubled because $Al(OH)_3$(s) also coprecipitates.

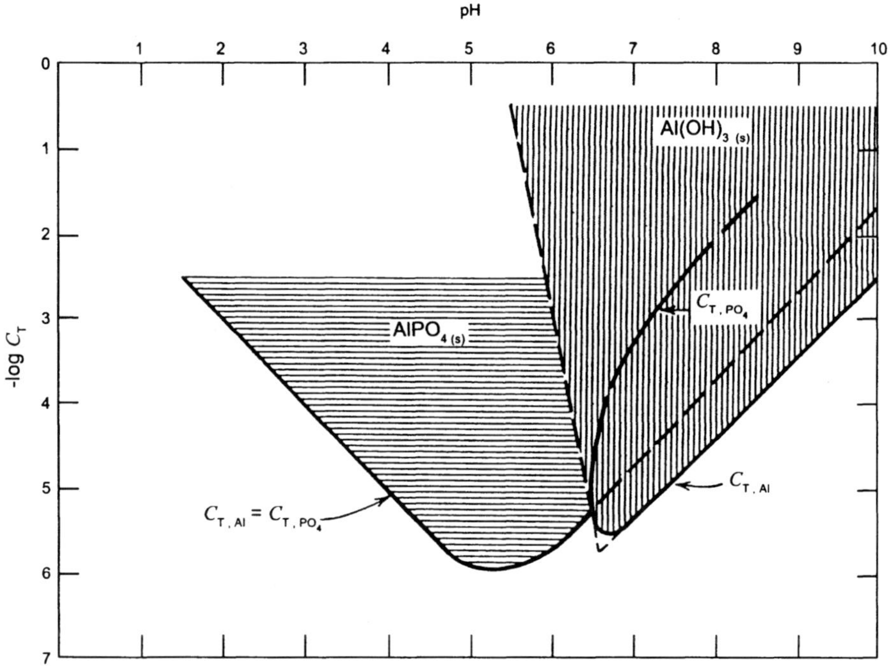

Figure 3-8. Concentration of phosphate and aluminum that can exist at various pH when $AlPO_4$(s) is precipitated from or dissolved in pure water. For pH lower than 5.4, concentrations of phosphate and aluminum are controlled by the solubility of $AlPO_4$(s). Above pH 5.4 $C_{T,\mathrm{Al}}$ is controlled by the solubility of $Al(OH)_3$(s). Thus, $C_{T,\mathrm{PO4}}$ concentrations in this range are greater than $C_{T,\mathrm{Al}}$ (Adapted from Ref. 2-1.)

3.3 LANGELIER INDEX

During chemical stabilization of water containing carbonates species, pH is often used to adjust the stability. The control is such that neither the established $CaCO_3$ is dissolved nor is any new $CaCO_3$ precipitated. The state of the water is such that the coating of $CaCO_3$ on the pipe is not removed so that the pipe surface will not corrode, but at the same time new deposits of $CaCO_3$ form that will clog the pipeline. We shall establish a criterion for this state. Remembering

$$CaCO_3(s) = Ca^{2+} + CO_3^{2-};\ K_s$$

$$CO_3^{2-} + H^+ = HCO_3^-;\ K_2^{-1}$$

addition of the above will yield

$$CaCO_3 + H^+ = Ca^{2+} + HCO_3^-;\ K_sK_2^{-1} \qquad [3\text{-}42]$$

or

$$((Ca^{2+})(HCO_3^-))/(H^+) = K_sK_2^{-1} = K \qquad [3\text{-}42a]$$

Let Q represent the products-to-reactants ratio of an actual concentration, or

$$Q = ((Ca^{2+})_{act}(HCO_3^-)_{act})/(H^+)_{act} \qquad [3\text{-}43]$$

The free energy of the reaction will be:

$$\Delta G = \Delta G^0 + RT \ln Q \qquad (a)$$

There will be an equilibrium concentration for all species involved:

$$K = [(Ca^{2+})_{eq}(HCO_3^-)_{eq}]/(H^+)_{eq} \qquad [3\text{-}44]$$

and at equilibrium, $\Delta G = 0$ and $Q = K$, thus,

$$\Delta G^0 = -RT \ln K \qquad (b)$$

Substituting into the original equation (a), then

$$\Delta G = RT \ln Q - RT \ln K = RT \ln (Q/K) \qquad (c)$$

Here if $Q < K$ or $Q/K < 1$, then ΔG, becomes a negative value, the reaction will proceed (dissolution) and the solution will be undersaturated; when $Q = K$, the system will be in equilibrium; and if $Q > K$ then ΔG will be positive, the solution oversaturated, and precipitation will occur. Hence,

$$Q < K,\ Q/K < 1;\ \text{undersaturated}$$

$$Q = K,\ Q/K = 1;\ \text{equilibrium}$$

$$Q > K,\ Q/K > 1;\ \text{precipitation} \qquad [3\text{-}45]$$

Going back to the equilibrium concentration, we qualify the manner in which the activities should be used,

$$K = [(Ca^{2+})_{eq}(HCO_3^-)_{eq}]/(H^+)_{eq}$$

or

$$K = [\gamma_{Ca^{2+}}(Ca^{2+})\gamma_{HCO_3^-}(HCO_3^-)]/(H^+)_{eq} \qquad [3\text{-}46]$$

Since $K = K_sK_2^{-1}$, we can rewrite the above in log form:

$$pH_{eq} = pK_2 - pK_s + pCa^{2+} + pHCO_3 - \log\gamma_{Ca^{2+}} - \log\gamma_{HCO_3^-} \qquad [3\text{-}47]$$

When examining [substituting Eqs. [3-43] and [3-44] into (c)],

$$\Delta G = RT\ln(\ (Ca^{2+})_{act}(HCO_3^-)_{act}(H^+)_{eq}\ )/K\ (H^+)_{act}$$

and

$$(Ca^{2+})_{act}(HCO_3^-)_{act} \cong K,\ \ \Delta G \propto \ln[(H)^+_{eq}/(H^+)_{act}] \propto pH_{act} - pH_{eq} \qquad [3\text{-}48]$$

Thus, the controlling factor is now on $(H^+)_{eq}$ and $(H^+)_{act}$, and again

$$pH_{act} < pH_{eq};\ \text{undersaturated}$$

$$pH_{act} = pH_{eq};\ \text{equilibrium}$$

$$pH_{act} > pH_{eq};\ \text{precipitation} \qquad [3\text{-}49]$$

Langelier developed an index in the pH range 6.5 – 9.5 that makes it possible to predict whether a given water will deposit or dissolve $CaCO_3(s)$. This index is termed the **Langelier index** or the **saturation index** (SI).

$$SI = pH_{act} - pH_{eq} = -;\ \text{undersaturated}$$

$$SI = 0;\ \text{equilibrium}$$

$$SI = +;\ \text{precipitation} \qquad [3\text{-}50]$$

In general, from Equation [3-47] and Alk = HCO_3^-,

$$SI = pH_{act} - pH_{eq} = pH_{act} - (pK_2 - pK_s) - pCa^{2+} - p(Alk) + \log\gamma_{Ca^{2+}} + \log\gamma_{HCO_3^-}$$

$$= pH_{act} - (pK_2' - pK_s') - pCa^{2+} - p(Alk) \qquad [3\text{-}51]$$

K_2' and K_s' are the equilibrium constants corrected for the activity coefficient. These are listed in Table 3-3.

[Example 3-4] Calculate the SI for a wastewater based on the following analytical data. pH = 7.9, T = 15°C. (The original problem only gives the two leftmost columns.)

Component	mg/L	MW	mM/L	μ (x 10^{-3})
Ca^{2+}	63.3	40.1	1.58	3.16
Mg^{2+}	14.8	24.3	0.61	1.22
Na^+	19.5	23.0	0.85	0.42
K^+	10.1	39.1	0.26	0.13
CO_3^{2-}	7.8	60.0	0.13	0.26
HCO_3^-	94.4	61.0	1.55	0.78
SO_4^{2-}	80.0	96.0	0.83	1.66
Cl^-	17.0	35.5	0.48	0.24
				$\Sigma\mu$ = 7.87

Solving this problem involves completing the values of the last three columns. From the total ionic strength of this water, we find in Table 19-3 that $pK_2' - pK_s' = 2.45$, $pCa^{2+} = 2.80$, $p(Alk) = p(2CO_3^{2-} + HCO_3^-) = 2.74$. Hence, SI $= 7.90 - 2.45 - 2.80 - 2.74 = -0.10$.

The water will not precipitate $CaCO_3$, but is slightly corrosive and will dissolve $CaCO_3$.

Table 3-3. Values of PK_2' and PK_3' at 25°C for Various Ionic strengths and the Difference ($PK_2' - PK_3'$) for Various Temperatures

Ionic Strength	Total Dissolved Solids	25°C			$PK_2' - PK_3'$									
	(mg/L)	PK_2'	PK_3'	$PK_2' - PK_3'$	0°C	10°C	20°C	30°C	40°C	50°C	60°C	70°C	80°C	90°C
0.000		10.26	8.32	1.94	2.45	2.23	2.02	1.86	1.68	1.52	1.36	1.23	1.08	0.95
0.001	40	10.26	8.19	2.07	2.58	2.36	2.15	1.99	1.81	1.65	1.49	1.36	1.21	1.08
0.002	80	10.25	8.14	2.11	2.62	2.40	2.19	2.03	1.85	1.69	1.53	1.40	1.25	1.12
0.003	120	10.25	8.10	2.15	2.66	2.44	2.23	2.07	1.89	1.73	1.57	1.44	1.29	1.16
0.004	160	10.24	8.07	2.17	2.68	2.46	2.25	2.09	1.91	1.75	1.59	1.46	1.31	1.18
0.005	200	10.24	8.04	2.20	2.71	2.49	2.28	2.12	1.94	1.78	1.62	1.49	1.34	1.21
0.006	240	10.24	8.01	2.23	2.74	2.52	2.31	2.15	1.97	1.81	1.65	1.52	1.37	1.24
0.007	280	10.23	7.98	2.25	2.76	2.54	2.33	2.17	1.99	1.83	1.67	1.54	1.39	1.26
0.008	320	10.23	7.96	2.27	2.78	2.56	2.35	2.19	2.01	1.85	1.69	1.56	1.41	1.28
0.009	360	10.22	7.94	2.28	2.79	2.57	2.36	2.20	2.02	1.86	1.70	1.57	1.42	1.29
0.010	400	10.22	7.92	2.30	2.81	2.59	2.38	2.22	2.04	1.88	1.72	1.59	1.44	1.31
0.011	440	10.22	7.90	2.32	2.83	2.61	2.40	2.24	2.06	1.90	1.74	1.61	1.46	1.33

3.4 COMPLEXATION-COORDINATION CHEMISTRY

As discussed in Section 3.2, metals can hydrolyze in water and become hydroxo or oxo complexes

$$Fe(H_2O)_6^{3+} = Fe(H_2O)_5OH^{2+} + H^+ = Fe(H_2O)_4(OH)_2^+ + 2H^+ = Fe(OH)_3(H_2O)_3(s) + 3H^+$$

Ion pairs may be responsible for the metal hydration complexes. Furthermore, metals can react with functional groups (coordination) of a molecule, and such molecules are called **ligands**. Vast forms of complexes are created in nature; therefore, complexation of trace metals is essential for microorganisms. Trace metal interaction with aquatic microorganisms may be viewed as organisms serving as assemblages of reactive ligands whose degrees of complexation with particular metals reflect different physiological effects. Various types of complexes are shown in Figure 3-9.

Briefly, **complexes** are formed from metals and ligands:

$$M + L = ML$$

Here, the ligand is defined as a complexing agent, and

$$(ML)/((M)(L)) = K_F$$

K_F is the **complex formation** or **stability constant**, and K_F^{-1} is referred to as the **instability constant**. Some ligands have more than one functional group site, thus a multidentate complex is formed,

$$ML + L = ML_2$$

$$ML_2 + L = ML_3$$

$$ML_{i-1} + L = ML_i$$

$$ML_{n-1} + L = ML_n$$

Hence two types of stability constants are often used:

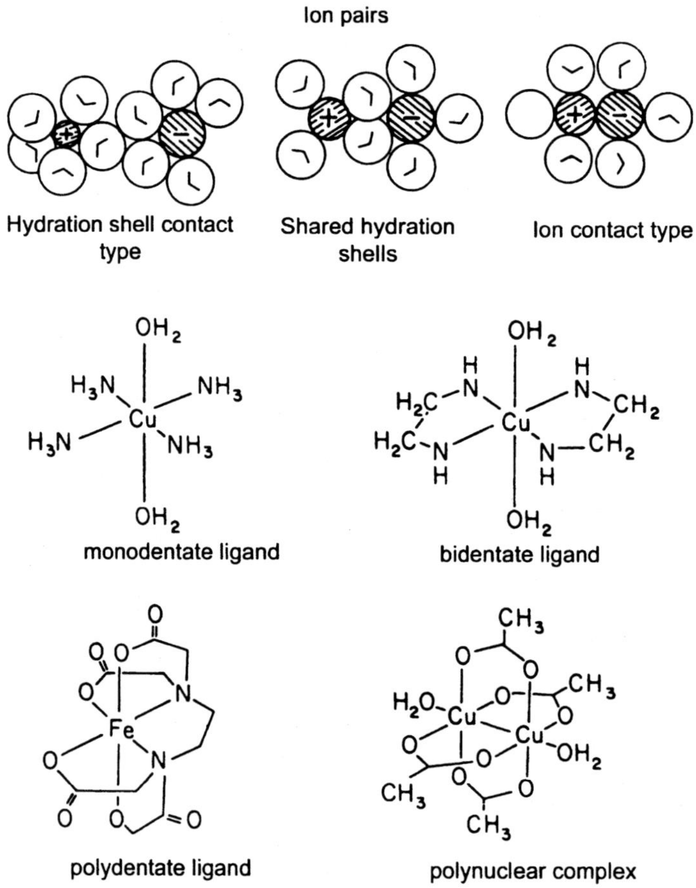

Figure 3-9. Various types of aqueous complexes. (From Ref. 3-1.)

$$K_i = (ML_i)/((ML_{i-1})(L)) \qquad [3\text{-}52]$$

and

$$\beta i = (ML_i)/((M)(L)^i) \qquad [3\text{-}53]$$

both of which are referred to in the literature as stability constants (see Table 3-4).

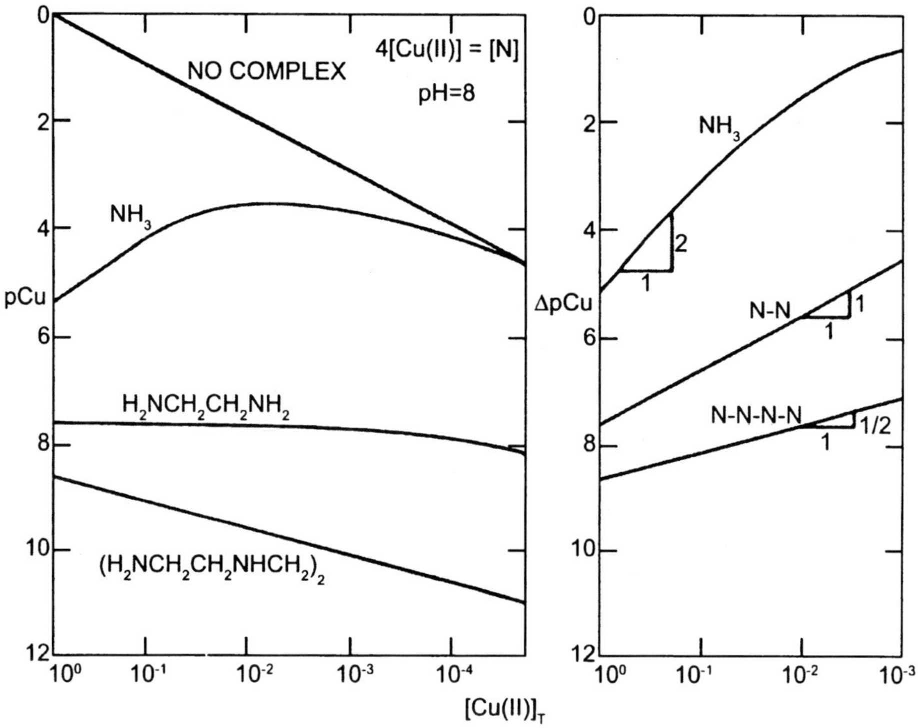

Figure 3-10. The chelate effect on complex formation of Cu·aq2+ with monodentate, bidentate, and tetradentate amines. pCu is plotted as a function of concentration in the left-hand diagram. On the right, the relative degree of complexation as measured by ΔpCu as a function of concentration is depicted. The extent of complexing is larger with chelate complex formers than with unidentate ligands. Unidentate complexes are dissociated in dilute solutions with chelates remains essentially undissociated at great dilutions. (From Ref. 1-6.)

As indicated in Figure 3-9, metals can form complexes with a single ligand (unidentate and monodentate) or multiple numbers of ligands (multidentate). In general, complexes formed with monodentate ligands are less stable than complexes with bidentate and tetradentate ligands. For example, different Cu complexes with N-bearing ligands are studied as a function of dilution, as shown in Figure 3-10. The quantitative degree of complexation as expressed in

$$\Delta p\,\mathrm{Cu} = \log \frac{\mathrm{Cu}_T}{\mathrm{Cu}^{2+}} = \log \mathrm{Cu}_T - \log\left(\mathrm{Cu}^{2+}\right) \qquad [3\text{-}54]$$

Table 3-4. Definition of Stability Constants

Mononuclear Complexes

Addition of ligand

$$M \xrightarrow[K_1]{L} ML \xrightarrow[K_2]{L} ML_2 \cdots \xrightarrow[K_i]{L} ML_i \cdots \xrightarrow[K_n]{L} ML_n$$

$$\text{———} \beta_2 \text{———}\rightarrow$$

$$\text{———} \beta_i \text{—————}\rightarrow$$

$$\text{———} \beta_n \text{————————}\rightarrow$$

$$K_i = \frac{[ML_i]}{[ML_{(i-1)}][L]}$$

$$\beta_i = \frac{[ML_i]}{[M][L]^i}$$

Addition of protonated ligand

$$M \xrightarrow[*K_1]{HL} ML \xrightarrow[*K_2]{HL} ML_2 \cdots \xrightarrow[*K_i]{HL} ML_i \cdots \xrightarrow[*K_n]{HL} ML_n$$

$$\text{———} *\beta_2 \text{———}\rightarrow$$

$$\text{———} *\beta_i \text{—————}\rightarrow$$

$$\text{———} *\beta_n \text{————————}\rightarrow$$

$$*K_i = \frac{[ML_i][H^+]}{[ML_{(i-1)}][HL]}$$

$$*\beta_i = \frac{[ML_i][H^+]^i}{[M][HL]^i}$$

Polynuclear Complexes

In β_{nm} and $*\beta_{nm}$ the subscripts *n* and *m* denote the composition of the complex M_mL_n formed. [If $m = 1$, the second subscript ($n = 1$) is omitted.]

$$\beta_{nm} = \frac{[M_mL_n]}{[M]^m[L]^n}$$

$$*\beta_{nm} = \frac{[M_mL_n][H^+]^n}{[M]^m[HL]^n}$$

The notation β_1 or K_1 is used indifferently to indicate the constants of formation of M_1L_1 complexes. β or K' indicate mixed acidity constants expressed as a function of the activity of H^+ and of the concentrations of other reactants and products. *Source*: Adapted from Ref. 1-6.

for different types of ligands is also expressed in Figure 3-10. The degree of complexation decreases more strongly with dilution for simple dentate complexes than for tetradentate complexes. Furthermore, the stability of a complex is not governed only by the template effect (podand) but also by the macrocyclic effect (Fig. 3-11A). The most stable configuration has cagelike clusters of ferredoxin (Fig. 3-11B). Another aspect of chelates is the difference in the metal; for example, if the tetradentate is porphyrin, the metallation/demetallation is governed by chemical stability as follows:

$$MP + 2\mathrm{H}^{+} = P\mathrm{H}_2 + M^{2+}$$

MP here is metalloporphyrinm and $P\mathrm{H}_2$ is the free base porphyrin. In general, the forward equation is demetallation, and the reverse is metallation. The disassociation of metalloporphyrin is governed by chemical stability.

Water (CO_2^{2-})	Ag(I), K(I), NA(I), Pb(II), Hg(II)
Dilute HCL	Zn(II), Cd(II), Mg(II), Fe(II)
Concentrate H_2SO_4	Cu(II), Mn(II), Co(II), Ni(II), Fe(III)
HBr-acetic acid	VO(IV)
Trichloroacetic acid	VO(IV)*

*Porphyrin is tetra-ms-4-biphenylporphin

From the above data, it is obvious that pheophytinization (removal of magnesium) is easy, whereas devanadylation (removal of VO) is difficult.

So far the most important groups in a complexing agent are the amino and the carboxylic functions. The use of a mixture of citric acid and trienthanoamine for derusting the bilges of naval ships has become traditional practice. The formation of a stable complex becomes a problem for treatment of this derusting water. Either biological or chemical degradation will lead to many dicarboxylic acid intermediates, and these intermediates still maintain their complexing capacities.

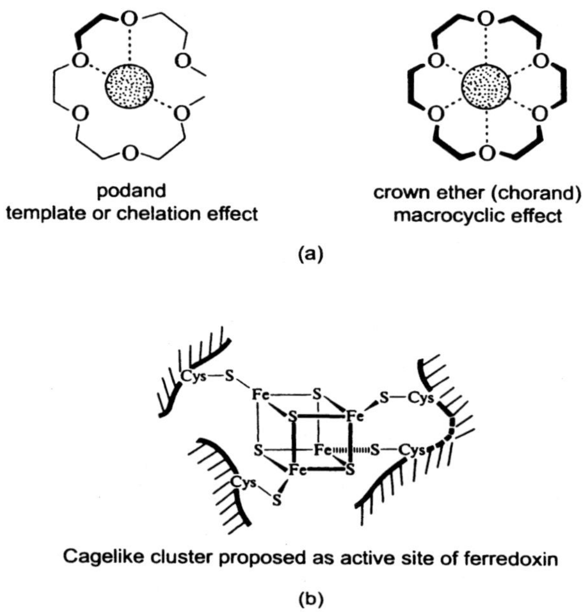

Figure 3-11. The macrocyclic effect.

In nature, both enterobactin and ferrichrome are strong complexing agents for Fe^{3+} (Fig. 3-12). The bacterial transport of iron and other metals hinges on these biological complexing agents. Note that they contain the hydroxamate functions. One widely used complexing agent is EDTA (ethylenediaminetetraacetic acid).

EDTA (H_4Y, see Table 3-5 for details) has four sites and can undergo deprotonation steps as follows:

$$H_4Y = H^+ + H_3Y^-$$

$$H_3Y^- = H^+ + H_2Y^{2-}$$

$$H_2Y^{2-} = H^+ + HY^{3-}$$

$$HY^{3-} = H^+ + Y^{4-} \qquad [3\text{-}55]$$

The last ionized product, Y^{4-}, can readily take part in a chelating reaction, for example, with Fe^{3+},

$$Y^{4-} + Fe^{3+} = FeY^- \qquad [3\text{-}56]$$

The utility and petroleum industries utilize **scale** (such as those caused by $BaSO_4$) **dissolvers**, DTPA (diethylenetrinitrilopentaaceticacid), DCTA (1,2-diaminocyclohexane tetraacetate), DOTA (1,4,7,10-tetraazacyclododecane *N*, *N'*, *N''*, *N''''* - tetraacetic acid), for keeping tubes from fouling. (The structures of most of these are shown in Fig. 3-13.) Table 3-6 provides a status report of the industry scale formation and abatement.

Another type of compound is the **scale inhibitor**. This type of compound is usually adsorbed onto the surface of a crystal in such a way as to prevent further growth or clustering of the crystal. A good example can be found with certain synthetic polypeptides (see Fig. 3-13 for structures). Figure 3-14 is a comparison of the degree of complex for Fe^{3+} in the presence of a number of complexing agents.

Table 3-5. Common Masking Agents

Masking Agent	Element Masked
F^-	Al, B, Be, Ca, Ce, Cr, Fe, Th, Ti, U, V, W, Zn, Ge, Mg, Mn, Mo, Nb, Ni, Sn
CN^-	Ag, Au, Cd, Co, Cu, Fe, Hg, Mn, Ni, V, Zn
CNS^-	Ag, Cd, Co, Cu, Fe, Mo, Ni, W, Zn
NH_3	Ag, Co, Cu, Fe, Ni, Zn
BAL	Al, As, Bi, Cd, Co, Cu, Fe, Hg, Mn, Pb, Sb, Sn, Zn
EDTA	Al, Ba, Bi, Ca, Cd, Ce, Co, Cr, Cu, Fe, Hg, Mg, Mn, Mo, Ni, Pb, Sb, Sr, Th, V, Zn
Citrate or tartrate	Al, Ba, Be, Bi, Ca, Cd, Ce, Co, Cr, Cu, Fe, Hg, Mn, Mo, Ni, Pb, Sb, Sn, Sr, Th, U, Zn
Oxalate	Al, Fe, Ge, Mg, Mn, Mo, Nb, Sn, U
$S_2O_3^{2-}$	Au, Cd, Co, Cu, Fe, Pb, Sb
H_2O_2	Co, Mo, Nb, Ti, U, V, W

BAL = 2,3-Dimercapto-1-propanol; EDTA = ethylenediaminetetracetic acid.

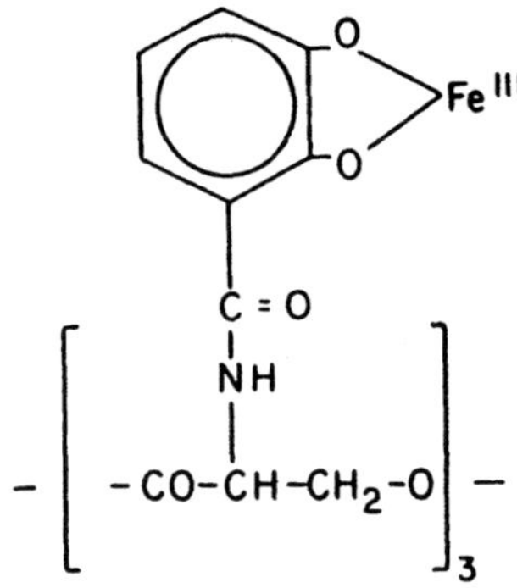

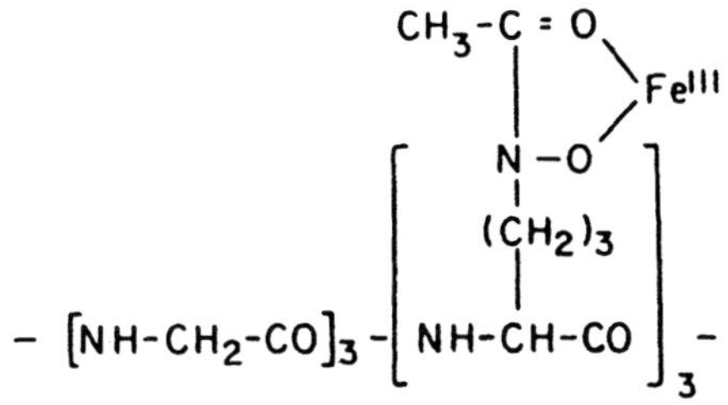

Enterobactin Ferrichrome

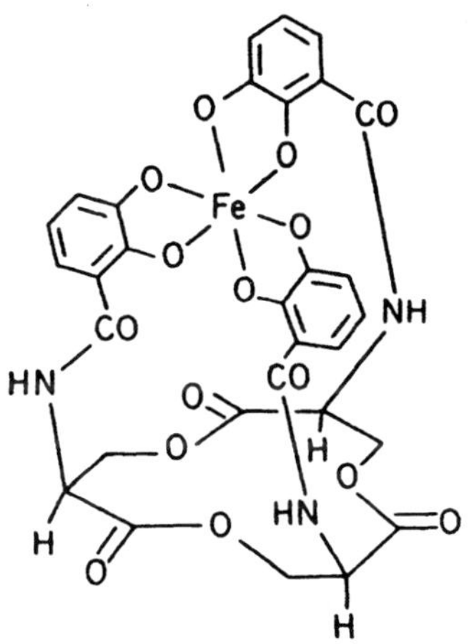

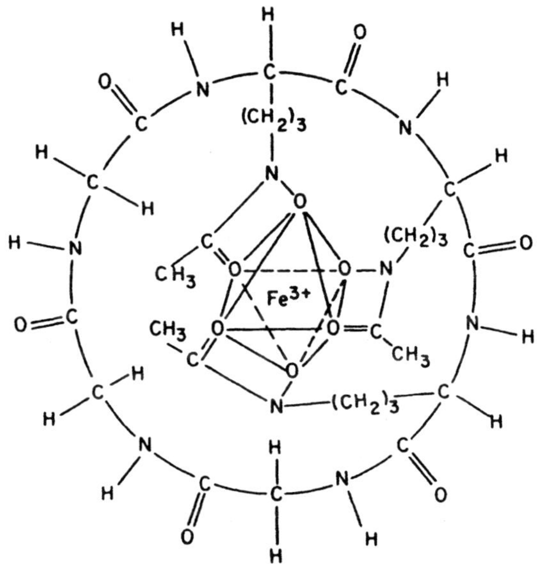

Figure 3-12. Structures of two archetypical siderophores: the catechol enterobactin and the hydroxamate ferrichrome. Both of these compounds and a large number of similar ones with equally strong affinity for Fe(III) ($K \cong 10^{20} - 10^{60}$) are produced by microorganisms for transporting — and competing for — iron. (From Ref. 3-1.)

EDTA

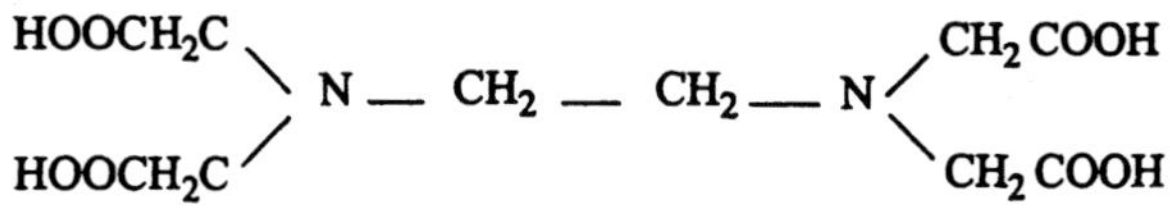

DTPA

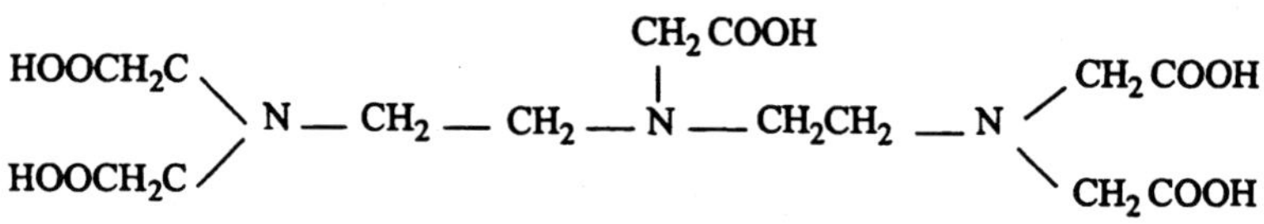

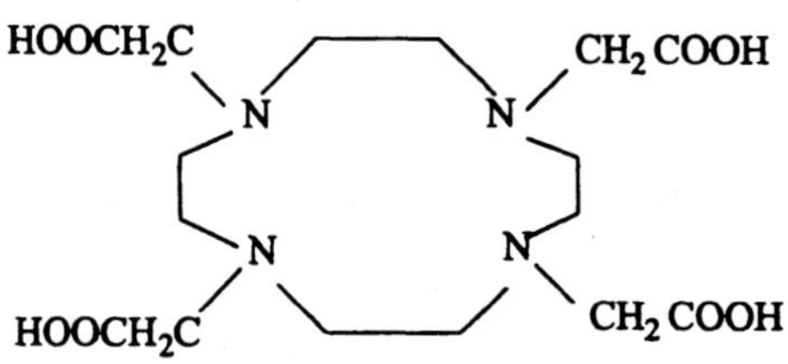

DOTA

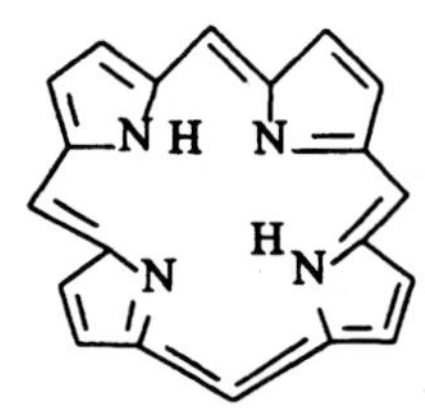

Porphyrin

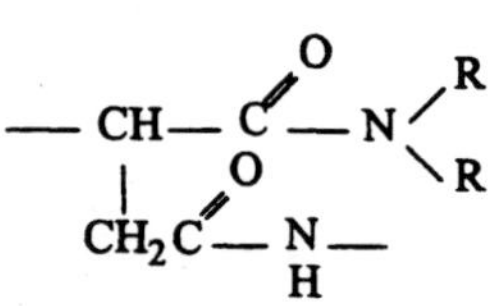

Polypeptide

Figure 3-13. Typical chelators.

Table 3-6. Common Scale Deposits in Petroleum Industry[a]

Deposit	Mineral	Most Frequent Causes of Scale Deposit	Removal Chemical
$CaCO_3$	Calcite	Mixing brines, changes in temperature and pressure	15% HCl
$CaSO_4 \cdot 2H_2O$ $CaSO_4$	Gypsum Anhydrite	Mixing brines, changes in temperature and pressure	Converting solutions, EDTA type dissolvers
$BaSO_4$	Barite	Mixing of brines	b
$SrSO_4$	Celestite	Mixing brines, changes in temperature and pressure	b
$FeCO_3$	Siderite	Mixing brines, changes in temperature and pressure	Sequestered acid
FeS	Trolite	Corrosion by sour crude or H_2S gas	Sequestered acid
Fe_2O_3 Fe_3O_4	Hematite Magnetite	Reactions of oxygen with dissolved ferrous ion	Sequestered acid
NaCl	Halite	Evaporation of water	Water or 1-3% HCl
$Mg(OH)_2$	Brucite	Excessive amounts of oxygen enter the well or alkaline fluids in well. High temperature.	15% HCl
Silicates		Cooling of hot brine high in dissolved silica	HCl:HF acid mixtures

[a]Source from P.J. Shuler of Chevron Petroleum Technology Co., 1997.

[b]Not developed. Maybe chelating agents, such as DTPA and DOTA will be used as dissolvers.

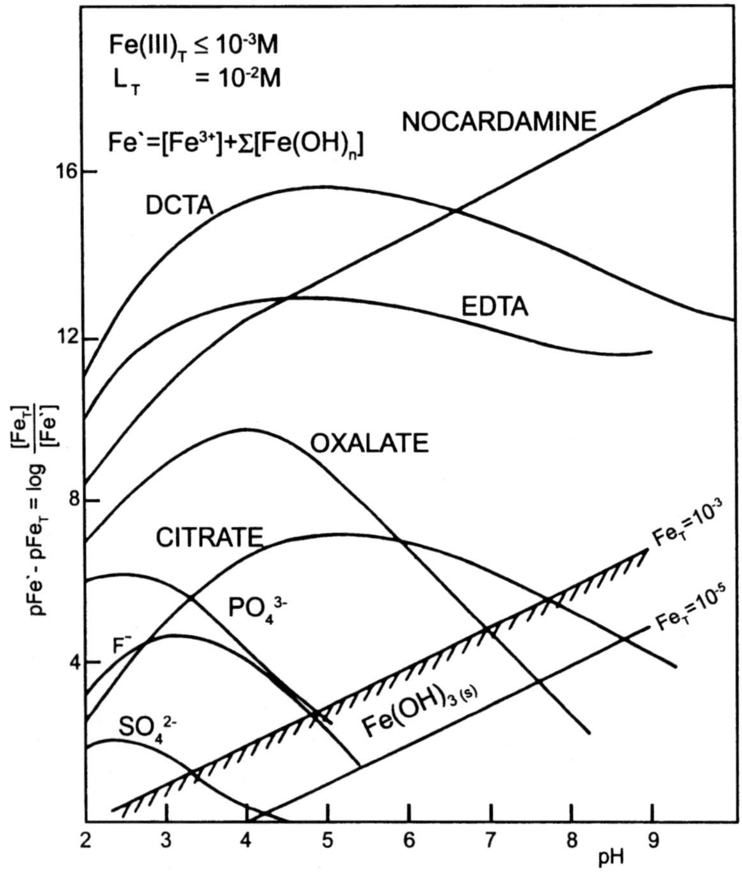

Figure 3-14. Complexing of Fe(III). The degree of complexation is expressed in terms of ΔpFe for various ligands (10^{-2} M). The competing effect of H^+ at low pH values and OH^- at higher pH values explains that effective complexation is strongly dependent on pH. Mono-, di-, and tridentate ligands are not able to keep a 10^{-3} M Fe(II) in solution at higher pH values. EDTA = Ethylenediaminetetraacetic acid; DCTA = 1,2-diaminocyclohexanetetraacetate. Nocordamine is a trihydroxamate. (Adapted from Ref. 17-6.)

[Example 3-5] Evaluate the stabilization of a fresh water system by the gradual addition of hydroxamate ferrichrome until a total of $10^{-5.8}$ M has been introduced. The fresh water contains 10^{-6} M Fe(III), 10^{-7} M Cu(II), $10^{-5.3}$ M CO_3^{2-}, $10^{-4.0}$ M SO_4^{2-}, $10^{-3.2}$ M Cl^-, $10^{-3.5}$ M Ca^{2+}, and $10^{-4.1}$ M Mg^{2+}, and the pH is 8.1. (Adapted from F.M.M. Morel).

A literature search reveals the following:

HY^-, $\log \beta_1 = 10.1$

H_2Y, $\log \beta_2 = 19.4$

H_3Y^+, $\log \beta_3 = 27.8$

CaY, $\log \beta_1 = 3.5$

MgY, $\log \beta_1 = 5.2$

FeY^+, $\log \beta_1 = 31.9$

$FeHY^{2+}$, $\log \beta = 32.6$

CuY, $\log \beta = 15.0$

$CuHY^+$, $\log \beta = 24.1$

CuH_2Y^{2+}, $\log \beta = 27.0$

$CuOH^+$, $\log \beta_1 = 6.5$

$Cu(OH)_2$, $\log \beta_2 = 11.8$

$$CuCO_3, \log \beta_1 = 6.7$$

$$Cu(CO_3)_2^{2-}, \log \beta_2 = 10.2$$

In the absence of a chelating agent, at pH 8.1 or $(OH^-) = 10^{-5.9}$, iron is precipitated as $Fe(OH)_3(s)$ and the free ferric ion concentration from Equation [3-20] is:

$$\left(Fe^{3+}\right) = \frac{10^{-38}}{\left(OH^-\right)^3} = 10^{-20.3} \qquad (a)$$

For free cupric ions, the concentration in the absence of a chelator can be approximated as shown below for the largest complexes of copper:

$$CuOH^+, \beta_1(OH^-) = 10^{+0.6}$$

$$Cu(OH)_2, \beta_2(OH^-)^2 = 10^{0.0}$$

$$CuCO_3, \beta_1(CO_3^{2-}) = 10^{1.4}$$

$$Cu(CO_3)_2^{2-}, \beta_2(CO_3^{2-})^2 = 10^{-0.4} \qquad (b)$$

$$Cu_T = (Cu^{2+})[1 + 10^{0.6} + 10^{0.0} + 10^{1.4} + 10^{-0.4}] = Cu^{2+} \times 10^{1.5} = Cu^{2+} \times 31.5 \quad (c)$$

Thus,

$$\frac{\left(Cu^{2+}\right)}{Cu_T} = \frac{1}{31.5} = 3\%$$

$$\frac{\left[Cu\left(OH^+\right)\right]}{Cu_T} = \frac{10^{0.6}}{31.5} = 13\%$$

$$\frac{[\mathrm{Cu(OH)_2}]}{\mathrm{Cu}_T} = \frac{10^{0.0}}{31.5} = 3\%$$

$$\frac{(\mathrm{CuCO_3})}{\mathrm{Cu}_T} = \frac{10^{1.4}}{31.5} = 80\%$$

$$\frac{[\mathrm{Cu(CO_3)_2}^{2-}]}{\mathrm{Cu}_T} = \frac{10^{-0.4}}{31.5} = 1\% \qquad \text{(d)}$$

Obviously the free cupric ion is only 3% of the total copper. The mole balance equation for the ferrichrome at low ligand levels can then be simplified.

$$Y_T = (Y^{2-})[1 + 10^{10.1}(\mathrm{H^+}) + 10^{19.4}(\mathrm{H^+})^2 + 10^{27.8}(\mathrm{H^+})^3 + 10^{3.5}(\mathrm{Ca^{2+}}) + 10^{5.2}(\mathrm{Mg^{2+}}) + 10^{31.9}(\mathrm{Fe^{3+}}) +$$

$$10^{32.6}(\mathrm{H^+})(\mathrm{Fe^{3+}}) + 10^{15}(\mathrm{Cu^{2+}}) + 10^{24.1}(\mathrm{H^+})(\mathrm{Cu^{2+}}) + 10^{27}(\mathrm{H^+})^2(\mathrm{Cu^{2+}})] \qquad \text{(e)}$$

All terms are negligible when compared to $\mathrm{Fe}Y^+$ concentration

$$Y_T = (\mathrm{Fe}Y^+) = 10^{31.9}(\mathrm{Fe^{3+}})(Y^{2-}) = 10^{11.:6}(Y^{2-}) \qquad \text{(f)}$$

As long as $Y_T < \mathrm{Fe}_T$, the ligand is entirely bound to iron and the free ion is controlled by $\mathrm{Fe(OH)_3(s)}$ precipitation. The copper remains chiefly in the copper carbonate complex (80%) [see Eq. (d)].

As soon as the ligand concentration exceeds the Fe_T, for example at $Y_T = 10^{-5.8}$ M, in addition to $\mathrm{Fe}Y^+$, the major species at pH 8.1 are still $\mathrm{H_2}Y$ and $\mathrm{H_3}Y^+$.

$$Y_T = (\mathrm{H_2}Y) + (\mathrm{H_3}Y) + (\mathrm{Fe}Y^+) = 10^{-5.8} \qquad \text{(g)}$$

We assume at this point that all the iron is bound to the organic ligand.

$$\mathrm{Fe}_T = (\mathrm{Fe}Y^+) = 10^{-6}$$

or

$$(\mathrm{H_2}Y) + (\mathrm{H_3}Y) = Y_T - \mathrm{Fe}_T = 10^{-6.23}$$

and

$$(Y^{2-})[10^{19.4}(H^+)^2 + 10^{27.8}(H^+)^3] = 10^{-6.23}$$

$$(Y^{2-})[10^{3.2} + 10^{3.5}] = 10^{-6.23}$$

$$(Y^{2-}) = 10^{-9.9} \qquad \text{(h)}$$

or

$$(FeY^{+}) = 10^{31.9}(Y^{2-})(Fe^{3+}) = Fe_T = 10^{-6} \qquad \text{(f)}$$

Thus, when compared with Equation (a),

$$(Fe^{3+}) = 10^{-28.0} < 10^{-20.3} \qquad \text{(i)}$$

and is dissolved.

3.4.1 Masking

Masking is used to prevent precipitation of a chemical species in a process. We will illustrate this principle with a Fe^{3+} solution in a basic medium. In basic conditions Fe^{3+} will form $Fe(OH)_3(s)$, and K_s $[Fe(OH)_3]$ will control the precipitation. One can first estimate a pFe^{3+} value from the solubility product constant, which is the principal reaction, $pFe^{3+} = pM_p$. From the masking reaction, one can estimate another $(pFe^{3+})_m$ value. If $(pFe^{3+})_m > pFe^{3+}$, then no precipitate of hydrous ferric oxide will be formed and the solution is masked. Expressed in general terms, a masking condition is

$$pM_m > pM_p \qquad [3\text{-}57]$$

Usually the metal ion concentration of the masking reaction and that of the principal reaction (from K_s) in Example 3-6 are compared, and a **masking ratio** larger than 7 is often found to be a stable system.

$$(pM_m)^2/pM_p > 7 \qquad [3\text{-}58]$$

[Example 3-6] Water that contains 0.1 M $FeCl_3$ is buffered by ammonia to an alkaline condition. If EDTA is introduced with a concentration in excess of 0.1 M, will there be any precipitation of $Fe(OH)_3(s)$? The buffering condition is: $(NH_4^+) = 0.4$ M, $(NH_3) = 0.04$ M.

First we write all the equations involved: the principal solubility product reaction (Eq. [3-20]):

$$K_{s,\mathrm{Fe(OH)_3}} = (\mathrm{Fe}^{3+})(\mathrm{OH}^-)^3 = 10^{-38} \qquad \text{(a)}$$

the buffering reaction (Eq. [3-9]):

$$K_a = ((\mathrm{H}^+)(\mathrm{NH_3}))/(\mathrm{NH_4^+}) = 10^{-9} \qquad \text{(b)}$$

the masking reaction:

$$K_m = ((\mathrm{Fe}^{3+})(Y^{4-}))/(\mathrm{Fe}Y^-) = 10^{-25} \qquad \text{(c)}$$

and the protonations of the ligand:

$$K_3 = ((\mathrm{H}^+)(\mathrm{H}Y^{3-}))/(\mathrm{H}_2Y^{2-}) = 6 \times 10^{-7} \qquad \text{(d)}$$

$$\mathrm{K}_4 = ((\mathrm{H}^+)(Y^{4-}))/(\mathrm{H}Y^{3-}) = 5 \times 10^{-11} \qquad \text{(e)}$$

From the buffering reaction (b), one can compute the pH of this solution; we found pH = 8. Substituting this value to the principal solubility product reaction (a),

$$\mathrm{pM_p} + 3\mathrm{pOH} = \mathrm{p}K_s$$

or

$$\mathrm{pM_p} = 20 \qquad \text{(f)}$$

From the protonation equations (d) and (e), and substituting pH = 8, we found that

$$(HY^{3-})/(H_2Y^{2-}) = 6 \times 10^{-7}/10^{-8}$$

$$(Y^{4-})/(HY^{3-}) = 5 \times 10^{-11}/10^{-8}$$

Thus,

$$(H_2Y^{2-}) : (HY^{3-}) : (Y^{4-}) = 1{:}60 : 0.30 \qquad \text{(g)}$$

In this manner

$$(Y^{4-}) = (0.3/61.3) \times 0.1 = 5 \times 10^{-4}$$

Substituting this value and (FeY^-) = 0.1 into the masking reaction (c), one finds (Fe^{3+}) = 2×10^{-23}, or $pM_m = 22.7$. Therefore, $10^{-22.7} < 10^{-20}$ and $pM_m > pM_p$, and there is no precipitate formed. The masking ratio of this case is 25.8, which is higher than 7. The state of this solution is stable.

Finally, having said that the polypeptide linkage can inhibit the growth of minerals (scale inhibitor), the dissolver also requires the imino or nitrilo functions with carboxylic acid functional groups. Proteins in living matters can initiate the growth of minerals. A simple example is the calcium carbonate shells in mollusks or silica gel formation in diatoms. This natural activity by selective extractions of certain ions from the suitable environment is termed **biomineralization**. Some plausible structures of how the different amino acids work as an anchor to mimic the shell growth are summarized in Figure 3-15.

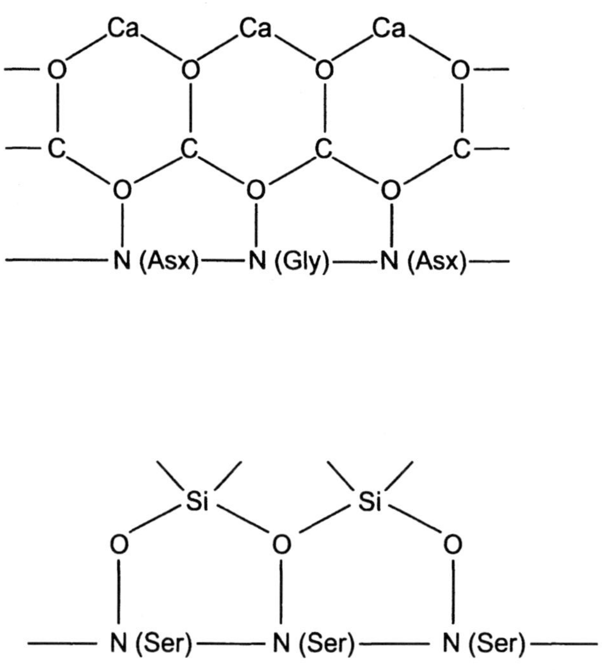

Figure 3-15. Plausible structures for biomineralization. Top: nucleation sites for calcium carbonate shell in mollusks. Lower: nucleation sites for silica gel in diatoms.

REFERENCES

3-1. F. M. M. Morel, *Principles of Aquatic Chemistry*, Wiley, New York, NY, 1983.

3-2. O. Söbnel and J. Gorside, *Precipitaion, Basic Principles and Industrial Applications*, Butterworth-Hernermann, Oxford, 1992.

3-3. W. F. Pickering, *Fundamental Principles of Chemical Analysis*, Elsevier, Amsterdam, 1966.

3-4. A. J. Rubin, *Aqueous-Environmental Chemistry of Metals*, Ann Arbor Science, Ann Arbor, Michigan, 1976.

3-5. A. Ringbow, *Complexation in Analytical Chemistry*, Wiley-Interscience, New York, 1963.

3-6. L. G. Sillen and A. C. Martell, *Stability Constants of Metal-Ion Complexes*, 2nd ed., special publication No. 17, The Chemical Society, London, 1964.

3-7. T. F. Yen, *The Role of Trace Metals in Petroleum*, Ann Arbor Science, Ann Arbor, Michigan, 1975.

PROBLEM SET

1. Find the Langelier saturation indexes at 25°C, 60°C, and the excess CO_2 (open and closed system) for the waters listed:

Ca, mg/L ($CaCO_3$)	Alkalinity, mg/L ($CaCO_3$)	Solids, mg/L	pH
(a) 252	164	284	7.9
(b) 17	8	37	6.8
(c) 40	32	60	10.1

2. Is it the $FeCO_3(s)$ or the $Fe(OH)_2(s)$ that controls the solubility of Fe(II) in anoxic water of alkalinity = 10^{-1} eq/L and pH = 6.8? You may estimate maximum soluble Fe(II) by considering the solubility equilibrium (25°C) with $FeCO_3(s)$ as well as that with $Fe(OH)_2(s)$.

CHAPTER 4

ION EXCHANGE

Ion exchange is a well-known process in nature. For example, clay in soil helps to transport ions to and from various sites. Silicate minerals are nature's exchangers, which provide for a number of geochemical processes, such as the enrichment of minerals. As we will learn, reaction and equilibrium processes can also be conducted in solid phases; in fact, heterogeneous reactions are quite common. The old statement *Corpora non agunt nisi fluida sive soluta* — that a substance does not react unless it is in a liquid or dissolved state — is no longer true.

In this chapter, we first explore the principles of ion exchange, such as exchange reactions and capacity. We also discuss the selectivity coefficient and exchange isotherm in terms of thermodynamics and equilibria. Later in the chapter, diffusion and transport within the context of exchange kinetics are illustrated. A number of examples of methods widely used in municipal and wastewater treatments are provided, including partial softening, desalination and demineralization, and reclamation of plating wastes. Finally, we conduct a study of ion chromatography, which makes up another useful tool in analyzing metals and other polar organic species in water for environmental purposes.

4.1 ION EXCHANGERS AND EXCHANGE REACTIONS

Ion exchange is the process in which ions in a liquid phase (an electrolyte solution) exchange with ions in a solid phase (a solid ion exchange medium, called an ion exchanger). Usually, ion exchangers are insoluble solid materials, which have fixed exchange sites and carry exchangeable cations or anions. These ions can be exchanged for a stoichiometric equivalent of other ions of the same sign when the exchanger is in contact with an electrolyte. Carriers of exchangeable cations are called **cation exchangers**, and the carriers of anions, **anion exchangers**. Materials that are capable of both cation and anion exchange are called **amphoteric** ion exchangers.

Some exchange reactions are given here:

1. Strong acid cation exchangers

Hydrogen form:

$$2RSO_3H + Ca^{2+} = (RSO_3)_2Ca + 2H^+$$

The product can be regenerated by a strong acid.

Sodium form:

$$2RSO_3Na + Ca^{2+} = (RSO_3)_2Ca + 2Na^+$$

This can be regenerated by a NaCl solution.

2. Weak acid cation exchangers

Hydrogen form:

$$2RCOOH + Ca^{2+} = (RCOO)_2Ca + 2H^+$$

The product can be regenerated by a strong acid.

Sodium form:

$$2RCOONa + Ca^{2+} = (RCOO)_2Ca + 2Na^+$$

This can be regenerated by a NaOH solution.

3. Strong base anion exchangers

Hydroxide form:

$$2RR'_3NOH + SO_4^{2-} = (RR'_3N)_2SO_4 + 2OH^-$$

The product can be regenerated by a NaOH solution.

Chloride form:

$$2RR'_3NCl + SO_4^{2-} = (RR'_3N)_2SO_4 + 2Cl^-$$

This can be regenerated by a NaCl or HCl solution.

4. Weak base anion exchangers

Free base or hydroxide form:

$$2RNH_3OH + SO_4^{2-} = (RNH_3)_2SO_4 + 2OH^-$$

The product can be regenerated by NaOH, NH_4OH, or Na_2CO_3.

Chloride form:

$$2RNH_3Cl + SO_4^{2-} = (RNH_3)_2SO_4 + 2Cl^-$$

This can be regenerated by an HCl solution.

In general, an ion exchanger containing counter ion *A* is placed in a solution containing counter ion *B*, and the counter ions are redistributed until equilibrium is attained.

$$\overline{AR} + BC = \overline{BR} + AC$$

A and *B* are **counter ions**; the resin, *R*, is always a solid phase (overlined) exchanger; and *C* is a **co-ion**, common to both counter ions. Both *BC* and *AC* are salts (Fig. 4-1).

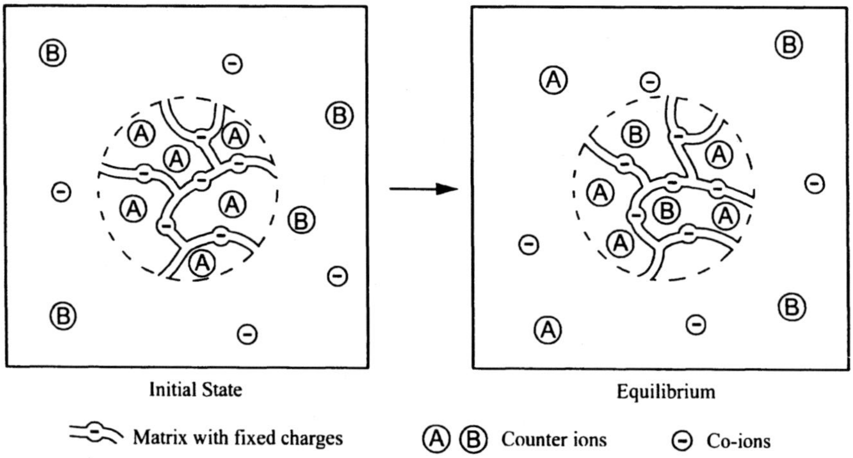

Figure 4-1. Ion exchange with a solution. A cation exchanger containing counter ions A is placed in a solution containing counter ions B (left). The counter ions are redistributed by diffusion until equilibrium is attained (right). (Adapted from Ref. 4-8.)

Various aluminosilicates such as zeolites are known to be exchangers. They usually have a rigid three-dimensional framework structure with cavities and chemicals in which the ions can move. Other materials, such as soils, humus, lignin, coal, and living cells, have been used as well. The vast exchangers are the organic ion exchanger resins, which consist of a three-dimensional cross-linked network with fixed sites of functional groups. The **resins** are cross-linked polyelectrolytes. An unlimited variety of resins with different fixed ionic groups of different compositions and varied degrees of matrix cross-linking can be synthesized. Figure 4-2 illustrates how the ion exchange resins are made. Both strong acids and weak acids are given. Also, some trade names of resin are shown.

4.1.1 Exchange Capacity

The **exchange capacity** of a resin is defined as the total number of ionic sites available for exchange per unit of mass. This value is usually used for characterization of a given ion exchanger. The unit is expressed as meq/g of dry resin in H^+ and Cl^- form. Occasionally the unit is expressed in volume, for example, eq/L of packed bed, eq/ft^3 bed, lb. CaO/ft^3 bed, etc. The value is determined by titration. For an acid resin,

$$\overline{RH} + \mathrm{NaOH} = \overline{RNa} + \mathrm{H_2O}$$

Figure 4-3 shows several titration curves of various cation exchange resins. The capacity unit expressed in weight, q_w, can be converted into a volume unit, q_v, as follows:

$$q_v = (1 - \beta)\,\rho\,(1 - w)q_w \qquad [4\text{-}1]$$

where β is the fraction void volume of the packing, ρ is the density of swollen resin in g/cm^3, and w is the water content in percentages.

For a weak acid or a weak base resin, the apparent pK value is used to obtain an apparent weight capacity number:

$$q_{\mathrm{app}} = \alpha q_w \qquad [4\text{-}1a]$$

Under experimental conditions, these values are found to be more useful. For a weak acid resin, $\overline{RH}$,

$$K = \frac{\left[\left(\overline{R^-}\right)\left(\overline{H^+}\right)\right]}{\left(\overline{RH}\right)} \qquad [4\text{-}2]$$

and the degree of dissociation, α, is defined as:

$$\alpha = \frac{\left(\overline{R^-}\right)}{\left[\left(\overline{R^-}\right)+\left(\overline{RH}\right)\right]}$$

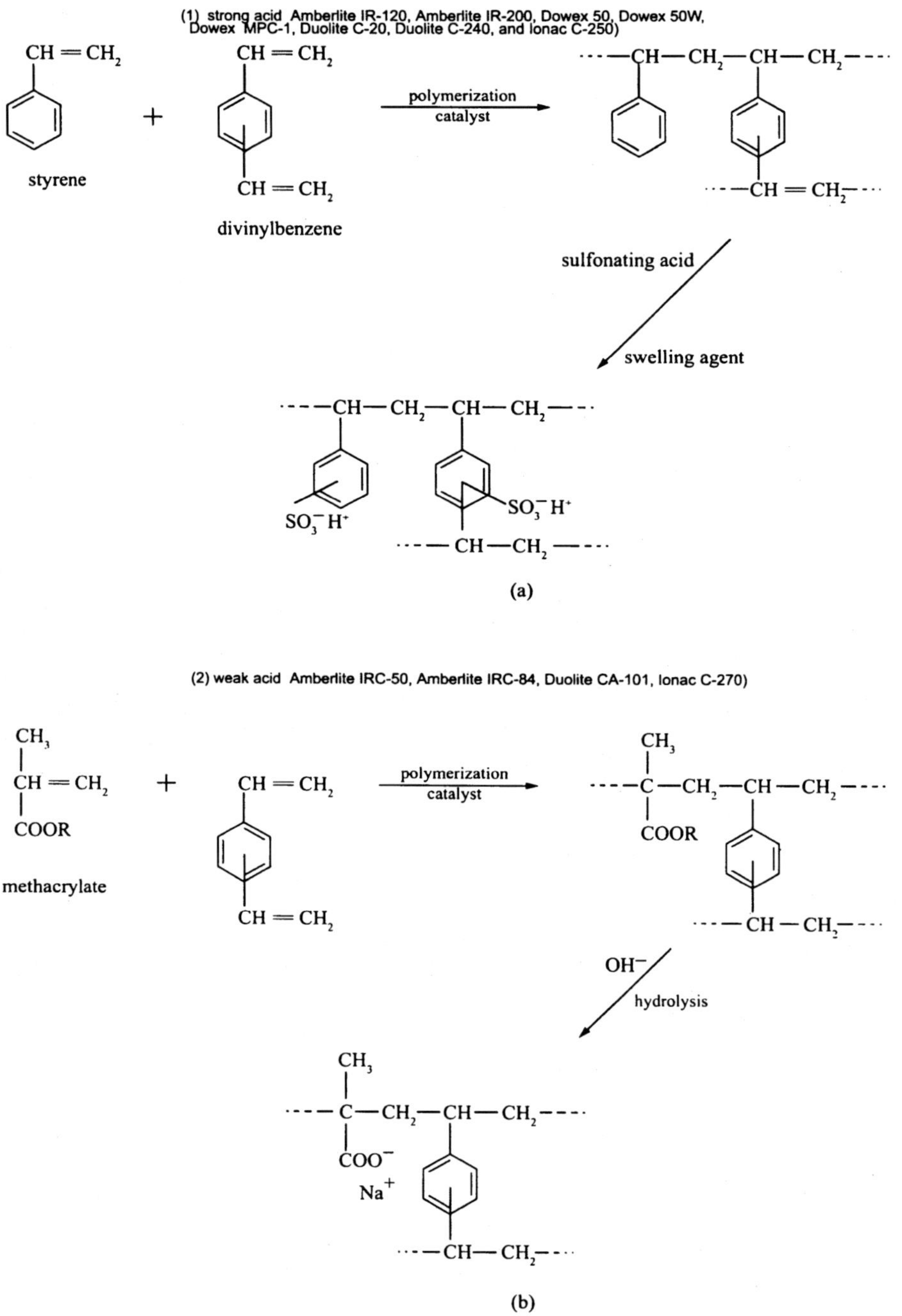

Figure 4-2. Preparation of ion exchange resin

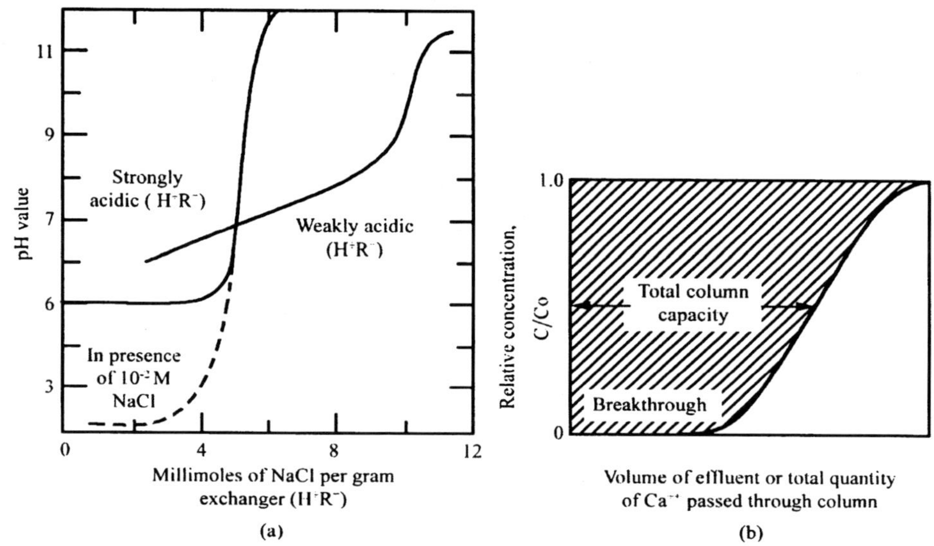

Figure 4-3. Determination of ion exchange capacity. (a) Alkalimetric titration of cation exchanger in the $\{H^+R^-\}$ form provides information about the acidity of the exchanger and permits the determination of the analytical exchanger capacity. (b) Determination of useful capacity of an ion exchange column.

Thus,

$$p\overline{H} = pK - \log\left[\frac{(1-\alpha)}{\alpha}\right] \qquad [4\text{-}3]$$

The second term vanishes when $\alpha = 0.5$. In weak acid resins, $\alpha = 0.5$, corresponding to a 50% conversion of resin from its H^+ form to the Na^+ form. Therefore,

$$pK = p\overline{H}\,(\alpha = 0.5) \qquad [4\text{-}4]$$

The pH of the solid phase should be related to the pH of the solution as a first approximation:

$$\frac{(Na^+)}{(H^+)} = \frac{(\overline{Na^+})}{(\overline{H^+})} \qquad [4\text{-}5]$$

At 50% conversion, the (Na^+) in the resin is:

$$(\overline{Na^+}) = \frac{(\overline{X})}{2} \qquad [4\text{-}6]$$

where $(\overline{X}) = (\overline{RH}) + (\overline{R^-})$ is the total concentration of dissociated and undissociated functional groups. Thus, from Equation [4-4],

$$pK = pH + \log(Na^+) - \log[(\overline{X})/2] \qquad [4\text{-}7]$$

and similarly, for a weak base anion exchanger,

$$pK = pH - \log(Cl^-) + \log[(\overline{X})/2] \qquad [4\text{-}7a]$$

Usually, the concentration of a univalent counter ion in a weak acid resin is:

$$(A^+) = \alpha(\overline{X}) \qquad [4\text{-}8]$$

and substituting Equation [4-3] into the degree of dissociation, we have:

$$\log[(1-\alpha)/\alpha^2] = pK - pH + \log(\overline{X}) - \log(A^+) \qquad [4\text{-}9]$$

[Example 4-1] Find q_w for a monosulfonated styrene-divinyl benzene copolymer. Convert it to volume units if, after being fully water-swollen, it has a density of 1.25 g/cm^3 and a water content of 46.8%.

The bulk of this polymer system is polystyrene sulfonate as indicated in Figure 4-4. The unit weight is $C_8H_8O_3S$ and is based on the relation q_w = 1000/184.2 = 5.43 meq/g. Actually, this copolymer also contains sulfonated divinyl benzene and ethylstyrene units with unit weights of 210.2 and 212.2. If the analysis shows that this resin has 8% DVB and 4% ethyl benzene units, then

q_w = 5.35 meq/g. Now the void volume of a packed bed may be taken as 40%. Hence,

$$q_v = (1 - 0.40)(1.25)(1 - 0.468)(5.4) = 2.2 \text{ eq/L bed}$$

Polystyrene
88%

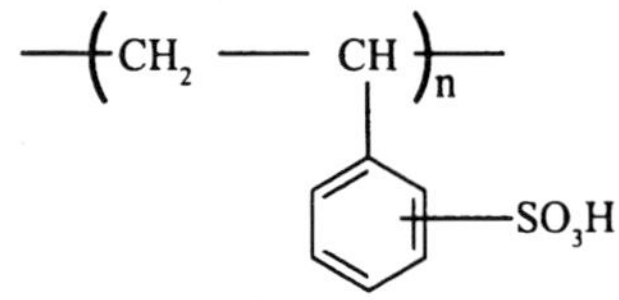

$C_8H_8O_3S$ MW 184.2

Polydivinylbenzene (DVB)
8%

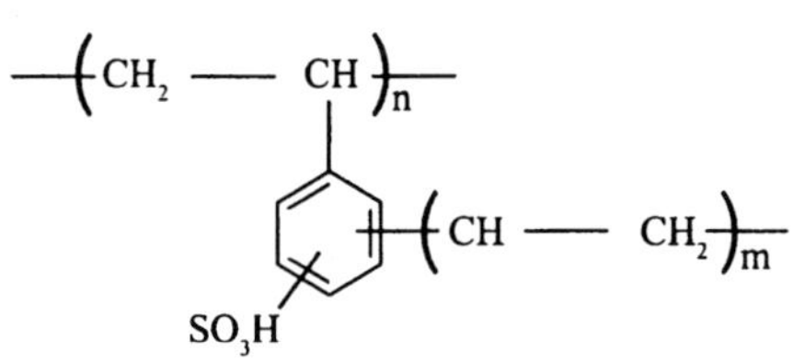

$C_{10}H_{10}O_3S$ MW 210.2

Polyethylstyrene
4%

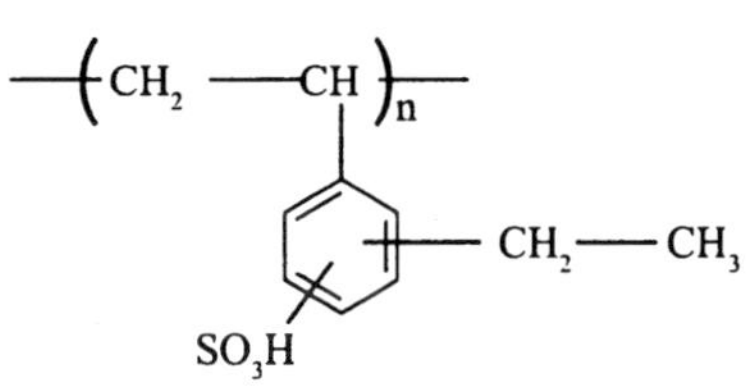

$C_{10}H_{12}O_3S$ MW 212.2

Figure 4-4. Monosulfonated copolymer of polystyrene.

[Example 4-2] A polymethacrylic acid resin with q_w = 9.0 meq/g, pK = 4.8, and a water content of 45% wt in H^+ form and 65% in Na^+ form is equilibrated with an excess of 0.02 M NaCl solution of pH = 6.0. Find the apparent capacity.

Since the average water content is (45 + 65)/2 = 55%, the concentration of the functional group in resin is $(\bar{X})$ = 9.0 × 45/55 = 7.4 meq/g water. Thus, $\log(1-\alpha)/\alpha^2 = 4.8 - 6.0 - \log(0.02) + \log(7.4) = 1.37$

Solving, $\alpha = 0.19$, $q_{app} = 0.19 \times 9.0 = 1.7$ meq/g.

4.2 Exchange Equilibria and Kinetics

Through the years, different theories and mechanisms have been advanced to account for the phenomenon of ion exchange. The **crystal lattice theory** attempted to account for the fixed sites of ions, and for many minerals the crystal lattice ions can be interchangeable. The **double layer theory** was developed for colloidal systems. The crystal lattice assumes a fixed number of exchange sites, which have to be satisfied regarding change of pH or concentration, whereas the double layer theory can accommodate a change in pH and concentration. Another explanation of ion exchange is the **Donnan membrane theory**. This theory pertains to the unequal distribution of ions on two sides of a membrane; one side contains an electrolyte, one of whose ions is not able to diffuse through the membrane. The salt concentration is greater on the side that is free of the nondiffusible ions. Both sides of the membrane will be balanced if another ion is added to the system, especially to the side that contains the nondiffusible ion. In this way, in the exchange, the colloidal micelle, whose top is attached to the exchangeable ion, is considered a nondiffusible ion. There are a number of proposed mechanisms. For example, resin swells as a spring in Hook's law so that ions will ultimately occupy the least volume. In all these different mechanisms, let us not forget that up until now we have found that ion exchange:

1. Is stoichiometric
2. Is reversible
3. Obeys the law of mass action
4. Behaves as an adsorption process
5. Involves diffusion and the transport process

4.2.1 Selectivity Coefficient and Ion Exchange Isotherm

For a general ion exchange reaction,

$$m\overline{BR_n} + nA^{m+} = n\overline{AR_m} + mB^{n+}$$

$$Q = \frac{\left(\overline{AR_m}\right)^n \left(B^{n+}\right)^m}{\left(\overline{BR_n}\right)^m \left(A^{m+}\right)^n} = \frac{\left[\frac{\left(\overline{AR_m}\right)}{\left(A^{m+}\right)}\right]^n}{\left[\frac{\left(\overline{BR_n}\right)}{\left(B^{n+}\right)}\right]^m} \qquad [4\text{-}10]$$

When n, $m = 1$ and $Q = \alpha$, the separation factor

$$\alpha = \frac{\left[\frac{\left(\overline{AR_m}\right)}{\left(A^{m+}\right)}\right]}{\left[\frac{\left(\overline{BR_n}\right)}{\left(B^{n+}\right)}\right]} \qquad [4\text{-}11]$$

If $n = m$ and $Q = \alpha^n$, after non-ideality correction,

$$K = Q\left(\frac{\gamma_B}{\gamma_A}\right)^n = \left[\alpha\left(\frac{\gamma_B}{\gamma_A}\right)\right]^n \qquad [4\text{-}12]$$

Both Q and K are **selectivity coefficients**; the separation factor pertains to the ratio of A and B.

Selectivity refers to the relative performance of resins for different types of ions. In general, the following rules are applicable to the ions:

1. Of larger volume:

$$PO_4^{3-} > SO_4^{2-} > Cl^-$$

$$Th^{4+} > Nd^{3+} > Ca^{2+} > Na^+$$

2. With smaller solvated volume (usually smallest solvated volume will have largest ionic diameter):

$$Cs^+ > Rb > K^+ > Na^+ > Li^+$$

$$CNS^- > ClO_4^- > I^- > NO_3^- > Br^- > CN^-$$

3. With greater ability to polarize:

$$NO_3^- > Br^- > NO_2^- > Cl^-$$

4. Of least participation with other ions to form complexes:

$$Cl^- > HCO_3^- > CH_3COO^- > OH^- > F^-$$

For a two-component system, such as a cation exchange,

$$A^n + n(R^-B^+) = nB^+ + R_n^-A^{n+}$$

or

$$A^n + n\overline{B}^+ = nB^+ + \overline{A}^{n+}$$

Suppose we ignore the activity coefficients and use ionic concentrations. At present, many **selectivity coefficients** have been estimated for the binary system, as indicated by Tables 4-1 and 4-2. Using the assumptions, cation exchange can be expressed as:

$$K_B^A = \frac{\left(\overline{B}^+\right)^n \left(A^{n+}\right)}{\left(\overline{A}^{n+}\right)\left(B^+\right)^n} \tag{4-13}$$

Table 4-1. Selectivity Scale for Cations on 8% Cross-Linked Strong-Acid Resin

Cation	Selectivity	Cation	Selectivity
Li^+	1.0	Zn^{2+}	3.5
H^+	1.3	Co^{2+}	3.7
Na^+	2.0	Cu^{2+}	3.8
NH_4^+	2.6	Cd^{2+}	3.9
K^+	2.9	Be^{2+}	4.0
Rb^+	3.2	Mn^{2+}	4.1

Table 4-1. Continued

Cation	Selectivity	Cation	Selectivity
Cs^+	3.3	Ni^{2+}	3.9
Ag^+	8.5	Ca^{2+}	5.2
UO^{2+}	2.5	Sr^{2+}	6.5
Mg^{2+}	3.3	Pb^{2+}	9.9
		Ba^{2+}	11.5

Source: Bonner and Smith, 1957.

Table 4-2. Approximate Selectivity Scale for Anions on Strong-Base Resins

Anion	Selectivity	Anion	Selectivity
I^-	8.0	OH^- (type II)	0.65
NO_3^-	4.0	HCO_3^-	0.40
Br^-	3.0	CH_3COO^-	0.20
HSO_4^-	1.6	F^-	0.10
NO_2^-	1.3	OH^- (type I)	0.06
CN^-	1.3	SO_4^{2-}	0.15
Cl^-	1.0	CO_3^{2-}	0.03
BrO_3^-	1.0	HPO_4^{2-}	0.01

Source: Peverson, 1953.

If expressed in fractions, then solution phase

$$X_{A^{n+}} = \frac{(A^{n+})}{C} \qquad X_{B^+} = \frac{(B^+)}{C} \tag{4-14}$$

where C is the total ionic concentration of solution in equivalents per liter. Similarly, in the resin phase,

$$\overline{X}_{A^{n+}} = \frac{(\overline{A}^{n+})}{\overline{C}} \qquad \overline{X}_{B^+} = \frac{(\overline{B}^+)}{\overline{C}} \tag{4-15}$$

$\overline{C}$ is now the total exchange capacity of resin per unit volume in equivalents per liter. Also,

$$X_{A^{n+}} + X_{B^+} = 1$$
$$\overline{X}_{A^{n+}} + \overline{X}_{B^+} = 1 \qquad [4\text{-}16]$$

Now Equation [4-13] becomes

$$\frac{\overline{X}_{A^{n+}}}{\left(1-\overline{X}_{A^{n+}}\right)} = K_B^A \left(\frac{\overline{C}}{C}\right)^{n-1} \frac{X_{A^{n+}}}{\left(1-X_{A^{n+}}\right)^n} \qquad [4\text{-}17]$$

When considering the monovalent exchange $n = 1$, C and $\overline{C}$ do not appear in the equation, while for other types of valence exchange, C and $\overline{C}$ become important in equilibrium concentration.

[Example 4-3] A monovalent exchange resin is used to remove nitrate from chloride well water. Given the water composition below and a total resin capacity of 1.4 eq/L, what is the maximum volume of water that can be treated per ft^3 of this strong base anion exchange resin?

Water Composition (in meq/L)

Cation		Anion	
Ca^{2+}	1.5	Cl^-	3.5
Mg^{2+}	1.0	SO_4^{2-}	0.0
Na^+	2.5	NO_3^-	1.5
Total Cation	5.0	Total Anion	5.0

The equivalent fraction of nitrate in solution is 1.5/5.0 = 0.30. From Table 4-2, $K_{Cl^-}^{NO_3^-} = 4$. Using Equation [4-17],

$$\frac{\overline{X}_{NO_3^-}}{1-\overline{X}_{NO_3^-}} = K_{Cl^-}^{NO_3^-} \frac{X_{NO_3^-}}{1-X_{NO_3^-}} = 4\frac{(0.3)}{(1-0.3)} = 1.71$$

or $\overline{X}_{NO_3^-} = 0.63$. Thus, 63% of resin sites can be loaded with nitrate ion from the well water. The maximum feed capacity is:

$$1.4\ \text{eq/L} \times 0.63 = 0.88\ \text{eq/L}$$

The maximum volume of water that can be treated per cycle is:

$$\frac{0.88\ \text{eq/L (resin)}}{1.5 \times 10^{-3}\ \text{eq/L (water)}} = 586\ \text{L of water/L of resin} = 4400\ \text{gal/ft}^3$$

Ion exchange can be viewed as an adsorption process. This occurs when a given amount of ions are within the resin and the ions are in solution when the resin is at equilibrium with the solution. These behaviors can be expressed by the Langmuir equation:

$$q_e = Q_0 bC/(1 + bC) \qquad [4\text{-}18]$$

where q_e is the amount of ion exchanged per unit mass, Q_0 is the limiting value of q_e, b is constant, and C is the ion concentration in the solution. Another important equation is the Freundlich equation,

$$q_e = K_F \ln C^{1/n} \qquad [4\text{-}19]$$

where both K_F and n are constants. A plot of q_e versus C will yield an **isotherm** (see Fig. 4-5). Of course the isotherm can be graphically reduced to a straight line, for example, for the Langmuir equation, q_e^{-1} versus C^{-1}, and for the Freundlich equation, log q_e versus log C.

We will illustrate how these isotherms can be derived from equilibrium considerations. For an exchange reaction,

$$A^+ + \overline{BR} = \overline{AR} + B^+$$

If we assume that the original amount of resin = b and the initial concentration of $A^+ = C_0$, then

$$\left(\overline{BR}\right) = b - \left(\overline{AR}\right)$$

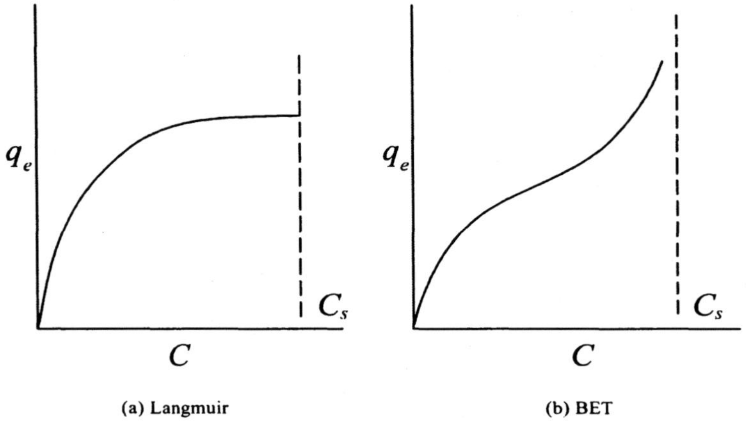

Figure 4-5. Typical isotherms for Langmuir and BET adsorption patterns.

$$\left(B^+\right) = C_0 - \left(A^+\right) \qquad [4\text{-}20]$$

Now,

$$Q = \frac{\left(\overline{AR}\right)\left[C_0 - \left(A^+\right)\right]}{\left[b - \left(\overline{AR}\right)\right]\left(A^+\right)} \qquad [4\text{-}21]$$

After rearrangement of the above, we have:

$$\frac{\overline{AR}}{b} = \frac{Q}{\left[\frac{\left(A^+\right)}{C_0}\right]^{-1} + (Q-1)} \qquad [4\text{-}22]$$

Here, we have a relation of the concentration of A^+ distributed in resin (solid) as $\frac{\left(\overline{AR}\right)}{b}$ and A^+ distributed in solution (liquid) as $\frac{\left(A^+\right)}{C_0}$. Figure 4-6 is a plot of $\frac{\left(\overline{AR}\right)}{b}$ versus $\frac{\left(A^+\right)}{C_0}$ with different Q values. It is evident that when $Q = 1$ for Equation [4-22], then

$$\frac{\left(\overline{AR}\right)}{b} = \frac{\left(A^+\right)}{C_0} \qquad [4\text{-}23]$$

Different Q values of the plot in Figure 4-6 will yield different isotherms. For $Q >> 1$, it suggests that most of the resin is consumed and only small portions of ions remain in the solution.

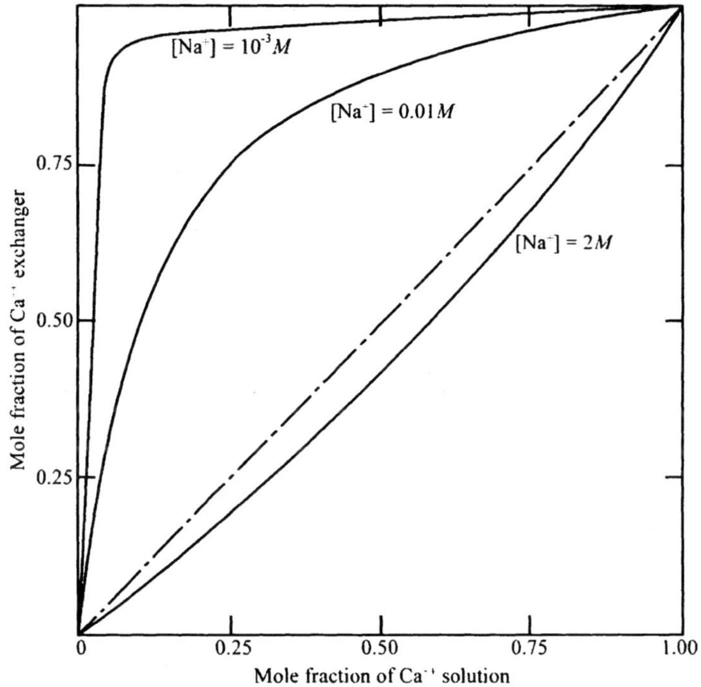

Figure 4-6. Exchange isotherms for the reaction $Ca^{2+} + \{Na^+R^-\} = \{Ca^{2+}R_2\} + 2\ Na^+$. In dilute solutions the exchanger shows a strong preference for Ca^{2+} over Na^+. This selectivity decreases with increasing ion concentration. The 45° line represents the isotherms for an exchanger with no selectivity. The exchanger can be regenerated with concentrated salt solutions.

4.2.2 Diffusion and Ion Exchange Kinetics

Ion exchange is a diffusion process. The mechanism is a redistribution of the counter ions by diffusion. The co-ion has relatively little effect on the kinetics and the rate of ion exchange. The rate-determining step in ion exchange is the interdiffusion of the exchange counter ions either within the ion exchanges themselves (particle diffusion) or within an adherent liquid film that is not affected by agitation of the solution (film diffusion). **Film diffusion** control is favored by a high capacity, a low degree of cross-linking of the resin, a weak agitation of the solution, and a low concentration of exchanging ions surrounding the resin particles. For film diffusion the appropriate expression for decreasing the rate of concentration C is:

$$\frac{-dC}{dt} = k_f \frac{a^{\circ}}{v}\left(C - C_e\right) \qquad [4\text{-}24]$$

Here, k_f is the film transfer coefficient, a° is the effective area for mass transfer, v is volume of the solution, and C_e is the equilibrated concentration in the solution. As for **particle diffusion**, the following spherical adsorbent equation can be used:

$$\frac{dC}{dt} = \frac{1}{r^2}\frac{1}{dr}\left(r^2 D_t \frac{dC}{dr}\right) - \frac{dq_e}{dt} \qquad [4\text{-}25]$$

where r is the resin particle radius and D_t is the diffusion coefficient. (For more discussion of diffusion, please see Chapter 7.)

The column packing can be illustrated by film diffusion control. Instead of using the rate of exchange of concentration with time, we use the rate of exchanging concentration for differential depth of resin used, dD, multiplying the latter by F_m, mass flow rate per unit area.

$$F_m\, dC/dD = (K^{\circ}a^{\circ})v^{-1}(C - C_e) = K^{\circ}\alpha\,(C - C_e) \qquad [4\text{-}26]$$

where $K^{\circ}\alpha$ is the overall mass transfer term. This expression can be further reduced to the change of mass of the resin in the column:

$$Q\, dC/dm = ((K^{\circ}\alpha)/(\rho_p\rho_l))(C - C_e) \qquad [4\text{-}27]$$

Here, Q is the flow rate in gallons/hr., C and C_e are quantified in meq/L, K° is quantified in meq/hr-ft^2, α is the surface area of resin per unit volume (of bed) in ft^{-1}, ρ_p is the packed density of resin in the column in g/cm^3, ρ_l is density of the liquid phase in g/cm^3, and m — required resin mass — is expressed in g and can be obtained from graphical integration:

$$m = \frac{\rho_p \rho_l Q}{K^\circ \alpha} \int_{C_2}^{C_1} \frac{dC}{(C - C_e)} \qquad [4\text{-}28]$$

The m calculated is usually the steady-state quantity of resin needed to accomplish the designed removal of certain ions from a water solution in contact with resin in a column. Data available for graphical integration can be arranged by plotting $(C - C_e)^{-1}$ versus C. The $C - C_0$ values can be obtained easily when the curve of q_w versus C_0 (supplied by manufacturer) and a line of q_w versus C are plotted in the same graph. The latter is usually constructed from the influent (one point) and effluent conditions (another point) of the solution whose ion is ready to be exchanged for a resin. The use of this method is better illustrated by the following example:

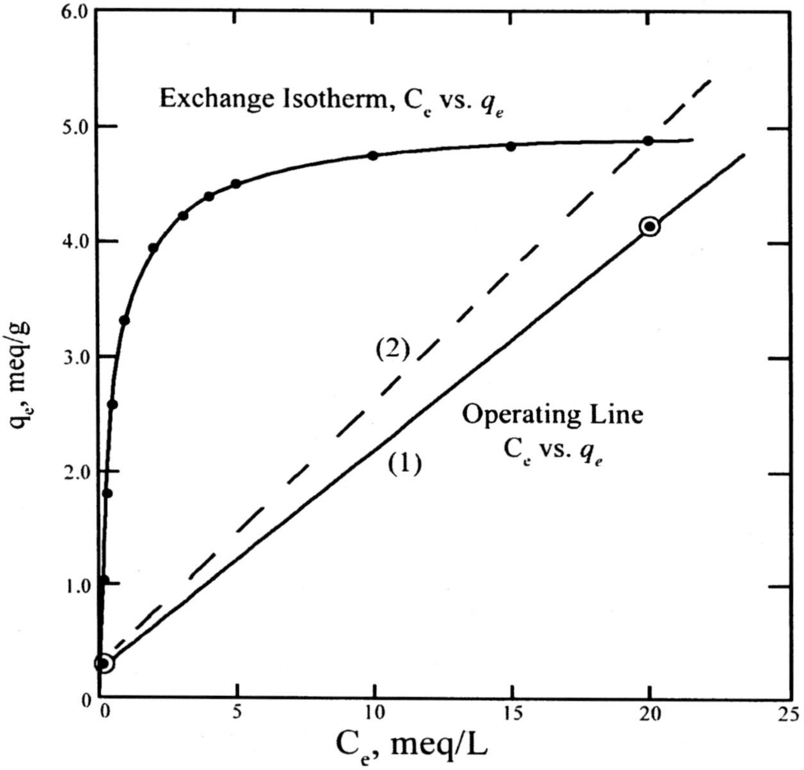

Figure 4-7. The operating line and the exchange isotherm of Cd^{2+}.

[Example 4-4] Industrial wastewater containing 20 meq of Cd^{2+}/L is allowed to be treated with an ion exchanger (for which the exchange capacity and corresponding equilibrium concentration are supplied by the manufacturer, as in the exchange isotherm data indicated in Fig. 4-7) in a countercurrent flow of a column where the resin is added continuously at the top to allow spent resin to withdraw at the bottom and where the input of the waste stream is from the bottom upward as shown in Figure 4-8. A removal of 99% of cadmium is required for a waste stream of 240,000 gpd. Given the overall mass transfer rate factor for a Cd^{2+} exchange reaction, $K^{\circ}a/\rho_s\rho_l = 2.0$ L/hr·g, the residual resin still contains 0.3 meq Cd^{2+}/g. Compute: (a) the minimum rate of addition of resin to the column; (b) same as (a) but providing a safety factor of design for 20% excess capacity; (c) the mass of required resin to maintain this removal. (After Weber)

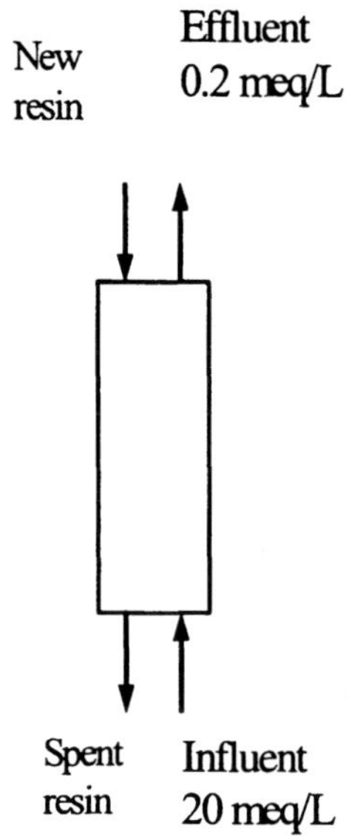

Figure 4-8. Countercurrent ion exchange column setup for Example 4-4.

(a) Influent concentration = 20 meq/L

Effluent concentration = 20 – 0.99(20) = 0.20 meq/L

Removal rate $= Q\Delta C$
= 24,000 gpd(20 – 0.20)meq/L
= 37850 L/hr (20 – 0.20)meq/L
= 750,000 meq/hr

Minimum resin utilization rate = removal rate/capacity
= [750,000 meq/hr]/(4.9 – 0.3)meq/g
= 163,000 g/hr

q_e at effluent concentration = 4.9 meq/g (from manufacture)
q_e at influent concentration = 0.3 meq/g (from regeneration data)

(b) Designing resin utilization rate (20% safety factor) = 1.2(163,000) g/hr = 196,000 g/hr

(c) First, construct an operating line (Fig. 4-7) for C versus q_e (dashed line). See operating line 1.

(output) Lower point $(C, q_e) = (0.2, 0.3)$

(input) Upper point $(C, q_e) = (20, 4.9^*)$

Since there is a 20% safety factor, a new q_e should be established for the upper point from mass balance:

$$750{,}000 \text{ meq/hr} = 196{,}000 \text{ g/hr } (q_e - 0.30)\text{meq/g}$$

Solving, $q_e = 4.1$ meq/g

Thus, an operating line is constructed with:

$$\text{Lower } (C, q_e) = (0.2, 0.3)$$
$$\text{Upper } (C, q_e) = (20, 4.1)$$

(See Fig. 4-7 for C versus q_e, solid line, operating line 2.)

With the two curves (exchange isotherm and operating line), construct a table of arbitrarily chosen C and read off the corresponding C_e. Next, plot $(C - C_e)^{-1}$ against C, and obtain the integrated area under the curve of 5.72 (see Table 4-3, Fig. 4-9).

Then, using Equation [4-28], the mass of resin required is:

$$m = \frac{\rho_p \rho_l Q}{K^\circ \alpha} \int_{C_2}^{C_1} \frac{dC}{(C - C_e)} = ((37850 \text{ L/hr})(5.72))/(2.0 \text{ L/hr}\cdot\text{g}) = 108{,}000 \text{ g} = 239 \text{ lb}$$

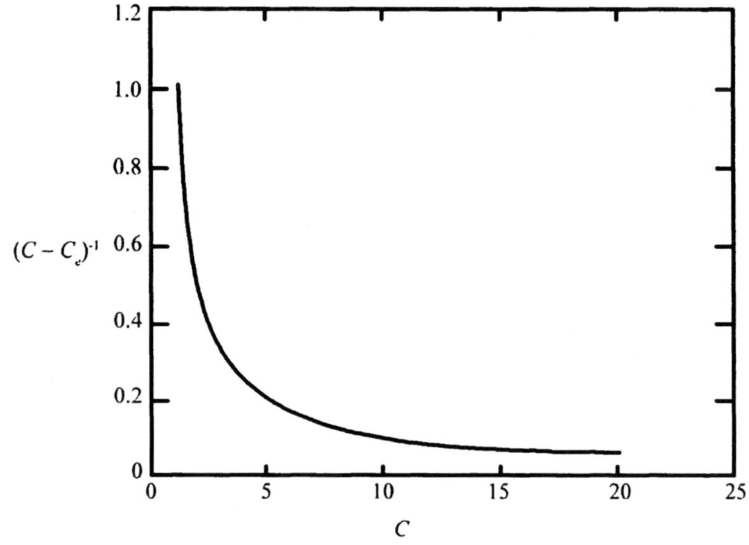

Figure 4-9. The quantity $(C - C_e)^{-1}$ as a function of C for Example 4-4.

Table 4-3. Ion Exchange Calculations

C	C_e	$(C - C_e)^{-1}$	C	C_e	$(C - C_e)^{-1}$
1	0.02	1.02	12	0.50	0.09
2	0.05	0.51	16	0.90	0.07
4	0.10	0.26	20	2.40	0.06
8	0.25	0.13			

4.3 Applications and Design of Exchangers

Either total or partial softening or total or partial demineralization can be achieved. For the tertiary treatment of water, zeolite (clinoptilotite) has been used to remove ammonia. Inorganic exchangers are also employed; for example, leucite (K_2O, Al_2O_3, $4SiO_2$) is used to remove the NaCl solution, which forms analcite (Na_2O, $Al_2O_3 \cdot 4SiO_2 \cdot 2H_2O$). Also, liquid exchangers exist. For example, heptadecyl phosphoric acid ($C_{17}H_{36}OPO(OH)_2$) is used to extract Th^{4+} from H_2SO_4, which is then to be stripped. Magnetic ion exchangers that are grafted polymers on γ Fe_2O_3 are used along with softening and classification in the process of water treatment. Ion exchange has been applied to nuclear waste treatment.

4.3.1 Adjustable Hardness

Since resin is virtually 100% complete in removing hardness, it is necessary to bypass a portion of water and then blend in order to obtain the desired final hardness. For the water-softening industry, the expression of hardness is in units of grains per gallon of water (gr/gal). It is useful to remember that 1 gr/gal = 17.1 mg/L. For fixed-bed exchangers, it is also important to determine the dimensions of the exchanger. Furthermore, it is essential to note that the exchange bed is usually separated into two regions: the exchange zone (50% utilized, between points of breakthrough and exhaustion, Figs. 4-10 and 4-11) and a remainder of the bed at the top (see Fig. 4-12). The total capacity must equal the two individual regions.

$$q = 0.5S\, zq_w + S(L - z)q_w$$

or

$$L = 0.5z + (q/Sq_w)$$

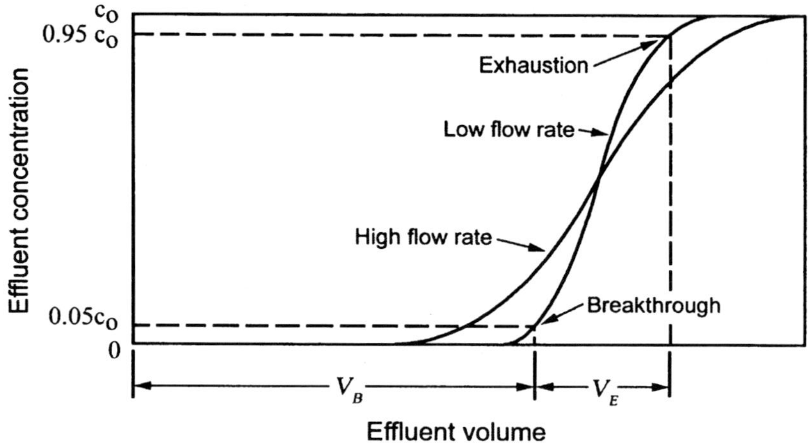

Figure 4-10. Typical breakthrough curves. (From Separation Processes in Practice, R. F. Chapman, ed., Reinhold, New York, 1961.)

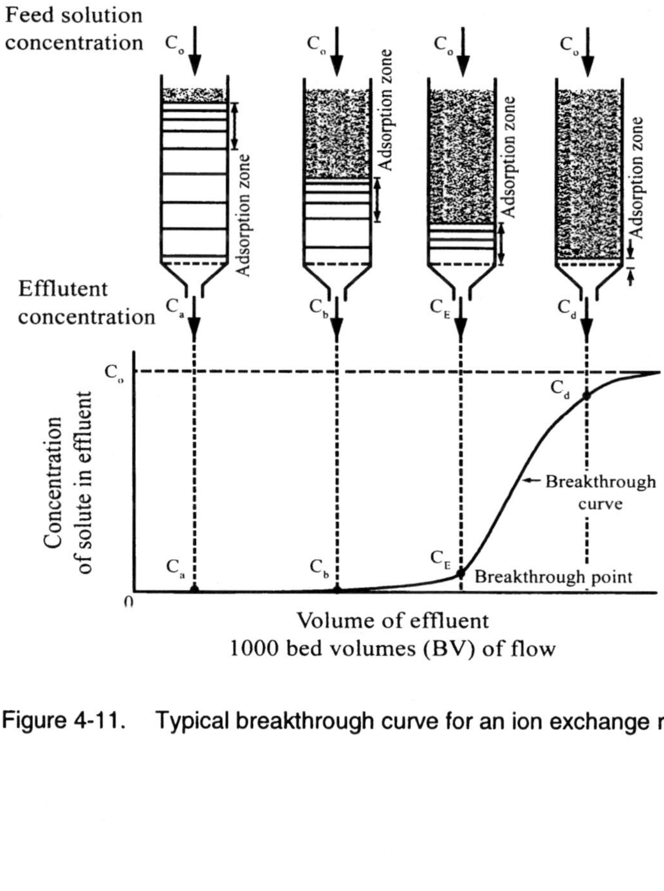

Figure 4-11. Typical breakthrough curve for an ion exchange resin.

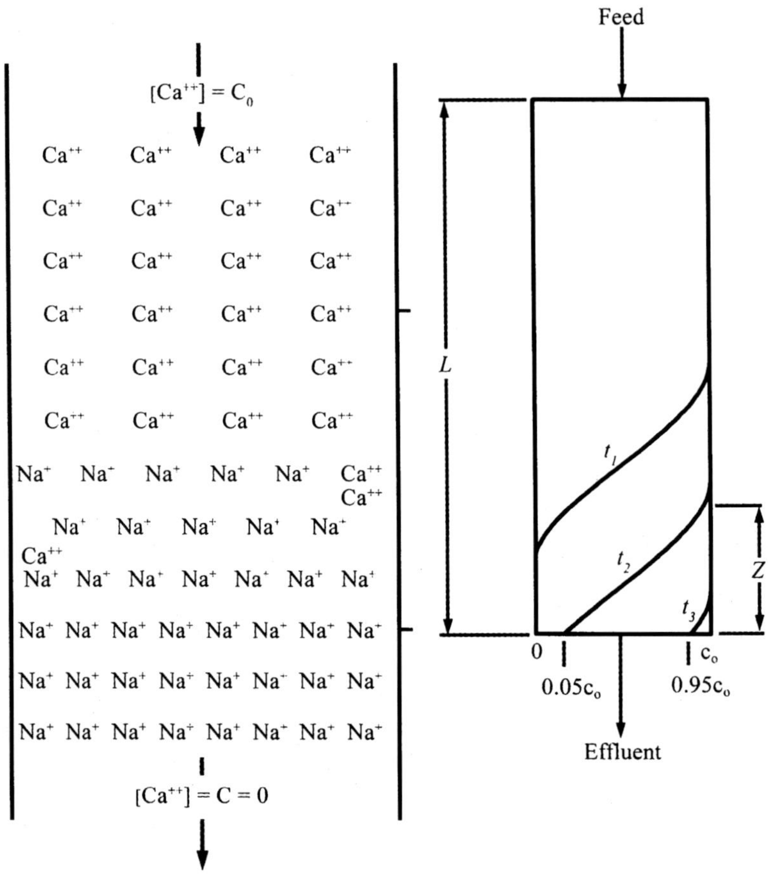

Figure 4-12. In breakthrough, the Ca^{2+} ion decreases, and the Na^{+} ion increases until $[Ca^{2+}] = C = 0$.

[Example 4-5] A home water softener has 0.1 m^3 of resin with an exchange capacity of 57 kg/m^3. The occupants of the home use 2000 L/day of water. If the water contains 280 mg/L of hardness (as $CaCO_3$) and it is desirable to soften it to 85 mg/L (as $CaCO_3$), find the bypassed amount. Also, determine the amount of time between regeneration cycles.

% Bypass = ((100)(85))/(280) = 30

Loading rate = (0.7)(280 mg/L)(2000 L/day) = 392,000 mg/day

Regeneration time = breakthrough time = (total capacity)/(loading rate)

Thus, $((57\ kg/m^3)(0.1\ m^3))/((392{,}000\ mg/d)(10^{-6}\ kg/mg)) = 14.5$ days

---

[Example 4-6] A fixed-bed exchanger is to be used to soften 20,000 gallons of water between regenerations. Hardness of the water averages 400 mg/L (as $CaCO_3$). If the capacity of the resin is 25 kgr (as $CaCO_3$)/ft^3 and the time between regeneration is 8 hours, calculate the dimensions of the exchanger.

Total capacity of the unit = $((400\ mg/L)(20{,}000\ gal))/((17.1\ mg\cdot gal/L\cdot gr)(1000\ gr/kgr))$

= 467 kgr (as $CaCO_3$)

A tentative volume of the exchanger required = (467 kgr)/(25 kgr/ft^3) = 18.7 ft^3.

Assuming a bed depth of 3 ft. (which is reasonable for most exchange units), the surface area is:

$$S = 18.7/3 = 6.23\ ft^2$$

Often, in figuring out the length of an exchange zone, a laboratory test is conducted for a given superficial velocity, U_s.

$$U_s = (20{,}000\ gal)/((6.23\ ft^2)(8\ hr)(7.48\ ft^3/gal)(3600\ sec/hr)) = 1.49 \times 10^{-2}\ ft/sec$$

This is within the range of $z = 1$ ft, and

$$L = 0.5\ (1\ ft) + (467\ kgr)/((25\ kgr/ft^3)(6.23\ ft^2)) = 3.5\ ft.$$

so the volume required for the exchange is:

$$V = LS = (3.5\ ft)(6.23\ ft^2) = 21.8\ ft^3$$

4.3.2 Desalination and Demineralization

The DESAL process, involving a three-step operation, is illustrated in Figure 4-13. In each step, different resins are used. A weak base anion exchanger, $R_3NH^+HCO_3^-$, is used in Column 1 to eliminate the Cl^- ion, and the resultant $R_3NH^+Cl^-$ is the spent resin. In Column 2, a weak acid cation exchanger, such as RCOOH, is used to eliminate the Na^+ ion. For the third column, a weak base anion exchanger, R_3N, is used to eliminate H_2CO_3 in saline water, and the resultant spent resin is $R_3NH^+HCO_3^-$, which is the resin to be used for Column 1. The equations involved are:

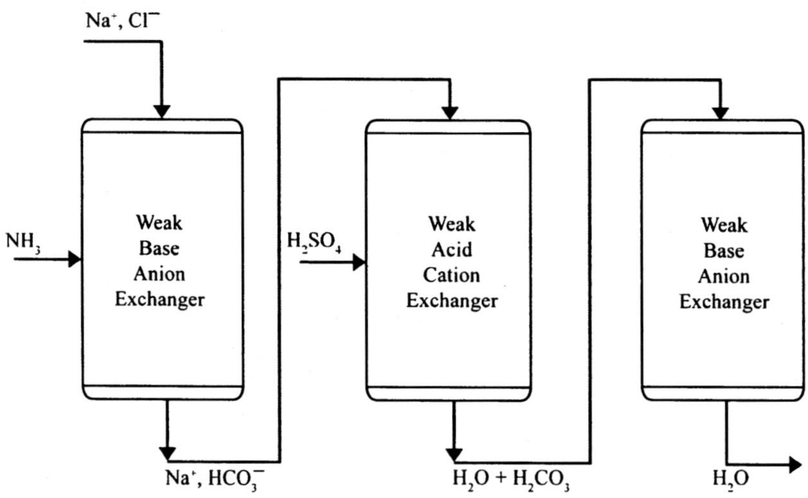

Figure 4-13. Schematic diagram of the DESAL ion exchange process.

Column 1:

Reaction	$R_3NH^+HCO_3^+ + Cl^- = R_3NH^+Cl^- + HCO_3^-$
Regeneration	$R_3NH^+Cl^- + NH_3 = R_3N + NH_4Cl$

Column 2

Reaction	$RCOOH + Na^+ = RCOONa + H^+$
Regeneration	$2RCOONa + H_2SO_4 = 2RCOOH + Na_2SO_4$

Column 3

Reaction	$R_3N + H_2CO_3 = R_3NH^+HCO_3^-$
Regeneration	to be used for Column 1.

Here, Columns 1 and 3 are in a cycle, since the regenerated resin from Column 1 can be used in Column 3, and the spent resin of Column 3 is used as the starting resin. The only chemicals spent in the DESAL process are NH_3 and H_2SO_4. Remember that in the regeneration of Column 1, we produce NH_4Cl. If we use lime, we can recover NH_3:

$$NH_4Cl + 1/2\ Ca(OH)_2 = NH_3 + 1/2\ CaCl_2 + H_2O$$

In this fashion, NH_3 can be produced from regeneration in Column 1. Therefore, with two heavy chemicals, lime and sulfuric acid, the remaining resins and chemicals are in a closed cycle.

For complete demineralization of water, a cation exchanger is generally used, followed by a weakly basic anion exchanger (see Fig. 4-14). If the cation resin is in H^+ form and the anion resin is in OH^- form, the unit will have water as its product and will remove all ionic species in water except carbon dioxide and silicate. If it is desirable to remove H_2CO_3, then a decarbonation unit that is a degasifier or vacuum deaerator can be installed at the end of the ion exchange cycle as in Figure 4-15. If both H_2CO_3 and silicate are objectionable, then an arrangement such as that shown in Figure 4-16 can be used. Notice that the degasifier in this case is in the middle. This is in an attempt to save the capacity for the strong base anion exchanger.

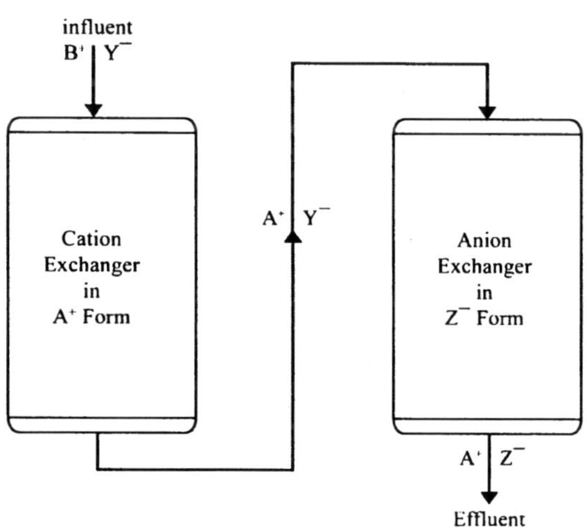

Figure 4-14. Two-stage ion exchange operation.

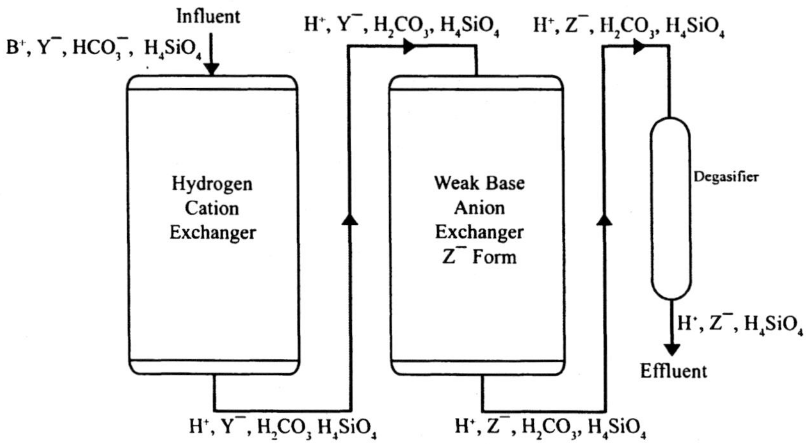

Figure 4-15. Two-stage operation with degasification for removal of carbon dioxide.

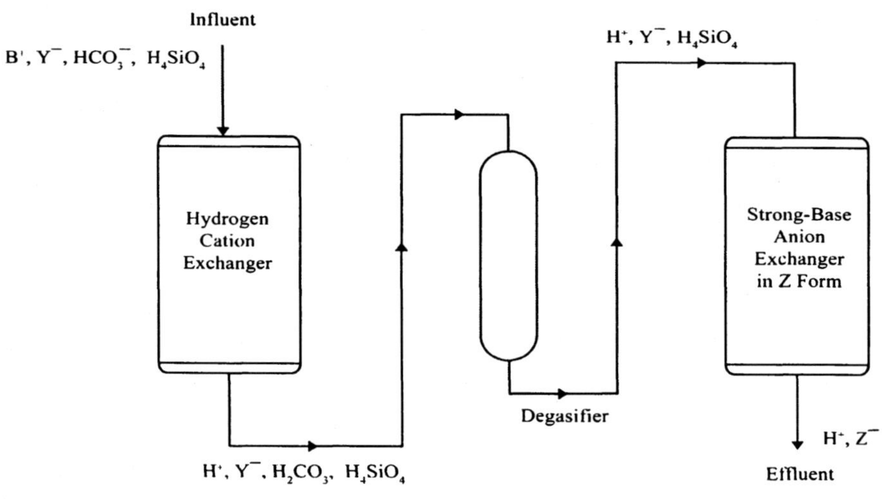

Figure 4-16. Two-stage operation for removal of both carbon dioxide and silica.

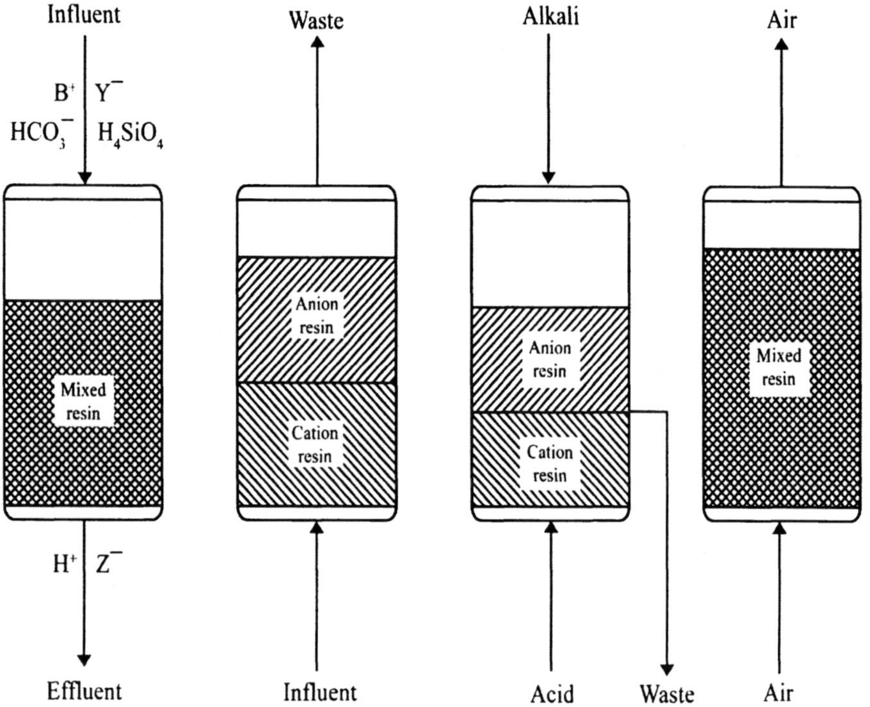

Figure 4-17. Mixed-resin exchange operation.

For the two-stage (two columns) operation, it is possible to use a mixed resin exchange operation as shown in Figure 4-17. In this unit, a cation resin is thoroughly mixed with an anion resin. For regeneration this mixed resin exchange operation must be separated by the density gradient (usually the cation and anion type chosen should have differences in density) for backwash. After regeneration, air is used for the mixing process. In order to have a continuously operating process, the columns can rotate in a merry-go-round fashion, as shown in Figure 4-18, with a liquid feed distributor above the columns divided into sections to accommodate all operations.

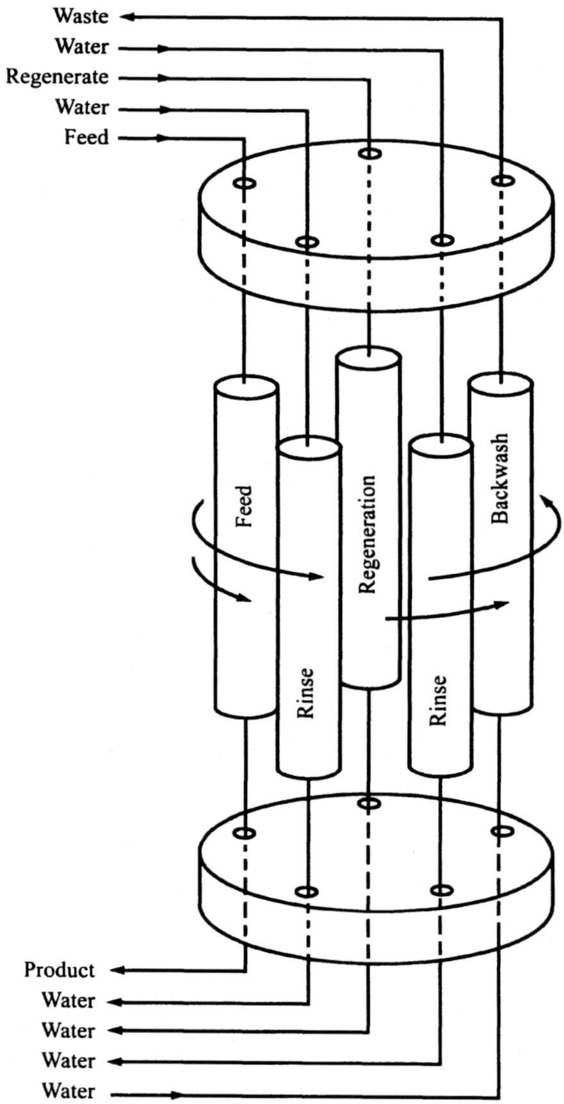

Figure 4-18. Aconex continuous ion exchange process. (From Ref. 4-19.)

4.3.3 Metal Recovery for Plating Wastes

Plating wastes that must be monitored contain Ni^{2+}, Zn^{2+}, Cu^{2+}, and Cd^{2+}, as well as CrO_4^{2-}. The divalent metals can be removed by a cation exchanger. However, the chromate CrO_4^{2-} has to be removed by an anion exchanger. A replacement for the Na_2CrO_4, typically another cation resin, is usually needed so that the final product is H_2CrO_4, (about 4–6%). This scheme is shown in Figure 4-19. The following example is a typical system for recovering plating waste from ion exchange.

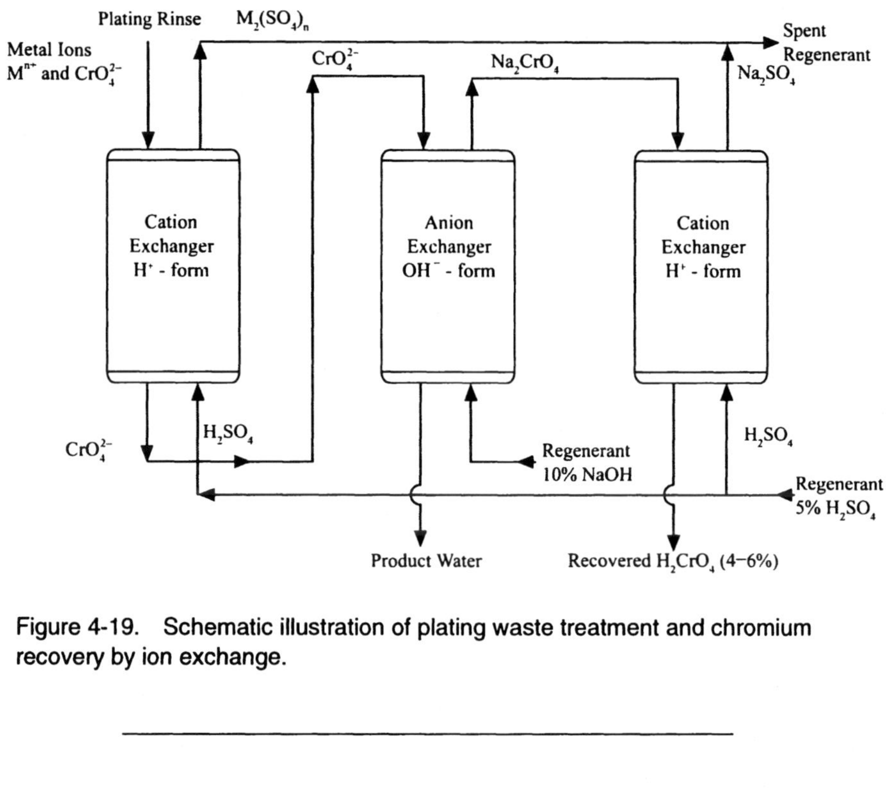

Figure 4-19. Schematic illustration of plating waste treatment and chromium recovery by ion exchange.

[Example 4-7] A metal plating plant generating 30,000 gpd (114 m^3 per day) of wastewater is intended for ion exchange. The composition of the analytes (both anions and cations) and the capacities of the resins and regeneration information are listed below. Assuming that resin is regenerated weekly, approximate the volume of resin in each exchanger and the amount that is regenerated.

Information of Analytes:

Cation	Concentration (mg/L)	Equivalent weight	meq/L
Ni^{2+}	25	29.4	0.85
Zn^{2+}	15	32.7	0.46
Cd^{2+}	35	56.2	0.62
Total cation			1.93
Anion CrO_4^{2-}	150	58	2.59
Total anion			2.59

Information of exchanger:

Resin	Capacity (eq/m^3)	Regenerant composition	Regenerant amount (kg/m^3 resin)
Cation	2650	5% H_2SO_4	190
Anion	1240	10% NaOH	80

For the cation exchanger,

$$RH_2 + M^{2+} = RM + 2H^+$$

$$RM + H_2SO_4 = RH_2 + MSO_4$$

Total equivalent removed per day = (1.93 meq/L)(1 eq/1000 meq)(114 m^3/day)(1000 L/m^3) = 220 eq/day, and the cation resin agent for a 7-day operation at 2650 eq/m^3 = (220 eq/day)(7 day/cycle)(1 m^3/2650 eq) = 0.58 m^3 resin/cycle. The H_2SO_4 needed for regeneration = (0.58 m^3)(190 kg H_2SO_4/m^3)(100 kg solution/5 kg H_2SO_4) = 2200 kg solution.

For the anion exchanger,

$$R(OH)_2 + CrO_4^{2-} = RCrO_4 + 2OH^-$$

$$RCrO_4 + 2NaOH = R(OH)_2 + Na_2CrO_4$$

Equivalents removed = (2.59)(1/1000)(114)(1000) = 295 eq/day. Anion resin for 7 days = (295)(7)(1/1240) = 1.66 m^3/cycle. NaOH needed = (1.66)(80)(100/10) = 1330 kg solution.

Finally, the chromate can be removed in another cation exchange:

$$RH_2 + Na_2CrO_4 = RNa_2 + H_2CrO_4$$

4.4 Ion Chromatography

Before we discuss ion chromatography, we should first understand the process of **ion exclusion**. Ion exclusion is a method for separating strong electrolytes from weak electrolytes and nonelectrolytes. No actual ion exchange occurs. The ion exchangers in this case merely work as adsorbents. A good example is shown in Figure 4-20. The electrolyte HCl can be separated from the non-electrolyte CH_3COOH. The Donnan membrane theory explains this behavior. The nonelectrolyte is strongly adsorbed and therefore strongly retained. A good electrolyte should readily pass through. Thus, if a solution containing ionic and nonionic components is placed on top of a resin bed and forced down through the column with water, the ionic material will be eluted first because it only displaces the solution in voids. In contrast, the nonionic solution must displace both the liquid in voids and the resin liquid, causing a delay in appearance.

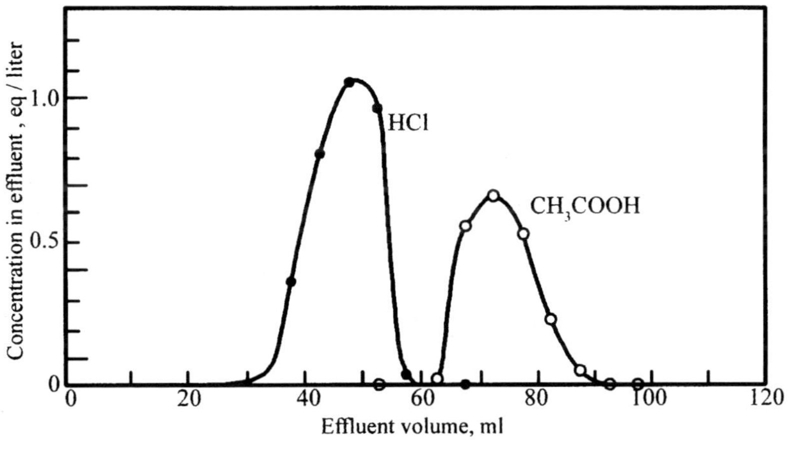

Figure 4-20. Experimental effluent history in separation of hydrochloric acid from acetic acid by ion exclusion. Feed: 15 ml 1.17 M HCl + 0.66 M CH_3COOH; resin: 100 ml Dowex 50-X8.

Another technique for separation is termed **ion retardation**, which involves the use of amphoteric ion exchange resins containing both anion and cation groups molecularly intertwined. Again the resin acts as a sorbent. Here the electrolyte sorption should be larger when compared with nonelectrolyte sorption due to the interaction of the functional groups with the electrolyte's charges. Hence, the opposite effect is observed, as shown in Figure 4-21.

Chromatography is a technique in which resolution of a mixture is achieved by virtue of the difference in migration rates of the components in a packed column. Some basic principles of chromatography are shown in Figure 4-22, indicating a sample of three components. Prior to injection of the sample, the packed column is equilibrated with an eluent (solvent), indicated by the dotted area in columns. When the sample arrives, it will begin to displace (ion exchange) the eluent from the column. Here, **retention**, **selectivity**, and **plates** are explained (Fig. 4-23).

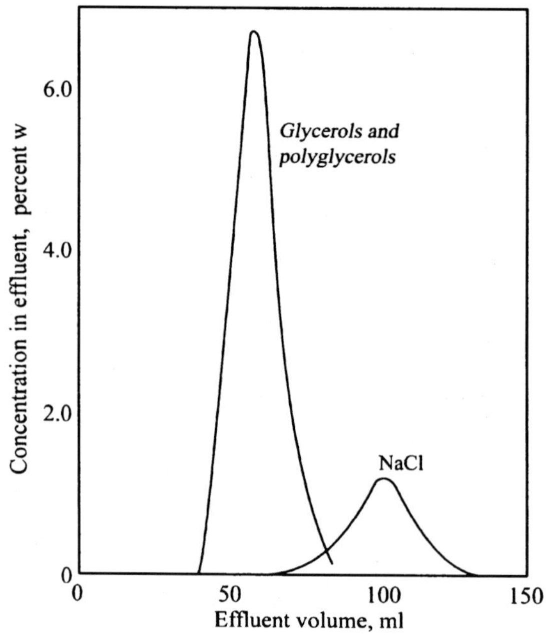

Figure 4-21. Experimental effluent concentration history in separation of NaCl from glycerol and polyglycerols by ion retardation. Feed: 12.5% wt glycerol, 6.2% wt polyglycerols, 6% wt NaCl; temperature 70°C; resin: Retardion 11 A 8.

The advantage of ion chromatography (IC) is that one can simultaneously quantify a number of ions. An example would be the use of an AS-1 (polystyrene–divinyl benzene–based resin) anion exchange column to simultaneously separate seven anions using a $NaHCO_3$–Na_2CO_3 solution as an eluent (see Fig. 4-24). Using the principle of ion exclusion in chromatography (or ion chromatography exclusion, ICE), separation of carbolytic acids and decarbolytic acids is achieved (see Fig. 4-25). Ion chromatography is comparable with other standard tools for analysis, such as atomic absorption (AA) and inductively coupled plasma (ICP) (see Table 4-4).

Table 4-4. Comparison Data for Ions

	Limit of detection		
Pollutant	IC	AA	ICP
Lead (Pb)	5	1	40
Copper (Cu)	0.3	4	5
Cadmium (Cd)	3	0.5	9
Zinc (Zn)	3	15 (flame)	4
Nickel (Ni)	3	75 (flame)	5

Values are in ppm; IC, ion chromatography; AA, atomic absorption; ICP, inductively coupled plasma.

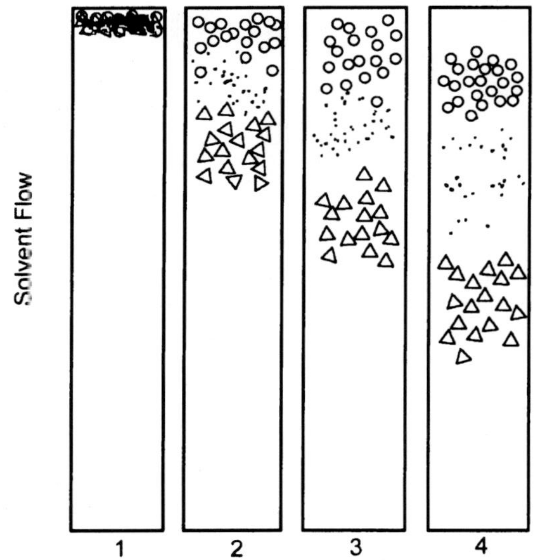

The dotted area represents the original solvent in the column which is being "displaced" during elution

Distribution coefficient: $K = \dfrac{[x]_s}{[x]_m}$

Retention: $k' = \dfrac{(V_i - V_o)}{V_o}$

(number of column volumes to elute a peak)

Selectivity: $\alpha = \dfrac{k_2'}{k_1'}$

(ratio of column volumes to elute peak 2 vs 1)

Plates: $N = 16\left(\dfrac{V_1}{W_1}\right)^2$

(column efficiency, i.e., narrowness of peaks)

Figure 4-22 Hypothetical separation of a three component mixture. Circles, squares and triangles represent three different components

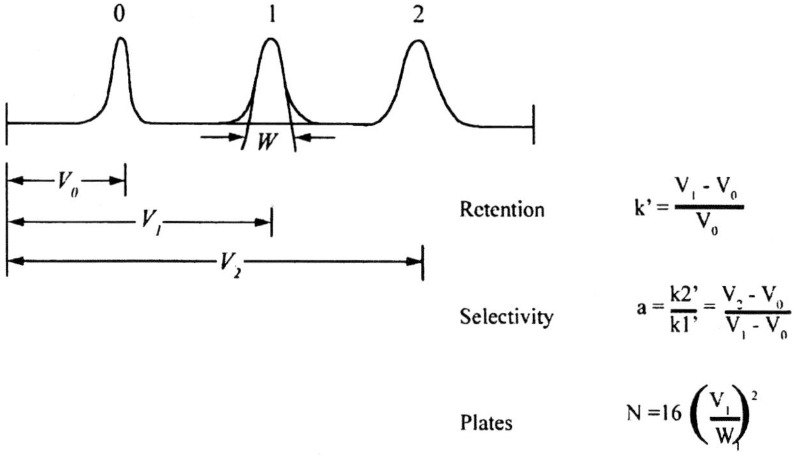

Figure 4-23. Equations for retention (also called capacity factor), selectivity, and plates (a measure of column efficiency).

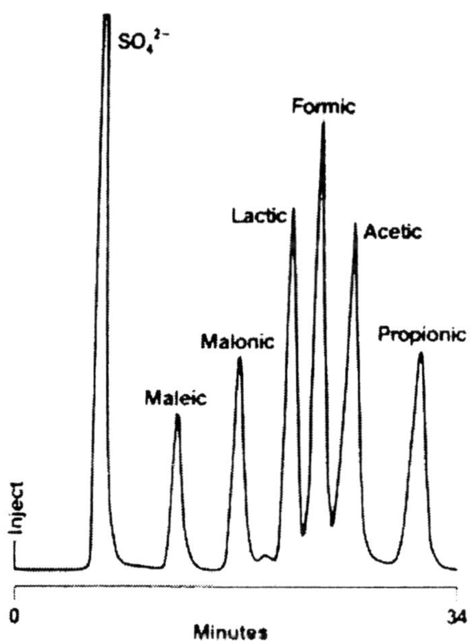

Figure 4-24. Standard ICE anion determination. Separation of organic acid anions on the first ICE column.

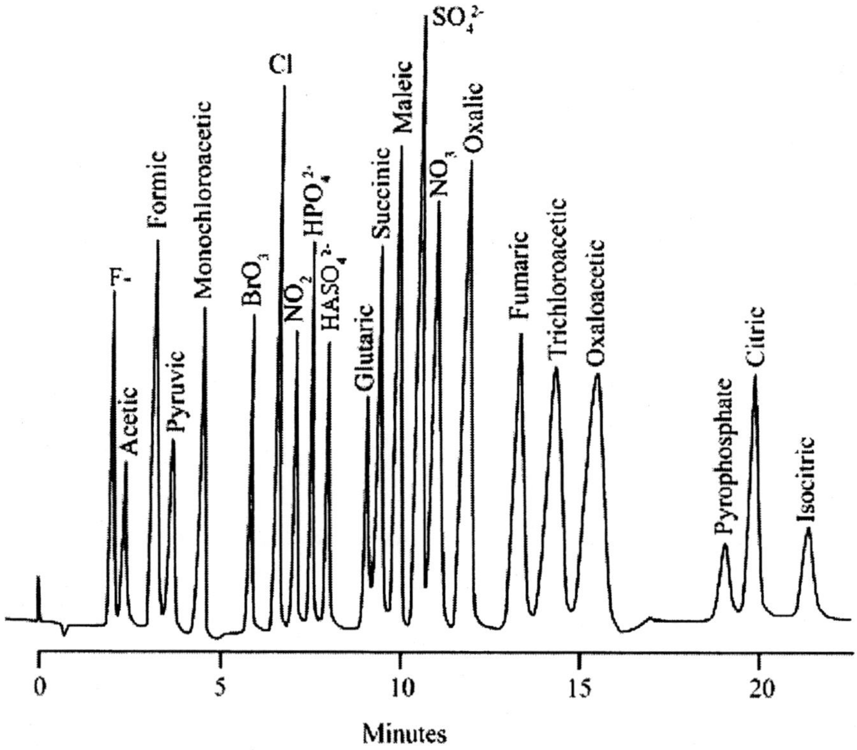

Figure 4-25. Gradient elution of anions.

REFERENCES

4-1. S. B. Applebaum, *Demineralization by Ion Exchange*, Academic Press, New York, 1968.

4-2. W. C. Bauman and Eichhorn, Fundamental properties of a synthetic cation exchange resin, *J. Amr. Chem. Soc.*, 69, 2830 (1947).

4-3. B.A. Bolto, Water treatment technology in Australia, in *Surface and Colloid Chemistry in Natural Waters and Water Treatment*, Plenum Press, 1990, pp. 87–101.

4-4. D.R. Dixon and L.O. Kolarik, The role of surface and colloid chemistry in the sirofloc process, in *Surface and Colloid Chemistry in Natural Waters and Water Treatment*, Plenum Press, 1990, pp. 103–118.

4-5. K. Dorfner, *Ion Exchangers—Properties and Applications*, Ann Arbor Science, Ann Arbor, Michigan, 1972.

4-6. D. J Downing, R. Kunin, and F. X. Pollio, DESAL process — economic ion exchange system for treating brackish and acid mine drainage effluents, *Water-1968*, Chemical Engineering Progress Symposium Series, 64, *Amr. Inst. Chem. Eng.*, 1968, p. 90.

4-7. T. J. Fadgen, A study of the possible use of ion exchange for recovery of metal from electroplating wastes, *Proc. 7th Ind. Waste Conf.*, Purdue University, 1952, p.6.

4-8. F. Helfferich, *Ion Exchange*, McGraw-Hill, New York, 1962.

4-9. R. J. Keating, R. Dvorin, and V. J. Calise, Applications of ion exchange to plating plant problems, *Proc. 9th Ind. Waste Conf.,* Purdue University, 1954, p.2.

4-10. R. Kunin, *Ion Exchange Resins*, 2nd ed., John Wiley and Sons, New York, 1958.

4-11. J. A. Marinsky, *Ion Exchange*, Vol. 1, Marcel Dekker, New York, 1966.

4-12. B. W. Mercer, L. L. Ames, C. J. Touhill, W. J. VanSlyke, and R. B. Dean, Ammonia removal from secondary effluents by selective ion exchange, *J. Water Pollut Contr. Fed.*, **42**, 2 (1970).

4-13. W. S. Midkiff and W. J. Weber Jr., Operating characteristics of strong-base anion exchange reactors, *Proc. 25th Ind. Waste Conf.*, Purdue University, 1970.

4-14. K. Odland and H. L. Pabich, Demineralizing brackish water, *Ind. Water Eng.*, 2, 12, (1965).

4-15. G. E. Osborn, *Synthetic Ion-Exchanger*, Chapman and Hall, London, 1961.

4-16. C. F. Paulson, Chromate recovery by ion exchange, *Proc. 7th Ind. Waste Conf.*, Purdue University, 36, 1952, p.6.

4-17. R. L. Sanks and W. J. Kaufman, Recycling of saline wastes, *Advan. Water Pollut. Res.*, Vol. 3, Proc. 3rd Internat. Conf. Water Pollut. Res., Water Pollut. Contr. Fed., Washington, DC, 1966.

4-18. V. L. Snoeynk and W. J. Weber Jr., Reaction of hydrated proton with active carbon, *Adsorption for Aqueous Solution*, American Chemical Society, Washington, DC, 1968, p.112.

4-19. R. M. Wheaton and A. H. Seamster, A basic reference on ion exchange, in *Encyclopedia of Chemical Technology*, 2nd ed., Vol. 11, 1966, pp.871–899.

4-20. B. A. Bolto and L. Pavlowski, *Wastewater Treatment by Ion-Exchange*, E & F. N. Spon, New York, 1987.

4-21. O. Samuelson, *Ion Exchange Separations in Analytical Chemistry*, Wiley, New York, 1963.

4-22. G. A. McKelvey, M. A. Thompson, and M. J. Parrotta, Ion exchange: A cost-effective alternative for reducing radium, *Am. Water Works Assoc. J.*, 61-66 (1993).

4-23. S. M. Robinson, J. M. Begovich, C. H. Brown, D. O. Campbell, and E. D. Collins, Treatment of radioactive wastewater by chemical precipitation and ion exchange, in *Recent Progress in Adsorption and Ion Exchange*, American Institute of Chemical Engineers, New York, 1978, pp. 55–65.

PROBLEM SET

1. In planning a softening plant for water containing 100 ppm calcium, 50 ppm magnesium, and 50 ppm bicarbonate alkalinity, all figured as $CaCO_3$, the following chemical cost data are available:

Lime (85% CaO)	\$32.00/ton
Soda ash (98% Na_2CO_3)	\$90.00/ton
Salt (100% NaCl)	\$44.00/ton

   Compute the chemical costs per million gallons for softening this water by (a) Lime-soda process, using 15% excess chemicals; (b) Zeolite softening — the zeolite to be used requires 0.35 lb salt per 1000 grains of hardness removed; (c) Lime-zeolite process, under the same conditions as in (a) and (b) (Note: 17.1 ppm = 1 grain per gallon)

2. Water containing 320 mg/L of hardness as $CaCO_3$ is to be softened by passage through a bed of ion-exchange resin at a rate of 3 gpm per square foot. The capacity of the resin is 2.0 meq/L bed volume, and 85% of the capacity is realized before breakthrough at the proposed rate of treatment. The bed is 30 inches in depth. How often will regeneration have to be carried out?

3. The exchange capacity of a strong cation exchange resin is reported as 37 kgr (as $CaCO_3$) per ft^3 of resin. The bulk density of the resin is 50 lb/ft^3 (0.8 g/ml). Determine the mass of resin needed to treat 10,000 gal (37.8 m^3) of water containing 30 mg/L of Cu^{2+}. (Note: 7000 grains = 1 lb)

4. A laboratory ion exchange column is treating a water stream with 37.8 mg/L Ca^{2+}, 5.2 mg/L Zn^{2+}, 3.6 mg/L Cr^{3+}, and 4.4 mg/L Fe^{3+}. The column contains 125 g of dry resin with an exchange capacity of 4.5 meq/g wet resin. Breakthrough data for the column are given in the following table, where C is the meq/L of exchangeable cations in the effluent and C_0 is the meq/L of total cations in the feed.

Concentration Ratio C/C_0	Effluent Volume (L)
~0	100
0.10	150
0.35	175
0.72	200
0.94	250
0.98	300
~1.0	400

Determine the actual quantity of cations exchanged using the breakthrough curve data, and compare it with the maximum exchange capacity of column.

5. Ammonia can be removed from wastewater by ion exchange on clinoptilotite, a natural zeolite. Consider an experiment where clinoptilotite is in contact with a solution containing 1.5×10^{-3} moles/L of Ca^{2+} and 1.2×10^{-3} moles/L of NH_4^+. The total capacity of the zeolite is 1.8 meq/g and the selectivity coefficient is 415. Determine the solid phase concentration $(Ca)_s$ and $(NH_4)_s$.

$$2NH_4^+ + Ca_s = 2(NH_4)_s + Ca^{2+}$$

$$K_{\frac{Ca}{NH4}} = \frac{\left(Ca^{2+}\right)\left(NH_4\right)_s^2}{\left(Ca\right)_s\left(NH_4^+\right)^2}$$

CHAPTER 5

AERATION AND STRIPPING

Gas transfer by water can be accomplished either by the addition of gas from outside the bulk phase to water (gas absorption) or by the removal of gas from water (gas desorption or stripping). The application of gas transfer in water treatment is essential to the extent that without it many of the treatment objectives cannot be met. As can be seen in Table 5-1, oxidation of reduced inorganic species and the destruction of harmful bacteria can be achieved by dispersing chlorine or ozone in water. Many oxidative reactions, especially biological oxidation by microorganisms, require an abundant supply of air or pure oxygen. Conversely, the removal of carbon dioxide or oxygen from groundwater can be helpful for the minimization of corrosion. The stripping of volatile organics from wastewater is also a common practice.

In this discussion, mass transfer is essential to an understanding of the underlying principle. We have spent some effort explaining the mass transfer coefficient in terms of its relationship with the dimensionless groups. From an analysis of similitude, we can comprehend a powerful tool such as dimensionless groups often used in environmental practice. We will start with Henry's law constants, which are used as the basis for the removal of undesirable contaminants and pollutants in water. Furthermore, we will discuss oxygen addition and oxygen removal in riverine waters based on the **Streeter-Phelps oxygen sag curve**. From mass transfer considerations; the technology of estuary and ocean disposal will be examined.

Table 5-1. Application of Gas Transfer in Water

Treatment	Water Treatment Objectives
Gas absorption	
O_2	Oxidation of Fe^{2+}, Mn^{2+}, S^{2-}; lake destratification
O_3	Disinfection, color removal, oxidation of selected organic compounds
Cl_2	Disinfection; oxidation of Fe^{2+}, Mn^{2+}, H_2S
ClO_2	Disinfection
CO_2	pH control
SO_2	Dechlorination
NH_3	Chloramine formation for disinfection
Stripping	
CO_2	Corrosion control
O_2	Corrosion control
H_2S	Odor control
NH_3	Nutrient removal
Volatile organics	Taste and odor control, removal of potential carcinogens (e.g., CH_4, $CHCl_3$, C_2HCl_3)

5.1 HENRY'S LAW

If H is expressed as a dimensionless distribution coefficient for the aqueous-gas equilibria,

$$A_{(aq)} \leftrightarrow A_{(g)}$$

then

$$H = \frac{C_{A(g)}}{C_{A(aq)}} \quad [5\text{-}1a]$$

There are two ways to express the Henry's law constant:

1. $$K = HRT\mathrm{M_W} \quad \text{(in atm.)} \quad [5\text{-}1b]$$

where M_w = the molecular weight of water and for different compound *I,* the K_i and H_i are used.

2. $$K_H = (HRT)^{-1} \quad (\text{in mole atm}^{-1}\ \text{L}^{-1}) \quad [5\text{-}1c]$$

Henry's constant is also discussed in Section 2.1.

Here we will use the first method. For a gas-liquid system at equilibrium, assuming ideal fluid behavior in each phase, the partial pressure of a substance in gas phase p_i is proportional to its concentration in liquid phase x_i (Henry's law).

$$p_i = K_i x_i \quad [5\text{-}1d]$$

where K_i is Henry's constant. When the gas phase is ideal, the total pressure of a gaseous mixture is the sum of the partial pressures of individual components (Dalton's law). Combining the two laws, we have:

$$y_i = \frac{K_i x_i}{P_T} \quad [5\text{-}2]$$

where y_i is the mole fraction of *i* in the gas phase and P_T is total pressure. The mole fraction x_i in the liquid phase can be expressed as:

$$x_i = \frac{n_i}{(n_i + n_j)} \quad [5\text{-}3]$$

For *i* species in an aqueous solution, the concentration in 1 L of water is 1000/18 = 55.6 moles of water. Some Henry's constants of commonly used gases and volatile organic compounds are listed in Tables 5-2 and 5-3. One of the properties of Henry's constants is their sensitivity to temperature changes, which follow the van't Hoff–type relation:

$$\log K_i = \frac{-\Delta H}{RT} + k \qquad [5\text{-}4]$$

Table 5-2. Henry's Constants and Temperature Dependence for Selected Volatile Organics in Water

		Henry's Constant	Temperature Dependence	
Gas	Formula	at 20 °C (atm)[a]	ΔH (kcal/kmole × 10^{-3}	k
Vinyl chloride	CH_2CHCl	3.55×10^5	—	—
Carbon tetrachloride	CCl_4	1.29×10^3	4.05	10.06
Tetrachloroethylene	C_2Cl_4	1.10×10^3	4.29	10.38
Trichloroethylene	$CCHCl_3$	550	3.41	8.59
Benzene	C_6H_6	240	3.68	8.68
Chloroform	$CHCl_3$	170	4.00	9.10
Bromoform	$CHBr_3$	35	—	—
Pentachlorophenol	C_6OHCl_5	0.12	—	—

[a] Computed from water solubility data and partial pressure of pure liquid at specified temperature.

Source: Landolt-Börnstein (1976).

Table 5-3. Henry's Constants and Temperature Dependence for Gases in Water

		Henry's Constant	Temperature Dependence	
Gas	Formula	at 20 °C (atm)[a]	ΔH (kcal/kmole × 10^{-3}	k
Oxygen	O_2	4.3×10^4	1.45	7.11
Nitrogen	N_2	8.6×10^4	1.12	6.85
Methane	CH_4	3.8×10^4	1.54	7.22
Ozone	O_3	5.0×10^3	2.52	8.05
Carbon dioxide	CO_2	1.51×10^2	2.07	6.73
Hydrogen sulfide	H_2S	5.15×10^2	1.85	5.88
Chlorine	Cl_2	5.85×10^2	1.74	5.75
Chlorine dioxide	ClO_2	54	2.93	6.76
Sulfur dioxide	SO_2	38	2.40	5.68
Ammonia	NH_3	0.76	3.75	6.31

[a] 1 atm = 101.3 kPa.

Source: Landolt-Börnstein (1976) and International Critical Tables.

5.2 GAS STRIPPING

Usually waste gases are removed from wastewater countercurrently in a packed tower. Figure 5-1 presents a gas-liquid contacting system where wastewater at a desirable pH is fed at the top and the stripping gas is introduced at the bottom. The overall material balance may be expressed as:

$$G(y_2 - y_1) = L(x_1 - x_2) \qquad [5\text{-}5]$$

where G and L are the loading rates of gas and liquid, respectively (moles area^{-1}time^{-1}), y and x are the mole fractions of solute in gas and liquid, respectively, and subscripts 1 and 2 are inlet and outlet conditions, respectively. Thus,

$$\frac{G}{L} = \frac{x_1 - x_2}{y_2 - y_1} \cong \frac{x_1 - x_2}{y_2} \qquad [5\text{-}6]$$

assuming that the desorbing or stripping gas contains no undesirable gas and $y_1 = 0$.

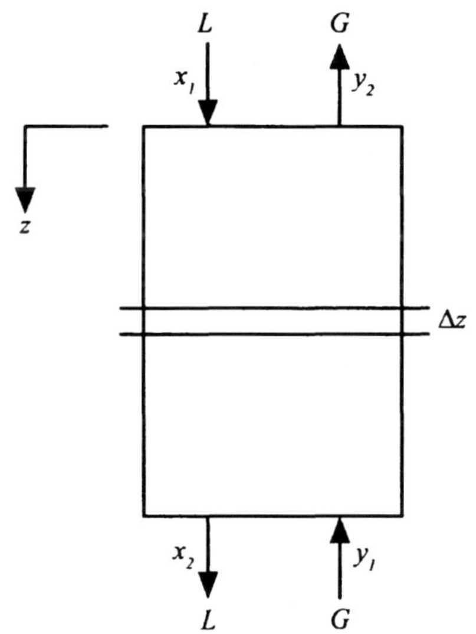

Figure 5-1. Notation for design of packed-column, gas-liquid transfer.

If Henry's law is applicable,

$$\frac{G}{L}=\frac{(x_1-x_2)P}{x_1 K} \qquad [5\text{-}7]$$

since y_2, the effluent gas concentration, is at equilibrium with the influent gas concentration, x_1 for the Henry's law. The **stripping ratio**, R, is usually defined as:

$$R = KG/L \qquad [5\text{-}8]$$

Some of the impurities commonly removed by air stripping are ammonia, hydrogen sulfide, and phenol. Common practice is to adjust the pH of the wastewater between 10.5 and 11.5 with lime for ammonia stripping. Conversely, for stripping of phenol or hydrogen sulfide, acid adjustment is used.

$$NH_3 + H_2O = NH_4^+ + OH^-, \quad pK' = 4.75$$

$$H_2S = HS^- + H^+, \quad pK' = 7.20$$

$$\phi - OH = \phi - O^- + H^+, \quad pK' = 9.89$$

and

$$(NH_3) = \left[1 + \frac{K(H^+)}{K_w}\right]^{-1}$$

Typical operating values for G/L values are 1000-6000 m^3 air/m^3 for wastewater that contains ammonia and 180–380 kg steam/m^3 for wastewater that contains H_2S. The depth of the tower for the stripping is usually 7–8 m.

[Example 5-1] If treated water in an enclosed reservoir is contaminated with chloroform at 500 μg/L, find the vapor pressure of chloroform in the headspace at 20°C and 10°C. What are their concentrations when compared to the NIOSH Occupational Exposure Standard of 10 ppm? How much effort is required to apply air stripping in order to meet this standard?

The mole fraction is:

$$x_i = \frac{c_i}{c} = \frac{(500\ \mu g/L)/(119\ g/mole)}{(55.6\ kmole/m^3)} \times 10^{-6}\ g/\mu g = 7.56 \times 10^{-8}$$

From Henry's law at 20°C,

$$y_i = \frac{Kx_i}{P_T} = \frac{170\ atm}{1 atm}\left(7.56 \times 10^{-8}\right) = 1.28 \times 10^{-5}$$

The partial pressure at 20°C is:

$$p_i = y_i\, p_T = 1.28 \times 10^{-5}\ atm = 1.28 \times 760 \times 10^{-5} = 9.73 \times 10^{-3}\ mmHg$$

The concentration at headspace

$$\frac{n_i}{V} = \frac{p_i}{RT} = \frac{(1.28 \times 10^{-5}\ atm)}{(0.08285 atm\text{-}L/mole\text{-}K)(293\ K)} = 5.27 \times 10^{-7}\ mole/L = 6.27 \times 10^{-5}\ g/L$$

Thus,

$$C_i = 62.7 \times 10^{-6}\ g/L = 62.7\ mg/m^3$$

Since density of air is 1.2 kg/m^3,

$$C_i = 52.3 \times 10^{-6}\ kg/kg\ dry\ air = 52.3\ ppm$$

For air stripping,

$$\frac{G}{L} = \frac{x_1 - x_2}{Kx_1} P_T = \frac{52.3 - 10}{170 \times 52.3} = 4.76 \times 10^{-3} = 5.71 \times 10^{-3}\ m^3, air(STP)/kg\ H_2O$$

This is equivalent to 5.71 L air/L H_2O or 9.15×10^{-2} ft^3/lb of water, which is a minor effort compared to a regular stripping unit.

At 10°C,

$$\log K = \frac{-(4 \times 10^3 \text{ kcal/kmole})}{(1.987 \text{ kcal/kmole})(283 \text{ K})} + 9.10$$

or

$$K = 97 \text{ atm}$$

Thus, $y_i = (97)(7.56 \times 10^{-8}) = 7.33 \times 10^{-6}$ and $p_i = 5.57 \times 10^{-3}$ mmHg at 10°C.

Similarly, at 10°C, where T=283°K,

$$n_i/V = 3.12 \times 10^{-7} \text{ mole/L}$$

$$C_i = 37.2 \times 10^{-6} \text{ g/L} = 37.2 \text{ mg/m}^3 = 31 \times 10^{-6} \text{ kg/kg dry air} = 31 \text{ ppm}$$

Similarly, at 10°C,

$$G/L = (31–10)/[(97)(31)] = 6.98 \times 10^{-3} = 8.38 \times 10^{-3} \text{ m}^3 \text{ air/kg H}_2\text{O}$$

$$= 8.38 \text{ L air/L H}_2\text{O} = 0.134 \text{ ft}^3 \text{ air/lb water}$$

Again, this is only a very minor effort when compared to a regular stripping unit.

5.3 Mass Transfer

Mass transfer often occurs at the interface between two phases. For aeration in biological treatment, the bulk phase of gaseous oxygen must first transfer into the liquid phase, which will proceed to the solid phase of biomass. A number of theories have been advanced, since even with the two-phase system of gas to liquid an interface is essentially involved. The simplest situation is illustrated in Figure 5-2 where Henry's law constant is used. The rate of mass transfer of species A per unit interfacial area, N_A, is dependant on various factors such as temperature, molecular diffusivity of A, and solubility of A in liquid. From Fick's law of diffusion, the turbulent rate of transfer from gas to liquid is:

$$N_A = k_G\left(P_A - P_{A_i}\right)$$

and

$$N_A = k_L\left(C_{A_i} - C_A\right) \quad [5\text{-}9]$$

where P is pressure, C is concentration, the subscript A is the A species in bulk, and A_i is the A species at the interface region. Lowercase k represents local rate transfer constants (mole area^{-1} time^{-1} pressure^{-1}), and K is the overall rate transfer constant.

$$N_A = K_G\left(P_A - P^*\right)$$

and

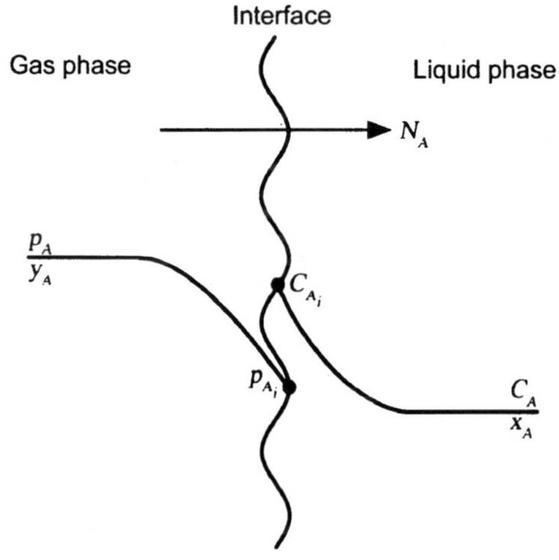

Figure 5-2. Mass transfer from gas to liquid phase.

$$N_A = K_L\left(C^* - C_A\right) \quad [5\text{-}10]$$

where the superscript asterisk depicts the hypothetical equilibrium values.

Therefore,

$$K_G\left(P_A - P^*\right) = k_G\left(P_A - P_{A_i}\right) \quad [5\text{-}11]$$

Assuming Henry's law holds, $P_A = HC^*$, $P_i = HC_i$, and $C_A = P^*/H$. After computation,

$$K_G^{-1} = k_G^{-1} + Hk_L^{-1}$$

$$K_L^{-1} = k_L^{-1} + Hk_G^{-1} \quad [5\text{-}12]$$

If the mass transfer is expressed by rate per unit volume (mole time^{-1} length^{-3}) and a is related to interfacial area per unit volume, then

$$M_A = K_G a\left(P_A - P^*\right) = K_L a\left(C^* - C_A\right) \quad [5\text{-}13]$$

Since it is difficult to measure interfacial areas in bubble aerations, the product $K_L a$ is most often used. It should be pointed out that K_L and k_L are interchangeable, although the former can be measured.

For vessels dealing with mass transfers, there are numerous variables, such as gas bubble diameter d_b, the bubble rise velocity v_b, power input to volume P/V, chemical reactions enhancement factor E, etc. Since the correlation among the variables is complex, some dimensionless groups are used (see following section for details). Notice that the symbols for units are different in different books (e.g., velocity can be v or u, or even V, which is not volume) as shown in the following section. We will discuss this more after we become acquainted with some of the dimensionless groups.

5.3.1 Similitude and Dimensional Analysis

The simplest case is as follows: If two systems are dynamically similar, the corresponding forces must occur in the same ratio. Forces for a fluid system such as rates are presented here (V represents velocity):

$$\begin{aligned}
&\text{Gravity}: \ F_G = mg = \rho L^3 g \\
&\text{Pressure}: \ F_p = \Delta pA = \Delta pL^2 \\
&\text{Viscosity}: \ F_v = \mu\left(\frac{du}{dy}\right)A = \mu\left(\frac{V}{L}\right)L^2 = \mu VL \\
&\text{Elasticity}: \ F_E = E_A = E_v L^2 \\
&\text{Surface tension}: \ F_T = \sigma L \\
&\text{Inertia}: \ F_I = ma = \rho L^3\left(\frac{L}{T^2}\right) = \rho L^4 T^{-2} = \rho V^2 L^2
\end{aligned} \tag{5-14}$$

Therefore,

$$F_G + F_P + F_V + F_E + F_V = \text{Resultant}$$

$$F_I = -\text{Resultant}$$

Thus,

$$\sum F = F_G + F_P + F_V + F_E + F_T + F_I = 0 \tag{5-15}$$

or for two systems, *j* and *k*,

$$\frac{F_{1j}}{F_{1k}} = \frac{F_{2j}}{F_{2k}} = \cdots \frac{F_{nj}}{F_{nk}} \tag{5-16}$$

The ratio of initial forces to viscous forces is called **Reynolds number**, Re:

$$\text{Re} = \frac{F_I}{F_V} = \frac{L^2 V^2 \rho}{LV\mu} = \frac{LV\rho}{\mu} = \frac{LV}{\upsilon} = \frac{dV\rho}{\mu} = \frac{dV}{\upsilon} \tag{5-17}$$

(e.g., d is the diameter of a pipe, V is kinetic viscosity, and $\rho V = \mu$).

Similarly, the ratio of inertia forces to gravity forces in square root form is termed the **Froude number**, Fr:

$$\mathrm{Fr}^2 = \frac{\rho L^2 V^2}{\rho g L^3} = \frac{V^2}{gL}$$
$$\mathrm{Fr} = \frac{V}{(gL)^{\frac{1}{2}}} \qquad [5\text{-}18]$$

Where compressibility is important, it is necessary to supply the ratio of the inertia forces to the elastic forces. The **Mach number**, Ma, is defined as:

$$\mathrm{Ma} = \left[\frac{\rho V^2 L^2}{E_V L^2}\right]^{\frac{1}{2}} = \frac{V}{\left[\frac{E_V}{\rho}\right]^{\frac{1}{2}}} = \frac{V}{c} \qquad [5\text{-}19]$$

where c is the sonic velocity and E_V is the volume modulus of elasticity in psi or N/m^2. (When Ma = 1, sonic; Ma < 1, subsonic; Ma > 1, supersonic; Ma >> 1, hypersonic.) For other types of flow, when surface tension becomes important, then the square root of the ratio of inertia forces to surface tension is known as the **Weber number**, We:

$$\mathrm{We} = \frac{V}{\left(\frac{\sigma}{\rho L}\right)^{\frac{1}{2}}} \qquad [5\text{-}20]$$

Sometimes a dimensionless quantity related to the ratio of the inertia forces to the pressure forces is known as the **Euler number**, Eu:

$$\mathrm{Eu} = \frac{V}{\left[\frac{\Delta p}{\rho}\right]^{\frac{1}{2}}} \qquad [5\text{-}21]$$

Dimensional analysis is a technique used to group the functionally dependent parameters in dimensionless combinations. For example, the drag force F_D exerted on a sphere may depend on the size of the sphere, the velocity of the sphere, and the medium properties that the sphere travels, e.g., the density and viscosity, or

$$F_D = f(d, V, \rho, \mu) \qquad [5\text{-}22]$$

in order to have the equivalence of dimension on both sides of the equation (C is a constant).

$$F_D = C d^a V^b \rho^c \mu^d \qquad [5\text{-}23]$$

using dimensions

$$\frac{ML}{T^2} = L^a \left(\frac{L}{T}\right)^b \left(\frac{M}{L^3}\right)^c \left(\frac{M}{LT}\right)^d \qquad [5\text{-}24]$$

To satisfy homogeneity, the exponents of each dimension must be identical. For M,

$$1 = c + d$$

For L,

$$1 = a + b - 3c - d$$

For T,

$$-2 = -b - d \qquad [5\text{-}25]$$

This is the case for equations with four unknowns. One must express a, b, c, in terms of d. From the above,

$$c = 1 - d$$

and

$$b = 2 - d$$

Thus,

$$a = 2 - d \qquad [5\text{-}26]$$

After substitution,

$$F_D = Cd^{2-d}V^{2-d}\rho^{1-d}\mu^{d} \qquad [5\text{-}27]$$

and grouping variables according to their exponents,

$$F_D = C_\rho (dV)^2 \left(\frac{Vd\rho}{\mu}\right)^{-d} \qquad [5\text{-}28]$$

which can be expressed as:

$$F_D = f'(\text{Re})\rho d^2V^2 \qquad [5\text{-}29]$$

A more generalized approach is the use of **Backnigham's π theorem,** which states that the number of dimensionless groups is equal to the number of variables minus the number of fundamental dimensions. If there are n variables of $Q_1, Q_2, \ldots Q_n,$

$$f(Q_1, Q_2, \ldots Q_n) = 0 \qquad [5\text{-}30]$$

Using m fundamental dimensions, there will be $(n - m)$ dimensionless groups:

$$f'(\pi_1, \pi_2, \cdots \pi_{n-m}) = 0 \qquad [5\text{-}31]$$

For the above illustration, there are five variables. Using three fundamental dimensions, M, L, and T, the result is $5 - 3 = 2$ dimensionless groups:

$$\Pi_1 = R, \Pi_2 = \frac{F_D}{\rho d^2V^2} \qquad [5\text{-}32]$$

[Example 5-2] It is found that the heat transfer rate per unit area q is dependent on the following:

viscosity, μ, which affects flow

density, ρ, which affects flow

specific heat capacity, C_P, of the fluid

thermal conductivity k of the fluid

linear dimension l of the surface

velocity u of the fluid over the surface

the temperature difference, ΔT

product of coefficient of cubical expansion and acceleration due to gravity, β_g, which can be a natural circulation effect caused by thermal expansion of the fluid on heating.

What are the dimensionless groups of this function?

$$f = \phi(u, l, \rho, \mu, C_p, \Delta T, \beta_g, k)$$

Express the variables in dimensions, L, M, T, θ (temperature), and H (heat). Now the dimension, D, of all the variables are:

$$DQ = HL^{-2}T^{-1}$$
$$Du = LT^{-1}$$
$$Dl = L$$
$$D\mu = ML^{-1}T^{-1}$$
$$D\rho = ML^{-3}$$
$$Dk = HT^{-1}L^{-1}\theta^{-1}$$
$$DC_P = HM^{-1}\theta^{-1}$$
$$D\Delta T = \theta$$
$$D\beta_g = LT^{-2}\theta^{-1}$$

Using the π theorem, the recurring set is l, ρ, μ, ΔT, k. Then the non-recurring variables are q, u, β_g, C_P.

Now for the recurring set, the major dimensions are:

$$L = l$$
$$M = \rho L^3 = \rho l^3$$
$$T = ML^{-1}\mu^{-1} = \rho l^3 l^{-1}\mu^{-1}$$
$$\theta = \Delta T$$
$$H = kLT\theta = kl\rho l^2\mu^{-1}\Delta T$$

Then the nonrecurring variables in the π groups are:

$$\pi_1 = q\,H^{-1}L^2T = q\,\kappa^{-1}\,l^{-3}\,\rho^{-1}\,\mu\,\Delta T^{-1}\,l\,\rho\,l^2\,\mu^{-1} = q\kappa^{-1}l\Delta T^{-1}$$

$$\pi_2 = u\,L^{-1}T = n\,l^{-1}\,\rho\,l^2\,\mu^{-1} = u\rho l\,\mu^{-1}$$

$$\pi_3 = C_P H^{-1}M\theta = C_P\kappa^{-1}l^{-3}\rho^{-1}\mu\Delta T^{-1}\rho l^3\Delta T = C_P\kappa^{-1}\mu$$

$$\pi_4 = \beta g\,L^{-1}T^2\theta = \beta g\,l^{-1}\,\rho^2\,l^4\,\mu^{-2}\,\Delta T = \beta g\Delta T\,\rho^2\,\mu^{-2}\,l^3$$

Therefore, the function can be expressed as:

$$\frac{ql}{k\Delta T} = \frac{hl}{k} = \phi\left(\frac{lu\rho}{m}\right)\left(\frac{C_P\mu}{\kappa}\right)\left(\frac{\beta g\Delta T l^3\rho^2}{m^2}\right)$$

which is $N = \phi\,(\text{Re})(\text{Pr})(\text{Gr})$

From the example above, one learns that there are three new dimensionless groups.

The **Nusselt group** Nu:

$$\text{Nu} = hl/\kappa \qquad [5\text{-}33]$$

where h is equivalent to $q/\Delta T$, which is the film coefficient for heat transfer.

The **Prandtl group**, Pr:

$$\text{Pr} = \frac{C_P\mu}{\kappa} = \frac{\nu}{\alpha} \qquad [5\text{-}34]$$

where α is thermal diffusivity, or

$$\alpha = \frac{\kappa}{\rho C_P}$$

The **Grashof group**, Gr:

$$\text{Gr} = \frac{\beta_g\Delta T\,l^3\,\rho^2}{\mu^2} = \frac{\beta_g\Delta T l^3}{\nu^2} \qquad [5\text{-}35]$$

Note that "group" and "number" are interchangeable.

For the condition in which only natural convection occurs, the velocity is dependent on the buoyancy effect, which is represented by the Grashof number, and in this case the Reynolds number can be omitted:

$$\mathrm{Nn} = f_1(\mathrm{Pr}, \mathrm{Gr}) \qquad [5\text{-}36]$$

When forced convection occurs, the Reynolds number becomes important and the natural convection, the Grashof number, is negligible.

$$\mathrm{Nu} = f_2(\mathrm{Re}, \mathrm{P}_\upsilon) \qquad [5\text{-}37]$$

Sometimes a **Stanton number** ($\mathrm{St} = \frac{h}{C_P \rho u}$) is used instead of the Nusselt number.

$$\frac{h}{C_P \rho u} = \frac{\left(\frac{hd}{k}\right)\mu}{ud\rho}\left(\frac{k}{C_P \mu}\right)$$

$$\mathrm{St} = \mathrm{Nu}\,\mathrm{Re}^{-1}\,\mathrm{Pr}^{-1} \qquad [5\text{-}38]$$

The **Peclet number**, Pe, is defined as convective mass transfer divided by diffusive mass transfer:

$$\mathrm{Pe} = \frac{dV}{D_A} \qquad [5\text{-}39]$$

The **Schmidt number**, Sc, is defined as momentum diffusivity divided by mass diffusivity:

$$\mathrm{Sc} = \frac{\mu}{\rho D_A} \qquad [5\text{-}40]$$

The **Sherwood number**, Sh, is defined as the total mass transfer divided by diffusive mass transfer:

$$\mathrm{Sh}=\frac{k_L d}{D_A} \qquad [5\text{-}41]$$

It should be pointed out that the above-mentioned numbers relate to the diffusion coefficient and that the Sherwood number is also related to mass transfer.

5.3.2 Correlation for Mass Transfer Coefficients

As we discussed in relation to mass transfer, the bubble diameter d_b affects the parameter a as well as the mass transfer coefficient K_L. Let us examine how K_La can be obtained from those relations. The bubble diameter is dependent upon the gas velocity in the orifice:

$$d_{\mathrm{b}}(\mathrm{cm})=0.736Q_f^n \qquad [5\text{-}42]$$

The Q_f^n is expressed as cm^3/sec/orifice, and n varies between 0.2 and 1, with a typical value of 0.4. For a bubble Reynolds number, $v_b d_b \rho_w/\mu_w$, of less than 300, the Levich correlation is used:

$$v_b=\frac{27.3\rho_w d_b^2}{\mu_w} \qquad [5\text{-}43]$$

For a Reynolds number between 300 and 4000, the **Mendelson correlation** is used:

$$v_b=\left[\frac{2\sigma}{d_b\rho_w}+\frac{gd_b}{2}\right]^{\frac{1}{2}} \qquad [5\text{-}44]$$

where σ is the surface tension between air and water. Of all the dimensionless groups, only the Sherwood number has a direct value of K_L. Other relations contain D_A, such as the Peclet and Schmidt numbers, and can link up with the Sherwood number. If the bubbles are rigid spheres, then the **Griffith equation**,

$$\mathrm{Sh}=2+0.57\,\mathrm{Re}^{0.5}\,\mathrm{Sc}^{0.35} \qquad [5\text{-}45]$$

is used for Re >> 1. If the bubbles are free flowing, then

$$\mathrm{Sh} = 2 + 0.31(\mathrm{Sc} \cdot \mathrm{Gr})^{\frac{1}{3}} \qquad [5\text{-}46]$$

Upon solving,

$$K_L = \frac{2D_A}{d_b} + 0.31\left[\frac{(\rho_g - \rho_w)\mu_w g}{\rho_w^2}\right]^{-\frac{1}{3}} \mathrm{Sc}^{-\frac{2}{3}} \qquad [5\text{-}47]$$

If the bubbles are wobbling, then a wide range of bubble Reynolds numbers are useful. For Re < 1,

$$\mathrm{Sh} = 0.65\mathrm{Pe}^{\frac{1}{2}};\ K_L = 0.65\left(\frac{D_A v_b}{d_b}\right)^{\frac{1}{2}} \qquad [5\text{-}48a]$$

For 10 < Re < 100,

$$\mathrm{Sh} = 0.65\mathrm{Pe}^{\frac{1}{2}}\left(1 + \frac{\mathrm{Re}}{2}\right)^{\frac{1}{2}} \qquad [5\text{-}48b]$$

For 100 < Re < 1000,

$$\mathrm{Sh} = 1.13\left(1 - 2.9\,\mathrm{Re}^{-\frac{1}{2}}\right)^{\frac{1}{2}} \mathrm{Pe}^{\frac{1}{2}} \qquad [5\text{-}48c]$$

And for Re > 1000,

$$\mathrm{Sh} = 1.13\mathrm{Pe}^{\frac{1}{2}};\ K_L = 1.13\left(\frac{D_A v_b}{d_b}\right)^{\frac{1}{2}} \qquad [5\text{-}48d]$$

Molecular diffusion coefficients can be estimated using the **Wilke-Chang** relation:

$$D_A = \frac{K_A T}{\mu_L} \qquad [5\text{-}49]$$

K_A here depends on the molar volume of contaminant A in water, and μ_L is viscosity. Some common organic compounds are listed in Table 5-4.

Going back to the $K_L a$ equation previously discussed, the overall mass transfer coefficient is used in order to specify the rate of aeration of wastewater.

$$\frac{dC}{dt} = K_L a\left(C^* - C\right) \tag{5-50}$$

After integration, $C = C^0$ at $t = 0$,

$$\ln\left(\frac{C^* - C^0}{C^* - C}\right) = K_L at \tag{5-51}$$

Table 5-4. Computation of Molecular Diffusion Coefficients Using the Wilke-Chang Relation

Compounds	Formula	K_A (× 10^8)	D_{AB}[a] at 20 °C × 10^{10} (m^2/sec)
Vinyl chloride	C_2H_3Cl	3.85	11.3
Methane	CH_4	6.18	18.1
Carbon tetrachloride	CCl_4	2.76	8.08
Tetra chloroethylene	C_2Cl_4	2.57	7.52
Trichlorethylene	$CCHCl_3$	2.86	8.37
Chloromethane	CH_3Cl	4.49	13.1
1,1,1-Trichloroethane	$C_2H_3Cl_3$	2.75	8.04
Benzene	C_6H_6	3.04	8.91
Chloroform	$CHCl_3$	3.12	9.15
1,2-Dichloroethane	$C_2H_4Cl_2$	3.10	9.08
Bromoform	$CHBr_3$	2.99	8.75

[a] $D_{AB} = K_A T/\mu_B$; T(°K), μ_B(cP); D_{AB} (ft^2/hr) = 3.875 × 10^4 D_{AB} (m^2/sec).

Normally, C^* is the saturated concentration of water with oxygen, or C^* = 8.65 mg O_2/L. In the laboratory, water samples are deoxygenated with sodium sulfide solution (using cobalt chloride as a catalyst), and aeration is introduced to obtain a series of C at different t. A simple semilog plot of $(C^* - C)$ to t will make it possible to determine the K_La value. As stated before, the unit of a is interfacial area per unit volume (cm^{-1}), while k_L is expressed as cm/time. Thus, the unit of K_La is time^{-1}. One method to determine a, the interfacial surface, is from the holdup ε and the bubble diameter as follows:

$$a = \frac{6\varepsilon}{d^b} \qquad [5\text{-}52]$$

The holdup can be determined from:

$$\frac{\varepsilon}{(1-\varepsilon)^4} = 0.0268 \nu_l^{-\frac{1}{6}} U_g \left(\frac{\sigma}{\rho_l}\right)^{-\frac{1}{8}} \qquad [5\text{-}53]$$

where ν_l is kinematic viscosity of liquid, U_g is superficial velocity of the gas (air), and σ and ρ_l are the surface tension and density. A correlation for K_L < 0.14 is:

$$K_L = \frac{1}{2} g^{\frac{5}{8}} D_A \rho^{\frac{3}{8}} \sigma^{-\frac{3}{8}} d_b^{\frac{1}{2}} \qquad [5\text{-}54]$$

or use of these relations, please see Example 5-3.

K_La can be related to the power consumption of agitated aerators. For example,

$$(K_L a)_{25^\circ C} = 21.67 \left(\frac{P}{V}\right)^{0.32} v_s^{0.46} \qquad [5\text{-}55]$$

where v_s is air velocity in m/hr and (P/V) is in kW/m^3 of liquid. K_La of waste-water can be evaluated by the α_K coefficient,

$$\alpha_K = \frac{(K_L a)_{\text{wastewater}}}{(K_L a)_{\text{purewater}}} \qquad [5\text{-}56]$$

K_La is also a factor in ionic strength.

$$\alpha_K = 1 + 3.78\mu \qquad [5\text{-}57]$$

Typical α values for the same wastewaters are listed in Table 5-5. K_La values for constituents other than oxygen are computed from the ratio of diffusivity of oxygen to the compound of interest. The ratio of diffusivities can be expressed by:

$$\beta = \left(\frac{D_{O_2}}{D_A}\right)^{\nu} \qquad [5\text{-}58]$$

where ν varies from 0.5 to 0.75 for a surface-aerated system. Reported values of β are listed in Table 5-6. Roberts has found that the K_La of volatile constituents are related to:

$$(K_L a)_{O_2}$$

and the value can be evaluated from D_A based on:

$$\frac{(K_L a)_A}{(K_L a)_{O_2}} = \left(\frac{D_A}{D_{O_2}}\right)^{0.66} \qquad [5\text{-}59]$$

[Example 5-3] A diffuser-supplied air bubble has a mean diameter of 0.3 cm in water at 20°C. Assuming Mendelson's correlation holds, find k_L (liquid mass transfer coefficient).

From Equation [5-44],

$$v_b = \left[\frac{2\sigma}{d_b \rho_w} + \frac{g d_b}{2}\right]^{\frac{1}{2}} = \left[\frac{(2)(72.75)}{(0.3)(1)} + \frac{(980)(0.3)}{2}\right]^{\frac{1}{2}} = 25.1 \text{ cm/sec}$$

$$\text{Re} = \frac{v_b d_b \rho_w}{\mu} = \frac{(25.1\, cm/\text{sec})(0.3\, cm)(1 g/cm^3)}{10^{-2}\, g/cm\text{-sec}} = 750$$

Table 5-5. Typical α Values for Some Specific Wastewaters

Unit	Waste	Alpha (α)	Conditions
Sparjers[a]	Activated sludge effluent	1.320	10–25 scfm/unit, 15 ft depth, 25 ft width
Plate tubes[a]	Activated sludge effluent	0.860	6–14 scfm/unit, 15 ft depth, 25 ft width
INKA system[b]	Kraft Black Liquor		
	20 ppm	0.875	6 ft depth, 6.8 ft width, 2.6 ft submergence
	100 ppm	0.750	6 ft depth, 6.8 ft width, 2.6 ft submergence
	200 ppm	0.625	6 ft depth, 6.8 ft width, 2.6 ft submergence
Small bubble diffuser[b]	Kraft Black Liquor		
	20 ppm	0.880	100 L volume, 3 ft depth, 10 × 10 cm bubble air diffuser
	100 ppm	0.813	100 L volume, 3 ft depth, 10 × 10 cm bubble air diffuser
	200 ppm	0.662	100 L volume, 3 ft depth, 10 × 10 cm bubble air diffuser
Lightning lab stirrer[c]	Nylon manufacturing wastewater		10 L volume, surface mixing
		0.560	212 rpm
		0.715	220 rpm
Aeration cone[d]	Tap water + 5 ppm anionic detergen	1.330	12,500 ft^3 volume, 6 ft diameter, 36 rpm

[a] From Barnhart (1965)

[b] From Bewtra (1964)

[c] From Randall (1971)

[d] From Eckenfelder (1968)

Table 5-6. Reported Values of β

Substance	β
1,1-Dichloroethylene[a]	0.62
Tetrachloroethylene[a]	0.55
Benzene[a]	0.53
Toluene[a]	0.53
1,2,4-Trimethylbenzene[a]	0.39
1,4-Dichlorobenzene[a]	0.39
Freon[b]	0.66
Chloroform[b]	0.56
1,1,1-trichloroethane[b]	0.61
Carbon tetrachloride[b]	0.617
Trichloroethylene[b]	0.615
Tetrachloroethylene[b]	0.608
Krypton[c]	0.82
Carbon dioxide[c]	0.89
Nitrogen[c]	0.91
Ethylene[d]	0.89
Propane[d]	0.72
Chloroform[e]	0.66
1,1-Dichloroethane[e]	0.71
Hydrogen[f]	1.3
Helium[f]	1.3

[a] From Matter-Müller (1980).
[b] From Roberts.
[c] From Tsivogloci (1973)
[d] From Rathbun (1973).
[e] From Smith and Bonnberger (1980).
[f] From Hutchinson (1932).

Therefore, we use Equation [5-48c]:

$$\mathrm{Sh} = 1.13\left(1 - 2.9\,\mathrm{Re}^{-\frac{1}{2}}\right)^{\frac{1}{2}} \mathrm{Pe}^{\frac{1}{2}} = 1.13\left[\left(1 - 2.9(750)^{-\frac{1}{2}}\right)^{\frac{1}{2}}\right]\left(\frac{0.3 \times 25.1}{2.21 \times 10^{-5}}\right)^{\frac{1}{2}} = 624$$

From Equation [5-41],

$$k_L = \frac{624 \times D_A}{d_b} = \frac{624 \times 2.2 \times 10^{-5}}{0.3} = 0.0459 \text{ cm/sec}$$

[Example 5-3] Air is used to generate ozone in a contact basin, which is a covered concrete tank 4.5 m deep with 10 min contact time. Air is introduced by porous diffusers at the bottom of the tank, producing 0.23 cm bubbles. Efficiency of the ozone generator is 1.5% from oxygen to ozone. The facility is designed to allow 2 mg/L of ozone to raw surface water to flow toward the basin at 0.2 m^3/sec. Estimate the overall mass transfer coefficient.

The area of the basin is:

$$A = \frac{Qt}{z} = \frac{\left(0.2\,\mathrm{m}^3/\mathrm{sec}\right)(10\,\mathrm{min})(60\,\mathrm{sec/min})}{4.5\,\mathrm{m}} = 2.67\,\mathrm{m}^2$$

Now, to find the air flow rate,

$$Q_w = \frac{\rho_A \cdot \%\mathrm{O}_3}{\text{dosage}} Q_A$$

or

$$Q_A = \frac{(2\,\mathrm{g/m}^3)(0.2\,\mathrm{m}^3/\mathrm{sec})}{(1208\,\mathrm{g/m}^3)(0.015)(0.21)} = 0.138\ \mathrm{m}^3/\mathrm{sec}$$

or

$$U_G = \frac{0.138\ \mathrm{m}^3/\mathrm{sec}}{26.7\ \mathrm{m}^2}(100\,\mathrm{cm/m}) = 0.4\,\mathrm{cm/sec}$$

To calculate the holdup from Equation [5-53],

$$\frac{\varepsilon}{(1-\varepsilon)^4} = 0.0268\,(0.01\ \text{cm}^2/\text{sec})^{-\frac{1}{6}}\,(0.4\ \text{cm/sec})\left(\frac{72\ \text{g/sec}}{1\ \text{g/cm}^3}\right)^{-\frac{1}{8}} = 0.0135$$

By trial and error, $\varepsilon = 0.014$. From Equation [5-52],

$$a = \frac{6\varepsilon}{d_b} = \frac{6\,(0.014)}{(0.2\ \text{cm})} = 0.42\ \text{cm}^2/\text{cm}^3$$

K_L can be obtained from Equation [5-54]:

$$K_L = 0.5\left(980\frac{\text{cm}^2}{\text{sec}}\right)^{\frac{5}{8}}\left(2.3\times10^{-5}\,\frac{\text{cm}^2}{\text{sec}}\right)\left(1\frac{\text{g}}{\text{cm}^3}\right)^{\frac{3}{8}}\left(72\frac{\text{g}}{\text{sec}}\right)^{-\frac{3}{8}}(0.2\,\text{cm})^{\frac{1}{2}} = 0.016\ \text{cm/sec}$$

Thus,

$$K_L a = (0.42\ \text{cm}^2/\text{cm}^3)\,(0.016\ \text{cm/sec}) = 0.0067\ \text{sec}^{-1}$$

5.4 Oxygen Sag Curve

In a natural water system, the concentration of dissolved oxygen drops below the equilibrium concentration, C^* or C_s. This difference is usually called the **oxygen deficit**, D, and is defined as:

$$D = C_s - C \qquad [5\text{-}60]$$

Since C_s is constant, the rate of change of deficit increases at the same rate that oxygen is used up

$$\frac{dD}{dt} = -\frac{dC}{dt} \qquad [5\text{-}61]$$

5.4.1 The Streeter-Phelps Equation

The Streeter-Phelps dissolved oxygen model of streams can be explained by defining the oxygen deficit in a stream as a function of both oxygen utilization and reaeration. The rate at which dissolved oxygen disappears from the stream actually corresponds to BOD exertion, thus,

$$y = L_0 - L_t$$

L_0, the ultimate BOD, is a fixed value, so

$$\frac{dy}{dt} = -\frac{dL_t}{dt}$$

From the BOD discussion, we have

$$\frac{dL_t}{dt} = -kL_t \qquad [5\text{-}62]$$

Therefore,

$$\frac{dD}{dt} = kL_t \qquad [5\text{-}63]$$

A more convenient expression is

$$r_D = k_1 L_t \qquad [5\text{-}64]$$

On the other hand, the rate of **reaeration (oxygen addition)** is a first-order reaction with respect to oxygen deficit, or

$$r_R = -k_2 D \qquad [5\text{-}65]$$

For overall consideration, both oxygen removal and oxygen addition have opposite effects on the deficit.

$$\frac{dD}{dt} = r_D + r_R = k_1 L_t - k_2 D \qquad [5\text{-}66]$$

This is illustrated graphically in Figure 5-3. The rate constants k_1 and k_2 can be formed, respectively, as shown in Tables 5-7 and 5-8. The actual oxygen concentration $C_t = C_s - D_t$ has a characteristic dip, as shown in Figure 5-3, resulting in the oxygen sag curve.

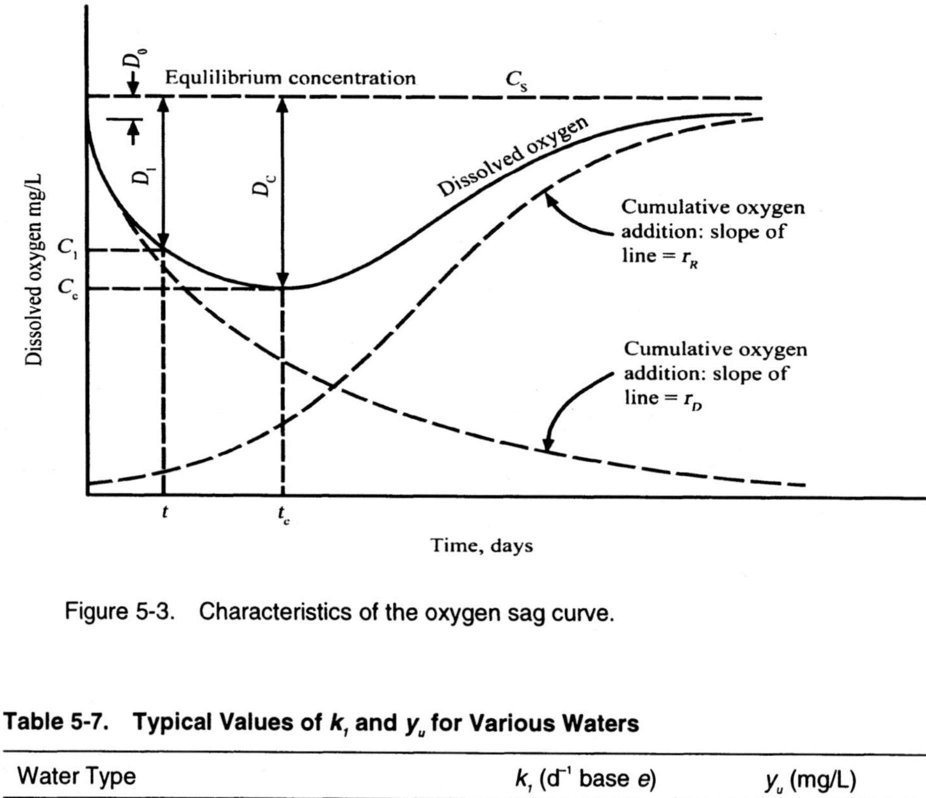

Figure 5-3. Characteristics of the oxygen sag curve.

Table 5-7. Typical Values of k_1 and y_u for Various Waters

Water Type	k_1 (d^{-1} base e)	y_u (mg/L)
Tap water	< 0.1	0–1
Surface waters	0.1–0.23	1–30
Weak municipal wastewater	0.35	150
Strong municipal wastewater	0.40	250
Treated effluent	0.12–0.23	10–30

Table 5-8. Reaeration Constants

Water Body	Ranges of k_2 at 20 °C, base *e*
Small ponds and backwaters	0.1–0.23
Sluggish streams and large lakes	0.23–0.35
Large streams of low velocity	0.35–0.46
Large streams of normal velocity	0.46–0.69
Swift streams	0.69–1.15
Rapids and waterfalls	> 1.15

Source: Adapted from Ref. 26-15.

Since

$$L_t = L_0 e^{-k_1 t}$$

the final expression of oxygen deficit is

$$\frac{dD}{dt} + k_2 D = k_1 L_0 e^{-k_1 t} \qquad [5\text{-}67]$$

Solving the differential equation (see Example 5-4), we have

$$D = \frac{k_1 L_0}{k_2 - k_1}\left(e^{-k_1 t} - e^{k_2 t}\right) + D_0 e^{-k_2 t} \qquad [5\text{-}68]$$

Equation 5-68 is the **Streeter-Phelps oxygen-sag equation**. In this equation, D_0 is initial oxygen deficit defined as $D = D_0$ when $t = 0$. If wastewater discharge is included, then

$$D_0 = (DO)_s - (DO)_{mix} \qquad [5\text{-}69]$$

and

$$(DO)_{mix} = \frac{(DO)_s Q_s + (DO)_w Q_w}{Q_s + Q_w} \qquad [5\text{-}70]$$

the subscript *s* stands for stream, while *w* represents wastewater.

The lowest point on the oxygen sag curve is (t_c, C_c) or critical time, critical concentration, and the corresponding D_c, **critical deficit**, is of great interest due to the minimum value of oxygen concentration, which represents the maximum impact on the *DO* as a result of wastewater discharge. At this maximum deficit, the rate of change of the deficit is zero, or from Equation [5-67],

$$0 + k_2 D_c = k_1 L_0 e^{-kt_{c1}}$$

or as rearranged,

$$D_c = \frac{k_1}{k_2} L_0 e^{-kt_{c1}} \qquad [5\text{-}71]$$

In order to find t_c, let us differentiate the D equation and set it equal to zero, since D_c is maximum at t_c.

$$0 = \frac{k_1 L_0}{k_2 - k_1}\left(-k_1 e^{-k_1 t_c} + k_2 e^{-k_2 t_c}\right) - k_2 D_0 e^{-k_2 t_c}$$

Rearranging the terms and solving for t_c,

$$t_c = \frac{1}{(k_2 - k_1)} \ln\left[\frac{k_2}{k_1}\left(1 - D_0 \frac{(k_2 - k_1)}{k_1 L_0}\right)\right] \qquad [5\text{-}72]$$

Both D_c and t_c can be determined. The t-scale on the x-axis can be replaced with distance x if the velocity is known, e.g., $tu = x$.

[Example 5-4] Solve the oxygen sag curve differential equation (Eq. [5-67]).

$$\frac{dD}{dt} + k_2 D = k_1 L_0 e^{-k_1 t}$$

This is a first-order differential equation of the form

$$\frac{dy}{dx} + Py - Q$$

Both P and Q are functions of x. Using the integration factor

$$e^{\int k_2 dt} = e^{k_2 t}$$

multiply both sides of the equation by the integration factor

$$e^{k_2 t}\frac{dD}{dt}+k_2 D e^{k_2 t}=k_1 L_0 e^{(k_2-k_1)t}$$

The two terms of the right side now are actually the derivative.

$$\frac{d}{dt}\left(De^{k_2 t}\right)$$

Therefore, integrating

$$\int dDe^{k_2 t}=k_1 L_0\int e^{(k_2-k_1)t}\,dt$$

which is

$$De^{k_2 t}=\frac{k_1 L_0}{(k_2-k_1)}e^{(k_2-k_1)t}+C$$

For boundary conditions, $D = D_0$ at $t = 0$, which is

$$D_0=\frac{k_1 L_0}{(k_2-k_1)}+C$$

Solving for C and substituting C in the integrated equation,

$$D\,e^{k_2 t}=\frac{k_1 L_0}{k_2-k_1}e^{(k_2-k_1)t}+D_0-\frac{k_1 L_0}{k_2-k_1}$$

After rearrangement,

$$D=\frac{k_1 L_0}{k_2-k_1}\left(e^{-k_1 t}-e^{k_2 t}\right)+D_0 e^{-k_2 t}$$

which is Equation [5-68].

5.4.2 The Biotic Index

Actually, the zone of critical oxygen deficit will have the worst aquatic community, e.g., rat-tailed maggot (see Fig. 5-4 a–e). According to the biotic index, the sag portion of the curve will have a score of 2 to 1 in comparison with scores for fresh water invertebrate species of 8 to 10 (see Table 5-9 and Figure 5-5).

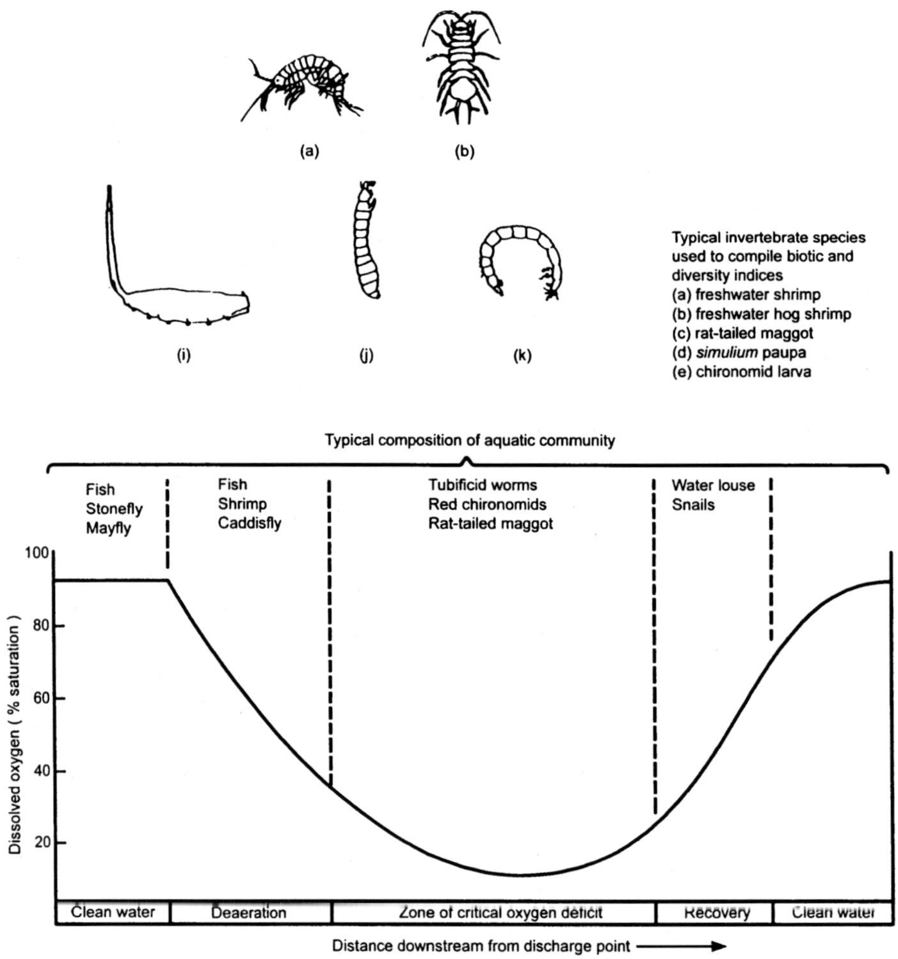

Figure 5-4. The effects of the oxygen sag resulting from a point source discharge on the composition of the aquatic community.

Table 5-9. Simplified Version of a Typical Biotic Index

Groups (families)	Score
Mayfly larvae (e.g., Euphemeridae, Ecdyonuridae) Stone-fly larvae (e.g., Leuctridae, Perlidae) Cased-caddis larvae (with stones and sand)	10(Good)
Damselfly and dragonfly larvae Caseless (free-living) caddis larvae (e.g., philopotamidae)	8
Mayfly larvae (Caenidae) Cased-caddis larvae (with plant debris)	7
Large freshwater mussels (60 × 100 mm) (Unionidae) Freshwater shrimp (Gammaridae)	6
Water bugs (e.g., Corixidae) Water beetles (e.g., Haliplidae, Elminthidae) Caseless caddis larvae (e.g., Hydropsychidae) Fly larvae (e.g., tipulidae, Simulidae) Flatworms (e.g., Planariidae)	5
Mayfly larvae (Baetidae) Alderfly larvae (Sialidae) Water mites	4
Snails (e.g., Lymnaeidae, Planorbidae, Physidae) Small freshwater bivalves (e.g., Sphaeriidae) Leeches (e.g., Glossiphoniidae) Water hog louse (Assellidae)	3
Fly larvae (midge) (Chironomidae)	2
Worms (e.g., tubificidae) Fly larvae (rat-tailed maggot)	1(Bad)

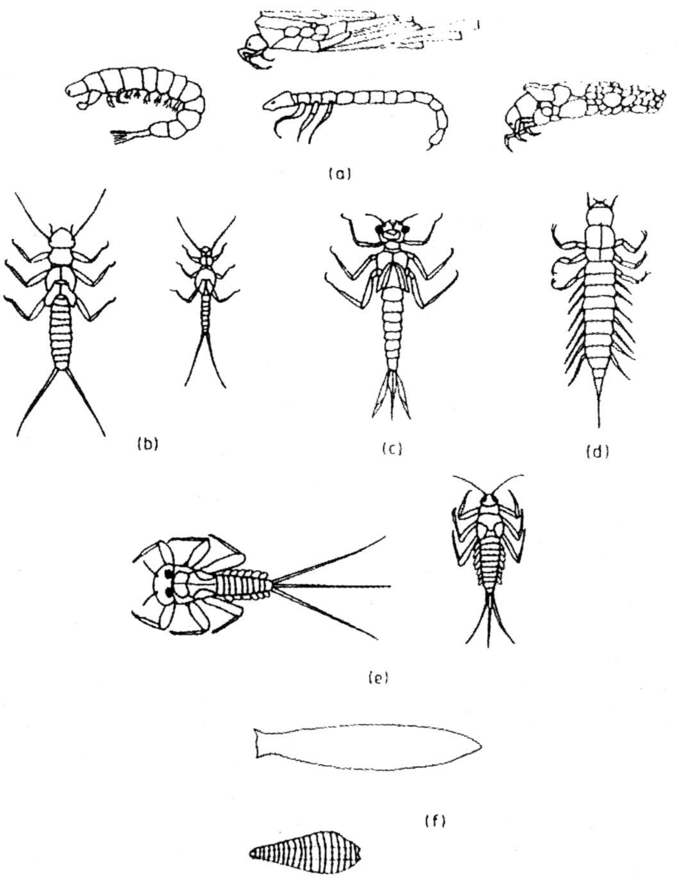

Figure 5-5. Fresh water invertebrate species: (a) caddis-fly larvae, (b) stonefly nymphs, (c) dragonfly nymph, (d) alderfly larva, (e) mayfly nymphs, (f) freshwater leaches.

5.4.3 Estuary and Ocean Disposal

As discussed previously, the general convective diffusion equation, such as Equation [5-63], can be expressed as (assuming a first-order reaction):

$$\frac{\partial C}{\partial t} = E\frac{\partial^2 C}{\partial x^2} - v\frac{\partial C}{\partial x} - k'C \qquad [5\text{-}73]$$

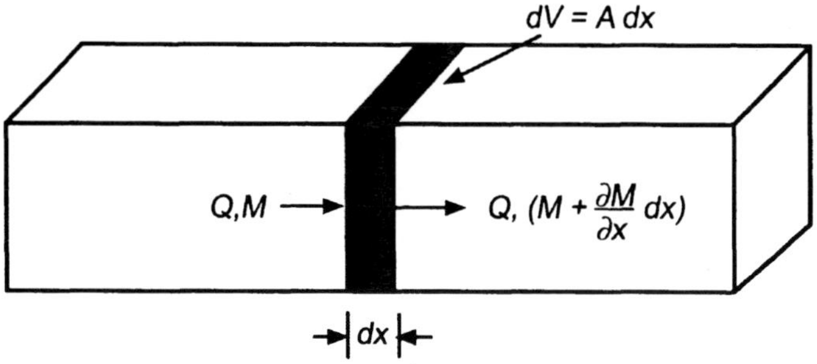

Figure 5-6. Definition sketch for estuary diffusion model.

Since mass flow due to eddy diffusion $\frac{\partial M}{\partial t}$ is proportional to concentration gradient $\frac{\partial C}{\partial x}$,

$$\frac{\partial M}{\partial t} = -EA\frac{\partial C}{\partial x} \qquad [5\text{-}74]$$

E in the above dispersion equation and the convective-diffusion equation is the coefficient of eddy diffusion, A is the cross-sectional area, and $v = Q/A$ (see Fig. 5-6). Integration of the diffusion equation at steady state,

$$\frac{\partial C}{\partial t} = 0$$

as shown in Figure 5-7.

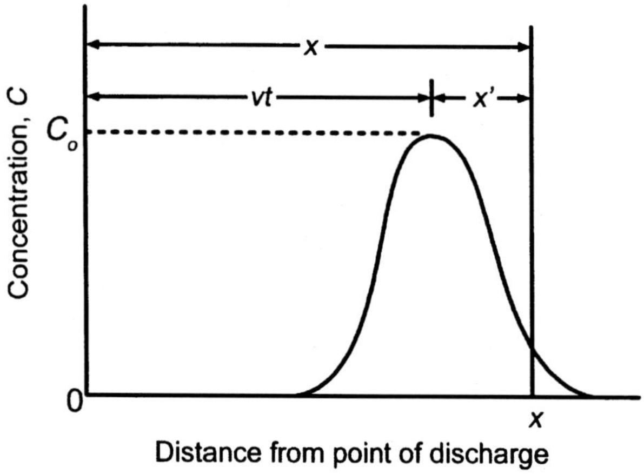

Figure 5-7. Definition sketch for estuary diffusion model for instantaneous dye release.

For instantaneous dye release at $x = 0$, $t = 0$ in a longitudinal cylinder,

$$C = C_0 e^{\left(-x^2 / Et\right)\left(-kt\right)} \tag{5-75}$$

where

$$C_0 = \frac{M}{2A\left(\pi E_t\right)^{\frac{1}{2}}}$$

M = mass discharged at $t = 0$, and $x' = x - vt$.

For continuous discharge of waste at a rate of W,

$$C = C_0 e^{jx}$$

where

$$C_0 = \frac{W}{Q}\left(1+\frac{4kE}{v^2}\right)^{\frac{1}{2}}$$

$$j = \frac{v}{2E}\left[1 \pm \left(1+\frac{4kE}{v^2}\right)^{\frac{1}{2}}\right] \qquad [5\text{-}76]$$

The positive root for j refers to upstream ($-x$) direction, and the negative root refers to downstream ($+x$) direction. Here, j can be approximated as V/E, and E can be measured.

For coefficient E, eddy diffusion can be evaluated from the following equation:

$$E = cnvR^{\frac{5}{6}} \qquad [5\text{-}77]$$

where $c = 77$, n = Manning's roughness coefficient, v = velocity in ft/sec, and R = hydraulic radius in ft. The E value is obtained in ft^2/sec. Other methods, such as the salinity intrusion method, are also available.

When waste is discharged to a diffuser, the velocity of the jet will cause turbulent mixing with the surrounding water. These are different ways to estimate the dilution effect:

$$D = D_1 D_2 D_3 \qquad [5\text{-}78]$$

D_1 is the **initial dilution** and is a function of depth of the discharge port, Y_0, the diameter of the discharge orifice d, and the Froude number Fr. From Equation [5-18],

$$\mathrm{Fr} = \frac{v_j}{\left[\left(\frac{\Delta S}{S}\right)gd\right]^{\frac{1}{2}}} \qquad [5\text{-}79]$$

where v_j is jet velocity, ΔS is the difference of density between the waste and surrounding seawater, S is the density of the waste, g is gravity acceleration, and d is discharge-jet diameter. Abraham made the correlation of Y_0/d to the Froude number with different D_1 numbers, as illustrated in Figure 5-8.

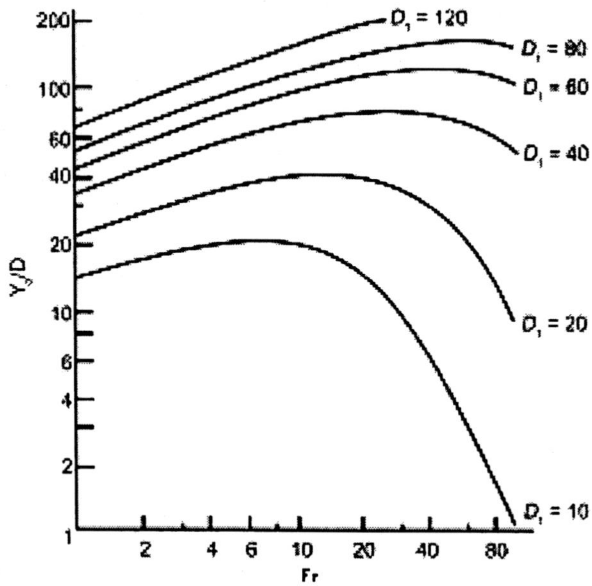

Figure 5-8. Initial dilution by turbulent jet mixing.

$$D_1 = \frac{v_x bd}{Q} \qquad [5\text{-}80]$$

where v_x is the current velocity, b is the diffuser length, d is the diffuser depth and Q is the waste water flow rate.

Dilution can be caused by **dispersion** due to eddy diffusion after the initial dilution. This dilution is represented by D_2 and can be evaluated through an error function:

$$D_2 = \frac{C_0}{C_t} = \left[erf \left(\frac{\frac{3}{2}}{\left[1 + \frac{2}{3}\beta\left(\frac{x}{b}\right)\right]^3 - 1} \right)^{\frac{1}{2}} \right]^{-1} \qquad [5\text{-}81]$$

where C_t is the maximum concentration of pollutant in time t, C_0 is the maximum pollutant concentration after initial dilution, x is the distance along plume concentration (ft), b is the effective diffuser system length (ft), β is a dimensionless number that equals 12 $E/V_x b$ (where $E = 1.0 \times 10^{-3}(b)^{4/3}$ ft^2/sec), and V_x is current velocity in ft/sec. A nomograph for the solution in both the U.S. and metric systems is shown in Figure 5-9.

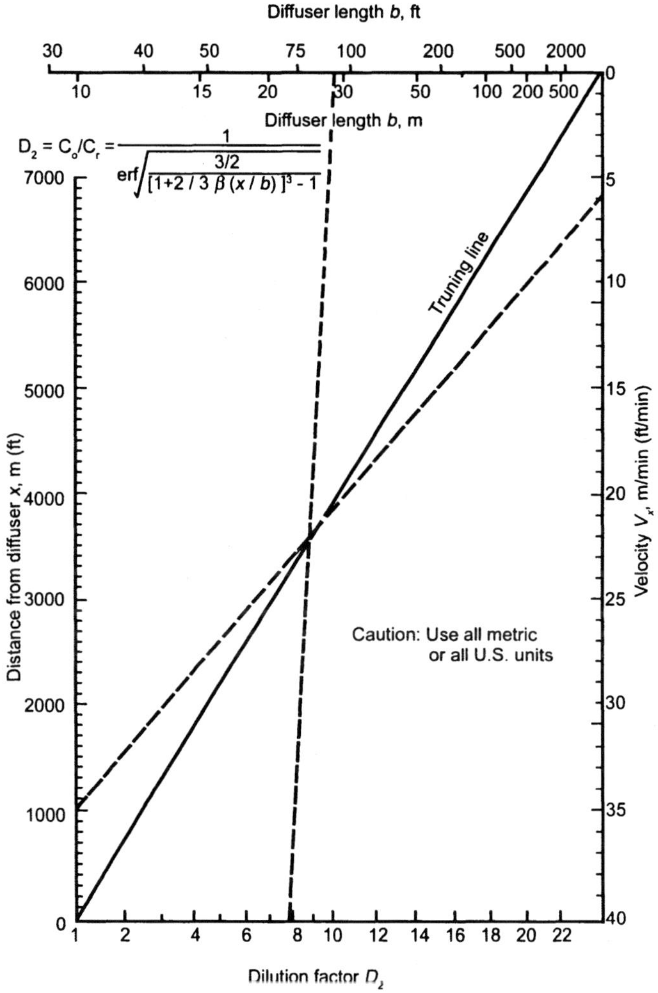

Figure 5-9. Nomograph for D_2. (The dashed line indicates an example with 10^3 m distance of diffuser and 6 m/min on-shore current; the diffuser length is 28 m.)

Another **dilution effect** is caused by **decay** of waste. Assuming first-order reaction,

$$r_c = -kC$$

$$C_t = C_0\, e^{-kt}$$

where C_t is initial bacterial concentration, C_0 is bacterial concentration after the initial dilution, and k is the bacterial decay constant. Usually T_{90}, the time in hours to achieve a 90% reduction in the bacterial number, is used.

$$D_3 = \frac{C_0}{C_t} = \exp\left[\frac{2.3x}{T_{90}\,60(V_x)}\right] = e^{\frac{2.3t}{T_{90}}} \qquad [5\text{-}82]$$

and V_x is velocity in m/min. A nomograph like the one shown in Figure 5-10 is designed for such use.

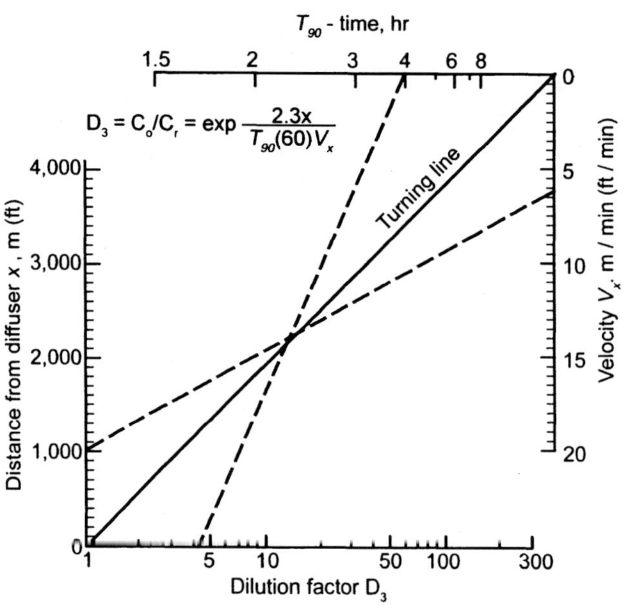

Figure 5-10. Nomograph for D_3. (The dashed line is the same as in Fig. 5-8, except T_{90} = 4 hr).

Usually, in order for an ocean outfall to meet a standard of 10 coliforms per mL, dilution has to be reduced by 75,000. This means that the length of the outfall has to be 2500 – 3000 m away from the shore (see Table 5-10). Table 5-11 outlines how different alternative treatments may require different lengths for the outfall.

Usually the outfall is used to convey the waste or pollutants (xenobiotics) to the diffuse section for disposal into the ocean. Since one fifth of the United States population lives near estuaries along the coastal periphery, the analysis and subsequent technological development are important. The same principle can also be applied to the Great Lakes area.

Table 5-10. Dilution Factors Versus Length of Outfall

Distance from Shore	Dilution Factor			
(m)	D_1	D_2	D_3	D_4
300	17	2.1	1.6	57
600	30	4.4	2.6	340
1000	48	7.9	4.9	1860
1500	70	13.0	11.0	10000
2000		1000[a]	24.0	24000
2500		1000	54.0	54000
3000		1000	120.0	120000

[a] Experience with existing outfall systems indicates that the maximum physical dilution achieved (initial dilution + transport dispersion) is in the neighborhood of 1000.

Note: m × 3.2808 = ft.

Table 5-11. Required Length of Outfall for Alternative Types of Treatment

Type of Treatment	Coliform Reduction (%)	Required Dilution to Meet Standard[a]	Length of Outfall Required (m)
Preliminary	25	75000	2700
Preliminary + Cl_2	25 + 98	1500	950
Activated sludge	90	10000	1500
Activated sludge + Cl_2	90 + 99	100	400

[a] Assumed to be 10 coliforms per mL.

Note: m × 3.2808 = ft.

REFERENCES

5-1. K. Akita and F. Yoshida, Bubble size interfacial area and liquid phase mass transfer coefficient in bubble columns, *Ind. Eng. Chem. Process Des. Develop.* 13, 1 (1974).

5-2. M. A. Churchill, H. L. Elmore, and R. A. Buckingham, Prediction of stream reaeration rates, *J. San. Eng. Divin.*, Proceeding of the American Society of Civil Engineers 88, SH4 (July 1962).

5-3. P. V. Danckenwerts, Significance of liquid-film coefficients in gas absorption, *Ind. Eng. Chem.* 43, 6, 1460 (1951).

5-4. T. H. Davies and S. Y. Ip, The droplet size and height effects in ammonia removal in a spray tower, *Water Res.*, 15(5), 521 (1981).

5-5. W. P. Isaacs and A. F. Gaudy, Jr., Atmospheric oxidation in a simulated stream, *J. Sanit. Eng. Div., Am. Soc. Civ. Eng.*, 94, SA2, 319 (1968).

5-6. D. Mackay, W. Y. Shiu, and R. P. Sutherland, Determination of air-water Henry's law constants for hydrophobic pollutants, *Environ. Sci. Technol.*, 13, 3 (1979).

5-7. K. H. Mancy and D. A. Okun, Effects of surface active agents on bubble aeration, *J. Water Pollut. Contr. Fed.*, 32, 4, 351 (1960).

5-8. K. H. Mancy and D. A. Okun, The effects of surface active agents on aeration, *J. Water Pollut. Contr. Fed.*, 37, 2, 212 (1965).

5-9. R. E. Pattle, The aeration of liquids. Part I. The solution of gas from rising bubbles, *Trans. Inst. Chem. Eng.*, (Br.), 28, 27 (1950).

5-10. R. L. Pigford and C. Pyle, Performance characteristics of spray-type absorption equipment, *Ind. End. Chem.*, 43(7), 1649 (1951).

5-11. P. Roberts, P. Dandliker, C. Matter, and C. Muniz, *Volatilization of Trace Organic Contaminants During Surface Aeration: Model Studies*, Tech. Report No. 257, Dept. of Civil Eng., Stanford University, 1981.

5-12. J. Smith and D. Bomberger, Prediction of volatilization rates of chemicals in water, in B. Afghan and D. Mackay (eds.), *Hydrocarbons and Halogenated Hydrocarbons in the Aquatic Environment*, Plenum Press, New York, 1980.

5-13. R. E. Treybal, *Mass Transfer Operations*, McGraw-Hill, New York, 1980.

5-14. W. Whipple, Jr., J. V. Hunter, B. Davidson, F. Ditmann, and S. Yu, *Instream Aeration of Polluted Rivers*, Water Resources Institute, Rutgers University, The State University of New Jersey, New Brunswick, NJ, August 1969.

5-15. Metcalf and Eddy, Inc., *Wastewater Engineering: Treatment, Disposal, Reuse*, 2nd ed., McGraw-Hill, New York, 1979.

5-16. J. M. Coalson, J.F. Richardson, J. R. Backhast, and J. H. Harker, *Chemical Engineering,* Vol. 1, Permagon Press, Oxford, United Kingdom, 1977.

PROBLEM SET

1. The following data were obtained on the unsteady-state reaeration of water using a surface aerator. Assuming C^* = saturation value = 9.1 ppm O_2, determine K_La using both the integral and difference methods.

2. Consider a bubble of pure oxygen leaving an orifice and having a d_b = 200 microns at the bottom of an activated sludge reactor 7 m deep. Considering mass transfer and hydrostatic effects, plot the bubble diameter and rise velocity as a function of distance from the bottom. Assume the bubble always remains pure oxygen. What fraction of the oxygen gas is transferred to the water for a temperature of 20°C and a dissolved oxygen concentration on the reactor of 2.0 mg/L?

3. Just below the point where a continuous discharge of pollution mixes with a river, BOD is 10.9 mg/L, and DO is 7.6 mg/L. The river and waste mixture has a temperature of 20 °C, a deoxygenation constant of 0.20/day, an average flow speed of 0.30 m/s, and an average depth of 3.0 m.

   (a) Find the time and distance downstream at which the oxygen deficit is a maximum.

   (b) Find the minimum value of DO.

4. An enclosed tank is contaminated with benzene of 300 μg/L. Find the vapor pressure of benzene in the headspace at 15 °C. What are their concentrations when compared to the NIOSH Occupational Exposure Standard of 20 ppm? How much effort is required to apply air stripping to meet this standard?

CHAPTER 6

COAGULATION

Suspended particles of colloidal ranges are present in almost all domestic and industrial wastewaters. These fine turbidity particles often cause taste, odor, and color problems. The nature of the chemical interactions that enable them to separate from an aqueous medium is the objective of analysis of this chapter. The coagulation process in water treatment is normally carried out prior to sedimentation and filtration. Usually chemicals (coagulants) are added to a flash mix and given a few minutes of agitation to disperse. Afterward, the solution is gently agitated between half an hour and an hour for floc formation (flocculation).

In this chapter we will first discuss destabilization and restabilization under the principles of colloidal and surface chemistry. Mechanisms in which destabilization occurs form the basis of the double layer. Hence, electrokinetic potentials of the DLVO theory will be explained, as well as the adsorption and entrapment of ions to hydrous metal oxides. Next, common coagulants together with a number of natural and synthetic polymers and polyelectrolytes will be reviewed. Rate processes such as perikinetic and orthokinetic flocculation are presented. Special attention is given to design criteria such as collision efficiency and velocity gradients. Finally, a typical coagulation tank with pedal-type agitators is illustrated and its power demand calculated.

6.1 Derjaguin-Landau-Verney-Overbeek (DLVO) Theory

Derjaguin and Landau together with Verney and Overbeek have independently developed a quantitative theory in which the stability of lyophobic sols is treated in terms of energy changes that take place when particles approach one another. The theory involves the estimation of repulsive energy due to the overlap of electric double layers and the attractive energy of the London–van der Waals force in terms of the interparticle distance. **Colloid stability** is often interpreted in terms of the nature of the summation of total energy as a function of interparticle distance.

Theoretical calculations have been made in different geometrical arrangements, for example, between two parallel plates or between two charged spheres. We will attempt to introduce the forces of repulsion first, followed by the forces of attraction. For the repulsive forces, a basic description is needed of the electric double layer and the associated electrokinetic phenomena. For the attractive forces, the concept of Hamaker constants should be understood. The combined summation of repulsive and attractive potential energies as a function of the distance between two interacting particles or surfaces can describe the colloid's stability. These types of diagrams, often referred to as **potential energy curves** as illustrated by Figure 6-1 or Figure 6-2, can be used to predict the critical coagulation concentration. The various studies of potential energy curve types can yield useful applications that are the essential component of the **DLVO theory**.

6.1.1. Electric Double Layer

When a substance is brought into contact with a polar (e.g., aqueous) medium, it generally acquires charges through ionization, ion adsorption, or ion dissolution. This acquisition of charges influences the charge distribution of nearby ions in the polar medium. Ions of opposite charge (counter-ions) are attracted toward the surface, while the ions of like charge (co-ions) are repelled away from the surface. In this manner, a shell of a diffuse **double layer,** as shown in Figure 6-3, will be formed. The first region is an inner region consisting of the original charge and the absorbed ions. In the second layer, however, the ions are distributed according to the influences of electric forces and random thermal motion. Gouy and Chapman independently have evaluated the potential of the double layer using the ion distribution function with the Poisson equation, which forms the basis of the Gouy-Chapman theory. Assuming that the electric potential is φ_0 at the surface and $\varphi(x)$ at a distance x from the surface in the electrolyte solution, then the Boltzmann equation is

$$n_i = n_{io} \exp\left[-\frac{z_i e\phi(x)}{kT}\right] \quad [6\text{-}1]$$

where

n_l = concentration of ions of kind i where the potential is $\varphi(x)$

n_{io} = concentration of ions in bulk solution

z_i = valency of ions of kind i

e = charge of an electron

A charge density ρ is used, which is the algebraic sum of the ionic charge per volume; that is,

$$\rho = \Sigma z_i e n_i \quad [6\text{-}2]$$

$$\rho = z e(n_+ - n_-)$$

$$= z e n_0 \left(\exp\left[\frac{-ze\phi}{kT}\right] - \exp\left[\frac{ze\phi}{kT}\right]\right)$$

$$= -2ze n_0 \sinh\left(\frac{ze\phi}{kT}\right) \quad [6\text{-}3]$$

The Poisson equation can relate the charge density to electric potential

$$\nabla^2 \varphi = -\frac{\rho}{D\varepsilon_0} \qquad [6\text{-}4]$$

where

D = dielectric constant

ε = permittivity of free space

and the n_i is defined by the previous equation (Eq. [6-1]).

The volume charge density can also be expressed as $n_i = n_+ - n_-$, such that for a symmetrical electrolyte, Equation [6-4] becomes

$$\nabla^2 \phi = -\frac{2zen_0}{D\varepsilon_0}\sinh\left(-\frac{ze\phi}{kT}\right) \qquad [6\text{-}5]$$

The solution of the preceding Poisson-Boltzmann equation is

$$\phi = 2\frac{kT}{ze}\ln\frac{1+\gamma\exp(-\kappa x)}{1-\gamma\exp(-\kappa x)} \qquad [6\text{-}6]$$

where κ is a unit of reciprocal length, and κ^{-1} is called the **Debye length.**

$$\kappa = \left(\frac{2n_0 ze}{D\varepsilon_0 kT}\right)^{\frac{1}{2}} = 3.288\, z\, c^{\frac{1}{2}} \text{ nm at } 25°\text{C} \qquad [6\text{-}7]$$

c = bulk electrolyte concentration in (m/L)

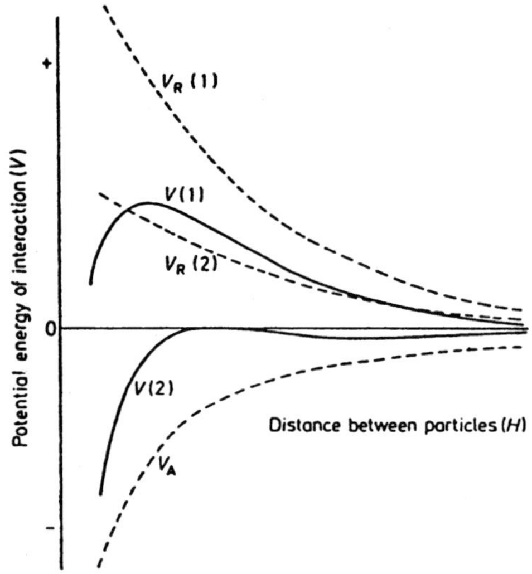

Figure 6-1. Total interaction energy curves, v(1) and V(2), obtained by the summation of an attraction curve, VA, with different repulsion curves, VR(1) and VR(2).

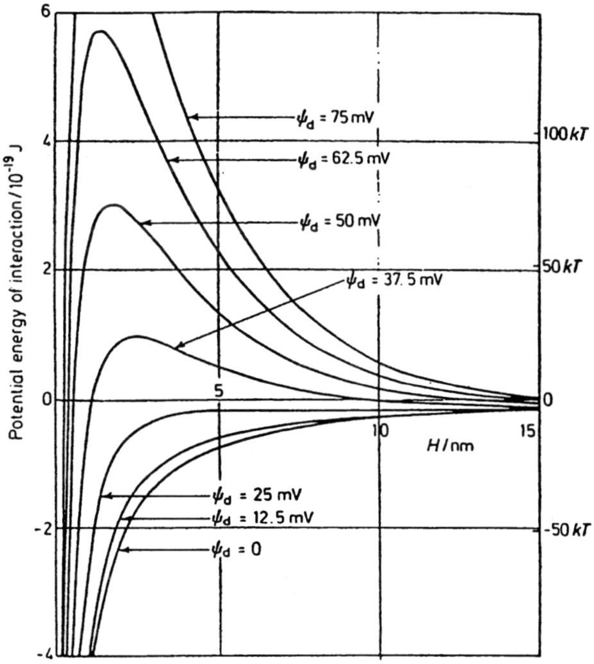

Figure 6-2. The influence of the Stern potential, ψd, on the total potential energy of interaction of two spherical particles: a = 10-7 m; T = 298 K; s =1; A11 = 2 x 10-19 J; A33 = 0.4 x 10-19 J; $\varepsilon/\varepsilon_0$ = 78.5; κ = 3 x 108 m-1 .

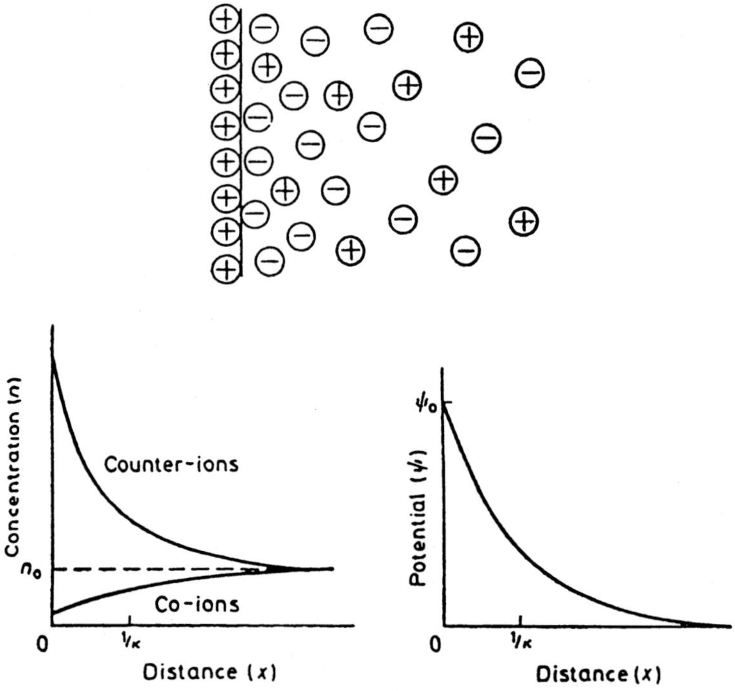

Figure 6-3. Schematic representation of a diffuse electric double layer.

The addition of electrolyte will increase the value of κ, and consequently will decrease the κ^{-1}, the **double layer thickness**, thereby compressing the double layer.

Also, γ in Equation [6-6] is

$$\gamma = \tanh \frac{ze\phi_0}{4kT} \tag{6-8}$$

or an approximation of the small surface potential is

$$\gamma \approx \frac{ze\phi_0}{4kT} \tag{6-9}$$

For large distances, the approximation of the solution from Equation [6-5] can be reduced to

$$\phi = \frac{4\gamma kT}{ze}\exp(-\kappa x) \qquad [6\text{-}10]$$

For calculation of the free energy of repulsion, ΔG_R,

$$\Delta G_R = \int_d^\infty \pi_x \, dx \qquad [6\text{-}11]$$

where d is the midpoint of the double layer on which the Maxwell stress is highest. If n here signifies the number of excess ions per unit volume, then using the van't Hoff equation,

$$\Pi = nkT \qquad [6\text{-}12]$$

Π is osmotic pressure and as before

$$n = 2n_0\left(\cosh\frac{ze\phi_d}{kT} - 1\right) \qquad [6\text{-}13]$$

or

$$\Pi_d = 2n_0 kT\left(\cos h\frac{ze\phi_d}{kT} - 1\right) \qquad [6\text{-}14]$$

After expanding the hyperbolic cosine and relating only the first term,

$$\Pi_d = 2n_0 kT\left(\frac{ze\phi_d}{kT}\right)^2 \qquad [6\text{-}15]$$

Integrating the ΔG_R and substituting, then

$$\Delta G_R = \frac{64 n_0 kT}{\kappa}\gamma^2 \exp(-2\kappa d) \qquad [6\text{-}16]$$

This is a key equation used for the DLVO theory.

6.1.2 Electrokinetic Phenomena

Four types of the electrokinetic phenomena are known which are caused by different field effects, such as electricity and so forth, at a different solid surface:

- Electrophoresis — electric field on particles with motion of dispersed phase
- Sedimentation Potential — gravitational field on particles with potential gradient
- Electroosmosis — electric field on tube wall or packed bed with motion of the medium
- Streaming Potential — motion of medium on tube wall or packed bed with potential gradient

The relative motion between a charged surface and the bulk solution causes a potential difference. In each of the four cases the shear plane between the double layer and the bulk medium is involved. The integration of net charge density out to infinity gives the total excess charge in solution per unit area. This charge is equal in magnitude but opposite in sign to the surface charge density, σ,

$$\sigma = -\int \rho \, dx$$

Another expression of Poisson's equation, in comparison with Equation [6-4], is

$$\nabla^2 \phi = -4\pi \frac{\rho}{D} \qquad [6\text{-}17]$$

due to the factor that

$$\varepsilon = \frac{q_1 q_2}{4\pi \, \varepsilon_0 D x}$$

In the last section, we solved ϕ as

$$\phi = \frac{4\gamma k T}{ze} \exp(-\kappa x) \qquad [6\text{-}18]$$

$$= \phi_0 \exp(-\kappa x)$$

After integration, Equation [6-17] and [6-18],

$$\sigma = \frac{D\kappa\,\phi_0}{4\,\pi} \qquad [6\text{-}19]$$

A **zeta potential**, ζ, is arbitrarily defined as the potential difference between a point away from the surface that is in the uniform diffuse medium and the plane of shear in relation to the double layer. To evaluate this potential, we use the same analogy as for the last equation, and therefore

$$\sigma = \frac{D\,\kappa\,\zeta}{4\,\pi} \qquad [6\text{-}20]$$

In electrophoresis, an electric potential gradient in solution is applied to determine the velocity of the charged particles. Then the force exerted on the surface per square cm is balanced by the viscous drag, and

$$\sigma F = \eta u \kappa \qquad [6\text{-}21]$$

where

F = field in voltz/cm

u = velocity

η = viscosity of the solution

or

$$u = \zeta DF/4\pi\eta \qquad [6\text{-}22]$$

In other words, the ζ potential or the $\sigma\kappa^{-1}$ — the electric moment per cm^2 — is proportional to the velocity per unit field. If the surface is located in a tube wall or packed bed, the mobile diffuse layer that moves under an electric field, will generate electroosmosis,

$$U = \pi\,r^2 u \qquad [6\text{-}23]$$

where

U = volume flow in mL/sec
r = radius of tube

then

$$U = \frac{r^2 \eta DF}{4\eta} \qquad [6\text{-}24]$$

Experimentally, a field of about 1500 V/cm created by electrodes is needed to produce a velocity of 1 cm/sec of air bubbles in the capillary at the return path for a zeta potential of 100 mV in water at 25°C. The preceding formula is used to determine the zeta potential. Zeta potential has many environmental applications; for example, the ζ values of bacteria can be an important factor for their transport in porous media as shown in Table 6-1.

The reverse of electroosmosis is used so that a potential can be induced, and this potential is termed the **streaming potential**, E. If a stream line of flow is assumed, then by Poiseuille's equation the velocity at a radius x from the center of the tube is

$$u = \frac{P(r^2 - x^2)}{4\eta l} \qquad [6\text{-}25]$$

where

P = pressure
l = length of the tube

The double layer is centered at

$$x = r - \kappa^{-1} \qquad [6\text{-}26]$$

and its velocity is

$$u_d = \frac{\kappa^{-1} rP}{2\eta l} \qquad [6\text{-}27]$$

Table 6-1. Zeta Potential of *Bacillus subtilis* and *Pseudonomas putida* (ATCC 12633 as determined from the electrophoretic mobility measurements)

Species	Cultural Age	Suspending Medium	Zeta Potential (mV)
Bacillus subtilis	18 hrs	el*	33.6
a spore-forming Gram positive	18 hrs	A**	27.1
facultative anaerobe	18 hrs	B***	37.8
	3 days	A	47.0
	3 days	G****	51.8
	3 days	B	63.28
Pseudomonas putida	12 hrs	el	53.9
a nonsporeforming Gram negative	12 hrs	B	52.4
aerobe	15 hrs	G	58.3
	15 hrs	A	58.7
	3 days	A	56.9

* Electrolyte solution containing 1,000 ppm NaCl

** Buffer solution containing 0.0663 M KH2PO4 and 0.0267 M NaOH, pH=7.0

*** Buffer solution containing 0.0127 M KH2PO4 and 0.0054 M NaOH, pH=7.0

**** Cells remained in growth medium without further separation and resuspension in synthetic solutions

Source: L.K. Jang and T.F. Yen, *International Bioresources J.*, **1** 226–246 (1985)

When the κ^{-2} term is neglected, the current due to the motion of the double layer is

$$i = 2\pi r \sigma u_d \qquad [6\text{-}28]$$

or

$$i = \frac{\pi r^2 \sigma \kappa^{-1} P}{\eta l} \qquad [6\text{-}29]$$

The conductance of the liquid is

$$C = \frac{\pi r^2 k}{l} \qquad [6\text{-}30]$$

and by Ohm's law, after substituting Equation [6-20],

$$E=\frac{i}{C}=\frac{\kappa^{-1}\sigma P}{\eta k}=\frac{\zeta PD}{4\pi\eta k} \qquad [6\text{-}31]$$

From streaming potential data, one can also determine the zeta potential.

6.1.3 Hamaker Constants

London has used a simple harmonic oscillator theory to treat the dispersion energy or the attraction between two molecules separated by large distances.

$$u(r)=-\Lambda_{ab}r^{-6}$$

and

$$\Lambda_{ab}=\frac{3}{2}\left(\frac{h\nu_a h\nu_b}{h\nu_a+h\nu_b}\right)\alpha_a\alpha_b \qquad [6\text{-}32]$$

where

ν = characteristic frequency

α = limiting polarizability

h = Planck's constant

Λ = London coefficient

Hamaker took London's expression of attraction for two molecules and integrated it for all the molecules in two separate particles to obtain the Gibb's free energy due to dispersion interactions,

$$\Delta G_{12}^{d}=-\int_{r_1}dv_1\int_{r_2}\frac{\Lambda_{12}\rho_1\rho_2}{r_{12}^{6}}dv_2 \qquad [6\text{-}33]$$

where

v = volume element

ρ = molecular number densities (molecules per unit volume)

For planer parallel plates (slabs) with slab thicknesses of b_1 and b_2 separated by a distance of H, the free energy per unit area is

$$\Delta G_{12}^{d} = -\frac{\pi \Lambda_{12} \rho_1 \rho_2}{12}\left[H^{-2} - (H + b_1 + b_2)^{-2} - (H + b_1)^{-2} - (H + b_2)^{-2}\right] \quad [6\text{-}34]$$

now set

$$A_{12} = \pi^2 \rho_1 \rho_2 \Lambda_{12} \quad [6\text{-}35]$$

If b_1 and b_2 are also large, then

$$\Delta G_{12}^{d} = -\frac{A_{12}}{12\pi H^2} \quad [6\text{-}36]$$

A_{12} is termed the Hamaker constant between two particles. If there is an intervening substance between two bodies, say 1 and 3, then by the Archimedean buoyancy principle,

$$A_{123} = A_{13} + A_{22} - A_{12} - A_{23} \quad [6\text{-}37]$$

If the intervening substance is between similar particles, then

$$A_{121} = A_{11} + A_{22} - 2A_{12} \quad [6\text{-}38]$$

From Berthelot's principle,

$$A_{12} = (A_{11} A_{22})^{\frac{1}{2}} \quad [6\text{-}39]$$

Therefore, you can also include

$$A_{123} = \left(A_{11}^{\frac{1}{2}} - A_{22}^{\frac{1}{2}}\right)\left(A_{33}^{\frac{1}{2}} - A_{22}^{\frac{1}{2}}\right) \qquad [6\text{-}40]$$

Some Hamaker constants A_{121} for nonretarded interaction across a film of water are illustrated by Table 6-2.

The summation of ΔG_{12} and ΔG_R — Equations [6-36] and [6-16] — is called the total potential energy that can be plotted in terms of H, the distance between two particles, a particle and a surface, or two surfaces. A typical example is demonstrated by Figure 6-1.

Table 6-2. Hamaker Constants A_{121} for the Nonretarded Interaction across a Film of Water

Substance	Hamaker Constant, A_{121} (J) ($\times 10^{20}$)
Gold	33.5
Silver	26.6
Germanium	16.0
Silicon	11.9
Selenium	4.77
Alumina	4.12
Magnesia	1.60
Crystalline quartz	1.70
Fused quartz	0.833
Fused silica	0.849
Calcite	2.23
Calcium fluoride	1.04
Sapphire	5.32
Ionic crystals	0.31 – 4.85
Oxides	1.76 – 4.17
Metals	3 – 33.4
Polymers	(0.35)
Polyisoprene	0.743
Polymethyl methacrylate	1.05
Polystyrene, average of 2	0.931
Polystyrene	0.27

Table 6-2. Continued

Substance	Hamaker Constant, A_{121} (J) ($\times 10^{20}$)
Polystyrene	0.35
Polytetrafluoroethylene	0.333
"Teflon FEP"	0.381
Polyvinyl chloride	1.30
Pentane	0.336
Hexane	0.360
Heptane	0.386
Octane	0.410
Nonane	0.435
Decane	0.462
Undecane	0.471
Dodecane	0.502
Tridecane	0.504
Tetradecane	0.514
Pentadecane	0.526
Hexadecane	0.540

6.2 DESTABILIZATION

Coagulation is a process in which chemicals are added to an aqueous system for the purpose of enhancing a rapid settling of aggregates out of finely divided, dispersed particulates with very slow or negligible settling velocities. The particulates are within the colloidal sizes, and their stabilities are governed by colloidal chemistry principles. These particulates possess the electrokinetic property causing double layer formation. Both hydrophilic and hydrophobic colloids have surface charges. **Hydrophilic colloids** are of biological origin (such as proteins) and usually consist of water-soluble functional groups, such as $-NH_2$, -COOH, -OH, SO_3H, and $-OPO_3H_2$. For example, for a protein particle,

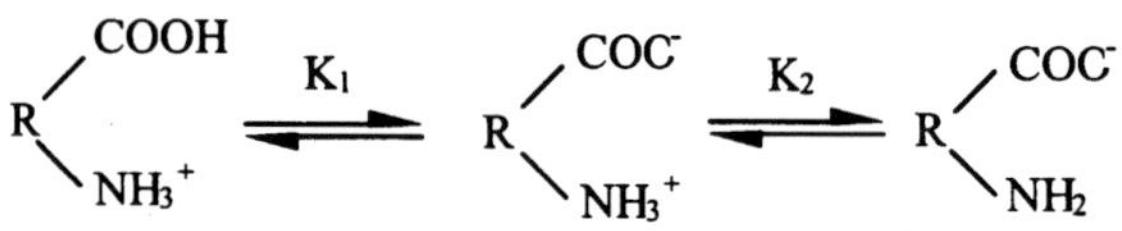

At low pH levels the surface is positively charged, while at high pH levels the surface is negatively charged. At some intermediate pH value, the net surface charge is zero. This pH value is usually referred to as the **isoelectric point** (IEP). It is routinely measured by adsorption of the net charges. For **hydrophobic colloids** such as clay, silt, and silica, surface charges occur at face boundaries or isomorphous replacements within the lattice due to lattice imperfections.

Evidently, if there is surface group of silanol, ≡SiOH, then

$$\equiv \mathrm{Si-OH_2^+} \xleftrightarrow{K_1} \mathrm{Si-OH} \xleftrightarrow{K_2} \mathrm{Si-O^-}$$

Most metal oxides and hydroxides are amphoteric, but they can adsorb H^+ or OH^- ions or their complexes. The pH at which that portion of the charges becomes zero due to H^+ or OH^- ions or their complexes is termed the **zero point of charge** (ZPC) and is represented by pH_{ZPC}. Usually, it is determined by alkalimetric-acidimetric titration curves. Some experimental values of pH_{ZPC} are listed in Table 6-3.

Surface charges can also originate from a solute coordinated to bind to a solid surface. For example,

$$\mathrm{Cu(s) + 2H_2S = Cu\ (SH)_2^{2-}\ (surface) + 2H^+}$$

Table 6-3. Zero Point of Charge

Material	pH_{zpc}	Material	pH_{zpc}
α-Al_2O_3	9.1	δ-MnO_2	2.8
α-$Al(OH)_3$	5.0	β-MnO_2	7.2
γ-AlOOH	8.2	SiO_2	2.0
CuO	9.5	$ZrSiO_4$	5
Fe_3O_4	6.5	Feldspars	2-2.4
α-FeOOH	7.8	Kaolinite	4.6
γ-Fe_2O_3	6.7	Montmorillonite	2.5
$Fe(OH)_3$ (amorph)	8.5	Albite	2.0
MgO	12.4	Chrysotile	>12

The values are from different investigators who have used different methods and are not necessarily comparable. They are given here for illustration.

$$MnO_2\,H_2O\,(s) + Zn^{2+} = MnOOHOZn^{+}\,(surface) + H^{+}$$

Surfactant ions can be adsorbed by a surface in this manner. These charges of particulates are essential to the understanding of colloidal stability. The following discussion will focus on the mechanism of destabilization.

6.2.1 Repression of Double Layer

From the DLVO model, the space charge density of ions in solution decreases rapidly with distance from the surface (Fig. 6-4), and the potential declines exponentially as a function of distance. From the Gouy-Chapman formulation of a single flat double layer, it is possible to derive a double layer equation from Poisson's equation with charges following Boltzmann's distribution.

$$\frac{d^2\psi}{dx^2} = \frac{\kappa^2 \sin h\left(\frac{zF\psi}{RT}\right)}{\frac{2F}{RT}} \qquad [6\text{-}41]$$

where ψ is potential in V, x is distance from the surface in cm, z is valence of the ion, and F is Faraday $[(6 \times 10^{23})(e) = 96490\ \text{C eq}^{-1}]$. The symbol κ represents the reciprocal Debye length, which is the reciprocal double layer thickness, such that

$$\kappa = \frac{e^2 \sum n_i z_i^2}{\varepsilon \varepsilon_0 kT} \qquad [6\text{-}42]$$

where e is electron charge, which is 1.6 × 10–19 C, ε is dielectric permittivity for water, which is 80, and ε0 is permittivity in a vacuum, which is 8.854 × 10–14 CV–1cm-1. At 20°C, this relation can be evaluated as κ–1 ~ (2.8 × 10–8)(μ–½) cm. For water and monovalent electrolyte, κ–1 is about 10 Å at 0.1 M and 100 Å for 1 × 10–3M. Some salt concentrations may be small enough for the two particles to approach each other and result in aggregation.

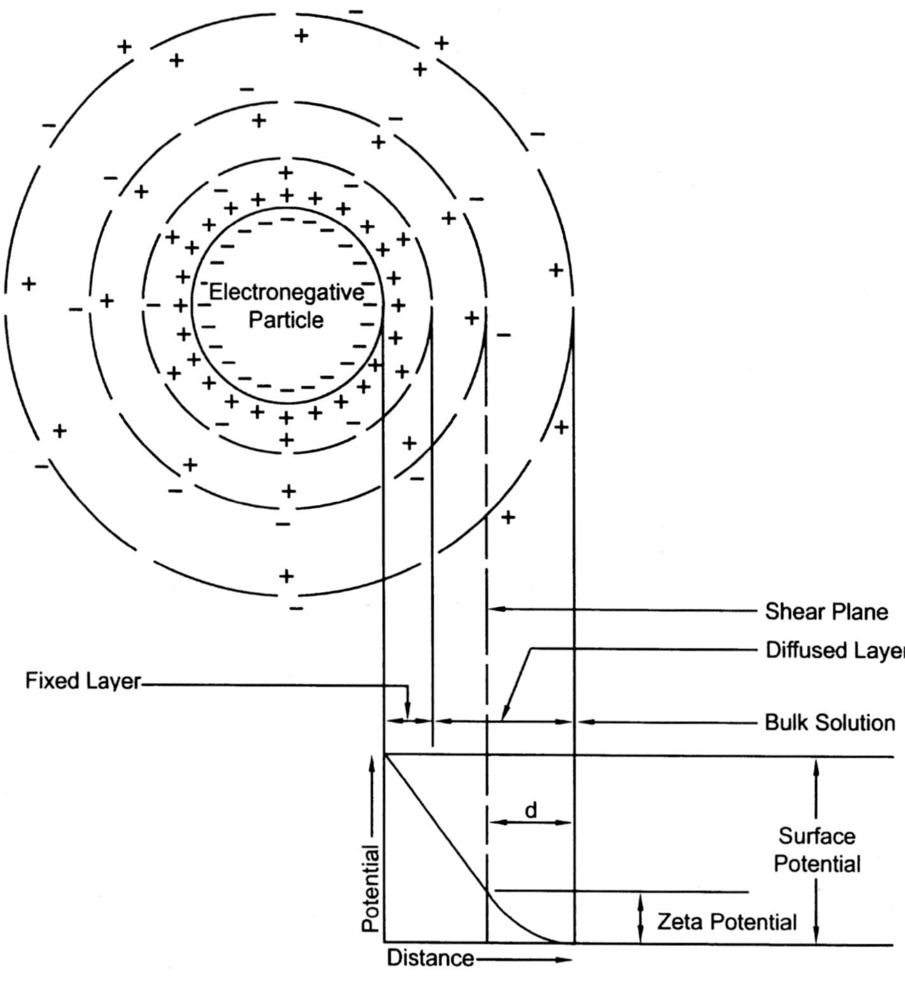

Figure 6-4. A negative colloidal particle with its electrostatic field. (Adapted from C.P. Preisting, A theory of coagulation useful for design, *Ind. Eng. Chem.* 54(8), 38(1962).)

The ionic strength is important, and since even the electrolyte concentration remains the same, the higher valence of counterions will shorten the double layer thickness. An empirical relation called the **Schulze Hardy rule** states that coagulation efficiency of counterions increases markedly with valency. This experimental observation is shown in Figure 6-5a. The trivalent ion is more efficient than the divalent ion, and the divalent ion is more efficient than the monovalent ion. Let us see if we can define this relationship more clearly.

Assume that there is van der Waals attraction energy, V_A, and double-layer Coulombic repulsive energy, V_R. The interaction energy is just a simple additive term, V_T, and:

$$V_T = V_R + V_A \leq 0 \qquad [6\text{-}43]$$

Furthermore, assuming that the approaching particulate is simply a flat plate interacting with a flat plate surface permit area, then

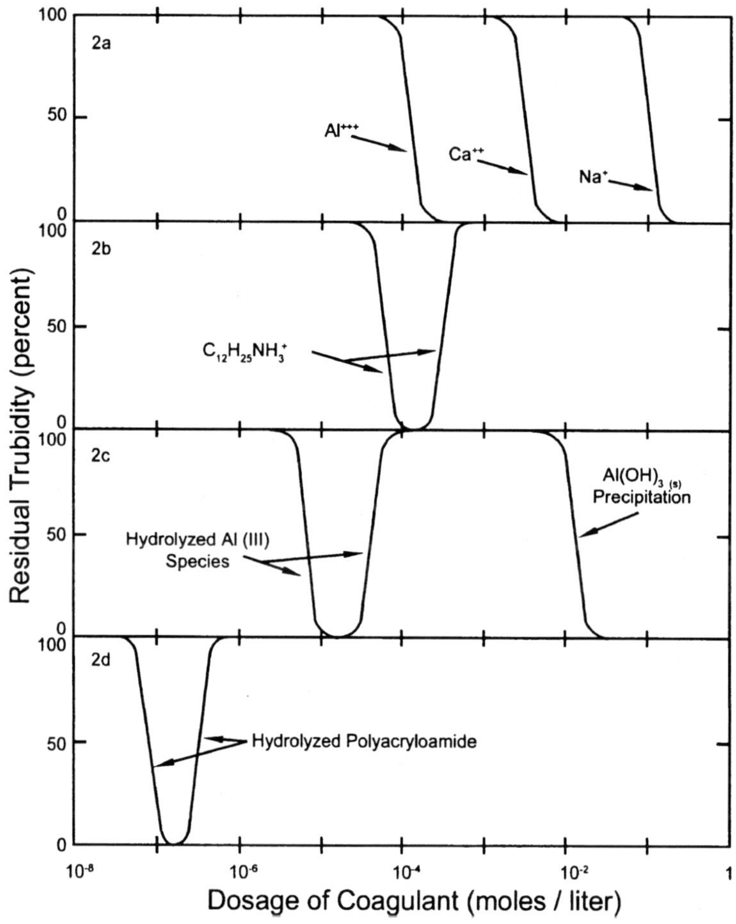

Figure 6-5. Schematic agglomeration curves for several different destabilizing agents. (From Ref. 6-1.) (x-axis is concentration)

$$V_R = \frac{64 n_s kT}{\kappa}\left[\tan h\left(\frac{ze\psi d}{4kT}\right)\right]^2 e^{-kd} \qquad [6\text{-}44]$$

$$V_A = -\frac{A_{11(2)}}{12\pi(d+2\delta)^2} \qquad [6\text{-}45]$$

Here, n_s is the number of molecules cm^{-3}, d is the interaction distance between two surfaces in nm, δ is the Stern layer thickness in nm, and $A_{11(2)}$ is the Hamaker constant that applies for interaction between surfaces 1 and medium 2 in J, or

$$A_{11(2)} = \left(A_{11}^{\frac{1}{2}} - A_{22}^{\frac{1}{2}}\right)^2 \qquad [6\text{-}46]$$

With a condition of minimum interaction with distance,

$$\frac{\partial V_T}{\partial d} = 0 \qquad [6\text{-}47]$$

we can evaluate the exact concentration of electrolyte required to coagulate colloids at 25°C, C_s, as

$$C_s = 8 \times 10^{-36} \frac{\left[\tan h\left(\frac{ze\psi d}{4kT}\right)\right]^4}{A_{11(2)}^2 z^6} \qquad [6\text{-}48]$$

or

$$C_s \propto z^{-6} \qquad [6\text{-}49]$$

From this simple relation, we can predict that the concentration of counterions necessary for coagulation with monovalent, divalent, and trivalent ions in order to achieve a similar effectiveness based on z^{-6} is mono:di:tri = 100:1.6:0.13. Again this simple relation has been verified in Figure 6-5a, as indicated by the concentration profile. In some hydrophilic colloids, the effectiveness of added electrolytes depends on the co-ions rather than the counterions. This phenomenon is known as salting-out. Instead of cations, the anions added are to compete for the charged particles for bound water formation. The Hofmeister series is an empirical rule, which lists the following anions in a descending order of effectiveness for coagulation: $SO_4^{2-} > Cl^- > NO_3^- > I^-$.

6.2.2 Charge Neutralization and Adsorption

For a typical monovalent counterion and a particle that has a potential across a diffusion layer of 100 mV, the Coulombic electrostatic potential involved is only about 2 kcal/mole. This is quite weak considering covalent, coordinate, or even hydrogen bonds. Therefore, events and interactions (e.g., as coagulant-colloid, coagulant-solvent, and colloid-solvent) become important when compared with the Coulombic energy. Such cases can be seen as surfactant-like molecules (e.g., dodecylammonium chloride) (Fig. 6-5b). As indicated, if it is based on the monovalent counter ion, the effectiveness of 3.5 orders of magnitude cannot be explained. This must be due to the long-chain tails of the counterion. When a sufficient number of counterions is adsorbed, charge reversal occurs and restabilization occurs. Note that destabilization-restabilization occurs within a short range of concentration. One way to view these zig-zag, long-chain counterions is that they attach to the colloid within the Stern layer, so that effective charge outside the shear layer is reduced, in contrast to the double-layer repression, which alters the charge distributions within the diffuse layer.

6.2.3 Entrapment in Precipitates

The Fe^{3+} and Al^{3+} salts are used as coagulants and destabilizers because they form polynuclear hydrolysis products, $M(OH)_n^{z+}$, which are adsorbed readily at particle water interfaces. The adsorption of cations and anions at hydrous metal oxide interfaces can be interpreted as surface complex formation and ligand exchange equilibrium. In other words, the hydrous metal oxides are amphoteric

$$\equiv M\mathrm{OH}_2^+ = \equiv M\mathrm{OH} + \mathrm{H}^+; \; K_{a_1} \qquad [6\text{-}50]$$

$$\equiv M\mathrm{OH} = \equiv M\mathrm{O}^- + \mathrm{H}^+; \; K_{a_2} \qquad [6\text{-}51]$$

upon reacting with cations and anions in a coordination reaction, as shown in Figure 6-6, such as:

$$\equiv MOH + M^{z+} = \equiv MOM^{z\text{-}1} + H^{+};\ {*K_1} \quad [6\text{-}52]$$

$$2 \equiv MOH + M^{z+} = (\equiv MO)_2 M^{z-2} + 2H^{+};\ {*\beta_2} \quad [6\text{-}53]$$

Similarly,

$$\equiv MOH + A^{z-} = \equiv MA^{z-1} + OH^{-};\ K_1 \quad [6\text{-}54]$$

$$2 \equiv MOH + A^{z-} = (\equiv MO)_2 A^{z-2} + 2OH^{-};\ \beta_2 \quad [6\text{-}55]$$

Noted also for protonated anions, the ligand exchange may be accompanied by deprotonation of the ligand at the surfaces:

$$\equiv MOH + HPO_4^{2-} = \equiv MHPO_4^{-} + OH^{-} = \equiv MPO_4^{2-} + H_2O \quad [6\text{-}56]$$

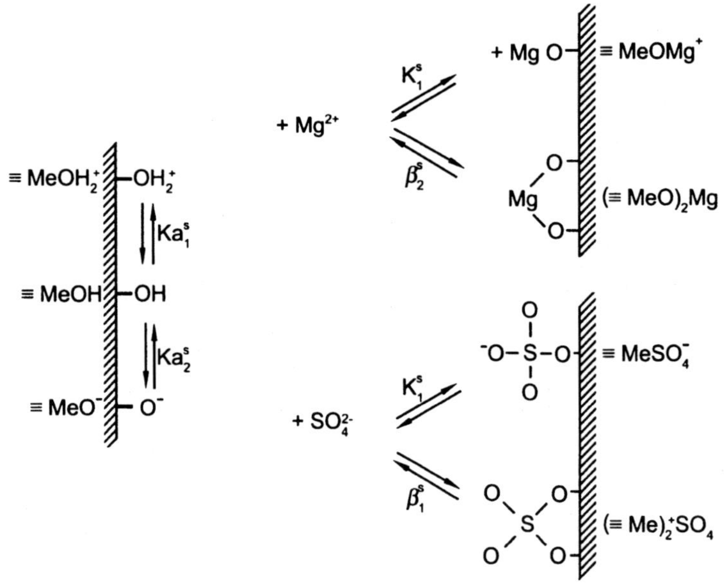

Figure 6-6. Interactions of hydrous oxides with cations and anions can be interpreted in terms of surface complex formation and ligand exchange equilibria. (After W. Stumm, "Chemical Interaction in Particle Separation," *Chemistry of Wastewater Technology*, A.J. Rubin, ed., Ann Arbor Science, MI, 1978.)

In summary, all the above reactions can be illustrated by Figure 6-7, and one can conclude that the adsorption of cations by hydrous metal oxides will lower the pH of the ZPC while simultaneously shifting the IEP to a higher pH value. On the other hand, the adsorption of anions will increase the pH of the ZPC while lowering the pH of the IEP. This finding is important since adsorption at the hydrous metal oxides sites will modify the net electric charges. At zero charge (which is at pH of the IEP), the maximum settling velocities of the suspended particles are observed. In this manner, the actual design of the experiment can be predicted (see Fig. 6-8 for the case of Mg^{2+} and SO_4^{2-} on Al_2O_3). Other charged particles can be similarly enmeshed in hydrous metal oxides. This is further established in Figure 6-5c, where high-dosage coagulation is brought out by the enmeshment of colloidal particles in the precipitate of $Al(OH)_3$.

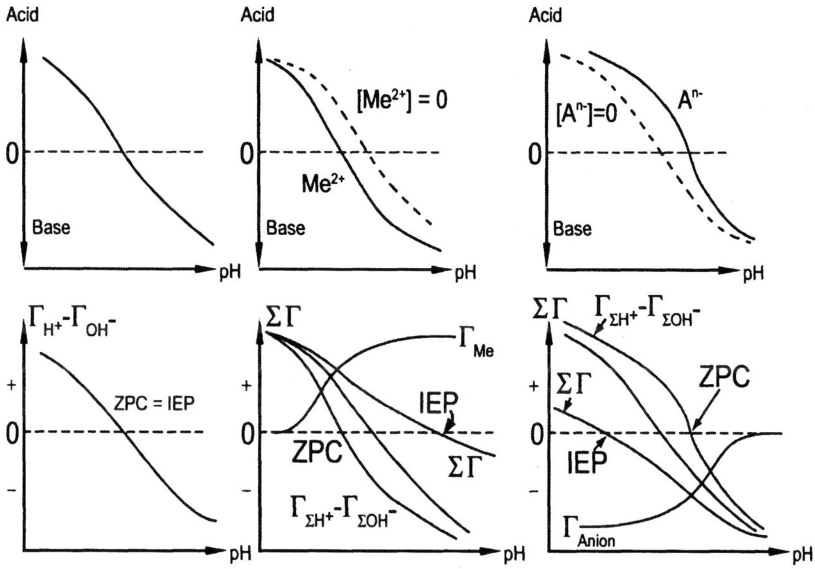

Figure 6-7. The net charge at the hydrous oxide surface is established by the proton balance (adsorption of H+ or OH– and their complexes) at the interface and specific bound cations and anions. This charge can be determined from an alkalimetric-acidimetric titration curve and from a measurement of the extent of adsorption of specifically adsorbed ions. Specifically adsorbed cations (anions) increase (decrease) the pH of the isoelectric point (IEP) but lower (raise) the pH of the zero point of charge (ZPC). (After Stumm)

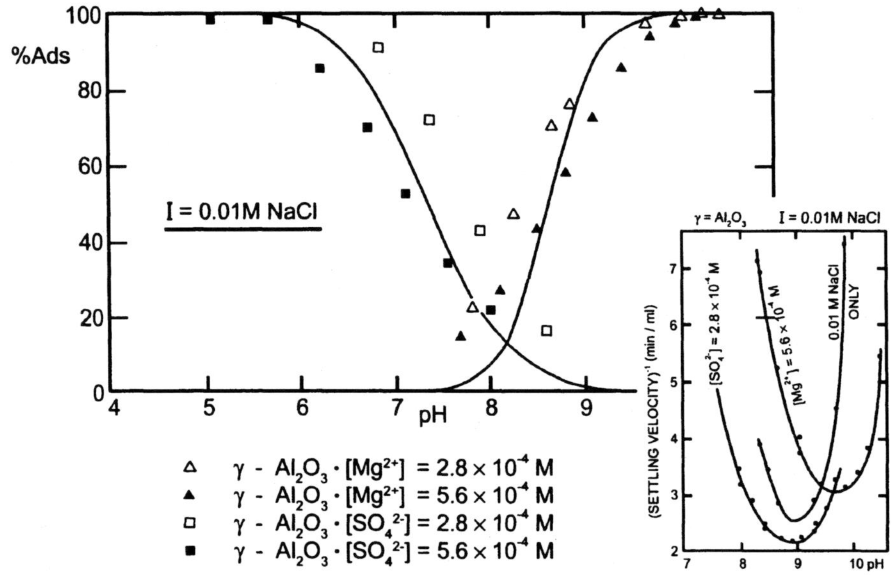

Figure 6-8. Extent of specific adsorption of SO_4^{2-} and Mg^{2+} on Al_2O_3. The drawn-out lines have been calculated from experimentally determined equilibrium constants for the adsorption from a 5.6 × 10^{-4} M $MgSO_4$ solution. As shown by the insert, Mg^{2+} causes an increase, and SO_4^{2-} a decrease in the isoelectric point (maximum settling rate).

6.2.4 Interparticle Bridging

Many synthetic polymers that bear ionizable functional groups and are usually soluble in water are termed polyelectrolytes. They can be either cationic, anionic, or ampholytic, but all have macromolecular structures with flexible chains and charges spreading over extensive areas. Humic acid acting as a naturally occurring polymer can also modify the surface of mineral particles (Fig. 6-9). The interaction with colloidal particles behaves as a sweeping action with subsequent floc formation. Restabilization can occur (see Fig. 6-5d), not through charge reversal, but with (a) excess polymer, (b) no particles available, (c) prolonged agitation, or (d) intramolecular adsorption by folding. All these possibilities are well illustrated in Figure 6-10 by the restabilization reactions.

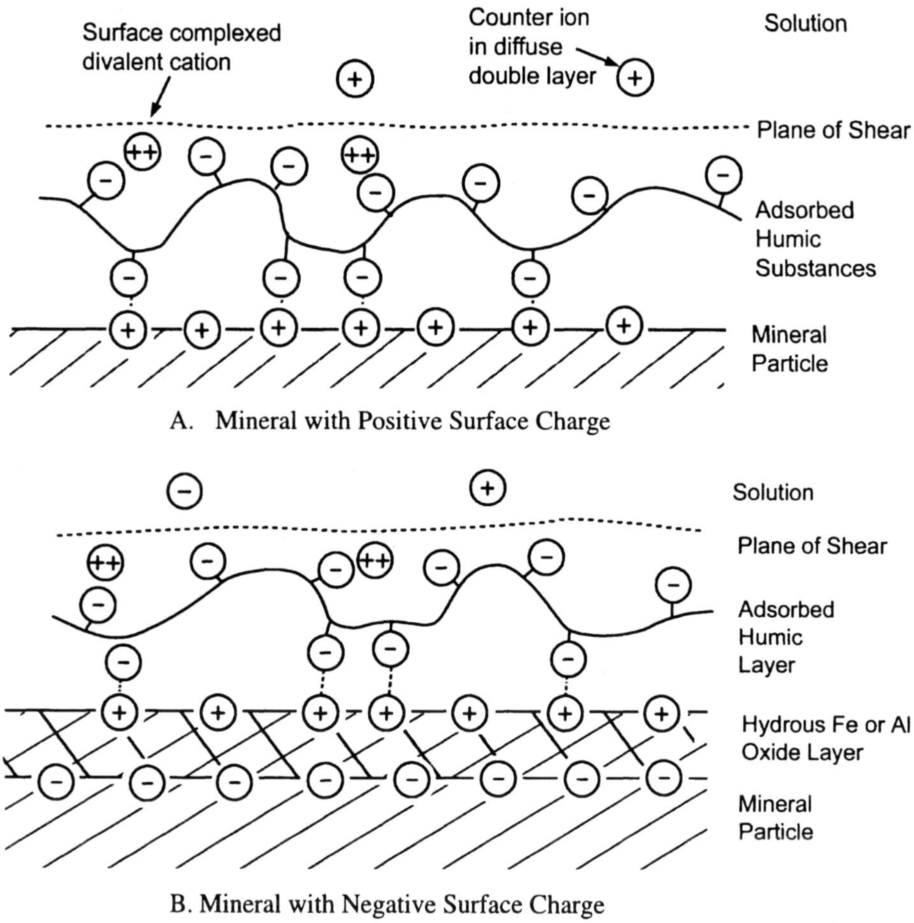

Figure 6-9. Schematic diagram depicting the role of adsorbed humic substances and divalent cations in modifying the surface charge of mineral particles.

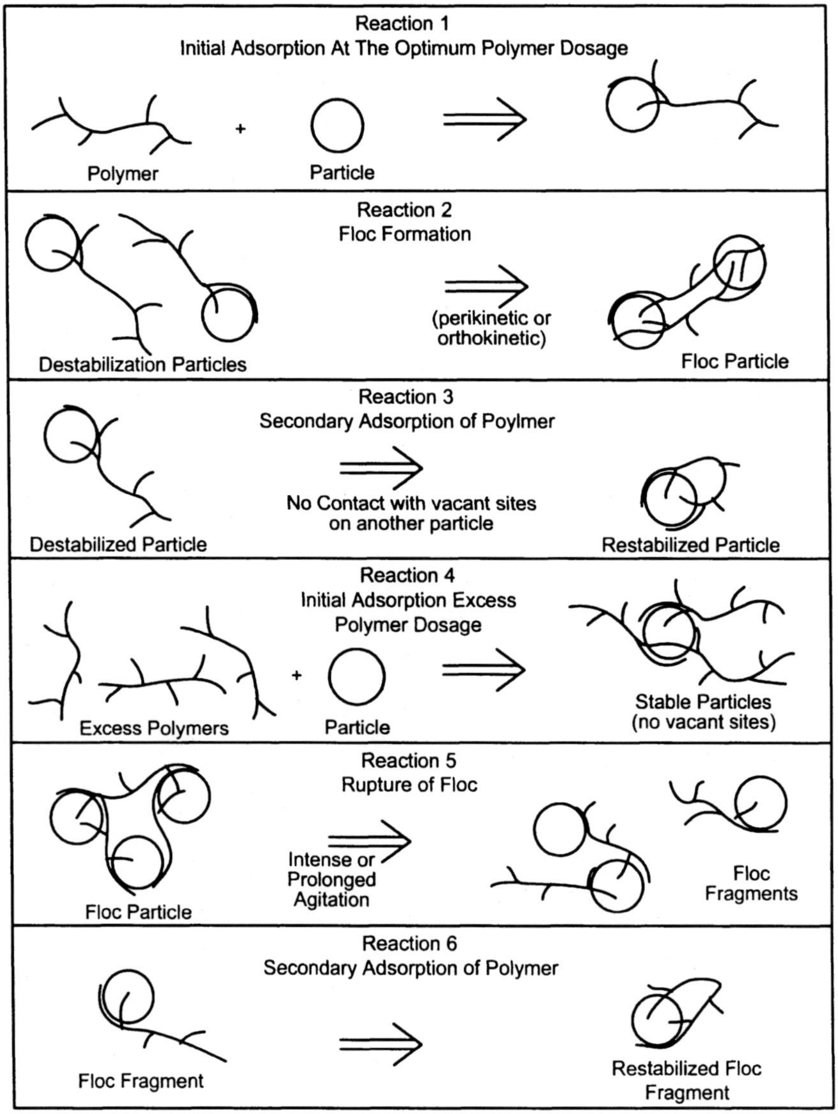

Figure 6-10. Schematic representation of the bridging model for the destabilization of colloids by polymers. (Adapted from C.R. O'Melia).

6.3 DESTABILIZATION CHEMICALS (COAGULANTS)

Filter alum (aluminum sulfate) is the most popular coagulant used for water treatment. In water the alum can use alkalinity as follows:

$$Al_2(SO_4)_3 \cdot xH_2O + 6HCO_3^- = 2\,Al(OH)_3 + 3\,SO_4^{2-} + 6CO_2 + xH_2O$$

The value of x for the filter alum can vary from 13 to 18. Note that in the above equations, for every mole of alum consumed six moles of carbon dioxide are produced. The net result is a reduction of pH. The decrease of pH will affect the promotion of dissociation of hydrous metal oxides due to the decrease of OH^- concentration by mass action:

$$M(\mathrm{OH})_n = M^{n+} + n\mathrm{OH}^-$$

The free cations enter the double layer of the particles in wastewater to reduce the potential. The colloidal particles thus neutralized can make contact, thereby causing flocculation. In the case where insufficient bicarbonate is present in water, alkalinity must be added for an alum reaction to occur. For example, if the added alkalinity is OH^-,

$$Al_2(SO_4)_3 \cdot x\,H_2O + 6OH^- = 2Al(OH)_3 + 3SO_4^{2-} + x\,H_2O$$

Copperas (ferrous sulfate) is another common coagulant. When copperas is introduced to water, the pH of water falls in the neutral range:

$$FeSO_4 \cdot 7H_2O + 2OH^- = Fe(OH)_2 + SO_4^{2-} + 7H_2O$$

The formation of ferrous hydroxide precipitate is obviously promoted by a pH increase. For copperas used in water treatment devoid of oxygen, the pH must be higher than 9.5. This effect is achieved by the addition of lime. However, when oxygen is present, the ferrous hydroxide is oxidized into ferric hydroxide, which is more insoluble:

$$4Fe(OH)_2 + O_2 + 2H_2O = 4Fe\,(OH)_3$$

In this manner, copperas is used as a coagulant when oxygen is sufficient. In the same way, chlorine is used as an oxidant:

$$6FeSO_4 \cdot 7H_2O + 3Cl_2 = 6Fe^{3+} + 6Cl^- + 42H_2O + 6SO_4^{2-}$$

and

$$Fe^{3+} + 3OH^- = Fe(OH)_3$$

Coagulants are able to remove phosphate as well as suspended solids. This can be done in reversal points, such as before the primary clarification. For example, alum, ferric chloride, and lime are used for phosphate removal:

$$Al^{3+} + PO_4^{3-} = AlPO_4 \downarrow, pk_s = 21$$

$$Fe^{3+}\ PO_4^{3-} = FePO_4 \downarrow, pk_s = 26$$

$$5Ca^{2+} + 3PO_4^{3-} + OH^- = Ca_5(OH)(PO_4)_3 \downarrow, pk_s = 57$$

The last compound is hydroxylapatite, which has the highest pk_s value.

Another important class of coagulants includes synthetic polymers and some natural polymers. The nonionic polymers include polyethylene oxide (PEO) and polyacrylamide (PAM):

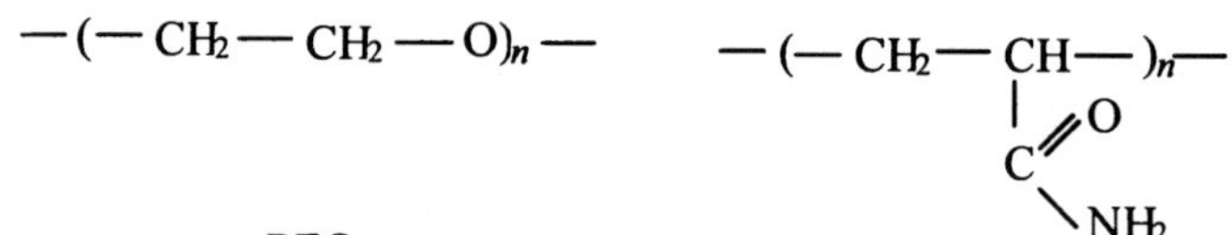

PEO PAM

The latter can be partially hydrolyzed in water to save polyacrylic acid (PAA). Usually the ratio of PAM to PAA can be controlled.

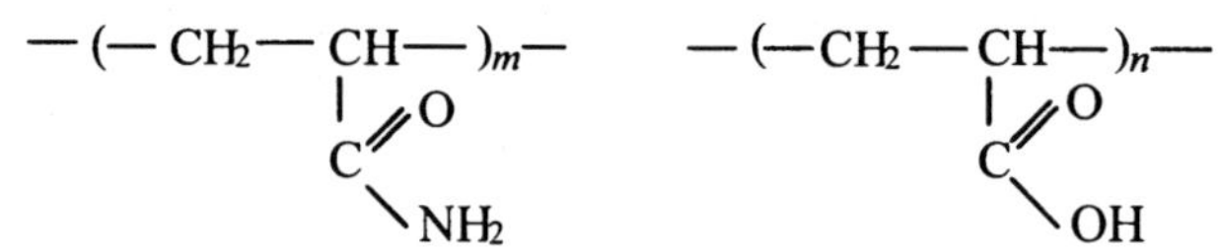

PAM PAA

PAA is an anionic polyelectrolyte. Polystyrene sulfonate (PSS) or lignin sulfonate also belongs to this class. The latter is a natural polymer. This is also true for humic substances that contain polycarboxylic acid groups. This is most important because humic substances are ubiquitous in all natural waters (see Fig. 6-9).

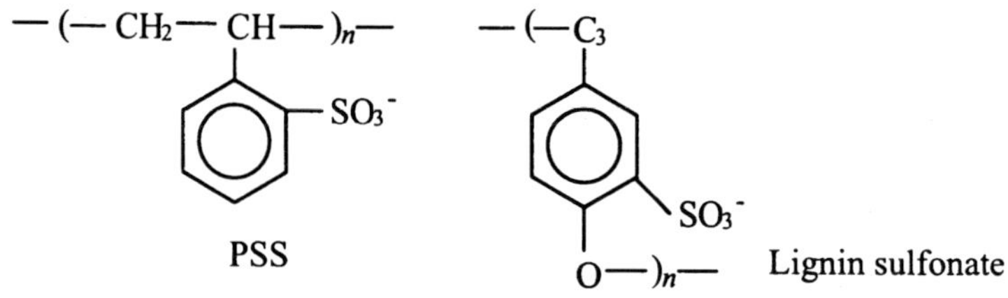

Humic acid is definitely linked with lignin (see Fig. 6-11). For some proposed structures of humic substance, see Figure 6-12.

The most important cationic polyelectrolyte is Cat-Floc, otherwise known as polydiallyl-dimethylammonium (PDADMA). Cat-Floc is formed by ring closure polymerization:

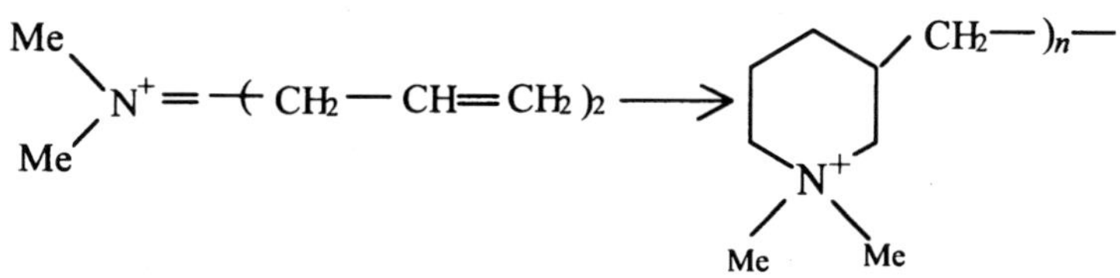

Others include the ionenes, which can be easily synthesized from:

$$Me_2{=}N{-}(CH_2)_n N{=}Me_2 + X(CH_2)_m X \longrightarrow -(-{}^{+}N{-}(CH_2)_n{-}N^{+}{-}(CH_2)_m{-})_p-$$

Difunctional tertiary amine and difunctional halide can be polymerized to yield *n,m*-ionene, such as 3,3-ionene and 3,4-ionene.

A new process called SIROFLOC has been developed that utilizes both inorganic and organic surfaces. In this process, magnetite particles are used as base destabilizers, while other polyelectrolytes are adhered to the surface of the base material as shown in Figure 6-13. Magnetite is used because it can be easily regenerated. This process is illustrated in Figure 6-14.

Polymers are used for evaluation, including polymers of the epichlorohydrin/amine type, which can have a fishnet structure due to cross-linking of the polymer.

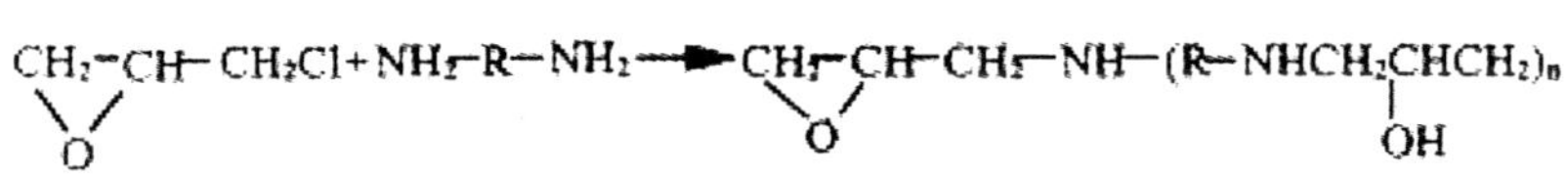

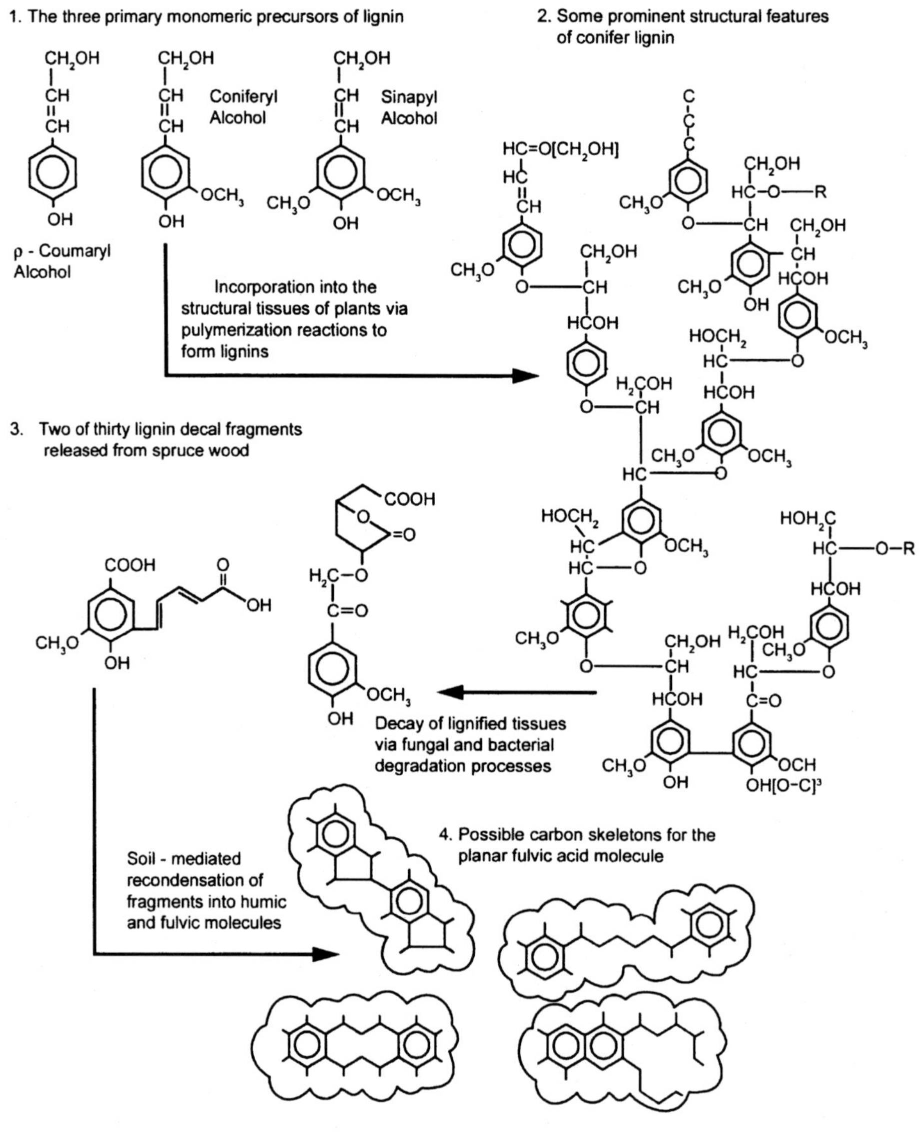

Figure 6-11. A possible schematic pathway for the formation of the terrigenous humic acids from lignins. Source: Crawford, 1981; Schnitzer and Khan, 1972; M.M. Morel, 1983.

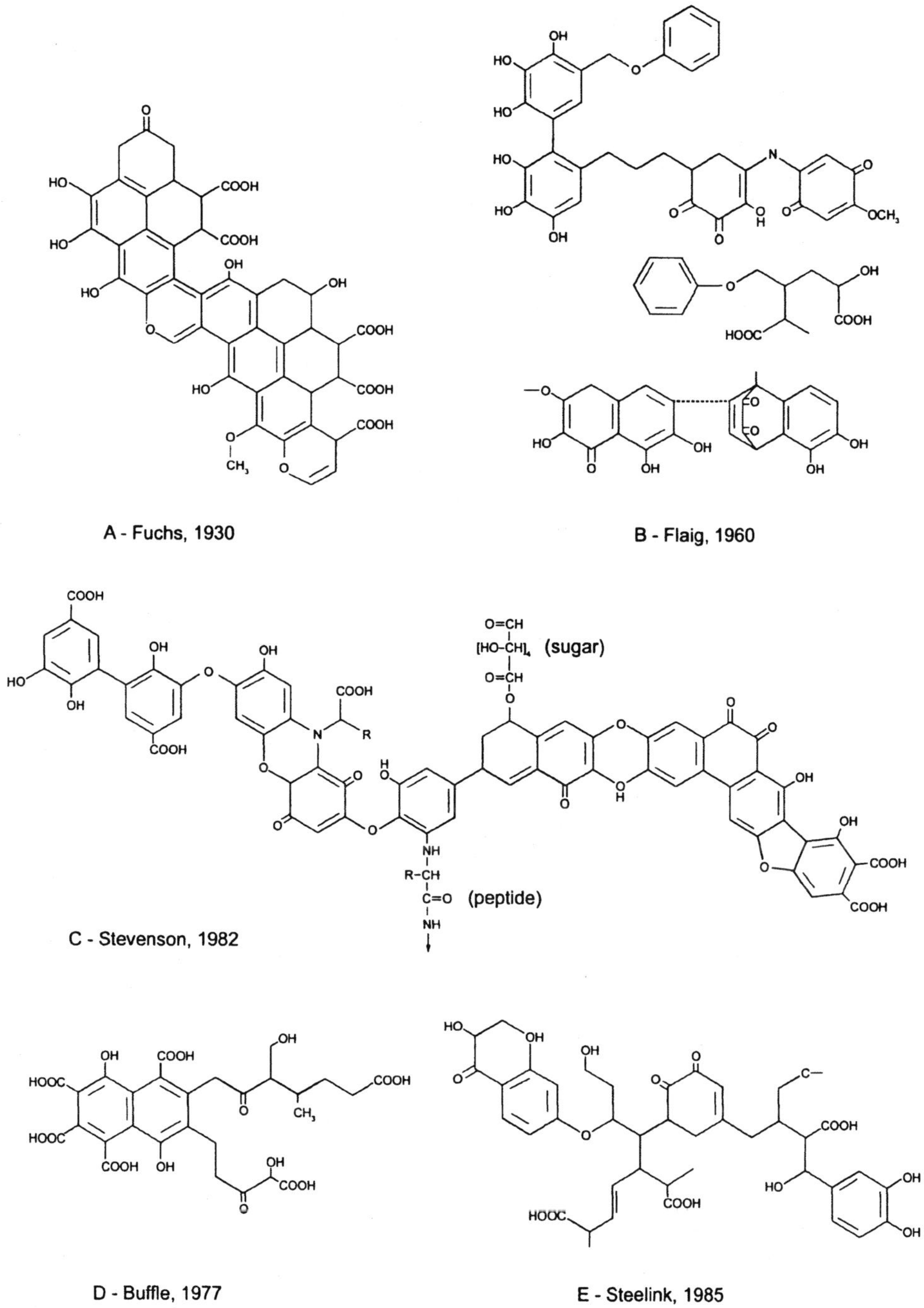

Figure 6-12. Structure of humic substances as proposed by various authors.

With a small amount of trifunctional amines, this type of epoxy resin can be a three-dimensional network or an extensive fishnet.

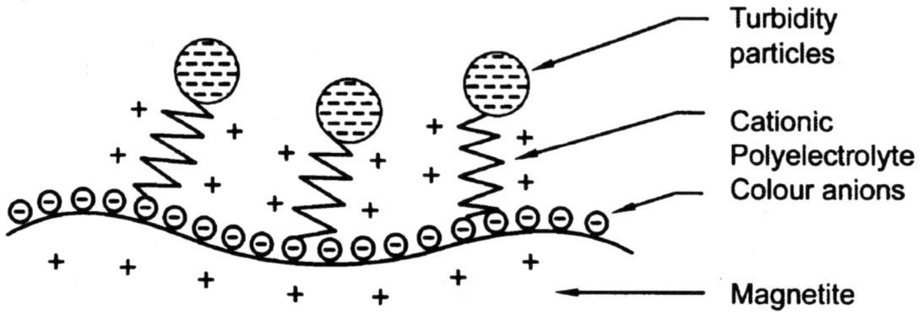

Figure 6-13. Diagrammatic representation of color and turbidity removal by magnetite.

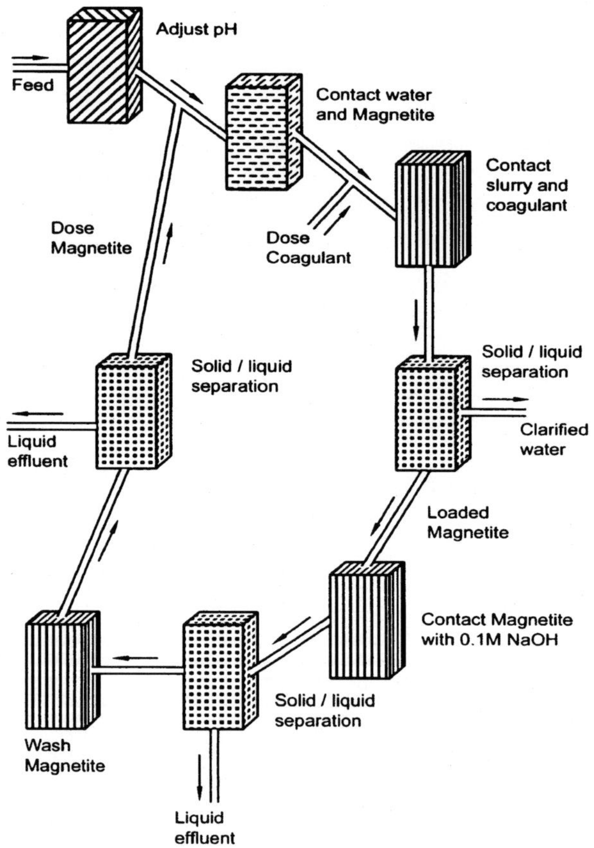

Figure 6-14. SIROFLOC process block diagram.

6.4 FLOCCULATION AND SEDIMENTATION

Often flocculation and coagulation are used interchangeably; nevertheless, a distinction should be made. **Coagulation** actually refers to the process of destabilization of colloidal particles, as discussed previously. During this process, chemicals such as coagulants are introduced. **Flocculation** refers to the postdestabilization process in which aggregations and large flocs are formed. The process will involve particle collision due to a rapid stirring either by Brownian motion or by velocity gradients. Both rapid stirring and slow stirring are required for a complete flocculation. **Sedimentation** refers to settling of the formed flocs without stirring for quiescence settling. All of these processes are depicted in Figure 6-15.

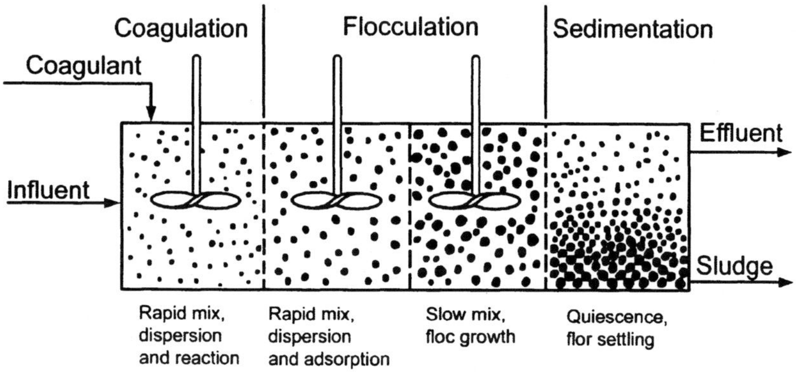

Figure 6-15. Coagulation/flocculation process sequence. (Adapted from Ref. 6-2.)

6.4.1 Orthokinetic Flocculation Versus Perikinetic Flocculation

For all other kinetic processes, electrokinetic flocculation is the result of the reduction of potential for the double layer. The process due to Brownian interparticle contact is called **perikinetic flocculation**. The process involving hydrodynamic fluid motion or agitation is called **orthokinetic flocculation**. For particles, the energy can be regulated by the Maxwell-Boltzmann distribution as:

$$N_j = N_0 e^{-\beta \varepsilon_f} \qquad [6\text{-}57]$$

The particle population is exponentially related to the energy. A second order rate law can express the time-dependent decrease in the concentration of particles due to collision by Brownian motion:

$$-dN/dt = k_p N^2 \qquad [6\text{-}58]$$

and

$$k_p = \alpha 4 D \pi d \qquad [6\text{-}59]$$

where α is the fraction of collision leading to agglomeration, D is the Brownian diffusion coefficient, and d is the particle diameter. Usually the Brownian diffusion coefficient can be expressed as:

$$D = kT/(3\pi\mu d) \qquad [6\text{-}60]$$

or

$$\left(\frac{dN}{dt}\right)_{pk} = -\left(\frac{4\alpha kT}{3\mu}\right) N^2 \qquad [6\text{-}61]$$

where μ is absolute viscosity. If one integrates the above equation,

$$N = N_0/[1 + (4\alpha kTN_0/3\mu)t] \qquad [6\text{-}62]$$

If

$$N = N_0 \,/\, [1 + t/t_{1/2}] \qquad [6\text{-}63]$$

then

$$t_{1/2} = 3\mu/(4\alpha kTN_0) = 1.6 \times 10^{11}/\alpha N_0) \qquad [6\text{-}64]$$

This simple relation indicates that for a poorly destabilized system (low α), $t_{1/2}$ becomes large. Even if a virus of 10^4/mL is completely destabilized ($\alpha = 1$), $t_{1/2}$ still requires 200 days.

Similarly, the orthokinetic flocculation rate can be expressed as:

$$\left(\frac{dN}{dt}\right)_{ok} = -\left(\frac{2\alpha G d^3}{3}\right)N^2 \qquad [6\text{-}65]$$

where G is the velocity gradient. Usually the volume fraction of a colloidal particle is used, so:

$$\phi = \pi d^3 N/6 \qquad [6\text{-}66]$$

After substitution,

$$\left(\frac{dN}{dt}\right)_{ok} = -\left(\frac{4\alpha G\phi}{\pi}\right)N \qquad [6\text{-}67]$$

and after integration,

$$\ln N/N_0 = -4\alpha\phi Gt/\pi \qquad [6\text{-}68]$$

Now let us compare the rates (Eqs. [6-61] and [6-65]):

$$\frac{(dN/dt)_{ok}}{(dN/dt)_{pk}} = \frac{\mu G d^3}{2kT} \qquad [6\text{-}69]$$

For water at 25°C, this ratio is about 1 when G = 10/sec. For a 0.1 μm particle, G is 10^4/sec, whereas for a particle of 10 μm, G is 10^2/sec. Thus, it is clear that small particles are not efficient for separation, but they should be aggregated into larger ones, i.e., close to 1 μm, in order for Browning collision to occur. In general, the overall rate of diminution of particles can be expressed as:

$$-\frac{dN}{dt} = 4\alpha N\left[\frac{kT}{3\mu}N + \frac{G\phi}{\pi}\right] \qquad [6\text{-}70]$$

It should be pointed out that α is important for both processes. In fact, the criterion for stability is $w = \alpha^{-1}$, since orthokinetic flocculation is always the control for the design of flocculation. It is suggested by Weber that the product

$\alpha G\phi t$ for wastewater treatment is in the order of 0.1, that is, G = 10/sec, $\phi = 10^{-4}$, $\alpha = 10^{-1}$, and $t = 10^3$ sec. Some of the collision efficiency factors are tabulated in Table 6-4.

Table 6-4. Observed Values of the Collision Efficiency Factor

Colloid	Coagulant	Type of Flocculation	G (sec^{-1})	η
Polystyrene latex spheres	NaCl	Orthokinetic	11	0.448
Polystyrene latex spheres	NaCl	Orthokinetic	45	0.344
Polystyrene latex spheres	Polyethylenimine	Orthokinetic	11	0.217
Polystyrene latex spheres	Polyethylenimine	Orthokinetic	45	0.063
Silica	Al(III)	Perikinetic	—	0.010
Silica	Al(III)	Orthokinetic	10	0.011
Oil	$Ca(NO_3)_2$	Perikinetic	—	0.355
Polystyrene latex spheres	NaCl	Perikinetic	—	0.375
Polystyrene latex spheres	NaCl	Orthokinetic	1-80	0.364

Source: Adapted from Ref. 6-1.

6.4.2 Comparison Between Filtration and Coagulation

The final process after coagulation, flocculation, and sedimentation is filtration, which involves separating the larger aggregates in wastewater. The similarities between filtration and coagulation are shown in Table 6-5.

Table 6-5. Comparison Between Coagulation and Filtration

Coagulation	α Collision efficiency	ϕ Volumetric concentraion of suspended particles	G Velocity gradient	t Time
Filtration	α	$1-f$ Volumetric concentration of filter medium	η Single collector efficiency (v,d) contact	L/d Number of collectors opportunities
Design and Operational Variables	Chemicals	Coagulation aids, media size, sludge recirculation	Energy input, Mass transport	Residence time, filter lengths, media diameter

The equation,

$$-\frac{dN}{dl}=\frac{3}{2}\frac{(1-f)}{d}\alpha\eta N \quad [6\text{-}71]$$

for packed-bed filtration is compared with orthokinetic flocculation:

$$-\frac{dN}{dt}=\frac{4}{\pi}\alpha G\phi N \quad [6\text{-}72]$$

where f is porosity, $1-f$ is the volume of filter media per unit volume of filter bed, L is bed depth, and η is a single collector efficiency reflecting the rate at which particle contacts occur between suspended particles and the filter bed.

In a three-dimensional plot for α, G, and ϕ, regions of natural water systems can be mapped for comparison to wastewater. Figure 6-16 applies the same concept of wastewater treatment to natural waters. Figure 6-17 indicates that the filtration rate can be increased by using the variables for improvement of collisions and decrease in contact opportunities. Figure 6-18 refers to the regions where particles can be removed in wastewater treatment. Generally, destabilization should be applied before filtration.

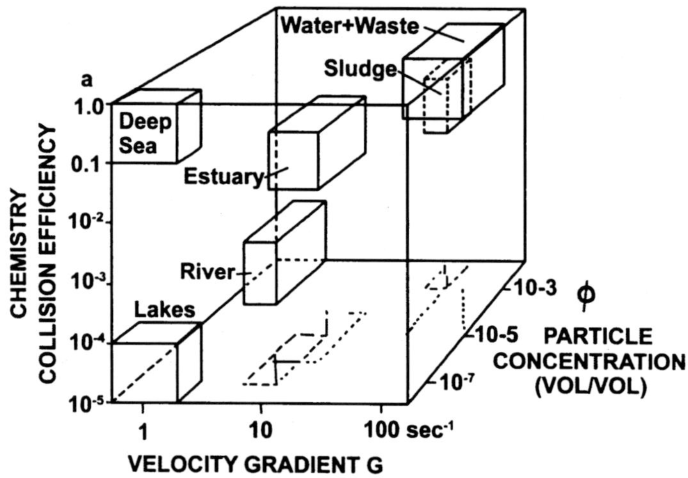

Figure 6-16. Variables that typically determine efficiency of coagulation in natural waters and in water and waste treatment systems. (W. Stumm, 1978)

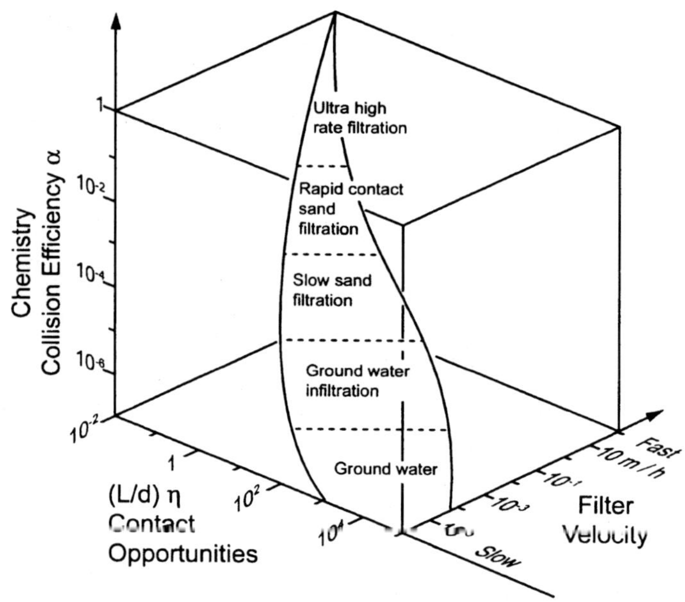

Figure 6-17. Filtration processes in natural waters and in water and wastewater systems. Similar efficiency in particle removal but marked increase in filtration rate, can be achieved by counterbalancing a reduction in contact opportunities by chemically improving the contact efficiency.

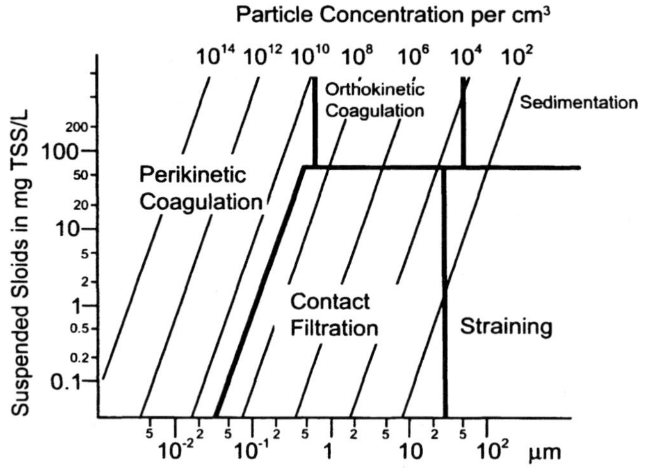

Figure 6-18. Most suitable ranges for application of various processes of particle removal in water and waste treatment.

6.4.3 Mean Velocity Gradient

Usually the velocity gradient can be related to the ratio of power input to fluid as follows:

$$G = \left(\frac{\overline{P}}{V\mu} \right)^{\frac{1}{2}} \qquad [6\text{-}73]$$

where V is flocculent volume of the reactor in m^3, μ is viscosity in (N)(sec)/m^2, and $\overline{P}$ is power requirement in W, or

$$Gt_d = \frac{V}{Q}\left(\frac{\overline{P}}{V\mu} \right)^{\frac{1}{2}} = \frac{1}{Q}\left(\frac{\overline{P}V}{\mu} \right)^{\frac{1}{2}} \qquad [6\text{-}74]$$

where t_d is retention time. In baffled flocculation chambers, interparticle unity can be based on:

$$\overline{P} = Q\rho g h_f \qquad [6\text{-}75]$$

where h_f is head loss of the tank, or

$$G=\left(\frac{Q\rho gh_f}{V\mu}\right)^{\frac{1}{2}}=\left(\frac{gh_f}{\nu t_d}\right)^{\frac{1}{2}} \qquad [6\text{-}76]$$

where ν is the kinematic viscosity. Interparticle contacts are accomplished by mechanical stirring with rotary pedals, and

$$F_D = C_D A\rho u^2/2 \qquad [6\text{-}77]$$

where F_D is drag force N, A is the cross-sectional area of the pedal, C_D is the drag coefficient, and u is the relative velocity of the pedal with respect to the fluid. Also, the power requirement can be related to the drag force:

$$\overline{P} = F_D u \qquad [6\text{-}78]$$

Difficulties are accounted for in the evaluation of u; it is estimated that u = 0.5 to 0.75 of u_P, or the pedal tip velocity.

[Example 6-1] Determine the power requirement and pedal area for coagulation if the G value in a tank of 3000 m^3 is 50/sec. The drag coefficient of the pedal is 1.8, and the pedal tip velocity is 0.6 m/sec.

$$\overline{P} = \mu G^2 V = (1.139 \times 10^{-3}\ \text{N·sec/m}^2)\ (50/\text{s})^2\ (3000\ \text{m}^3) = 8.543\ \text{kW}$$

$$A=\frac{2\overline{P}}{C_D\rho u^3}=\frac{\left(2\times 8543\ \text{kg/m}^2\cdot\text{sec}^2\right)}{(1.8)\left(999.1\ \text{kg/m}^3\right)\left(0.75\times 0.6\ \text{m/sec}\right)^3} = 104.3\ \text{m}^2$$

REFERENCES

6-1. C. R. O'Melia, in *Physicochemical Processes for Water Quality Control* (W. J. Weber Jr., ed.) Wiley-Interscience, New York, 1972.

6-2. S. Daniels, A survey of flocculating agents — process descriptions and design considerations, *AIChE Symp. Ser.*, 70, 266–281 (1974).

6-3. W. E. Neuman, Optimizing coagulation with pilot filters and zeta potential, *Am. Water Works Assoc. J.*, 83, 73 (1991).

6-4. S. D. Faust and O. M. Aly, *Chemistry of Water Treatment*, Butterworth, Boston, 1983.

6-5 M. J. Hammer, *Water and Wastewater Technology*, 2nd ed., Wiley, New York, 1986.

6-6. R. Leu and M. M. Ghosh, Polyelectrolyte characteristics and flocculation, *Am. Water Works Assoc. J.*, 80, 159 (1988).

6-7. S. Kawamura, Effectiveness of natural polyelectrolyte in water treatment, *Am. Water Works Assoc. J.*, 83, 88-91 (1991).

6-8. T. S. Tanaka and M. Pirbazari, Effects of cationic polyelectrolyte on the removal of suspended particulate during direct filtration, *Am. Water Works Assoc. J.*, 78, 59-64 (1986).

PROBLEM SET

1. The orthophosphate (PO_4^{3-}) concentration in a neutral wastewater is 30 mg/L. Estimate the concentration of PO_4^{3-} remaining in solution after the addition of 50 mg/L of CaO. Consider the following equilibrium:

$$Ca_5(PO_4)_3(OH)_{(s)} = 5\ Ca^{2+} + 3\ PO_4^{3-} + OH^-;\ \ pK = 55.6$$

2. An acid mine drainage $(H_2SO_4)_T = 10^{-4}$M, $PCO_2 = 10^{-1.5}$] mixes 1:2 with a stream in which $PCO_2 = 10^{-1.5}$ at pH = 6.0. What are the alkalinity and pH of (a) the acid mine drainage, (b) the stream, (c) the mix at the point of confluence, (d) the mix after cascading down the hill and equilibrating with the atmosphere, and (e) the mix after standing in a small lake and supporting the growth of 3 mg/L of algal biomass (NO_3^- as N source)?

CHAPTER 7

DISINFECTION

The total destruction and removal of microorganisms from a treated medium is termed **sterilization**, whereas the process in which only pathogenic microorganisms are destroyed is termed **disinfection**. In a broad sense, disinfection is used to destroy a particular species at some stage of its development. This includes the treatment of water, food, beverages, and air.

The mechanism of disinfection depends on the nature of the disinfectants as well as the types of microorganisms. Halogen disinfectants such as chlorine and chlorine-containing compounds are thought to act by penetrating through the cell to deactivate the essential enzyme. The oxidation potential correlates directly with the disinfecting capabilities. Since the cytochemistry of pathogens reflects variations among many types of cells, such as spores, cysts, and fungi, different disinfection methods are available. Nonchemical disinfections, such as heat and ionizing radiation, play an important role as well.

In this chapter, we will first discuss the kinetics of disinfectants. Examples of the use of steam in a closed system and irradiation by ultraviolet light are then illustrated, followed by the chemistry of chlorine, as it is still the primary chemical used in water disinfections. A number of problems in dealing with chlorination will be examined thoroughly in the discussion, followed by a discussion of the breakpoint curve as a guideline for chlorination. In the last section, we will briefly discuss disinfection in the water distribution system, making use of the basic principles of fluid mechanics and convective diffusion.

7.1 DISINFECTION KINETICS

The destruction of pathogens depends upon contact time, temperature, pH, concentration of the disinfectant, concentration of the organisms, and the presence of any interfering substances. Disinfection is carried out over a span of time, and the rates are determined by diffusion and chemical kinetics. Since the mechanisms of disinfection are not well known, many of the expressions are still empirical.

For a given disinfectant concentration, the rate of kill for bacteria can be described by Chick's law, which is first-order:

$$-dN/dt = kN \qquad [7\text{-}1]$$

where N is the concentration of living microorganisms at time t and k is the first-order rate constant. Assuming the initial concentration of microorganism is N_0, after integration:

$$\ln N/N_0 = -kt \qquad [7\text{-}2a]$$

or

$$N/N_0 = e^{-kt} \qquad [7\text{-}2b]$$

In practice N_0 can never be zero, and N is expressed in terms of N_0. For example, for a 90% kill, 10% remains, and $N = 0.1\ N_0$; for a 99% kill, 1% remains, and $N = 0.01\ N_0$, and so on. Chick's law is illustrated in Figure 7-1.

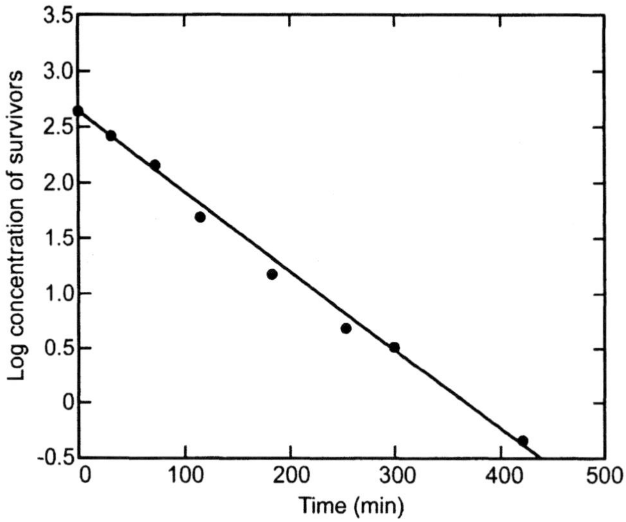

Figure 7-1. Disinfection of anthrax spores with 5% phenol. (Adapted from Ref. 7-4.)

[Example 7-1] The bacterial count of a sewage effluent is 10^9/mL; in order to meet the bacteria discharge standard of 1/100 mL, how large a tank is needed for the disinfection of 1 mgd of the effluent, assuming a kinetic constant of 0.5 min^{-1}?

For the solution, the time required for treatment is:

$$t = \left(\frac{2.303}{0.5}\right)\left(\log\frac{10^9}{10^{-2}}\right) = 50.6\,\text{min}$$

Tank size can be calculated based on $V = Qt$, which is 134 m^3, since 1 mgd = 0.044 m^3/sec.

Often, deviation from Chick's law is incurred. In Figure 7-2, when the log of the concentration of survivors is plotted versus time, cases of retardation or acceleration occur. For the cases of retardation, they will fit

$$-dN/dt = kN/(1 + \alpha_r t) \qquad [7\text{-}3]$$

where α_r is the retardative coefficient. For the cases of acceleration, the autocatalytic model is used:

$$-dN/dt = k_1N + k_2(N_0 - N)N \qquad [7\text{-}4]$$

in which k_2 is called the velocity acceleration constant. Anomalies are presented in Figure 7-2. Curve B is the original Chick's law, while curve A is represented by:

$$-dN/dt = \frac{kN}{(1+\alpha_r Ct)} \qquad [7\text{-}3a]$$

where C is concentration. Curve C can be a case of Equation [7-4]. Curve D has two steps, each following a different rate of kill. A typical survival curve (Fig. 7-3) actually involves an initial lag as curve B (Fig. 7-2). It will then follow Chick's law, and, finally, it may have a decline rate as either curve C or curve D. Selleck has developed a model to explain and fit the experimental data.

First, if $-\log(N/N_0)$ is plotted versus log Ct for the k in Chick's law, the following can be expressed:

$$k = 0 \text{ for } Ct < \tau$$

$$k = k' \text{ for } Ct = \tau$$

$$k = k'/b(Ct) \text{ for } Ct > \tau \qquad [7\text{-}4a]$$

No disinfection occurs until t exceeds a minimum time, t'. After integration,

$$\frac{N}{N_0} = 1 \text{ for } Ct < \tau$$

$$[7\text{-}4b]$$

$$\frac{N}{N_0} = \left(\frac{Ct}{\tau}\right)^{-n} \text{ for } Ct > \tau$$

where $n = k'/b$. Physically, τ and n can be evaluated from the graph for a number of disinfectants (Table 7-1 and Fig. 7-4).

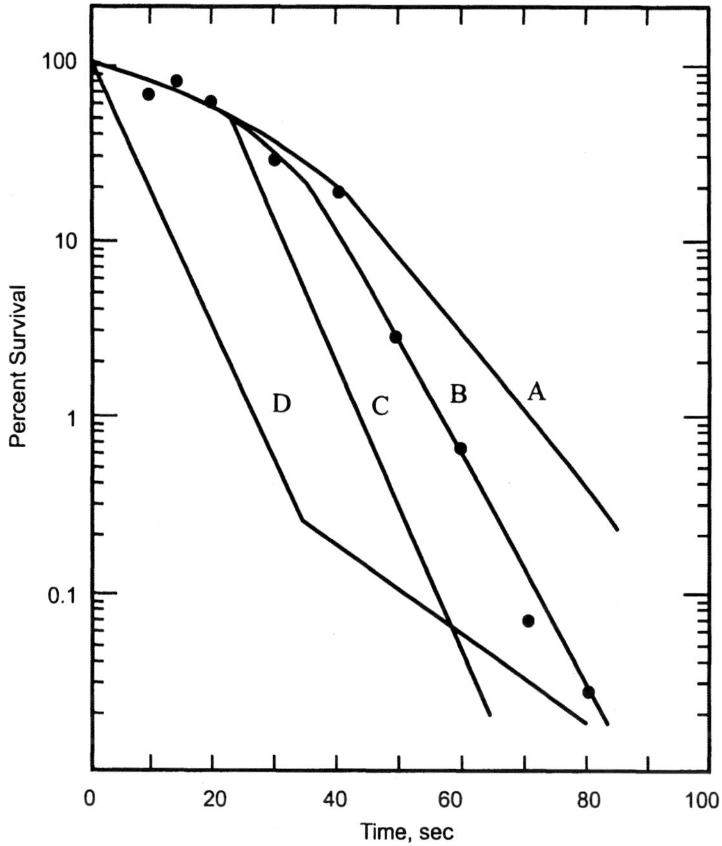

Figure 7-2. Anomalies and deviation from Chick's law. A, Retardation; B, Chick's law; C, acceleration; D, two rates. Data points are E. coli vs. ClO2 (From P. V. Scarpino et al., Proc. AWWA Tech. Conf., Kansas City, MO, 1977.)

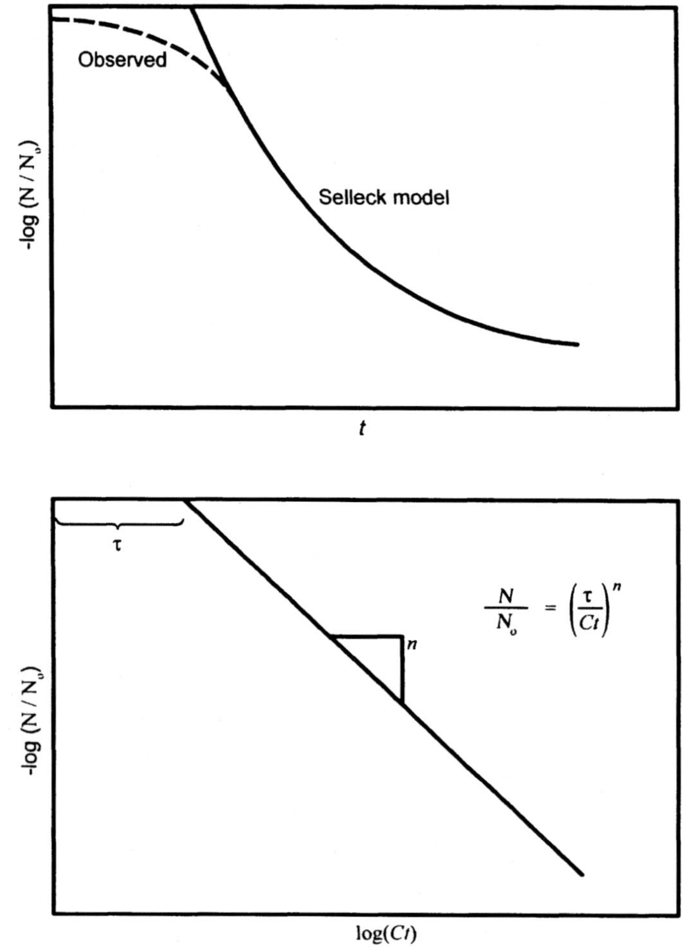

Figure 7-3. Plot of Selleck model: (a) versus time; (b) versus product of concentration and time.

Furthermore, the departure of Chick's law sometimes takes the form:

$$\ln N/N_0 = -kt^m \qquad [7\text{-}5]$$

where m is a constant. If $m < 1$, the kill will decrease with time; if $m > 1$, then the kill will increase with time. Graphically, m can be obtained by plotting the double log if the extent of bacteria survived versus the log of time is:

$$\log(-\ln N/N_0) = \log k + m \log t \qquad [7\text{-}6]$$

Table 7-1. Applications of Selleck's Model

	τ	n	r^a	Organism
Chlorine				
Secondary effluent (combined residual)	4.06	–2.82	0.86	Coliform
Nitrified effluent (free residual)	1.33	–2.40	0.87	Coliform
Nitrified effluent (combined residual)	6.74	–3.17	0.91	Coliform
Nitrified, filtered (free residual)	0.59	–2.12	0.69	Coliform
Chlorine dioxide				
Secondary effluent	0.89	–2.18	0.78	Coliform
Nitrified effluent	0.78	–2.06	0.75	Coliform
Nitrified, filtered	0.003	–1.13	0.81	Coliform
Ozone				
Secondary, filtered[b]	8.65	–4.46	0.96	Coliform
Ozone-demand free water[c]	0.002	–2.51	0.76	*E. Coli*
UV				
Secondary effluent	8.01	–1.99	0.70	Fecal coliform

[a] Correlation coefficient.
[b] Results obtained using bubble diffuser ozone contactor and naturally occurring coliforms
[c] Results obtained using batch reactor and laboratory-grown *E. coli* at 1°C

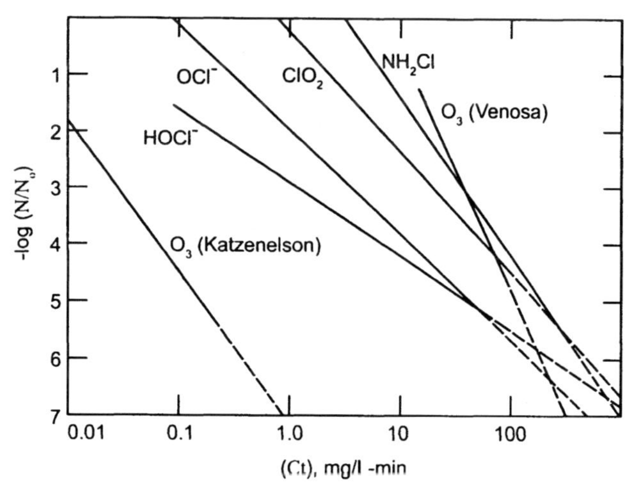

Figure 7-4. Application of model comparison of disinfectants.

At times even empirical formulas such as the following are used (see Fig. 7-4b):

$$N/N_0 = kt^m \qquad [7\text{-}7]$$

$$N/N_0 = (1+0.23\ Cl_r t)^{-3} \qquad [7\text{-}8]$$

Here, Cl_r is the residual chlorine concentration. In selecting a chemical agent for disinfection, it is important to consider the concentration of that chemical agent.

$$C^n t_p = \text{constant} \qquad [7\text{-}9]$$

where C is the concentration of a disinfectant and t_p is time required to effect a constant percentage of kill.

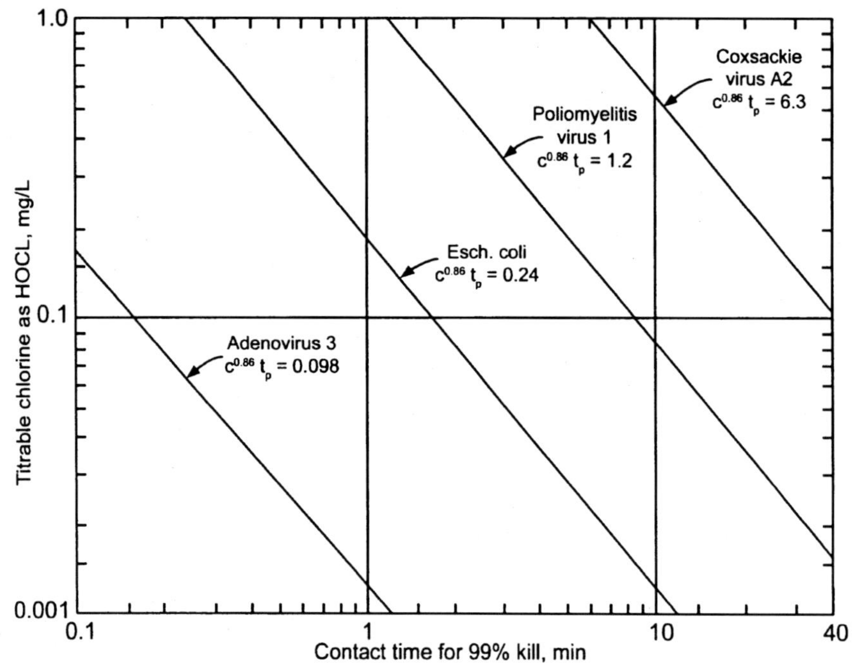

Figure 7-5. Time-concentration relationships in disinfection. Concentration of chlorine as HOCL required for 99% kill of *E. coli* and three enteric viruses at 0 – 6°C. (Adapted from Ref. 7-16.)

Also,

$$C^q N_p = \text{constant} \tag{7-10}$$

where N_p is the concentration of organisms reduced at a given percentage in a given time. In the above equations, n and q are constants. All these expressions are empirical. The time-concentration relationship is illustrated by Figure 7-5, where n is evaluated as 0.86. However, for a number of viruses and *E. coli,* the constant value varies.

In conjunction with this is the multi-hit concept developed by Kimball and Gurian:

$$P_t = (1 - e^{-kt})^m \tag{7-11}$$

where P_t represents the probability that the organism or the vital center within it will be destroyed before or at time t, k is still the rate constant, and m is the number of vital sites. For simplicity,

$$P_t \cong 1 - me^{-kt}$$

$$\ln (1 - P_t) = \ln m - kt \cong \ln (\ln N/N_0) \tag{7-12}$$

which has just been discussed in Equation [7-2a].

For some disinfectants, a second-order reaction is usually used:

$$-dN/dt = kNC \tag{7-13}$$

where C is the concentration. The solution of this type is:

$$\frac{1}{C_0 - N_0} \ln \frac{C}{C_o} \cdot \frac{N_0}{N} = kt \tag{7-14}$$

If $C_0 >> N_0$, then it will reduce to the first-order reaction:

$$\ln N/N_0 = kC_0 t \tag{7-2a}$$

which is Chick's law. For chlorine and its derivatives, the following equation is actually used:

$$-dN/dt = kNt \tag{7-15}$$

Here is an example of the common dechlorination of water.

[Example 7-2] For chlorine disinfection, what length of discharge pipeline with 0.5 m diameter will be needed in order to kill 99% of the bacteria in 1 mgd of wastewater?

Since the rate constant is not known, one has to conduct a laboratory experiment to determine it. Suppose a sample of this water killed 99.9% of all bacteria in 10 minutes when treated with 5 mg/L of chlorine:

$$-dN/dt = kNt$$

Then

$$-\ln\frac{N_2}{N_1}=\frac{k}{2}\left(t_2^2-t_1^2\right)$$

If t_1=0, then

$$\frac{N_2}{N_1}=\frac{1-0.999}{1}=10^{-3}$$

and

$$k = 0.138\ \text{min}^{-1}$$

Many water chlorination designs are of a plug-flow type; the reaction changes with the axial direction, x, of flow, and the velocity component is v_x. The conservation equation is:

$$v_x dN - r_N dV = 0$$

which can also be written as:

$$QdN - r_N dV = 0$$

where the reaction is:

$$-r_N = kN\theta$$

and $\theta = V/Q$, which is the retention time, and:

$$d\theta = dV/Q$$

Therefore,

$$dN + kN\theta d\theta = 0$$

$$-\ln\frac{N_2}{N_1} = \frac{k\theta^2}{2}$$

For 99% kill,

$$-\ln\frac{N_2}{N_1} = -\ln\left(\frac{10^{-3}}{1}\right) = 2.3 \times 3$$

$$\theta = \left(\frac{2.3\times 6}{0.138}\right)^{\frac{1}{2}} = 10\,\text{min}$$

Thus, $V = Q\theta = 0.044\ \text{m}^3/\text{sec} \times 10 \times 60 = 26.4\ \text{m}^3$, and

$$L = \frac{V}{A} = \frac{26.4}{\frac{\pi}{4}(0.5)^2} = 134\ \text{m}$$

7.1.1 Disinfection Using Heat and Ultraviolet Light

In general, wet heat has been employed in foods and beverages for disinfection. Usually a pasteurizer is employed with heating and cooling cycles, as shown in Figure 7-6. There is also a hold period in which the temperature is kept constant.

Not all bacteria can be destroyed by heat. The relative resistance of microorganisms to sterilization by moist heat is illustrated in Table 7-2. Some thermophilics have been known to survive at 120°C. Usually, when using a batch autoclave pasteurizer for processing beverage products, tests must be made for evaluation of *k*, especially at the holding temperature. These can be obtained using the Arrhenius equation.

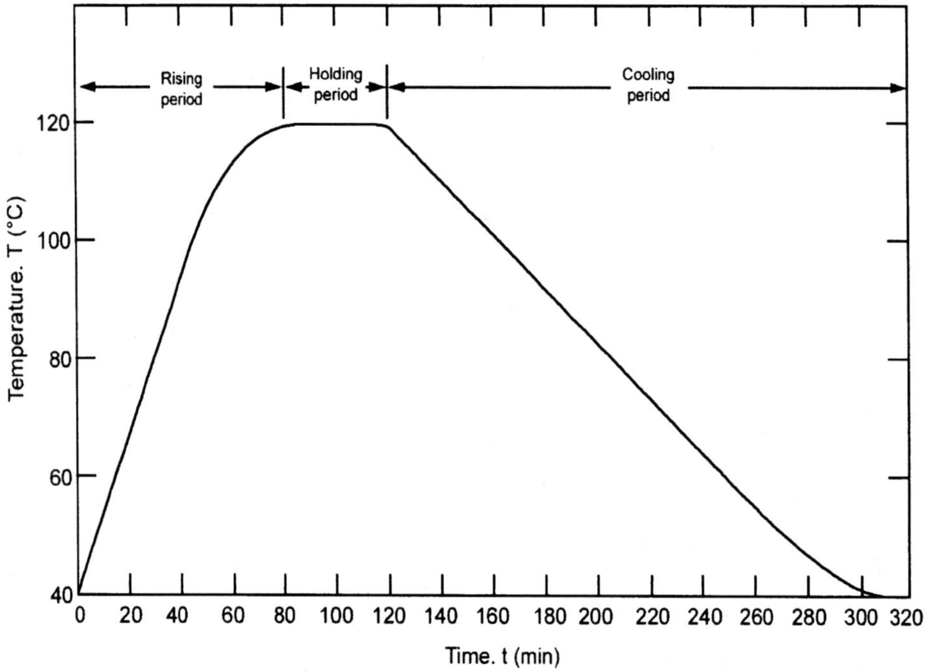

Figure 7-6. Heating cycle in a batch disinfection process.

Table 7-2. Relative Resistances of Microorganisms to Sterilization by Moist Heat

Organism	Relative Resistance
Escherichia coli	1
Bacterial spores	3,000,000
Mold spores	2–10
Viruses and bacteriophages	1–5

Source: O. Rahn, Physical methods of sterilization of microorganisms, *Bacteriol. Rev.*, 9 1–47 (1945).

[Example 7-3] Using a batch pasteurizer as shown in Figure 7-6, a beverage containing 10^7 spores/mL is exposed to moist heat. For a test run of 6 minutes at 80° and 110°C, 10 and 50% of the spores are destroyed, respectively. How long must the holding period be in order for the product to ensure 99 out of 100 batches are free of spores? Assume that the minimum lethal temperature is 80°C. (This example is adapted from Ref. 7-15.)

Compute the k values at 80° and 110°C:

$$k_{80} = 2.3 \log (100/90)/6 = 0.0169 \text{ min}^{-1}$$

$$k_{110} = 2.3 \log (100/50)/6 = 0.115 \text{ min}^{-1}$$

An Arrhenius plot is constructed in which the temperature is in absolute scale (Fig. 7-7). From the time levels of Figure 7-6, the temperature is determined, and from each of these temperatures the rate constants are found in Figure 7-7. All these data are entered in Table 7-3. The rate constants are plotted as a function of time in Figure 7-8 and graphically integrated.

$$k_{av} = \frac{\int_{t_1}^{t_2} kdt}{\Delta t} = \frac{\left(\sum kdt\right)_{\text{rising}} + \left(\sum kdt\right)_{\text{cooling}}}{t_2 - t_1} = \frac{5.32 + 6.84}{50 + 90} = 0.0868 \text{ min}^{-1}$$

Using Chick's law to determine the number of surviving spores,

$$\log N_2 = \log N_1 - (tk/2.3) = \log 10^7 - [(50{+}90) \times 0.0868/2.3] = 1.72$$

$$N_2 = 52.7$$

Using $k = 0.178$ min^{-1} at a holding period (120°C), determine the length of the holding period required to reduce the remaining number of spores to 0.01:

$$t = 2.3 \log (52.7/0.01)/0.178 = 48 \text{ min}$$

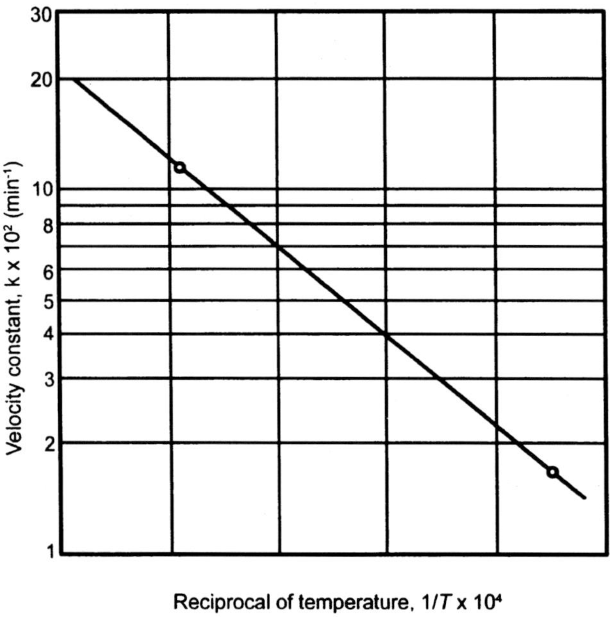

Figure 7-7. Velocity constant as a function of reciprocal temperature.

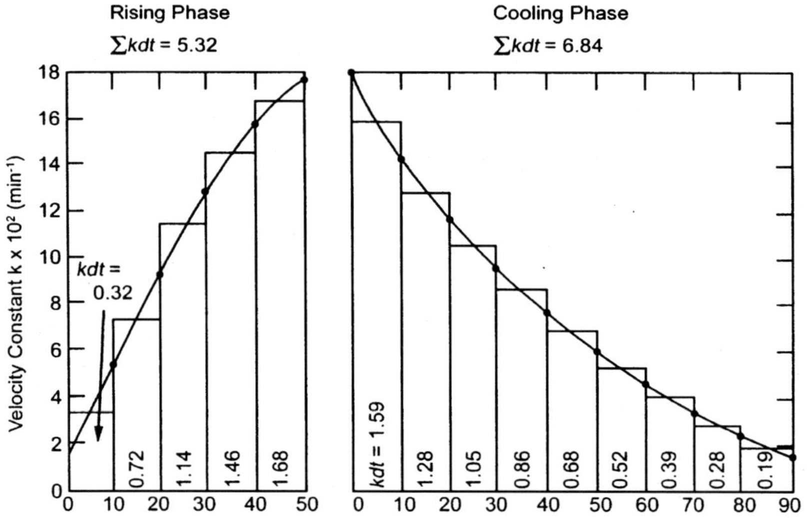

Figure 7-8. Graphical integration of velocity constant–time curves for rising and cooling periods.

Table 7-3. Velocity Determined from Figure 7-7

t (min)	T (°C)	$1/T$	k
		Rising Phase	
0[a]	80	125.0×10^{-4}	1.7×10^{-2}
10	95	105.3	5.1
20	106	94.3	9.5
30	113	88.5	13.2
40	117	85.5	15.6
50	120	83.3	17.8
		Cooling Phase	
0	120	83.3	17.8
10	115	87.0	14.2
20	111	90.2	12.0
30	106	94.3	9.5
40	102	98.0	7.8
50	97	103.0	5.8
60	93	107.5	4.5
70	88	113.6	3.2
80	84	119.0	2.4
90	79	126.3	1.6

[a] Pasteurization begins and ends at 80°C, the minimum lethal temperature.

The continuous process pasteurizing of beverages has become common practice. For the disinfection of large amounts of flowing water, ultraviolet radiation is used. The germicidal ultraviolet wavelength ranges from 200 to 295 nm, with a maximum effect at 253.7 nm. Similar to a disinfection dosage, the rate of kill is proportional to the light intensity, I:

$$k = KI^n \qquad [7\text{-}16]$$

In an ordinary case, n is close to unity (K and n are constants). The intensity in a medium is:

$$I = I_o e^{-ax} \qquad [7\text{-}17]$$

where x is the distance between I and I_0, and a is coefficient of absorption (see Table 7-4 for some common media).

For water treatment, the average value of the rate constant is used since the rate of kill varies throughout the depth of flow:

$$k_{av} = \frac{\int_0^x k dx}{x} = \frac{KI_0}{x} \int_0^x e^{-ax} dx = \frac{KI_0}{x}\left[\frac{1}{a} - \frac{e^{-ax}}{a}\right] \qquad [7\text{-}18]$$

$K = 1740$ if I and k have the units of cal/cm^2/min and min^{-1}, respectively.

Table 7-4. Coefficients of Absorption for Various Waters, λ = 2537 Å

Water Source	*A* (per cm)
Distilled	0.008
Swimming pool	0.031
Cleveland tap	0.050
Drilled well	0.056
Fish pool	0.070
Lake Erie	0.083
Concrete cistern	0.297

Source: M. Luckiesh and L.L. Holiday, Disinfecting water by means of germicidal lamps, *General Electric Rev.*, 47, 45–50 (1944).

[Example 7-4] Well water containing 1000 bacteria per ml is irradiated with 610 μW/cm^2 intensity as the water flows through a concrete channel 10 feet long and 2 feet wide at a depth of 3 inches. Find out the rate of flow necessary to get a 1000-to-1 assurance that all the bacteria are destroyed (Adapted from Ref. 7-15).

The absorption coefficient can be obtained from Table 7-4 as 0.056 cm^{-1}, and

$$k_{av} = \frac{1740\left(610 \times 10^{-6}\right)\left(13.34\right)}{7.62}\left[\frac{1}{0.056} - \frac{e^{-0.056(7.62)}}{0.056}\right] = 12.6 \text{ min}^{-1}$$

The time of exposure is:

$$t = \frac{2.3}{12.6}\log\left(\frac{1000}{0.001}\right) = 1.09\,\text{min}$$

and the flow can be evaluated as follows:

$$Q = \frac{V}{t} = \frac{(2 \times 10 \times 0.25)(7.48)}{1.09} = 34.3\,\text{gpm}$$

7.2 Chlorine Chemistry

The use of chlorine as a disinfectant dates back to 1904, the year when the London Metropolitan Water Board decided to use chlorine for the emergency control of water-bearing diseases. Subsequently, in 1910 the New Jersey Supreme Court took similar action. The use of chlorine in water became widespread up until 1912, when commercial liquid forms of chlorine were made available.

The practice of chlorinating water has many advantages:

1. From the public health perspective, water-borne diseases have decreased significantly. It has been claimed that the death rate caused by typhoid declined from 40/100,000 in 1900 to zero in 1924 (see Fig. 7-9).
2. The vast availability of heavy chemicals such as chlorine makes it the most economical method.
3. This method has a good retention of chlorine as residue.
4. The method can also remove toxic ions — ferrous, manganous, sulfide, sulfite, etc.

On the other hand, chlorination of water also yields many disadvantages:

1. Trihalomethane (THM) and chloroform are produced, which have been proven carcinogenic to animals.
2. Chlorine can induce odor and taste problems when it reacts with phenols or humic acids existing in natural waters.
3. Toxic chloro-substituted purines and pyrimidines are formed because their parent nitrogenous compounds are part of the decomposition byproducts.

4. For over half a century, chlorine has been manufactured with sodium hydroxide through the electrolysis of brine. In this chloro-alkali process (see Chapter 12), mercury is used in the membrane for partition. In this manner, mercury escapes into the environment. Increased production of chlorine would promote mercury pollution. In recent years, the process has been modified to exclude the use of mercury, although the mercury used in the previous process remains in the environment.

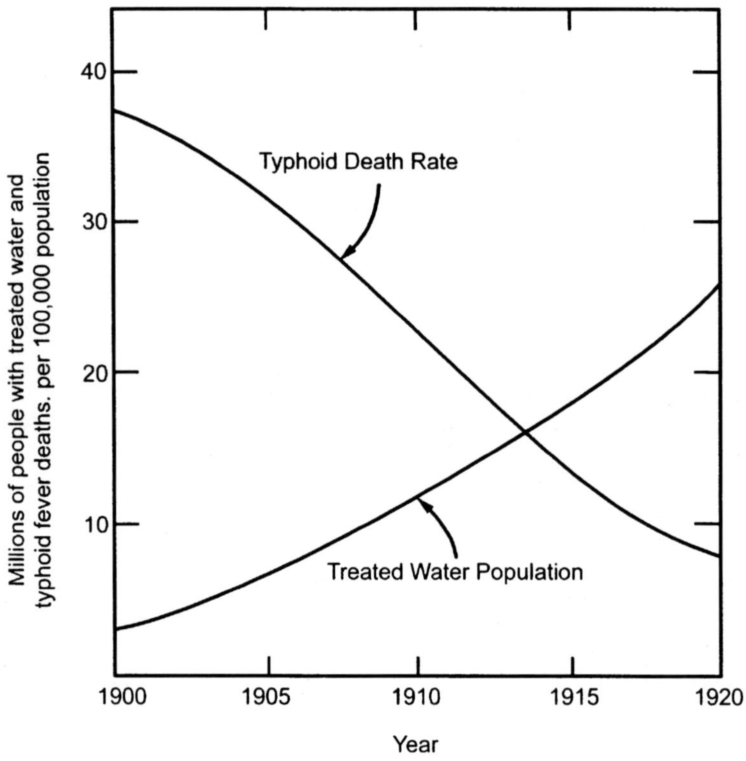

Figure 7-9. Typhoid fever and treated water supplies during two decades.

Chlorine is a yellow-green gas that dissolves in water according to the following hydrolysis equation:

$$Cl_2 + H_2O = HOCl + HCl$$

Its K_h value is:

$$K_h = \frac{(\mathrm{HOCl})(\mathrm{H^+})(\mathrm{Cl^-})}{(\mathrm{Cl_2})} = 4.5 \times 10^{-4}\ (\mathrm{M/L})^2 \text{ at } 25^\circ\mathrm{C} = 1.5 \times 10^{-4}\ (\mathrm{M/L})^2 \text{ at } 0^\circ\mathrm{C}$$

For example, when pH is 5 and the Cl^- concentration is 350 mg/L, then

$$\frac{(\mathrm{HOCl})}{\mathrm{Cl_2}} = \frac{K_h}{(\mathrm{H^+})(\mathrm{Cl^-})} = 4500$$

The overwhelming concentration of the hyperchlorous acid and the amount of Cl_2 is less than 0.00022 at 25°C. In this instance, more than 99.97% of the chlorine is hydrolyzed. Thus, when pH > 3, the reaction always moves to the right. In seawater, Cl^- = 19,000 mg/L. Thus, at pH > 6, hydrolysis is greater than 99.6%. In ordinary water with pH = 7.0 and Cl^- = 50 mg/L, the hydrolysis is 99.9999%. Analytically, Cl_2 and HOCl are not easy to distinguish. However, Cl_2 is soluble in CCl_4, whereas HOCl is not.

Also, hypochlorous acid is readily ionizable into the hypochlorite ion as:

$$\mathrm{HOCl = H^+ + OCl^-}$$

with a pK_a at 25°C of 7.54. We can readily write:

$$\frac{K_a}{(\mathrm{H^+})} + 1 = \frac{(\mathrm{OCl^-}) + (\mathrm{HOCl})}{(\mathrm{HOCl})} = \frac{T}{(\mathrm{HOCl})} \qquad [7\text{-}19]$$

$$(\mathrm{HOCl}) = T/[1 + K_a/(\mathrm{H^+})] \qquad [7\text{-}20]$$

$$(\mathrm{OCl^-}) = T/[1 + (\mathrm{H^+})/K_a] \qquad [7\text{-}21]$$

where T is the free available chlorine. At pH 7.54, (HOCl) = (OCl^-). As indicated in Figure 7-10, the HOCl concentration is very sensitive to pH. For example, at pH 6.54, 91% of the species is HOCl, whereas at pH 8.54, only 9% of the species is HOCl. The fraction of the free available form of chlorine, HOCl, is a stronger disinfectant than OCl^-. Therefore, the germicidal effectiveness is greater at lower pH levels. Other oxidant reactions for chlorine in aqueous systems are listed in Table 7-5.

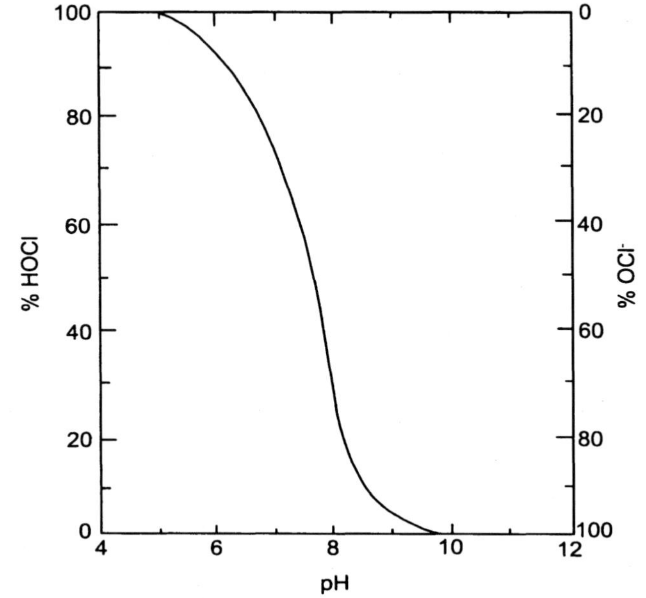

Figure 7-10. Relative amounts of HOCl and OCl^- in water at 20°C.

Table 7-5. Free Oxidant Reactions: Principal Reactions of Environmental Concern for Chlorine in Aqueous Solutions

Reaction Type	Example
Water	$Cl_2 + H_2O = HOCl + HCl$
Ammonia	
Substitution	$NH_3 + HOCl \rightarrow NH_2Cl + H_2O$
Oxidation	$2NHCl_2 + H_2O \rightarrow N_2 + HOCl + 3H^+ + 3\ Cl^-$
Inorganic oxidation	$Mn^{2+} + HOCl + 2H_2O \rightarrow MnO(OH)_2 + 3H^+ + Cl^-$
Disproportionation	$3OCl^- \rightarrow 2Cl^- + ClO^{3-}$
Decomposition	$2HOCl \rightarrow 2H^+ + 2Cl^- + O_2$
Organic reactions	
Oxidation	$RCHO + HOCl \rightarrow RCOOH + H^+ + Cl^-$
Addition	$RC{=}CR' + HOCl \rightarrow RC(OH)C(Cl)R'$
Substitution	
N–Cl bond	$RNH_2 + HOCl \rightarrow RNHCl + H_2O$
C–Cl bond	$RCOCH_3 + 3HOCl \rightarrow RCOOH + HCCl_3 + 2H_2O$

Table 7-6. Actual and Available Chlorine in Pure Chlorine-Containing Compounds

Compound	Molecular Weight	Chlorine Equivalent[a]	Actual Chlorine (%)	Available Chlorine (%)
Cl_2	71	1	100	100
Cl_2O	87	2	81.7	163.4
ClO_2	67.5	5	52.5	260
NaOCl	74.5	2	47.7	95.4
CaClOCl	127	1	56	56
$Ca(OCl)_2$	143	2	49.6	99.2
$NaClO_2$	90.5	4	39.2	157
HOCl	52.5	2	67.7	135.4
$NHCl_2$	86	2	82.5	165
NH_2Cl	51.5	2	69	138

[a] Valency of Cl reduced to −1.

Hypochlorites such as NaOCl, $Ca(OCl)_2$, and CaCl(OCl) (bleaching powder) also yield dissolved chlorine species such as molecular chlorine.

$$Ca(OCl)_2 = Ca^{2+} + 2\ OCl^-$$

$$OCl^- + H^+ = HOCl$$

The buffering capacity of these solutions will determine the effectiveness of chlorine. Sodium hypochloride is produced by electrolysis of brine where Cl_2 is anodically generated and sodium hydroxide is cathodically generated upon interaction:

$$Cl_2 + 2\ NaOH = NaCl + NaOCl + H_2O$$

On the other hand, the chlorination of lime will produce bleaching powder:

$$Ca(OH)_2 + Cl_2 = CaCl(OCl) + H_2O$$

usually a mixture consisting of 20% $Ca(OH)_2$ and 10% of other carbonates.

The actual and available chlorine in a number of pure chloro-containing compounds are calculated readily in the following manner:

$$\%(Cl_2)_{act} = \frac{(\text{No. of Cl in compd})(\text{At. Wt. of Cl})}{(\text{MW of compd})} \times 100 \qquad [7\text{-}22]$$

$$\%(Cl_2)_{avl} = (\text{Cl eqv.})\,[\%(Cl_2)_{act}] \qquad [7\text{-}23]$$

where (Cl_{eqv}) is the value of valency of chloride in the compound reduced to –1. Table 7-6 gives the actual and available chlorine values for a number of commonly used compounds.

The widespread use of chlorine in the sanitation and wastewater treatment areas is illustrated by Table 7-7.

7.3 Breakpoint Reaction

If ammonia is present in water, hypochlorous acid will react to form chloroamines:

$$HOCl + NH_3 = NH_2Cl + H_2O$$

$$HOCl + NH_2Cl = NHCl_2 + H_2O$$

$$HOCl + NHCl_2 = NCl_3 + H_2O$$

The relative concentrations of the chloroamines depends to a great extent on the Cl_2/NH_3 ratio as well as the pH value. The formation of monochloroamines is favored by a high pH and a low Cl_2/NH_3 ratio. It retains the oxidizing capacity of free chlorine but less oxidizing occurs. Therefore, its germicidal property is less effective than HOCl (see Fig. 7-11). Dichloroamine is an unstable species. The formation of trichloroamine is favored by a low pH and a large ratio of Cl_2 to NH_3. It has a pungent odor and is irritating to the eyes and nose. It is not produced in significant amounts at a pH greater than 6 unless the Cl_2/NH_3 ratio is much higher.

In practice, when chlorine is added to a body of water, it readily oxidizes all the water-bearing, reducible species. Many ions, such as Mn^{2+}, Fe^{2+}, NO_2^-, Br^-, CN^-, and SO_3^{2-}, are readily consumed due to the added dosage of chlorine. A conceptual model with a degree of experimental verification is shown as the breakpoint curve in Figure 7-12.

Table 7-7. Additional Chemical Applications in Wastewater Collection, Treatment, and Disposal

Application	Chemicals used[a]	Remarks
Collection		
Slime-growth control	Cl_2, H_2O_2	Control of fungi and slime-producing bacteria
Corrosion control (H_2S)	Cl_2, H_2O_2	Control brought about by destruction of H_2S in sewers
Odor control	Cl_2, H_2O_2, O_3	Especially in pumping stations and long, flat sewers
Treatment		
Grease removal	Cl_2	Added before preaeration
BOD reduction	Cl_2, O_3	Oxidation of organic substances
Ferrous sulfate oxidation	Cl_2[b]	Production of ferric sulfate and ferric chloride
Filter-ponding control	Cl_2	Residual at filter nozzles
Filter-fly control	Cl_2	Residual at filter nozzles, used during fly season
Sludge-bulking control	Cl_2, H_2O_2, P	Temporary control measure
Digester supernatant oxidation	Cl_2	
Digester and Imhoff tank foaming control	Cl_2	
Ammonia oxidation	Cl_2	Conversion of ammonia to nitrogen gas
Odor control	Cl_2, H_2O_2, O_3	
Oxidation of refractory organic compounds	O_3	
Disposal		
Bacterial reduction	Cl_2, H_2O_2, O_3	Plant overflows, storm water
Odor control	Cl_2, H_2O_2, O_3	

[a] Cl_2 = Chlorine; H_2O_2 = hydrogen peroxide; O_3 = ozone; P = polymer.

[b] $6FeSO_4 \cdot 7H_2O + 3Cl_2 \rightarrow 2FeCl_3 + 2Fe_2(SO_4)_3 + 42H_2O$.

Many of the reducing species existing in natural water do not have the ability to trap and retain residual chlorine as is the case with ammonia and other nitrogenous compounds. These reducing species have little or no residual chlorine (AB in the breakpoint curve).

At point B In Figure 7-12, after meeting the demand exerted by reducing agents, the reaction and consequent formation of residual chlorine starts with ammonia or other nitrogenous organics. Substitution reactions of chlorine with other organic compounds can yield combined residual chlorine, which can be utilized, such as chloroamines. As chlorine content increases to the point where NH_2Cl forms stoichiometrically over a period of several hours, concurrent losses of oxidizing chlorine and ammonia occur. Point C represents the maximum reserve for chlorine in combined form, and it is consumed. It is also at this point that initiation of oxidation of the trapping species of nitrogen to gaseous form starts. For example,

$$2NH_2Cl + HOCl \rightarrow N_2 + H_2O + 2HCl$$

$$NH_2Cl + NHCl_2 \rightarrow N_2 + 3HCl$$

$$NH_2Cl + NHCl_2 + HOCl \rightarrow 2NO + 5HCl$$

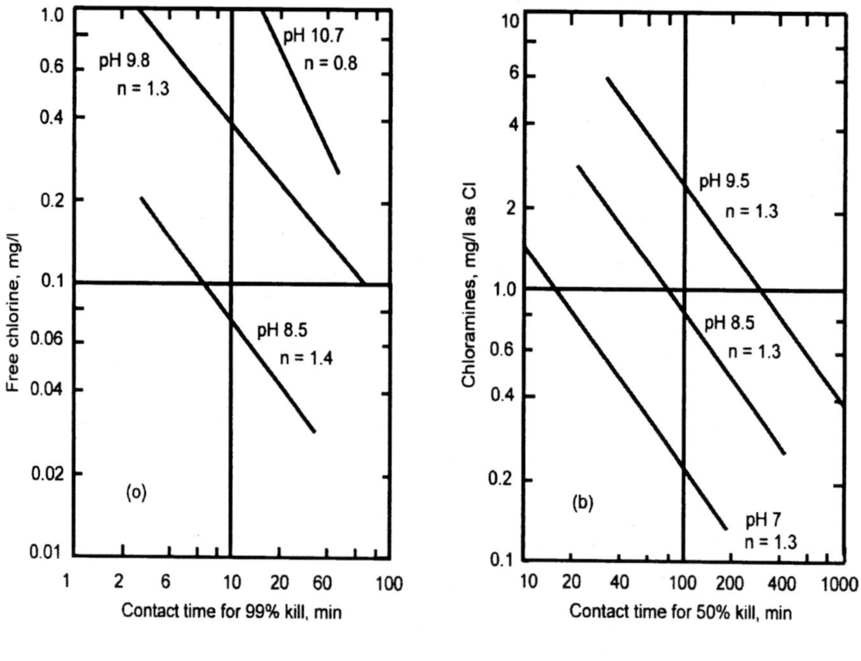

Figure 7-11. Time-concentration relationships in disinfection: (a) concentration of free available chlorine required in order to kill 90% of E. coli at 25°C; (b) concentration of combined available chlorine required in order to kill 50% of E. coli at 25°C. (After Fair et al., 1968.)

All these oxidations occur at a chlorine to nitrogen ratio higher than that of the stoichiometric ratio. In this manner, the residual chlorine is spent and the demands of all oxidizable chlorination will increase the free residual of chlorine. The minimum point D of this curve is called the breakpoint, since further residual accumulates beyond this point. Usually, the chlorine demand is anywhere in the breakpoint curve, the difference between applied chlorine dosage and residual chlorine: chlorine demand = applied chlorine dosage – residual chlorine (see Fig. 7-12).

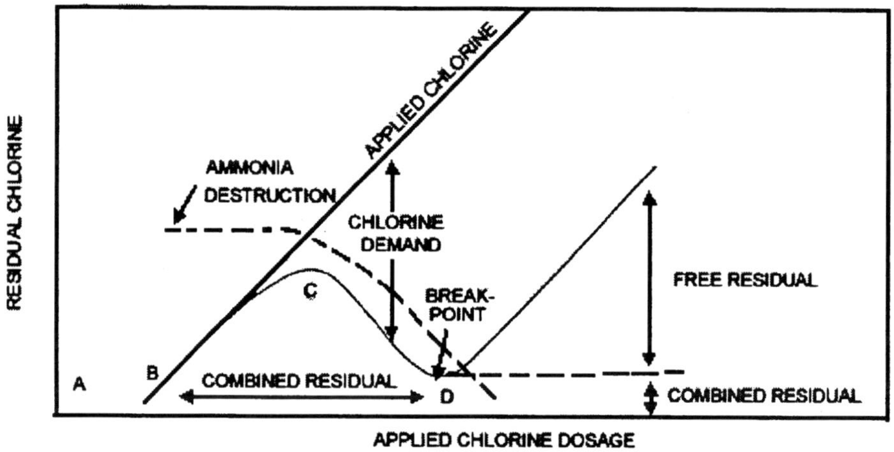

Figure 7-12. Graphic description of the breakpoint phenomenon. The combined residual after breakpoint consists essentially of the organic chloramines not destroyed by free chlorine.

7.3.1 Trihalomethane and Haloacetonitrile Formation

Trihalomethanes have the general form CHX_3, where X can be Cl, Br, or I. Chloroform, or $CHCl_3$, is particularly important because it is a suspected carcinogen. Proposed regulations recommend a maximum contaminant level of 100 μg/L (ppb) for THM in purified drinking water. Even at this level it has been estimated there will be 0–400 more tumors per year from exposure to THM. The origin of THM is traced back to the acetyl group in humic substances (the structure of marine humic material is shown in Fig. 7-13) contained in water. A proposed mechanism is shown in Figure 7-14. The overall equation is:

$$RC(O)\text{-}CH_3 + 3HOCl \rightarrow CHCl_3 + RCOOH + 2H_2O$$

The reaction proceeds more rapidly at a higher pH than at a lower pH during water treatment. Treated water often contains not only chloroform but other combinations of THM, such as $CHCl_2Br$, $CHClBr_2$, and $CHBr_3$. Since natural water contains inorganic bromide, it is easily oxidized:

$$HOCl + Br^- \rightarrow HOBr + Cl^-$$

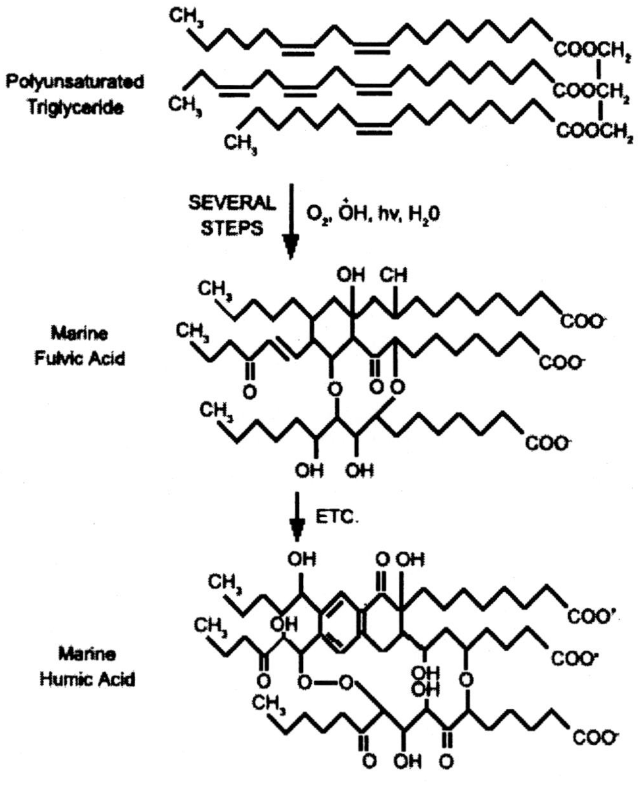

Figure 7-13. A possible pathway for the formation of marine humic acids from a triglyceride

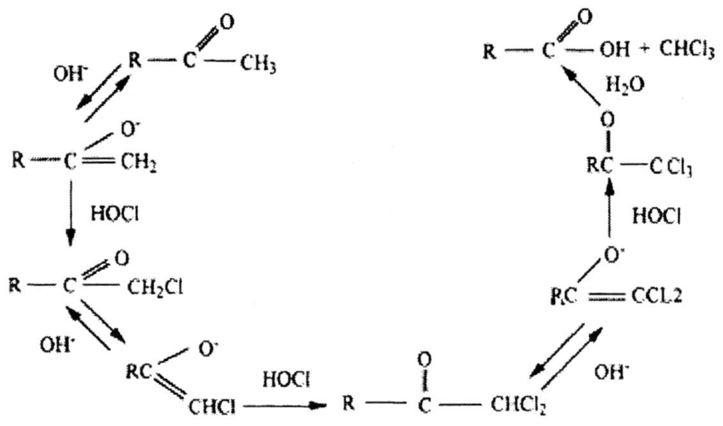

Figure 7-14. Mechanism of THM formation based on enolization and chlorination.

Thus, the enolization and halogenation of the mechanism in Figure 7-15 will give the anticipated mixed products. All intermediates listed in the scheme are found in treated water, including trichloroacetone (TCA). If iodide is present in water, HOCl will also oxidize to give an additional species, with $CHCl_2I$ being one of the species commonly observed for this case.

Similar to ammonia, the amino group of amino acids can form monochloroamino and dichloroamino functions and can serve as residual sources. For alanine, valine, and leucine, deamination occurs with the formation of α-keto carboxylic acids. For glycine, the product is HCN and CO_2, and at the end of the breakpoint curve region,

$$2NH_2CH_2COOH + 9HOCl \rightarrow N_2 + 4CO_2 + 9HCl + 5H_2O$$

In addition, most amino acids upon deamination will give chloroaldehydes and chloronitriles as intermediates. The discovery in drinking water of a new class of anthropogenic compounds generally termed haloacetylnitriles (HANs) is important (see Fig. 7-16). Their risk has not yet been thoroughly studied, and their formation is believed to involve THM (see the mechanism in Fig. 7-17).

Table 7-8 lists the purgeable THMs found in drinking waters worldwide. Efforts have been conducted to remove the THMs in drinking water by some polishing steps. So far column adsorption as well as some advented oxidation processes, such as UV-H_2O_2 or ultrasound-H_2O_2, have been explored with the possibility of abatement. In the latter process, chloroform is decomposed to CO_2:

$$CHCl_3 + 2OH \rightarrow CO_2 + 3Cl^- + 3H^+$$

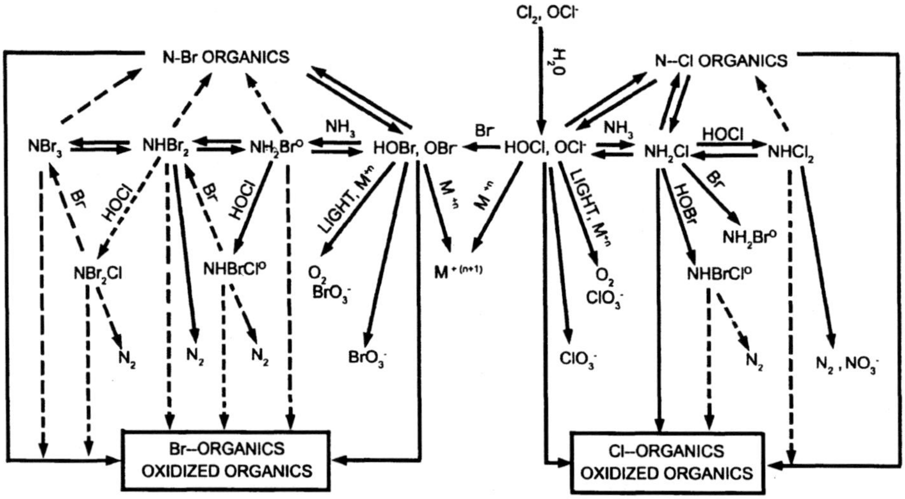

Figure 7-15. Principal pathways for reaction, degradation, and environmental fate of free available chlorine in the aquatic environment. Presumed pathways (not yet proven) are shown by dashed arrows. Compounds formed at different places in the reaction scheme are designated with superscripts a and b to assist in understanding interrelationships. Halides (Cl- and Br-) are not depicted but are products of many of the chemical pathways.

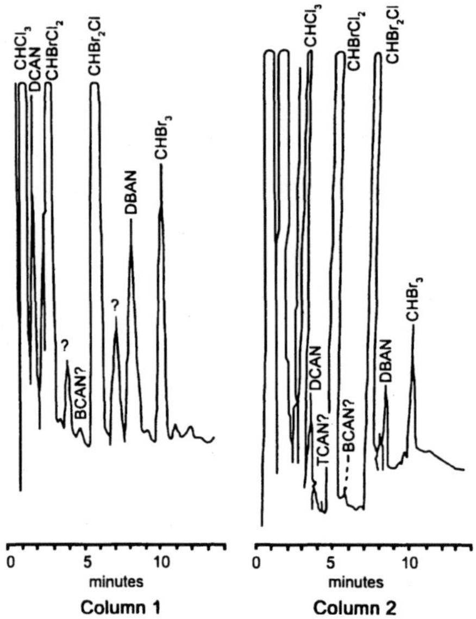

Figure 7-16. Analysis of site A drinking water using two-dimensional PDPR gas chromatography. DCAN = Dichloroacetonitrile; TCAN = Trichloroacetonitrile; DBAN = Dibromoacetonitrile; BCAN = Bromoacetonitrile. (From Ref. 7-18.)

Table 7-8. Purgeable Orfanics Found in International Drinking Waters

Description	Southern China	Northern Taiwan	Southern Philippines	Northern Philippines	Northern Egypt	Southern Indonesia
Water Source	Local	Well	River/Wells	River	River	River
Water Treatment	Sed/Filt/Cl_2	None	Sed/Filt/Cl_2	Sed/Filt/Cl_2	Sed/Filt/Cl_2	Sed/Filt/Cl
Methylene chloride	ND	4.4	ND	ND	1.6	ND
Trichlorofluoromethane	ND	ND	ND	ND	0.3	ND
Trans-1,2-Dichloroethee	ND	ND	ND	ND	0.5	ND
Chloroform	3.4	0.05	4.9	1.8	ND	6.8
Carbon tetrachloride	ND	ND	ND	ND	ND	ND
Dichlorobromomethane	7.6	ND	2.3	1.7	ND	3.0
Trichloroethene	ND	ND	0.03	0.01	1.2	ND
Dibromochloromethane	13	ND	1.2	1.5	ND	0.7
Benzene	ND	ND	0.12	ND	ND	ND
Bromoform	6.3	ND	ND	ND	ND	ND
Tetrachloroethene	0.2	ND	ND	ND	0.3	ND
Toluene	0.7	0.7	1.5	0.5	1.1	1.4
Ethyle benzene	ND	ND	0.2	ND	0.02	ND

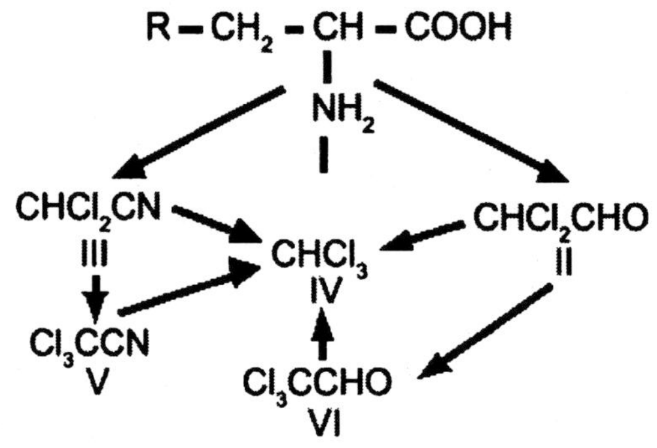

Figure 7-17. Potential HAN-THM interrelations. (From Ref. 7-17.)

7.3.2 Cyanide Conversion and Phenol Reaction

Alkaline chlorination of the CN^- ion is of great importance, since it is a detoxification procedure. At first, cyanogen chloride is formed:

$$CN^- + Cl_2 = CNCl + Cl^-$$

Further CNCl is oxidized to cyanate under pH ≈ 8:

$$CNCl + 2OH^- = CNO^- + Cl^- + H_2O$$

Finally, conversion to CO_2 and N_2 occurs:

$$2CNO^- + 4OH^- + 3Cl_2 = 2CO_2 + N_2 + 6Cl^- + 2H_2O$$

Chlorine readily undergoes electrophilic substitution to the *para-* and *ortho-* positions of phenol, resulting in dichloro (2,4 and 3,6)- and trichloro (2,4,6)-phenols, as shown in Figure 7-18. Because several of these are odorous and bitter and this type of phenol is present in water resulting from industrial discharge or natural decay (see Fig. 7-18), their formation has become of concern in water treatment.

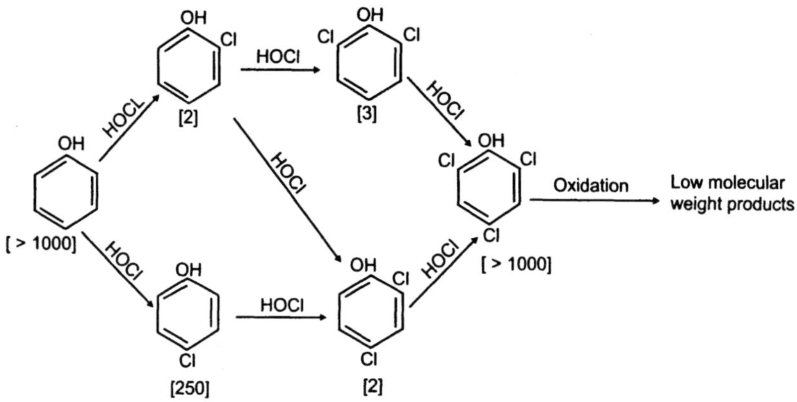

Figure 7-18. Reaction scheme for the chlorination of phenol. [ ] indicate odor threshold concentration (μg/L). (From R. H. Burttschell, A. A. Rosen, F. M. Middleton, and M. B. Ettinger, Chlorine derivatives of phenol causing taste and odor, J. Am. Water Works Assoc., 51, 205–214 (1959).

7.3.3 Specialized Disinfectants

Chloro-substituted amines bearing adjacent carbonyl groups (strong electron-withdrawing groups) can easily release HOCl upon contact with water. A good example is urea, which is known as a Cl reservoir. *N*-Bromo succinimide and dactin solids can release HOCl in water readily. Halazone tablets are used in emergencies for camping applications. "Pool aid" used in swimming pools is actually isocyanauric acid. It can readily either take up or release chlorine. Army tablets of tetraglycine hydroperiodide are used against cysts (see Fig. 7-19 for structures). The effect of bromine chloride is far better than that of chlorine (see Tables 7-9 and 7-10). Usually an overdose of chlorine (superchlorination) will create a number of chlorinated organics in all types of wastewater (see Tables 7-11 and 7-12). The environmental fate of free, available chlorine in an aquatic system is of concern.

Table 7-9. Properties of Bromine Chloride

Property	Br_2	BrCl	Cl_2
FP (°C)	-7.3	–66	-101
BP (°C)	58.8	5	–34
Density (20)°C	3.12	2.34	1.40
Vapor pressure psig (20°C)	0	30	100
Solubility (g/100g water)	3.3	8.5	0.75

Table 7-10. Disinfection Efficiency of BrCl Compared to Secondary Effluent

		Total Coliforms[a] (Average No./100ml)	
Disinfectant Dosage (mg/L)	Contact Time (min)	BrCl	Cl_2
1.0	5	< 100	TNT
	15	12	> 5,000
1.5	5	< 100	TNT
	15	24	> 4,000
2.0	5	25	~ 8,000
	15	20	280
2.5	5	20	3,000
	15	8	20

[a] Total coliforms of untreated effluent were 6×10^5/100 ml TNT (too numerous to count), suspended solids 18 mg/L, pH 7.3, temperature 57°F.

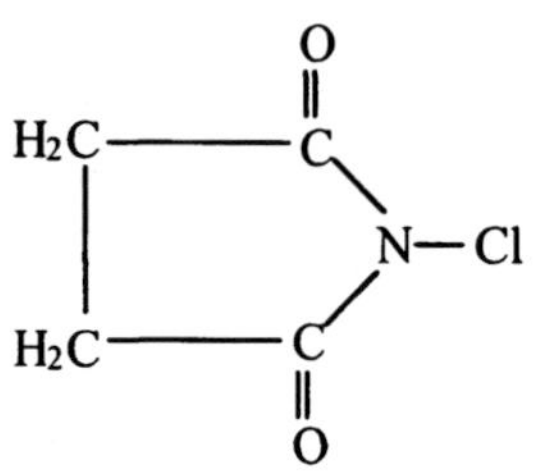

NCA (*N*-Chloroacacetamide)

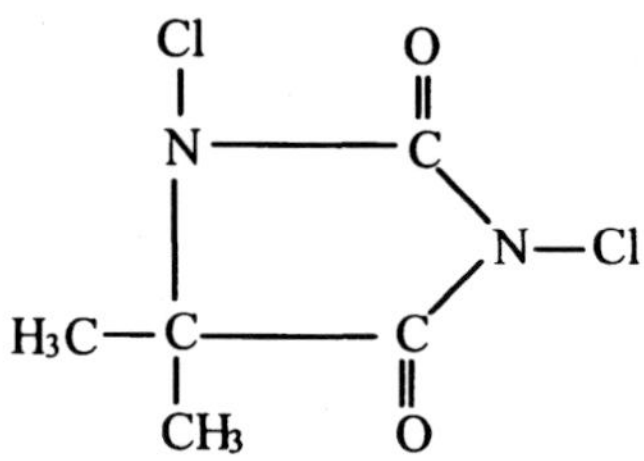

Dactin (*N,N*-Dichlorodimethyl hydantoin)

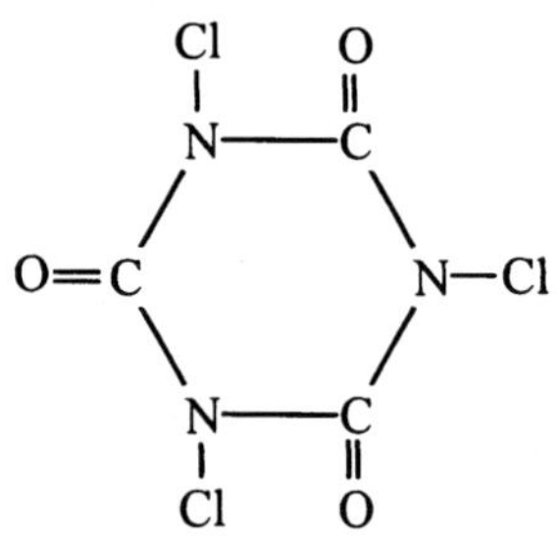

Trichloroisocyanuric acid

$[K(NH_3^+CH_2COC^-)_4]I_3^-$

Tetraglycine hydroperiodide

Figure 7-19. Some specialized disinfectants.

Table 7-11. Summary of New Chlorinated Organics Found in "Superchlorinated" Municipal Wastewater

Compound Number	Compound Name	Concentration (μg/L)
1	Chloroform	—
2	Dibromochloromethane	—
3	Dichlorobutane	27
4	3-Chloro-2-methylbutlene	285
5	Chlorocyclohexane (118)	20
6	Chloroalkyl acetate	—
7	*o*-Dichlorobenzene	10
8	*p*-Dichlorobenzene	10
9	Chloroethylbenzene	21
10	Tetrachloroacetone	11
11	Pentachloroacetone	30
12	Hexachloroacetone	30
13	Trichlorobenzene	—
14	Dichloroethylbenzene	20
15	Chlorocumene (154)	—
16	*N*-Methyl-trichloroaniline (209)	10
17	Dichlorotoluene	—
18	Trichlorophenol	—
19	Chloro-α-methyl benzyl alcohol	—
20	Dichloromethoxytoluene	32
21	Trichloromethylstyrene (220)	10
22	Trichloroethylbenzene (208)	12
23	Dichloro-α-methyl benzyl alcohol (190)	10
24	Dichloro-bis(ethoxy)benzene (220)	30
25	Dichloro-α-methyl benzyl alcohol (190)	—
26	Trichloro-*N*-methylanisole	—
27	Trichloro-α-methyl benzyl alcohol	25
28	Trichloro-α-methyl benzyl alcohol	25
29	Tetrachlorophenol	30
30	Trichloro-α-methyl benzyl alcohol	50

Table 7-11. Continued

Compound Number	Compound Name	Concentration (μg/L)
31	Trichlorocumene (222)	—
32	Tetrachloroethylstyrene (268)	—
33	Trichlorodimethoxybenzene (240)	—
34	Tetrachloromethoxytoluene (258)	4
35	Dichloroaniline derivative (205)	13
36	Dichloroaromatic derivative (249)	15
37	Dichloroacetate derivative (203)	20
38	Trichlorophthalate derivative (296)	—
39	Tetrachlorophthalate derivative (340)	—

Values in parentheses are mass number observed.

If ozone is used as an disinfectant for drinking water, one of the concerns is that the bromide ion (from marine origin), which intrudes the groundwater, will be oxidized to bromate ion. The bromate ion is known to be detrimental to human health.

7.3.4 Fluoridation, Defluoridation, and Dechlorination

Low levels of fluorine in drinking water often cause caries, and excessive amounts often result in mottled tooth enamel. Recommended and approved fluoride limits of 1 mg/L will give consumers a maximum reduction in tooth decay with no aesthetically significant dental fluorosis. The fluoride compounds commonly applied in water treatment are sodium fluoride (NaF), sodium silicofluoride (Na_2SiF_6), and fluosilicic acid (H_2SiF_6). On the other hand, defluoridation is accomplished through a packed tower of activated Al_2O_3, which can be regenerated with caustic soda. Another method employs bone char, which can be reused by backwash with a 1% NaOH solution.

In the destruction of odors and tastes in drinking water, dechlorination is conducted after superchlorination. Usually a reducing agent such as SO_2, SO_3^{2-}, HSO_3^-, or C is used. The last is activated carbon.

$$SO_2 + Cl_2 + 2H_2O \rightarrow H_2SO_4 + 2HCl$$

$$NaHSO_3 + Cl_2 + H_2O \rightarrow NaHSO_4 + 2HCl$$

$$2Na_2S_2O_3 + Cl_2 \rightarrow Na_2S_4O_6 + 2NaCl$$

$$C + 2Cl_2 + 2H_2O \rightarrow CO_2 + 4HCl$$

Table 7-12. Chloro-organics in Cooling Waters and Process Effluents

	Sewage Effluent (ppb)	Watts Bar Lake Sample (ppb)	Mississippi River Sample (ppb)
Nucleoside			
5-Chlorouridine	1.7	0.6	7
Purine			
8-Chlorocaffeine	1.7	1.1	6
6-Chloro-2-aminopurine	0.9	1.0	3
8-Chloroxanthine	1.5	3	—
Pyrimidine			
5-Chlorouracil	4	0.6	3
Aromatic acid			
2-Chlorobenzoic acid	0.3	1.1	10
3-Chlorobenzoic acid	0.6	0.2	8
4-Chlorobenzoic acid	1.1	0.3	8
3-Chloro-4-hydroxybenzoic acid	1.3	0.8	3
4-Chloromandelic acid	1.1	1.8	6
4-Chlorophenylacetic acid	0.4	3	20
5-Chlorosalicylic acid	0.2	3	18
Phenol			
4-Chloro-3-methylphenol	1.5	0.2	0.7
2-Chlorophenol	1.7	0.2	4
3-Chlorophenol	0.5	0.2	6
4-Chlorophenol	0.7	0.2	2
4-Chlororesorcinol	1.2	0.5	7

7.4 DISINFECTION IN WATER-DISTRIBUTION SYSTEMS

After chlorine is dispensed as a disinfectant in a water supply system, its effectiveness as it is transported with water in the conduit for pipeline distribution must be assured. Apparently, deactivation of residual chlorine in a pipeline delivery system may occur due to chemical reaction with the reducing agents such as the divalent metal ions or the net ions of mass transfer. Convective diffusion is essential in estimating the extent of chlorine deactivation through the long distance of pipelines. Furthermore, the laminar flow and the turbulent flow will affect deactivation differently. We will introduce some basic concepts in fluid mechanics relevant to this topic followed by some examples of its estimation.

Mass transfer and fluid mechanics are essential for environmental practices in various areas. The understanding of mass transfer and fluid mechanics will help in evaluating oxidation and reduction, electrochemistry, aeration and stripping, dispersion and tracer analysis, trickling filters and rotating biological contactors, and carbon adsorption.

7.4.1 Fluid Mechanics and Flow

Incompressible, viscous flows are described by the Navier-Stokers equation:

$$\rho\frac{\partial u}{\partial t}+\rho u\cdot\nabla u=-\nabla p+\mu\nabla^2 u+\rho g \qquad [7\text{-}24]$$

where ρ is the fluid density, u is the fluid velocity, t is time, p is pressure, μ is the absolute viscosity, and g is gravity force. This nonlinear, momentum differential equation, with the removal of the dot product term

$$\rho\frac{Du}{Dt}=-\nabla\rho+\mu\nabla^2 u+\rho g \qquad [7\text{-}25]$$

and

$$\frac{D}{Dt}=\frac{\partial}{\partial t}+u_1\frac{\partial}{\partial x_1}+u_2\frac{\partial}{\partial x_2}+u_3\frac{\partial}{\partial x_3} \qquad [7\text{-}26]$$

Another convention for the Navier-Stokes equation is the tensor notation:

$$\rho \frac{\partial u_i}{\partial t} + \rho u_k \frac{\partial u_i}{\partial x_k} = -\frac{\partial p}{\partial x_i} + \mu \frac{\partial^2 u_i}{\partial x_j \partial x_j} + \rho g_i \qquad [7\text{-}27]$$

The time derivative on the left of the equation is called the local acceleration, the first term on the right is the pressure gradient term, the second term involves viscosity and the Laplacian of velocity representing viscous stresses, and the final term on the right is the gravity term. The first two terms on the right deal with the surface force of the differential fluid element. The last term on the right is known as the body force of the fluid element. The simplest application of the Navier-Stokes equation is when there is no fluid motion, or $u = 0$, in which case

$$\nabla P = \rho g \qquad [7\text{-}28]$$

Using a coordination system of x_1, x_2, and x_3 as shown in Figure 7-19, the gravity has only one component in the negative x_2-direction. Therefore, $\partial p/\partial x_1 = 0$, and $\partial p/\partial x_3 = 0$:

$$\partial p/\partial x_2 = -\rho g_2 = -\rho g \qquad [7\text{-}29]$$

After integrating,

$$p = -\rho g x_2 + \text{constant} \qquad [7\text{-}30]$$

Let $p = 0$ at the fluid surface ($x_2 = 0$). We find that the integrating constant $= 0$. Replacing x_2 with depth D, we have:

$$p = \rho g D \qquad [7\text{-}31]$$

This is the hydrostatic equation stating that the hydrostatic pressure increases with depth linearly for a constant density and gravity.

Batchelor modified the total pressure P in the Navier-Stokes equation as:

$$P = p_0 + \rho g x + p \qquad [7\text{-}32]$$

where p_0 is constant pressure (such as atmospheric pressure), the second term is the fluid static as the equation is hydrostatic pressure, and the third term is fluid pressure.

Then take the gradient operation:

$$\nabla P = \rho g + \nabla p \qquad [7\text{-}33]$$

Substituting in the Navier-Stokes equation,

$$\rho \frac{Du}{Dt} = -\nabla p + \mu \nabla^2 u \qquad [7\text{-}34]$$

The gravity term is dropped, and the result represents viscous flow with no free surface. Let us consider the solution of unidirectional flow, particularly the plane couette flow as shown in Figure 7-20. In this case the dot product of the nonlinear convective acceleration terms is zero. Also there is only one velocity derivative in the *y*-direction, and this is a no-free-surface flow. Thus, the x_1 component is:

$$\frac{dp}{dx} = \frac{\mu d^2 u_x}{dy^2} \qquad [7\text{-}35]$$

Integrating once,

$$\left(\frac{dp}{dx}\right) y = m\left(\frac{du_x}{dy}\right) + A \qquad [7\text{-}36]$$

Since $y = 0$, then $du/dy = 0$. Thus, $A = 0$. Integrating again,

$$\frac{dp}{dx}\frac{y^2}{2} = \mu u_x + B \qquad [7\text{-}37]$$

For boundary condition at $y = a$ (or $y = -a$). If $u = 0$ and $B = (dp/dx)(a^2/2)$, we can use substitution and rearrange the equation to get:

$$u_x = -\frac{1}{2\mu}\frac{dp}{dx}\left(a^2 - y^2\right) \qquad [7\text{-}38]$$

This equation states that velocity in annular space is inversely proportional to the viscosity of the annulus squared when $y = 0$. This form is the well-known Poiseuille equation.

If now we define a term involving the gradient of stress tensor, τi_j, then the Navier-Stokes equation can be written as:

$$\rho \frac{\partial u_i}{\partial t} + \rho u_k \frac{\partial u_i}{\partial x_k} = -\frac{\partial p}{\partial x_i} + \frac{\partial \tau_{i_j}}{\partial x_i} + \rho g_i \qquad [7\text{-}39]$$

The stress tensor of a Newtonian fluid for three-dimensional, rectangular coordinates is:

$$\tau_{z_x} = \mu\left(\frac{\partial u_y}{\partial x} + \frac{\partial u_x}{\partial y}\right) \qquad [7\text{-}40]$$

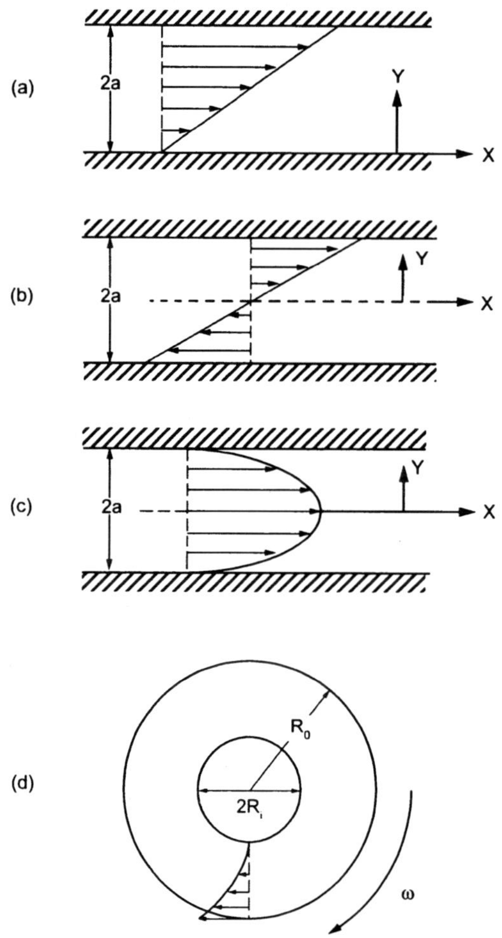

Figure 7-20. (a) Plane couette flow with one wall moving; (b) plane couette flow with walls moving at the same speed in opposite directions; (c) plane couette flow due to pressure gradient; and (d) couvette flow in circular annulus with one wall moving.

Applying this equation to Fig. 7-20(a), there is no y-direction component of velocity, or

$$\tau_{z_x} = \mu \frac{du_x}{dy} \qquad [7\text{-}41]$$

This equation is Newton's viscosity relation, which defines the stress, the viscosity, and the strain rate (du_x/dy).

Now let us reexamine the unidirectional flows in a nonsteady system with a boundary condition of time variance. Omitting gravity and pressure gradient terms (u is not function of x), the Navis-Stokes equation for Figure 7-21(c) can be written as

$$\frac{\partial u}{\partial t} = \nu \frac{\partial^2 u}{\partial y^2} \qquad [7\text{-}42]$$

where ν is kinematic viscosity, $\nu = \mu/\rho$. This well-known differential equation can be solved with the error function. The boundary conditions are:

$$u(0,t) = \begin{bmatrix} 0 \text{ for } t \le 0 \\ U_{\max} \ \text{ for } t > 0 \end{bmatrix} \qquad [7\text{-}43]$$

The solution is:

$$u(y,t) = u_{\max}\left[1 - erf\left(\frac{y}{2(\nu t)^{\frac{1}{2}}}\right)\right] \qquad [7\text{-}44]$$

In reality, the moving boundary at three successive times, t_1, t_2, and t_3, seems to diffuse out into the fluid. To check the solution of error function, the following procedure is used. When the function:

$$\left[\frac{y}{2}(\nu t)^{\frac{1}{2}}\right] = \frac{3}{2}$$

then the value of error function is 0.96, or $u(y,t)$ = 0.04 U_b. In this case, the boundary layer thickness, δ,must be:

$$\delta = 3(\nu t)^{\frac{1}{2}} \qquad [7\text{-}45]$$

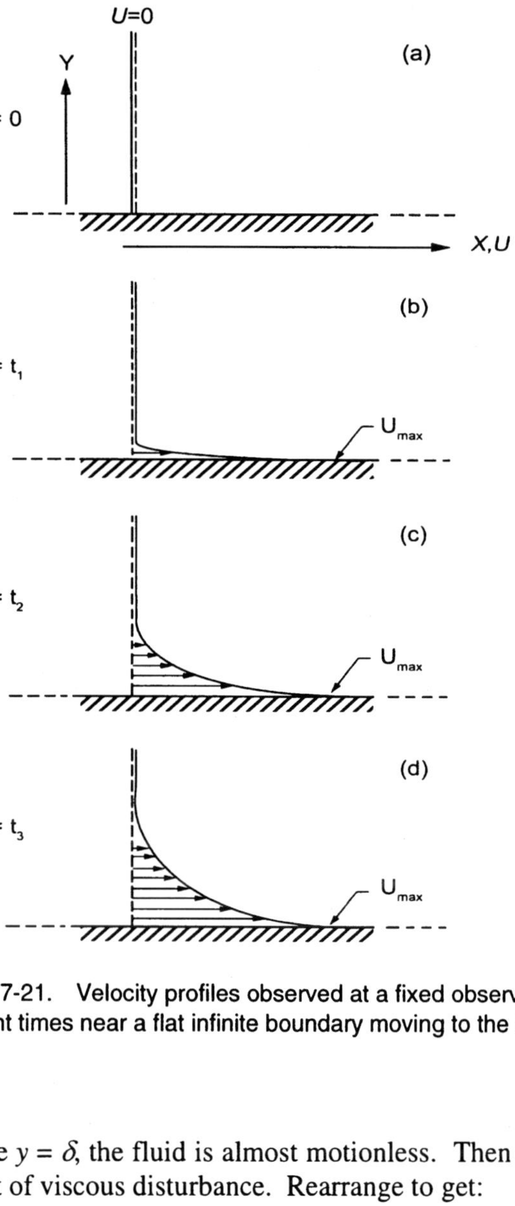

Figure 7-21. Velocity profiles observed at a fixed observation point and at four different times near a flat infinite boundary moving to the right with constant velocity Umax.

Above $y = \delta$, the fluid is almost motionless. Then the above equation gives the element of viscous disturbance. Rearrange to get:

$$t = \frac{\delta^2}{9\nu} \qquad [7\text{-}46]$$

which states that the time for viscous disturbance is proportional to the distance squared and inversely proportional to the kinematic viscosity.

The Navier-Stokes equation can be written as a nondimensional form. There are four independent variables:

$$\tilde{u} = \frac{u}{U}$$
$$\tilde{x} = \frac{x}{L}$$
$$\tilde{t} = \frac{\nu}{L^2}t \qquad [7\text{-}47]$$
$$\tilde{p} = \frac{L}{\rho \nu U}p$$

Thus,

$$\frac{\partial \tilde{u}}{\partial \tilde{t}} + Re\left(\tilde{u} \cdot \tilde{\nabla}\,\tilde{u}\right) = -\tilde{\nabla}\tilde{p} + \tilde{\nabla}^2\tilde{u}$$

If $\tilde{\nabla} = \nabla L$,

$$\text{Re} = \frac{\rho UL}{\mu} = \frac{UL}{\nu} \qquad [7\text{-}48]$$

When the Reynolds number becomes very small, Re → 0, the convection product term may be dropped:

$$\frac{\partial u}{\partial t} = -\frac{1}{\rho}\nabla p + \nu\nabla^2 u \qquad [7\text{-}49]$$

For steady flow and taking the divergence of above equation,

$$\nabla \cdot \nabla p = \mu\nabla \cdot \nabla^2 u \qquad [7\text{-}50]$$

Using identities,

$$\nabla^2 p = \mu\nabla^2(\nabla \cdot u) \qquad [7\text{-}51]$$

but

$$\nabla \cdot u = 0$$

for incompressible liquids. Therefore,

$$\nabla^2 p = 0 \qquad [7\text{-}52]$$

This is the Laplace equation for low Reynolds number for creeping flow.

7.4.2 Convective Diffusion

Let us first review diffusive transport. Frick's first law of diffusion states that the diffusive molar flux of component *A* relative to coordinates moving around with local molar average velocity is:

$$J_A = -CD_{AB}\nabla X_A \tag{7-53}$$

where C is total fluid molar concentration, D_{AB} is the diffusion coefficient of component *A* in component *B*, and X_A is the mole fraction of component *A*:

$$X_A = \frac{C_A}{C} \tag{7-54}$$

If the expression is in terms of mass, then

$$j_A = -\rho D_{AB}\nabla W_A \tag{7-55}$$

where ρ is the fluid density and W_A is the mass fraction of component *A*.

$$W_A = \frac{\rho_A}{\rho}$$

$$\begin{aligned} N_A &= C_A u - CD_{AB}\nabla X_A \\ n_A &= \rho_A u - \rho D_{AB}\nabla W_A \end{aligned} \tag{7-56}$$

A simplified version of Frick's first law assumes that $u = 0$.

$$\begin{aligned} N_A &= J_A = -D_{AB}\nabla C_A \\ n_A &= j_A = -D_{AB}\nabla \rho_A \end{aligned} \tag{7-57}$$

For Frick's second law, the mass-average velocity is zero and the diffusion coefficient and density are constant:

$$\frac{\partial \rho_A}{\partial t} + \nabla \cdot j_A = 0 \tag{7-58}$$

Substituting in the previous equation,

$$\frac{\partial \rho_A}{\partial t} = D_{AB} \nabla^2 \rho_A \qquad [7\text{-}59]$$

or, divided by molar mass of species *A*,

$$\frac{\partial C_A}{\partial t} = D_{AB} \nabla^2 C_A \qquad [7\text{-}60]$$

If u is not equal to zero and a chemical reaction occurs, then

$$\frac{\partial \rho_A}{\partial t} + \nabla \cdot n_A = r_A \qquad [7\text{-}61]$$

where r_A is the rate of chemical reaction ($\mathrm{mL^{-3}T^{-1}}$). Similarly, R_A is the molar equivalence, which has a dimension of $\mathrm{mol \cdot L^{-3}L^{-1}}$. Substituting for n_A, then

$$\frac{\partial \rho_A}{\partial t} + u \nabla \rho_A + \rho_A (\nabla \cdot u) = \nabla \cdot (\rho D_{AB} \nabla W_A) + r_A \qquad [7\text{-}62]$$

Since $\nabla \cdot u = 0$ and ρ is constant,

$$\frac{\partial \rho_A}{\partial t} + u \cdot \nabla \rho_A = D_{AB} \nabla^2 \rho_A + r_A$$

or

$$\frac{\partial C_A}{\partial t} + u \cdot \nabla C_A = D_{AB} \nabla^2 C_A + R_A \qquad [7\text{-}63]$$

This is a general version of a convective-diffusion equation containing a term on the left side involving the dot-product similar to the Navier-Stokes equation. This term is referred to as the diffusive-transportation term.

Now express the convective-diffusion equation in cylindrical coordinates. The dot product can be expanded:

$$\frac{\partial C_A}{\partial t} + u_r \frac{\partial C_A}{\partial r} + u_\theta \frac{1}{r} \frac{\partial C_A}{\partial \theta} + u_z \frac{\partial C_A}{\partial_z} = D_{AB} \left[\frac{1}{r} \frac{\partial}{\partial r} \left(r \frac{\partial C_A}{\partial r} \right) + \frac{1}{r^2} \frac{\partial^2 C_A}{\partial \theta^2} + \frac{\partial^2 C_A}{\partial z^2} \right]$$

[7-64]

This is known as Poiseuille flow. Consequently, there is no r or θ component of velocity. Thus,

$$\frac{\partial C_A}{\partial t}+u_z\frac{\partial C_A}{\partial z}=D_A\left[\frac{1}{r}\frac{\partial}{\partial r}\left(r\frac{\partial C_A}{\partial r}\right)+\frac{\partial^2 C_A}{\partial z^2}\right] \quad [7\text{-}65]$$

For larger velocity and/ or low D_{AB}, the last term (streamwise diffusion) may be dropped.

$$\frac{\partial C_A}{\partial t}+u_z\frac{\partial C_A}{\partial_z}=D_{AB}\left[\frac{1}{r}\frac{\partial}{\partial r}\left(r\frac{\partial C_A}{\partial r}\right)\right] \quad [7\text{-}66]$$

At a steady state,

$$u_z\frac{\partial C_A}{\partial z}=D_{AB}\left[\frac{1}{r}\frac{\theta}{\partial r}\left(r\frac{\partial C_A}{\partial r}\right)\right] \quad [7\text{-}67]$$

In this equation, the cross-stream component is important when compared with the diffusion arise component. Using appropriate boundary conditions, this equation has been solved. Assuming that:

$$\tilde{C}_A=\frac{C_A-C_{A,S}}{C_{A,i}-C_{A,S}} \quad [7\text{-}68]$$

where $C_{A,i}$ is the uniform inlet concentration and $C_{A,s}$ is the steady concentration, the axial coordinate z is also transformed:

$$\tilde{z}=\frac{\frac{z}{a}}{\text{ReSc}} \quad [7\text{-}69]$$

and r is nondimensional with pipe radius:

$$\tilde{r} = \frac{r}{a}$$
$$\tilde{C}_A(\tilde{r},0) = 1$$
$$\tilde{C}_A(1,\tilde{z}) = 0 \qquad [7\text{-}70]$$
$$\frac{\partial \tilde{C}_A(0,\tilde{z})}{\partial \tilde{r}} = 0$$
$$\tilde{C}_A(\tilde{r},\infty) = 0$$

The solution is:

$$\tilde{C}_A(\tilde{r},\tilde{z}) = \sum_{n=0}^{\alpha} C_n R_n(\tilde{r}) \exp\left(-\lambda_n^2 \tilde{z}\right) \qquad [7\text{-}71]$$

In the above solution, C_n is an infinite series of constants, and R_n represents a function that depends only on the dimensionless coordinate r. Both r and λ_n are the eigenvalues. For pipe-flow problems, we generally adopt an exit concentration, C_A, as an average weighted by the local velocity.

$$C_{A,av} = \frac{2\pi \int_0^a C_A(r,\tilde{z})\, u(r)\, r dr}{\pi a^2 u_{av}} = \frac{2}{a^2 u_{av}} \int_0^a C_A(r,z)\, u(r)\, r dr \qquad [7\text{-}72]$$

Friedlander as well as Skelland treated the problem as:

$$\frac{C_{A,i} - C_{A,av}}{C_{A,i} - C_{A,S}} = \frac{C_{A,av} - C_{A,i}}{C_{A,S} - C_{A,i}} = 1 - 8 \sum_{n=0}^{\infty} \frac{G_n}{\lambda_n^2} \exp\left(-\lambda_n^2 \tilde{z}\right) \qquad [7\text{-}73]$$

where $G_n = (C_n/2)\, R'n(1)$ and the solutions for the first five constants are:

n	λ^2_n	G_n
0	7.312	0.749
1	44.62	0.544
2	113.8	0.463
3	215.2	0.414
4	348.5	0.382

If $n > 4$, then

$$\lambda_n = 4n + \frac{8}{3} \qquad [7\text{-}74]$$
$$G_n = 1.01276\lambda_n^{-1/3}$$

[Example 7-5] After the directing water leaves a treatment plant, chlorine is used as a disinfectant for a 75-m length of pipeline. Suppose the initial concentration of Cl_2 is 0.5 mg/L, the pipe diameter is 5 cm, average velocity of flow is 1 cm/sec, and the temperature is 20°C. Find the maximum decrease in chlorine concentration at the end of the pipeline, assuming that chlorine is only consumed at the pipe surface (after M. M. Clark).

Rearranging Equation [7-73],

$$\frac{C_{A,av}}{C_{A,i}} = 8\sum_{n=0}^{\infty}\frac{G_n}{\lambda_n^2}\exp\left(-\lambda_n^2\tilde{z}\right)$$

The Reynolds number is:

$$\mathrm{Re} = \frac{2aU_{av}}{\nu} = \frac{5\mathrm{cm}\,(1\mathrm{cm/sec})}{0.01\,\mathrm{cm}^2/\mathrm{sec}} = 500$$

The Schmidt number is:

$$\mathrm{Sc} = \frac{\nu}{D_{AB}} = \frac{0.01\ \mathrm{cm}^2/\mathrm{sec}}{1.22\times10^{-5}\ \mathrm{cm}^2/\mathrm{sec}} = 820$$

The dimensionless axial coordinate is:

$$\tilde{z} = \frac{z/a}{\mathrm{ReSc}} = \frac{(75\ \mathrm{m})(100\ \mathrm{cm/m})/2.5\mathrm{cm}}{(500)(820)} = 7.32\times10^{-3}$$

Using λ^2_n and G_n values of five terms,

$$\frac{C_{A,av}}{C_{A,i}} = 0.866$$

The maximum decrease along the pipe is 13%.

In this section we have learned the momentum transfer from the fluid mechanics point of view where the momentum flux,

$$J = \mu du/dz \qquad [7\text{-}75]$$

becomes important. There are other approaches in which mass transfer and heat transfer are equally important. For example, the fluxes can be represented as:

$$J_A = -D_{AB}\frac{dC_A}{dz} \qquad [7\text{-}76]$$

$$q = -k\frac{dT}{dz} \qquad [7\text{-}77]$$

Here, mass transfer (Eq. [7-76]) is analogous to heat transfer (Eq. [7-77]). Actually, the latter is Fourier's law of heat conduction, where k is thermal conductivity (energy/$TL°kC$) and

$$k = \alpha C_p \rho \qquad [7\text{-}78]$$

where α is thermal diffusivity, C_p is the specific heat of the medium, and ρ is fluid density. Similarly, the diffusion coefficient, D_{AB}, can be related to the mass transfer coefficient, k_c.

$$k_c = \frac{D_{AB}}{\Delta z} \qquad [7\text{-}79]$$

thus

$$J_A = k_c C_A^* \qquad [7\text{-}80]$$

where C_A^* is the concentration difference.

A discussion of this concept can be found in Chapter 5 for the dimensionless groups. The analogy has been studied by Chilton and Colburn for mass transfer and heat transfer. For example,

$$\frac{k_c}{u}\mathrm{Sc}^{\frac{2}{3}} = \frac{h}{\rho u C_p}\mathrm{Pr}^{\frac{2}{3}} \qquad [7\text{-}81]$$

This analogy can be extended to momentum transfer.

For a disinfection system, although laminar flow is common in smooth pipe, turbulence may be anticipated at certain distances. For turbulent flow, Brodkey and Hershey developed a correlation: for 10,000 < *Re* < 100,000, and 432 *Sc* < 97,600,

$$\mathrm{Sh} = 2\frac{ak_c}{D_{AB}}0.0096\,\mathrm{Re}^{0.913}\,\mathrm{Sc}^{0.346}$$

Also, in a number of cases chemical loss occurs due to chemical reaction along distribution distances. If the chemical reaction is disconnected, a pseudo plug-flow condition can be adopted (see Example 7-2).

[Example 7-6] For the chlorine disinfection of drinking water, a pipeline is 200 m long with a diameter of 20 cm and an initial Cl_2 concentration of 0.5 mg/L. The velocity of flow is 0.5 m/sec, and the temperature is 20°C. Find the decrease in Cl_2 concentration, assuming the concentration at pipe surface is zero. (Adapted from M. M. Clark).

$$\mathrm{Re} = 2\frac{au_{av}}{\nu} = \frac{20\text{ cm }(0.5\text{ m/sec})(100\text{ cm/m})}{0.01\text{ cm}^2/\text{sec}} = 100{,}000$$

$$\mathrm{Sc} = \frac{\nu}{D_{AB}} = \frac{0.01\text{ cm}^2/\text{sec}}{1.22\times10^{-5}\text{ cm}^2/\text{sec}} = 820$$

Using the Brodkey-Hershey correlation,

$$k_c = \frac{D_{AB}\mathrm{Sh}}{2a} = D_{AB}(0.0096)\mathrm{Re}^{0.913}\frac{\mathrm{Sc}^{0.346}}{2a}$$

$$= \frac{1.22\times10^{-5}\text{ (cm}^2/\text{sec)}(0.0096)(10^5)^{0.0913}\,820^{0.346}}{2(10\text{ cm})} = 2.2\times10^{-3}\text{ cm/sec}$$

The flux to the pipe, as shown in Figure 7-22 from Equation [7-80], is:

$$J_A = k_c C_A$$

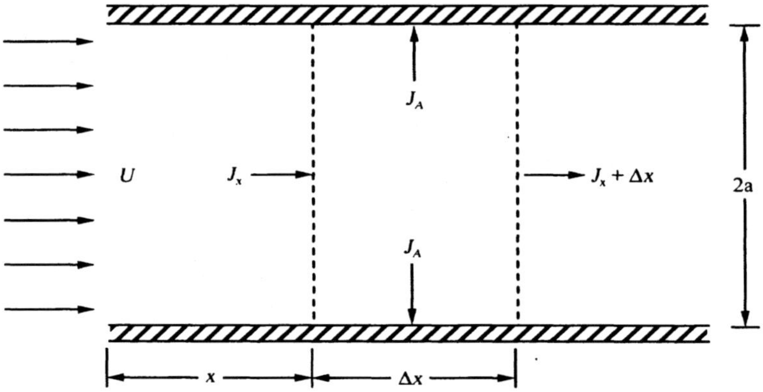

Figure 7-22. Flow diagram for Example 7-5, showing the flux of chlorine in water within a pipe.

Assuming the uniform concentration is radial direction and plug flow along the pipe, then

$$V\frac{\partial C_x}{\partial t} = J_x A - J_{x+\Delta x} A - 2\pi a \Delta x J_A$$

where A (area) = πa^2 and V (volume) = $\pi a^2 \Delta x$. Let $J_x = UC_x$, then

$$J_{x+\Delta x} = U\left(C_x + \frac{\partial C_x}{\partial x}dx\right)$$

The above equation becomes:

$$\frac{\partial C_x}{\partial t} + U\frac{\partial C_x}{\partial x} = \frac{2J_A}{a}$$

In a steady state,

$$U\frac{dC_x}{dx} = -\frac{2k_c C_x}{a}$$

The initial condition is $C_x = C_{x=0}$ at $x = 0$. Integrating,

$$C_x = C_{x=0}\exp\left(-\frac{2k_c}{aU}x\right)$$

or

$$C_{x=200\,\mathrm{m}} = C_{x=0} \exp\left[-\frac{2(2.2 \times 10^{-5}\,\mathrm{m/sec})\,200\,\mathrm{m}}{(0.1\,\mathrm{m})(0.5\,\mathrm{m/sec})}\right] = 0.84 C_{x=0}$$

$$C_{x=200} = 0.42\ \mathrm{mg/L}$$

The decrease in Cl_2 concentration is 16%.

REFERENCES

7-1. R. Beckett, *Surface and Colloid Chemistry in Natural Waters and Water Treatment*, Plenum Press, 1990.

7-2. R. Beckett, The surface chemistry of humic substances in aqueous systems, in *Surface and Colloidal Chemistry in Natural Waters and Water Treatment*, Plenum Press, New York, 1990, pp. 3–20.

7-3. R. E. Smith, *Ion Chromatography Applications*, CRC Press, Boca Raton, Florida, 1988.

7-4. H. Chick, An investigation of the laws of disinfection, *J. Hyg.*, 8, 698 (1908).

7-5. N. A. Clarke and P. W. Kabler, The inactivation of purified coxsackie virus in water by chlorine, *Amr. J. Hyg.*, 59, 119 (1954).

7-6. B. M. Cow and P. V. Roberts, Halogenated byproduct formation by ClO_2 and Cl_2, *J. Div. Eng., Proc. Am. Soc. Civ. Eng.*, 107(EE4), 609–618 (August 1981).

7-7. A. Kelner, Effects of visible light on the recovery *tomyces Griseus Conidia* from ultraviolet radiation injury, *Proc. Natl. Acad. Sci.*, 35, 73–79 (1949).

7-8. C. A. Lawrence and S. S. Block, *Disinfection, Sterilization and Preservation*, Lea and Febiger, Philadelphia, 1968.

7-9. J. R. Neefe, J. B. Baty, J. J. Reinhold, and J. Stokes, Jr., Disinfection of water containing causative agent of infectious (epidemic) hepatitis, *J. Amr. Med. Assoc.*, 128, 1076 (1945).

7-10. O. K. Scheible and C. D. Bassell, Ultraviolet disinfection of a secondary wastewater treatment plant effluent, Municipal Environmental Research Laboratory, Cincinnati, EPA 600/2-81-152, August 1981.

7-11. G. Sykes, *Disinfection and Sterilization*, D. Van Nostrand Co., Inc., New York, 1958.

7-12. G. D. White, *Disinfection of Wastewater and Water for Re-use*, Van Nostrand Reinhold Company, New York, 1978.

7-13. E. M. Witkin, Ultraviolet repair in *Escherichia coli*, *Bacteri. Rev.*, 40, 869 (December 1976).

7-14. H. W. Wolf, A. C. Petrasek, Jr., and S. E. Esmond, Utility of the UV disinfection of secondary effluent, in *Proceedings of the National Symposium, Progress in Wastewater Disinfection Technology*, Cincinnati, Ohio, September 1978, EPA-600/9-79-018, June 1979.

7-15. L. G. Rich, *Unit Processes of Sanitary Engineering*, Wiley, New York, 1963.

7-16. G. Berg, The virus hazard in water supplies, *J. N. Eng. Water Works Assoc.*, 78, 79 (1964).

7-17. M. C. Trehy and T. I. Bieker, Detection, identification and quantitative analysis of dihaloacetonitriles in chlorinated natural waters, in *Advances in Identification and Analysis of Organic Pollutants in Water*, Ann Arbor, Michigan, 1981, pp. 941–978.

7-18. L. H. Keith, R. C. Hall, R. C. Hanisch, J. E. Henderson, and R. G. Landolt, New methods for gas chromatographic analysis of water pollutants, in *Water Chlorination, Environmental Impact and Health Effects*, Vol. 4, Book 1 (R. L. Jolley, etc.) Ann Arbor, Michigan, 1983, pp. 563–582.

7-19. A. R. Trussell, J. L. Cromer, M. D. Umphres, P. E. Kelley, and J. G. Moncur, Monitoring of volatile halogenated organics: a survey of twelve drinking waters from various parts of the world, in *Water Chlorination, Environmental Impact and Health Effects*, Vol. 3 (R. L. Jolley, W. A. Brungs, R. B. Cunning, and V. A. Jacobs, Ed.), Ann Arbor, Michigan, 1980, pp. 39–53.

7-20. R.A. Granger, *Fluid Mechanics*, Dover, Mineola, New York, 1995.

7-21. E.A. Bryant, G.P. Fuelton, and G.C. Budd, *Disinfection Alternatives for Safe Drinking Water*, Van Nostrand Rienhold, New York, 1992.

7-22. S. Miller, Disinfection products in water treatment, *Environ. Sci. Technol.*, 27 (1993).

7-23. D. Grasso, Ozonation of drinking water, *Ozone Sci. Eng.*, 9, 109–124 (1987).

7-24. S. W. Krasner, M. J. McGuire, J. G. Jacangelo, and N. L. Patania, The occurance of disinfection by-products in U.S. drinking water, *Am. Water Works Assoc. J.*, 81, 41–53, (1989).

7-25. R. W. Gilpin, S. B. Dillon and P. Keyser, Disinfection of circulating water systems by ultraviolet light and halogenation, *Water Res*. 19(7), 839–848 (1985).

7-26. H. Hajenian, Alternatives to chlorine wastewater disinfection in *Viruses and Wastewater Treatment*, Pergamon, Oxford, 1981, pp. 171–174.

7-27. H. Cember, *Introduction to Health Physico*, Mc-Graw Hill, New York, 1983.

7-28. C. A. Lawrence and S. S. Block, *Disinfection, Sterilization and Preservation*, Lea and Febiger, Philadelphia, 1968.

7-29. S. Farooq, T. D. Kurucz, and T. D. Warte, Disinfection of wastewaters: high-energy electron vs. gamma irradiation, *Water Res.*, 1177–1184 (1993).

7-30. E. Blatchley, Disinfection and antimicrobial processes, *Water Environ. Res.*, 361–368 (1994).

7.31. M. Luckiesh, *Application of Germicidal, Erythermal and Infrared Energy*, D. Van Nostrand, New York, 1946.

PROBLEM SET

1. Results of a chlorine demand test on a raw water are as follows:

Sample Number	Chlorine Dosage (mg/L)	Residual Chlorine after 10-min Contact (mg/L)
1	0.20	0.19
2	0.40	0.36
3	0.60	0.50
4	0.80	0.48
5	1.00	0.20
6	1.20	0.40
7	1.40	0.60
8	1.60	0.80

Sketch a chlorine demand curve. What is the breakpoint dosage, and what is the chlorine demand at a dosage of 1.2 mg/L?

2. A dosage of 0.60 mg/L liquid chlorine is needed to maintain a residual of 0.20 mg/L in the distribution system. What dosage rate in kg/day is needed to treat 40,000 m^3/d?

3. How many grams of available chlorine are contained in 1.0 L of sodium hypochlorite with a strength of 15%? How many liters would be required for a dosage of 0.6 mg/L to 4000 m^3/day of water?

4. A new main is disinfected with water containing 50 mg/L of chlorine by feeding a 1.0% solution of chlorine to the water entering the pipe. (a) How many kilograms of dry hypochlorite powder containing 70% available chlorine must be added to 160 L of water to make 1.0% solution? (b) At what rate should this 1.0% solution be applied to the water entering the pipe to provide a concentration of 50 mg/L?

CHAPTER 8

COMBUSTION AND INCINERATION

Incineration is an old practice employed in various parts of the world for the conversion of organic matter containing moisture. Some configurations vent or dump into the open sea (Vulcanus), while others make use of a platform (Old Rig). However, the recent use of catalytic incineration has greatly changed approaches to the air emission problem.

This chapter will first review combustion chemistry. Flame retardants, soot, and toxic substances (such as dioxin) will be discussed. Then the thermochemistry dealing with incineration will be examined.

8.1 COMBUSTION CHEMISTRY

The rapid oxidation process of fuel, accompanied by exothermic steps, with conduction, convection, and radiation of thermal energy is usually termed **combustion**. The combustion process involves mobile, stationary, and industry (roasting) methods. The fuels involved range from gas to liquid to solid, such as LPG, JP-4, diesel, oil, coal, and propellants. They can be single-component (such as CH_4) or multicomponent (such as petroleum) fuels. Gases can even range from producer gas ($CO + N_2$) to coal gas ($H_2 + CH_4$) to water gas ($CO + H_2$). In order to regulate the rate of the oxidation reaction, a diluent such as N_2 is used. For example, the combustion of CH_4 (a fuel) by oxygen occurs too fast, so air is used (with approximately 79% nitrogen) to adjust the combustion rate. Fundamentally, there are three distinctly different components in a combustion system: the fuel, the oxidant, and the diluent.

A fuel containing energy-rich bonds, such as carbon-carbon, carbon-hydrogen, and hydrogen-hydrogen bonds, is a common source of chemical potential energy. During combustion, such fuels are oxidized and their chemical potential energy is transformed into thermal energy. The oxidant is most commonly oxygen, which reacts with the fuel during combustion. A diluent is a substance that does not participate chemically but often influences the combustion process by acting as a thermal sink to limit the temperature increase achieved by combustion. Nitrogen, as mentioned above, is the most common diluent. A combustion process can be expressed as:

$$\text{Fuel} + \text{oxidant} + \text{diluent} \rightarrow \text{products}$$

8.1.1 Equivalence Ratio

A stoichiometric mixture of fuel and oxidant is one in which the quantity of oxidant present is just sufficient to completely oxidize the fuel. For example, the simplest hydrocarbon is methane. Stoichiometric combustion requires two moles of oxygen for each mole of methane.

$$CH_4 + 2\,O_2 \rightarrow CO_2 + 2\,H_2O$$

A generalized formula for the combustion of hydrocarbon in air is:

$$CH_y + nO_2 + n\left(\frac{79}{21}\right)N_2 \rightarrow aCO_2 + (1-a)CO + bH_2O + \left(\frac{y}{2} - b\right)H_2 + \left[n - a - \frac{1-a}{2} - \frac{b}{2}\right]O_2 + n\left(\frac{79}{21}\right)N_2$$

A combustible mixture that contains excess oxidants is referred to as "lean." In the reverse case, where there are insufficient oxidants for complete oxidation of the available fuel, the mixture is referred to as "rich." It is important to know not only whether a combustible mixture is rich, lean, or stoichiometric, but also how rich or lean it is. Fuel:air or air:fuel ratios are the most common ways of quantitatively expressing the richness or leanness. If n_f is the mole of fuel, M_f is the molecular weight of fuel, and n is the number of moles of oxygen per mole of fuel in the above equation, then the fuel:air ratio of the mixture is either:

$$\left(\frac{F}{A}\right)_{mol} = \frac{n_f}{n\left[1 + \left(\frac{79}{21}\right)\right]} \qquad [8\text{-}1]$$

or

$$\left(\frac{F}{A}\right)_{wt} = \frac{n_f M_f}{n\left[1 + \left(\frac{79}{21}\right)\right]29} \qquad [8\text{-}2]$$

where the molecular weight of air is 29. Air:fuel ratios by nature are reciprocals of fuel:air ratios.

Often it is desirable to compare the richness or leanness of combustion for different fuels. The equivalence ratio, Φ, is convenient for this type of comparison, and it is defined as the quotient of the actual air: fuel ratio divided by the stoichiometric air: fuel ratio.

$$\Phi = \frac{\left(\frac{A}{F}\right)_{\text{actual}}}{\left(\frac{A}{F}\right)_{\text{theoretical}}} \qquad [8\text{-}3]$$

Diagrams relating the products derived at different values of Φ are discussed further in the Air Pollution chapter of the Environmental Chemistry series.

8.1.2 Free Radicals Process

Free radicals, an important factor in combustion reactions, proceed with the ordinary initiation, propagation, and termination reactions. A number of researchers have investigated methane-air flame at atmospheric pressure. To summarize, some of the events are presented below (following the tradition of combustion chemistry, the dot sign of the free radical is omitted):

1. Formation of CH_3:

$$CH_4 + O_2 \rightarrow CH_3 + HO_2$$

$$CH_4 + H \rightarrow CH_3 + H_2$$

$$CH_4 + OH \rightarrow CH_3 + H_2O$$

$$CH_4 + M \rightarrow CH_3 + H + M$$

2. Interaction with O to form CO and CO_2:

$$CH_3 \xrightarrow{O} CH_2O \xrightarrow{O} CHO \xrightarrow{O_2} CO \xrightarrow{OH} CO_2$$

3. Conversion of CO to CO_2:

$$CO + OH \rightarrow CO_2 + H$$

4. Chain-branching step:

$$H + O_2 \rightarrow OH + O$$

A more detailed mechanism of the CH_4 - O_2 system is presented in Figure 8-1.

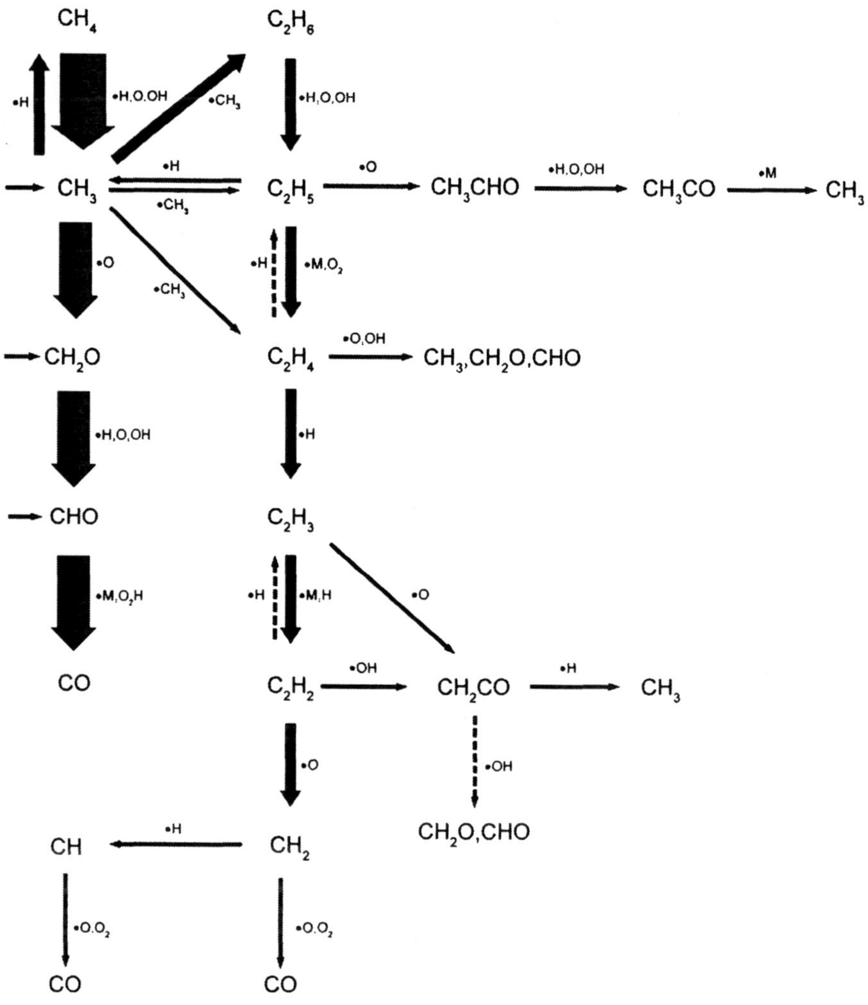

Figure 8-1. Flow diagram for the oxidation of CH_4 in stoichiometric methane-air flames at $P = 1$ bar, $T_u = 298$ K. The calculations were done with a one-dimensional laminar flat flame model. The thickness of the arrows is proportional to the reaction rates integrated over the whole flame front.

8.1.3 Zeldovich Mechanism

The formation of nitrogen oxide results from the high-temperature combination of nitrogen and oxygen present in the air. This is often called the **Zeldovich mechanism**.

$$N_2 + O \rightarrow NO + N$$

$$N^{\cdot} + O_2 \rightarrow NO + O$$

It is also possible that:

$$N + OH \rightarrow NO + H$$

Furthermore, the following reaction has been studied:

$$CH + N_2 \rightarrow HCN + N$$

Therefore in rich flames, hydrogen cyanide is probably the source of nitric oxide (for details see Fig. 8-2).

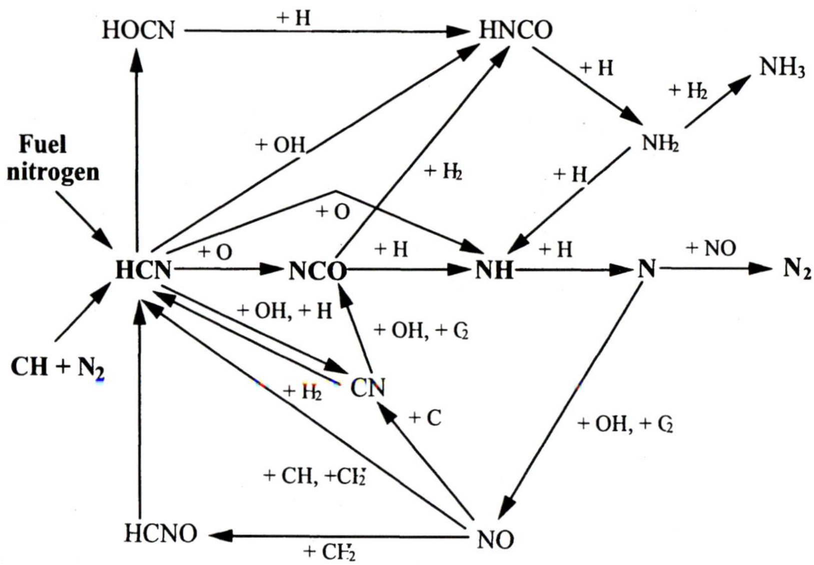

Figure 8-2. Hydrogen cyanide as the source for nitric oxide.

There are two basic flame types, the **premixed** and the **non-premixed**. Both types can be turbulent or laminar in fluid motion. An example of the turbulent premixed flame type is the spark-ignited gasoline engine (Otto engine) and for the laminar, the Bunsen flame. For the second flame type, the Diesel engine is an example of the turbulent case, while the candle flame or wood fire is an example of the laminar case.

8.1.4 Flame Retardants

Flame retardants are chemicals added to a material that resist burning when subjected to a relatively low-energy ignition source, such as a match, candle, cigarette, or stove burner. Some flame retardants are fire retardant but rarely become fireproof. Flame retardants are used in insulation material, cables, automobile seat cushions, protective coatings, carpets, draperies, and even garments such as children's sleepwear. The chemical classes are phosphorus, chlorine and bromine, antimonyl trioxide, alumina trihydrate, and boron. These functional groups or compounds can be incorporated into textiles by a treatment procedure. Some phosphate-containing flame retardants for the treatment of fabrics are shown in Figure 8-3. Phosphorus or boron is used to greatly increase the conversion of organics to char during flame. Consequently, the amount of flammable carbon-containing gas is greatly decreased to inhibit combustion. The function of bromine or chlorine is to remove the hydrogen free radical from the flaming gas, so that combustion is interrupted.

For example,

$$HBr + H \rightarrow H_2 + Br$$

$$H + Br + M \rightarrow HBr + M$$

Also, the hydroxyl free radicals are reduced:

$$HBr + OH \rightarrow H_2O + Br$$

The antimony trioxide is usually added to the chlorinated or brominated material. Upon burning, the released HCl or HBr will react with Sb_2O_3 to form antimony chloride ($SbCl_3$) or antimony oxychloride ($SbOCl_3$). These species are gaseous and will participate in the reactions of removing both H and OH radicals.

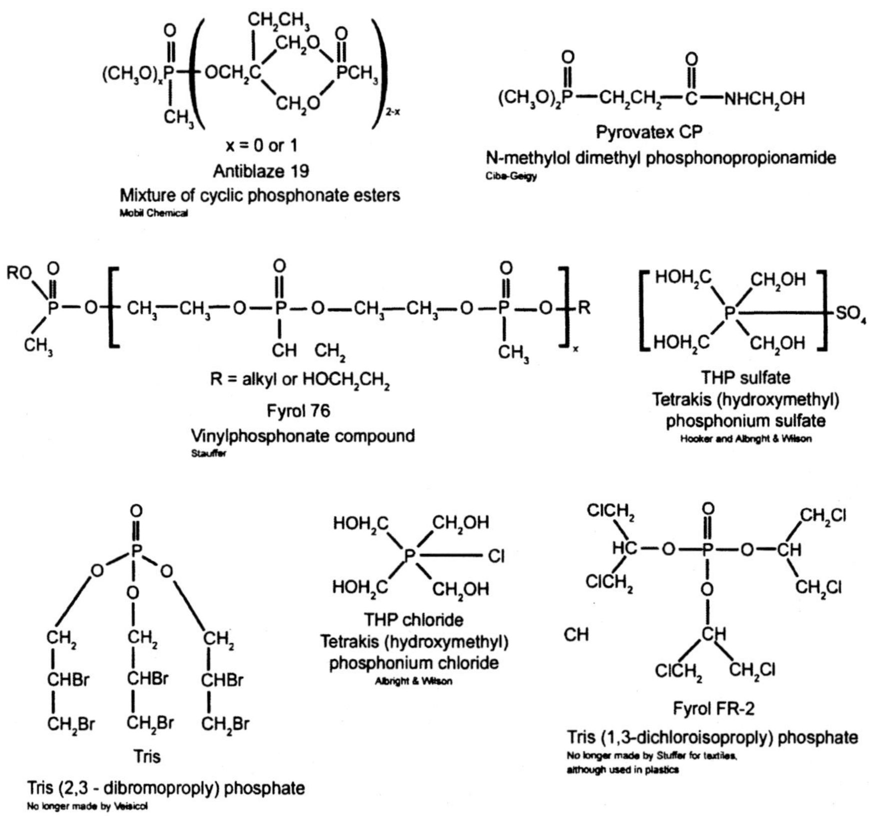

Figure 8-3. Some flame retardants now or previously used to treat textiles.

8.2 THERMOCHEMISTRY AND INCINERATION

The quantity of heat absorbed or liberated in a chemical reaction is usually termed the heat of reaction, ΔH: positive values signify absorption of heat, whereas negative values indicate that heat is liberated. The heat of reaction can be determined from either the heat of formation or the heat of combustion, as illustrated by Figure 8-4. For example,

$$CO_2(g) + H_2(g) = CO(g) + H_2O(g); \quad \Delta H_{25} = 9.838 \text{ kcal}$$

From Figure 8-4,

$$\Delta H = \Sigma(\Delta H_f)_P - \Sigma(\Delta H_f)_R \qquad [8\text{-}4]$$

The terms *P* and *R* represent products and reactants, respectively. It can be calculated from the heat of formation that:

$$CO_2(g) = -94.052 \text{ kcal}$$

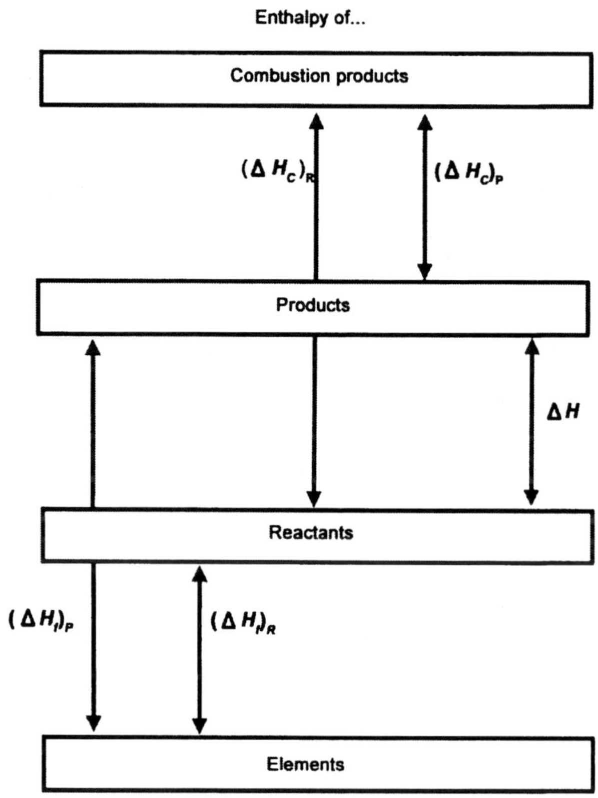

Figure 8-4. Enthalpy levels. [From T.E. Corrigan, Chemical engineering fundamentals, chemical equilibrium — 1, *Chem. Eng.*, 61, 222(1954)].

$$H_2(g) = 0$$

$$CO(g) = -26.416 \text{ kcal}$$

$$H_2O(g) = -57.798 \text{ kcal}$$

Hence, a value of 9.838 kcal is obtained. Alternatively, the heat of reaction can be evaluated from the heat of combustion, again from Figure 8-4.

$$\Delta H = \Sigma(\Delta H_c)_R - \Sigma(\Delta H_c)_P \qquad [8\text{-}5]$$

Similarly, the heats of combustion are:

$$CO_2(g) = 0$$

$$H_2(g) = -57.798 \text{ kcal}$$

$$CO(g) = -67.636 \text{ kcal}$$

$$H_2O(g) = 0$$

when $\Delta H = 9.838$ kcal.

A number of volume-reduction techniques are needed for resource recovery. The incinerator is one of the earliest workable machines that employs a reduction technique. The incinerator must meet all pollution-control requirements, and in particular, its volatile organic carbon (VOC) must be completely destroyed. Modern incinerators are modular in design; for example, they are equipped with a catalyst performance with which most of the VOC can be removed (see Table 8-1). They can be used in the open sea (Vulcanus) or mounted on a platform (Old Rig); they do not need post-treatment of the exhaust (Catcom). Most modular units have the following features: (a) hot effluent is used for the generation of steam; (b) the unit is portable; (c) it is also equipped with a control device for pollution; and (d) the quality of ash can be used for landfills. Normally, a conventional traveling grate incinerator (Fig. 8-5) with auxiliary firing will be able to reduce the bulk volume of raw refuse to 90%. Solid waste after thermal treatment still retains some heating value, and it is termed **refuse derived fuel** (RDF).

The most important principle is a computation of mass balance and energy balance. In order to achieve an autogenous (self-sustaining with auxiliary firing) state, the heat of combustion of refuse, the free value of refuse, or its related materials, become important. Usually this value can be read from a graph (Fig. 8-6). It can be approximated empirically from:

$$h = 127R + 400 \text{ (in cal/g)} \quad [8\text{-}6]$$

where

$$R = 0.25\,(2.66\%\text{C} + 7.94\%\text{H} - \%\text{O}) \quad [8\text{-}7]$$

for any organic sources containing C,H, and O; h can be expressed in Btu/lb simply by multiplying it with (450 × 3.97)/1000 (empirical method 1).

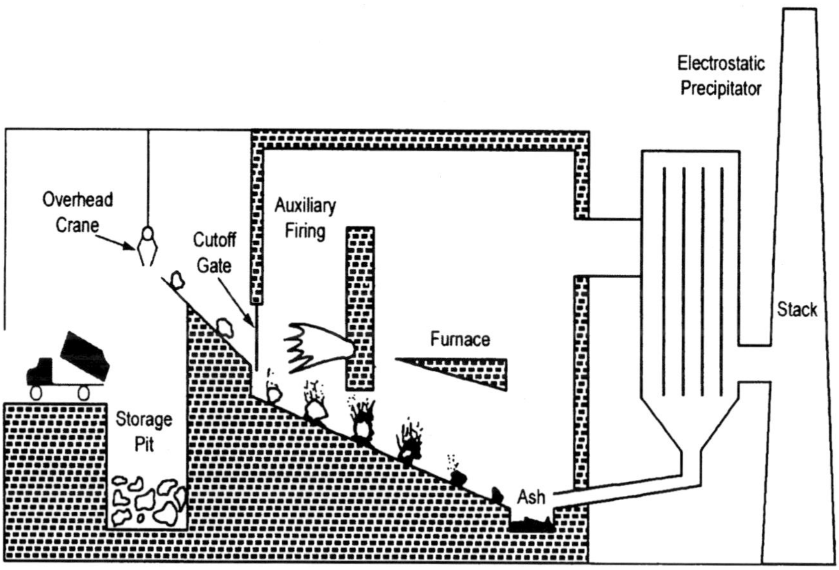

Figure 8-5. Schematic of conventional traveling grate incinerator.

Another empirical method for determining heat of combustion is:

$$h = (\Delta(O_2)/MW) \times 435 \text{ KJ/g} \quad [8\text{-}8]$$

where $\Delta(O_2)$ is the number of moles of oxygen required to burn any mole of organic compound. For example, in the case of cellulose or solid waste:

$$C_6H_{10}O_5 + 6O_2 = 6CO_2 + 5H_2O$$

Table 8-1. Catalyst Performance in Japan

Name of Constituent	Chemical Formula	Formula Weight	Temp. Rise at 1000 ppm (°C)	Operating Temp. (°C)	Concentration Before Treatment (ppm)
Styrene	$C_6H_5CHCH_2$	104.15	138	250	310
Acetaldehyde	CH_3CHO	44.05	35	350	240
Benzene	C_6H_6	78.12	103	210	380
Toluene	$C_6H_5CH_3$	92.14	123	210	320
m-Xylene	$C_6H_4(CH_3)_2$	106.17	143	210	270
Phenol	C_6H_5OH	94.11	101	300	380
Formaldehyde	HCHO	30.03	17	150	410
Acrolein	CH_2CHCHO	56.06	51	180	500
Acetic acid	CH_3COOH	60.05	26	350	590
Butyric acid	C_3H_7COOH	88.11	66	250	370
Acetone	CH_3COCH_3	58.08	57	350	410
Methyl ethyl ketone	$CH_3COC_2H_5$	72.11	74	220	380
Ethyl acetate	$CH_3COOC_2H_5$	88.11	68	350	350
Butyl acetate	$CH_3COOC_4H_9$	116.16	108	350	480
Methyl alcohol	CH_3OH	32.04	21	150	830
Ethyl alcohol	C_2H_5OH	46.07	44	350	550
Isopropyl alcohol	C_3H_7OH	60.10	64	280	230
Butyl alcohol	C_4H_9OH	74.12	84	260	330
Carbon monoxide	CO	28.01	9	150	4000
Methyl cellosolve	$HOCH_2CH_2OCH_3$	76.10	55	300	110
Ethyl cellosolve	$HOCH_2CH_2OC_2H_5$	90.12	76	300	80
Butyl cellosolve	$HOCH_2CH_2OC_4H_9$	118.18	118	300	50

Source: Sumitomo Metal Mining Co., Ltd., brochure 83.6.05 NM.

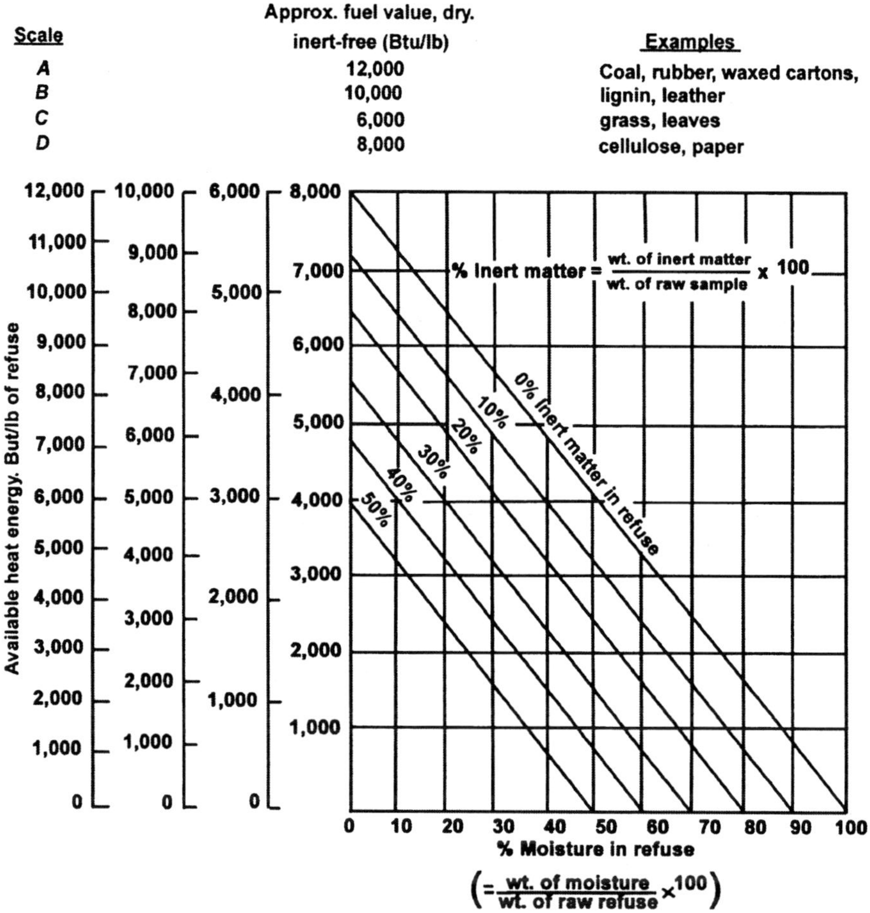

Figure 8-6. Relationship between heat, energy, and moisture content as a function of the inert matter in raw refuse. (From Municipal Incineration, SERP, Bulletin No. 5, University of California, 1951.)

the $\Delta(O_2)$ value is 6 and the molecular weight (MW) value is 435 (empirical method 2). Table 8-2 shows this method of estimation applied to a number of polymers including cellulose. For example, a solid waste containing 20% moisture and 10% inert materials can be approximated from Figure 8-6 as

5500 Btu/lb. Approximating from empirical method 1 (assuming 70% of $C_6H_{10}O_5$), the heat of combustion is 5470 Btu/lb. Approximating from empirical method 2, the heat of combustion is 4840 Btu/lb. If one assumes MW to be $C_6H_{10}O_5$, method 2 will yield 6900 Btu/lb.

Table 8-2. Specific Heat of Combustion of Some Polymers

Polymer	Elementary Composition of Structural Unit	*M*	$\Delta(O_2)$	Δh_{comb} (kJ/g)	
				Calc.	Exp.
Polyformaldehyde	CH_2O	30.0	1.0	14.5	16.7
Poly(methyl methacrylate)	$C_5H_8O_2$	100.1	6.0	26.1	—
Polyacrylonitrile	C_3H_3N	53.1	3.75	30.8	30.6
Polyethylene	C_2H_4	28.1	3.0	46.3	46.5
Polypropylene	C_3H_6	42.1	4.5	46.5	46.5
Polyisoprene	C_5H_8	68.1	7.0	44.7	44.9
Polybutadiene	C_4H_6	54.1	5.5	44.2	45.2
Polystyrene	C_8H_8	104.1	10.0	41.7	41.5
Cellulose	$C_6H_{10}O_5$	162.2	6.0	16.1	16.7
Poly(ethylene terephthalate)	$C_{10}H_8O_4$	192.2	10.0	22.7	22.2
Poly(vinyl alcohol)	C_2H_4O	44.1	2.5	24.7	25.1
Nylon-66	$C_{12}H_{22}O_2N_2$	226.3	16.5	31.7	31.4
Polycarbonate	$C_{16}H_{14}O_3$	254.3	18.0	30.8	31.0
Nomex ®	$C_{14}H_{10}O_2N_2$	238.3	15.5	28.3	28.7
Polychloroprene (Neoprene ®)	C_4H_5Cl	88.5	5.0	24.5	24.3
Poly(vinyl chloride)	C_2H_3Cl	62.5	2.5	17.5	18.0
Poly(vinylidene chloride)	$C_2H_2Cl_2$	97.0	2.0	9.0	10.45

Another factor to be considered is that if a given material is heated, the quantity of heat for the energy value consists of two types: **sensible heat** and **latent heat**. The sensible heat requirement is:

$$Q_s = C_P W (T_2 - T_1) \qquad [8\text{-}9]$$

where C_P is the specific heat, W is the weight of the material, and T_1 and T_2 are the initial and final temperatures of the material. Latent heat evolves in the transitions between states or phases. For example, the latent heat of vaporization between water and steam is:

$$Q_l = \lambda W \qquad [8\text{-}10]$$

where λ is the unit latent heat; the latent heat of water is 975 Btu/lb. For refuse incineration, the C_P values expressed as Btu/lb/°F are as follows:

Material	Specific Heat (Btu/lb/°F)
Solid in refuse	0.25
Water	1.00
Steam	0.5
Air	0.25
Ash	0.2

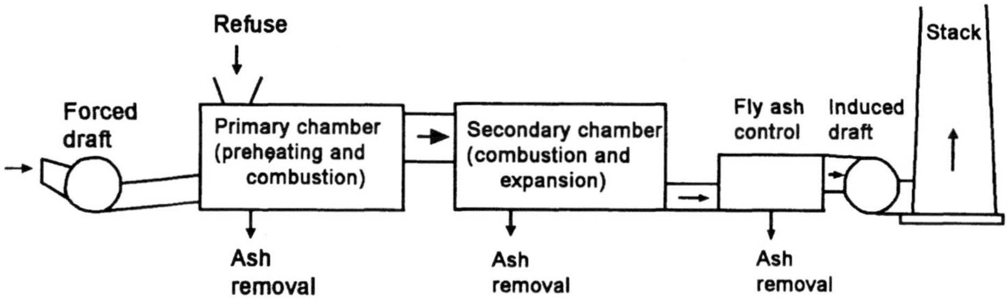

Figure 8-7. Schematic drawing of a refuse incinerator. (From Municipal Incinerator, SERP, Bulletin No. 5, University of California, 1951)

The specific heat of any solid waste or organic can be approximated from an analysis of the different functional components of a given material. Using Table 8-3, we can evaluate the specific heat of cellulose as the repeating units of:

5 (–C–H) = 5(15.6)
2 (–O–) = 2(16.8)
3 (–OH) = 3(17)
1 ($-CH_2-$) = 25.35
Total = 188 J/mol/°K

This value can easily be converted into 0.27 Btu/lb/°F, which is close to the value of 0.25 Btu/lb/°F.

[Example 8-1] Compute the waste heat available for incineration of a municipal solid waste (MSW) (Fig. 8-5).

First a mass balance is performed and followed by a heat balance. The results on a per pound basis are listed in Figure 8-8. The detailed calculation is included in Table 8-4 (Mass Balance) and Table 8-5 (Energy Balance).

Table 8-3. Group Contributions to Molar Heat at 25°C (J/mol·K)

Group	C_p^s (Satoh)	C_p^l (Shaw)	C_p^s / R per atom	C_p^l / R per atom
$-CH_3$	30.9	36.9	0.92	1.10
$-CH_2-$	25.32	30.4	1.01	1.21
$-\overset{\vert}{C}H-$	15.6	20.95	0.93	1.25
$-\overset{\vert}{\underset{\vert}{C}}-$	6.2	7.4	0.74	0.88
$=CH_2$	22.6	21.8	0.90	0.87
$=CH-$	18.65	21.4	1.11	1.28
$=\overset{\vert}{C}-$	10.5	15.9	1.25	1.90
$-CH_2-$(5 ring)	19.9	26.4	0.79	1.05
$-CH_2-$(6 ring)	18.0	26.4	0.71	1.03
$>CH_{ar}$	15.4	22.2	0.92	1.33
$>CH_{ar-}$	8.55	12.2	1.02	1.45

Continued next page

Table 8.3. Continued

Group	C_p^s (Satoh)	C_p^l (Shaw)	$C_p^s\,/\,R$ per atom	$C_p^l\,/\,R$ per atom
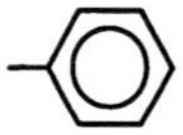	85.6	123.2	0.94	1.35
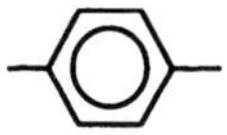	78.8	113.1	0.95	1.36
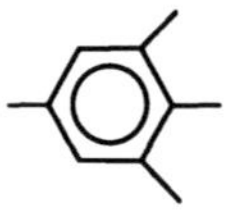	65.0	93.0	0.98	1.40
–F	(21.4)	(21.0)	2.55	2.50
–Cl	27.1	(39.8)	3.23	4.75
–Br	26.3	—	3.14	—
–I	22.4	—	2.67	—
–CN	(25)	—	1.50	—
–OH	17.0	44.8	1.01	2.68
–O–	16.8	35.6	2.01	< 4.25
–CO–	23.05	52.8	1.38	3.15
–COOH	(50)	98.9	1.50	2.95
–COO–	(46)	65.0	1.83	2.58
$-NH_2$	20.95	—	0.83	—
–NH–	14.25	(31.8)	0.85	1.90
>N–	17.1	(44.0)	2.04	5.25
$-NO_2$	41.9	—	1.67	—
–CONH–	(38–54)	(90.1)	1.12–1.63	2.68
–S	24.05	44.8	2.37	5.35
–SH	46.8	52.4	2.78	3.12
$-SO_2-$	(50)	—	2.00	—

Source: D. W. Van Krevelen, *Properties of Polymers*, Elsevier, Amsterdam, 1976.

Assumptions are made concerning this MSW in comparison to pyrolysis. Table 8-6 lists a number of other fossil fuel feed stocks and MSW that can obtain similar results under heat treatment.

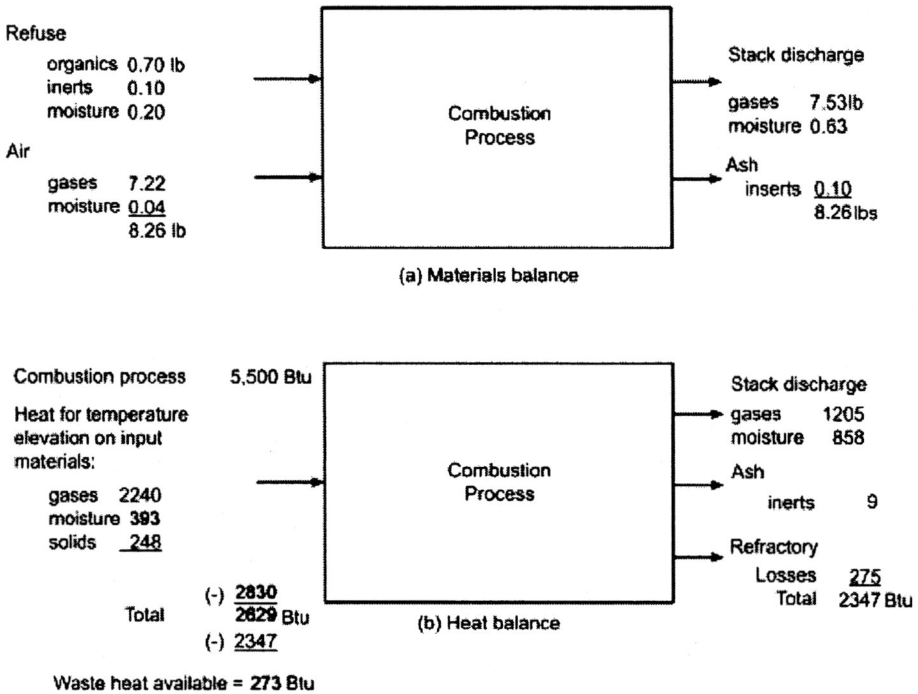

Figure 8-8. Materials and heat balances across a combustion process (on the basis of 1 lb of moist refuse).

Table 8-4. Material Balance on Refuse[a]

Main equation: $C_5H_{10}O_5 + 6O_2 = 6CO_2 + 5H_2O$

Input mass	Pound base (lb.)
A. Refuse	
1. Organics	0.7
2. Inerts (inorganics)	0.1
3. Moisture	0.2
B. Air	

1. Gases Oxygen requirement $= \left(\frac{0.70}{162}\right)(6 \times 32) = 0.83$ lb.

Assuming 200% excess oxygen, then dry air required $= \left(\frac{2(0.83)}{0.23}\right) = 7.22$

2. Moisture

From humidity chart (Fig. 7-15) 50% relative humidity at 60°F = 0.005 lb. moisture/lb. dry air. Thus, moisture = 7.22(0.005) = 0.036

Output mass	Pound base (lb.)
A. Stack discharge	

1. Gases

CO_2 produced $\left(\frac{0.70}{162}\right)(6 \times 44) = 1.14$ lb.

The discharge dry gases = gas input + CO_2 produced in system – oxygen utilized

$= 7.22 + 1.14 - 0.83 = 7.53$

2. Moisture

Moisture produced in system

Total moisture in stack discharge $= \left(\frac{0.70}{162}\right)(5 \times 18) = 0.39$ lb.

= refuse moisture + air moisture + water produced by combustion

$= 0.20 + 0.036 + 0.39 = 0.626$

B. Ash

1. Inerts	0.1

Input – Output = 8.25 – 8.256 = 0

[a] One pound moist refuse.

Table 8-5. Energy Balance on Refuse[a]

Energy budget

A. Heat of combustion 5500 Btu

B. Heat of refuse and air from 60°F to 1300°F

1. Moisture

sensible heat

$$Q_s = 1\,(0.20 + 0.036)(212 - 60) = 31 \text{ Btu}$$

latent heat

$$Q_l = 975\,(0.20 + 0.036) = 230 \text{ Btu}$$

further sensible heat

$$Q_s = 0.5\,(0.20 + 0.036)(1300 - 212) = 128 \text{ Btu}$$

total moisture heating

$$= Q_s\,(\text{water}) + Q_l\,(\text{water}) + Q_s\,(\text{steam})$$

$$= 31 + 230 + 128 = 388 \text{ Btu}$$

2. Gases

$$Q_s = 0.25\,(7.22)\,(1300 - 60) = 2238 \text{ Btu}$$

3. Solids

$$Q_s = 0.25\,(0.1 + 0.7)\,(1300 - 60) = 248 \text{ Btu}$$

C. Stack discharge

1. Moisture (from 60°F to stack temperature of 700°F)

$$Q_s\,(\text{water}) + Q_l\,(\text{water}) + Qs\,(\text{steam})$$

$$= 1\,(0.626)\,(212 - 60) + 975\,(0.626) + 0.5\,(0.626)\,(700 - 212)$$

$$= 95 + 610 + 153 = 858 \text{ Btu}$$

2. Gases

$$Q_s = 0.25\,(7.53)\,(700 - 60) = 1205 \text{ Btu}$$

D. Ash (Ash temperature = 500°F)

$$Q_s = 0.2\,(0.1)\,(500 - 60) = 9 \text{ Btu}$$

E. Refractory loss (50%)

$$Q_s = (5500)\,(0.05) = 275 \text{ Btu}$$

Available waste = A – (B + C + D + E) = 279 Btu

[a] One pound moist refuse.

Table 8-6. Typical Ultimate Analyses of Solid Wastes and Some Fossil Fuel Feed Stocks for Conversion to Pipeline Gas

	Lignite— N. Dakota, Beulah Seam	Oil Shale— Colorado, Green River Deposit	Bituminous Coal— Pennsylvania Pgh. Seam	Subbituminous Coal— Wyoming, Monarch Seam	Solid Waste Average Municipal Refuse
Carbon	42.4	23.8[a]	76.6	54.6	35.4
Hydrogen	6.7	2.6	5.2	6.4	4.4
Oxygen	43.3	12.3	6.2	33.8	28.2
Nitrogen	1.7	0.5	1.6	1.0	0.4
Sulfur	0.7	1.0	1.3	0.4	0.2
Ash	6.2	59.8	9.1	3.8	31.4[b]
Total	101.0(sic)	100.0	100.0	100.0	100.0
Moisture	34.8	0.2	2.6	23.2	20.7

[a] Includes 4.6% carbon as carbonate.

[b] Includes ash, metals, glass, and ceramics.

8.2.1 Catalytic Incineration and Wet Combustion

Incineration processes usually burn the organic portion of sludge into ash. Multiple hearth incinerators are the most common type of combustion equipment (Fig. 8-9). They contain several refractory hearths with solids located alternately at the center and the periphery of adjacent hearths. Rakes attached to the arms traverse the surface. The wet sludge moves in stages countercurrently to the ascending air and combustion products. Another incinerator type that increases thermal efficiency is called the fluidized-bed incinerator, which contains a suspension of heated sand fluidized by an upward flow of air. The water-wall combustion process usually takes advantage of water-filled tubes for the generation of additional steam for energy recovery.

Although pollutants can be generated from incineration, catalytic incineration can convert the fugitive air pollutants into nonhazardous products at a reduced temperature. Many VOCs or their precursors, as well as their intermediates, can be transformed (Table 8-7). The ignition temperature (or flashpoint), t_f, usually can be approximated from its boiling point, t_b, as follows:

$$t_f = 0.683\, t_b - 119 \text{ (in °F)} \qquad [8\text{-}11]$$

Table 8-7. Ignition Temperatures

	Catalytic[a]		Thermal[b]	
	°F	°C	°F	°C
Benzene	400	204	1040	560
Toluene	400	204	896	480
Formaldehyde	300	149	806	430
Methyl ethyl ketone	475	246	960	515
Carbon monoxide	320	160	1128	609

[a] The ignition temperature, defined as the point of transition from the kinetic to the diffusion regime, was taken between 50 to 75% conversion on the S-shaped conversion vs. temperature curve and does vary with catalyst test conditions.

[b] NFPA NO. 325M.

In most cases, the catalytic incinerator will automatically take care of the pollutant central devices. For example, in a rich-burn engine for NO,

$$2NO + 2H_2 \xrightarrow{Cu\,-\,Ni} N_2 + 2H_2O$$

$$2NO + 5H_2 \xrightarrow{Cu\,-\,Cr} 2NH_3 + 2H_2O$$

Finally, the wet combustion, or Zimmerman process (see Fig. 8-10), is a chemical oxidation process under pressure in a water solution or under suspension for organic material. Injection of compressed air supports the combustion process. As the temperature increases from 100 to 374°C (the critical temperature of water), oxidation will increase, and processes carried out at less than 374°C will reach the oxidation equilibrium in a relatively short time. Regardless of the initial concentration of the input, the final concentration of organic material remains very low (see Figs. 8-11 and 8-12).

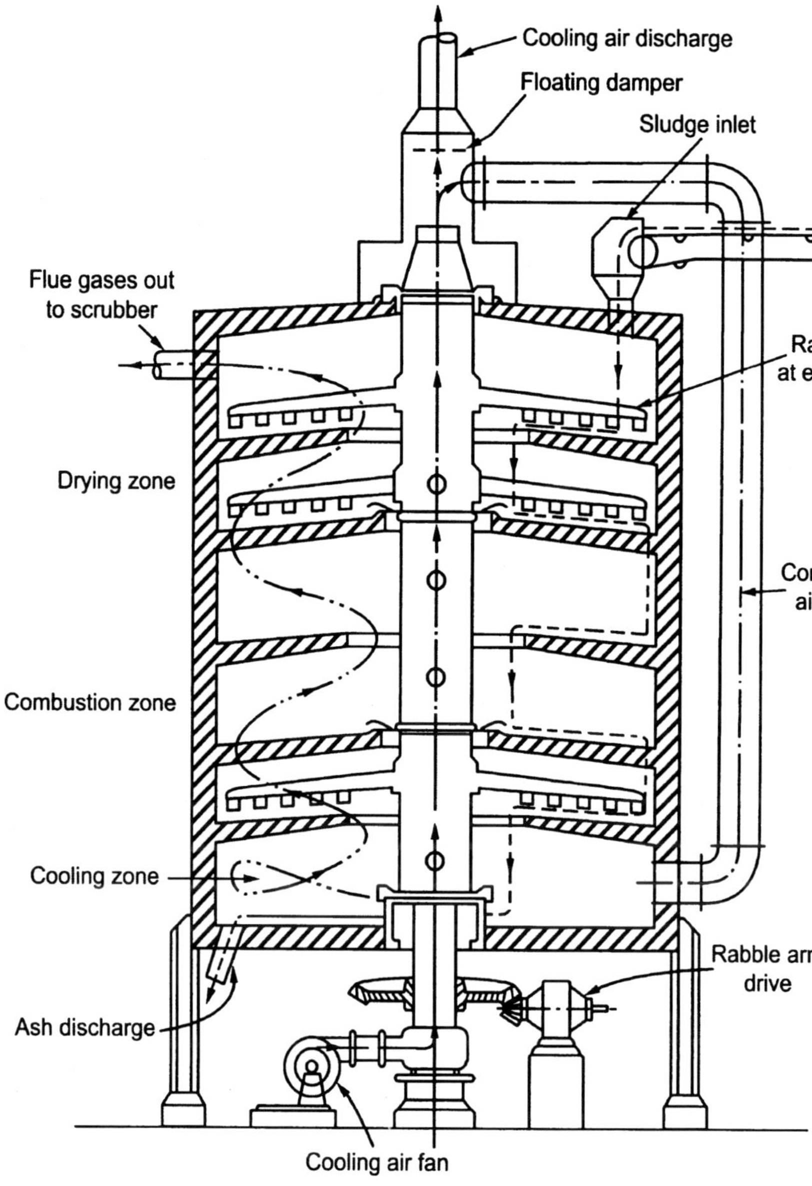

Figure 8-9. Cross section of a typical multiple-hearth incinerator.

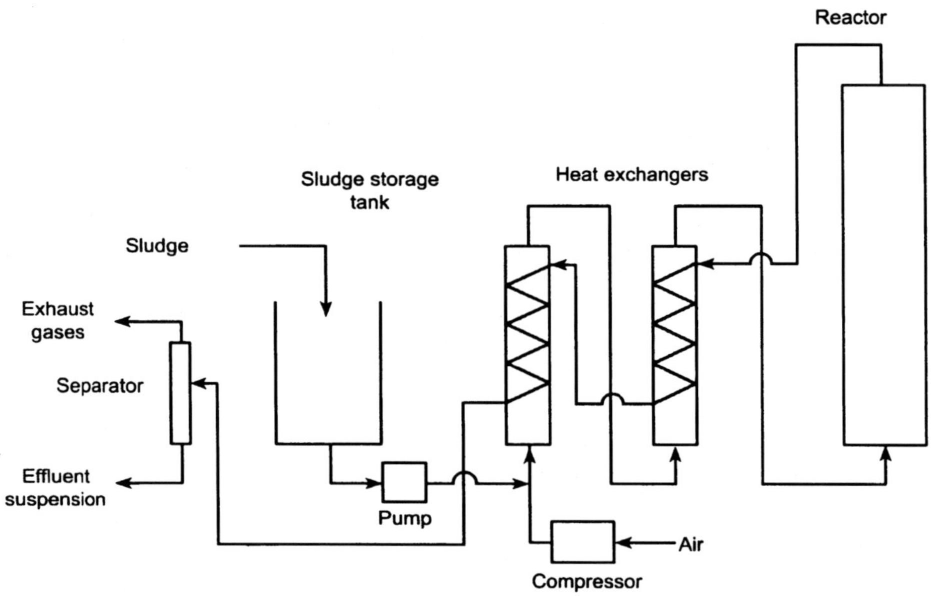

Figure 8-10. Schematic diagram of a process for the wet combustion of an organic sludge. [Adapted from Wet air combustion, Industrial Water and Wastes, 6, 102–106, (1961)].

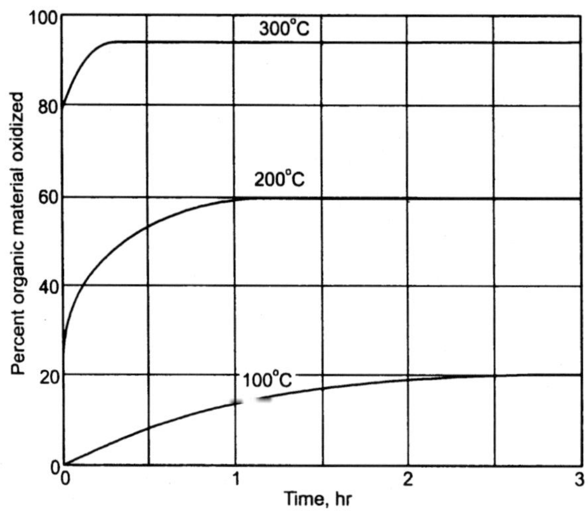

Figure 8-11. Effect of temperature upon time required to reach oxidation equilibrium. [Adapted from E. Huriwitz and W. A. Dundas, Ultimate disposal of sewage sludge by wet oxidation, Proc. Fourteenth Industrial Waste Conference, Purdue University, (1959).]

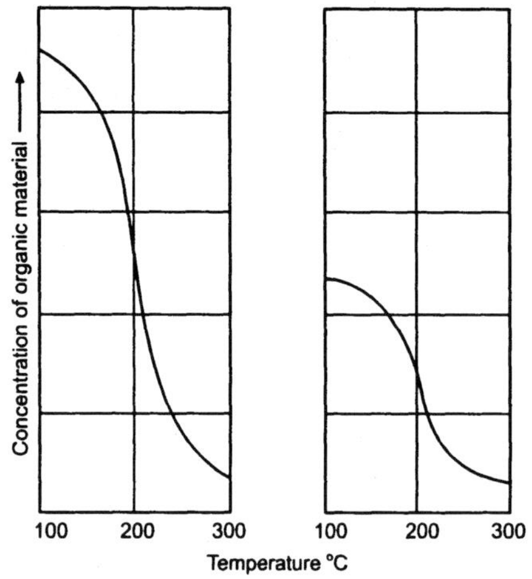

Figure 8-12. Effect of temperature upon oxidation of organic material at equilibrium. [Adapted from Wet Combustion, *Industrial Water and Wastes*, 6, 102–106, (1961).]

8.3 Formation of Polycyclic Aromatic Hydrocarbons (PAH) and Soot

Fuel is usually converted to C1 and C2 hydrocarbons. Higher hydrocarbons are formed from recombination of these fragments. Acetylene formed in high amounts under fuel-rich conditions (both premixed and non-premixed flames). Figure 8-13 is an example of the formation of acetylene.

The formation of acetylene to benzene is as follows:

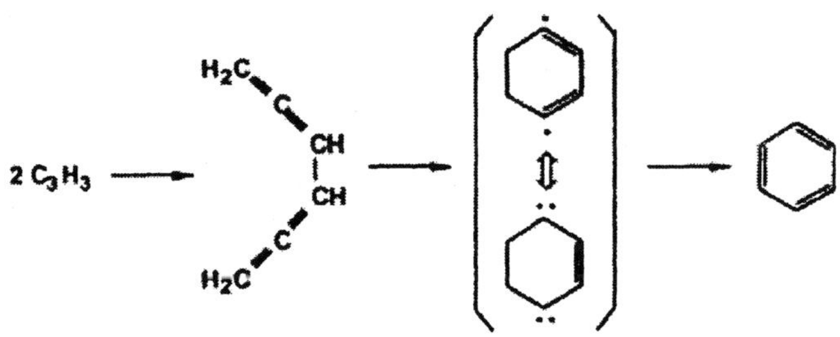

The addition of C_2H_2 to phenyl radical can form styryl radical; additional C_2H_2 can form naphthalene. Furthermore alternating radical formation and H attack is possible.

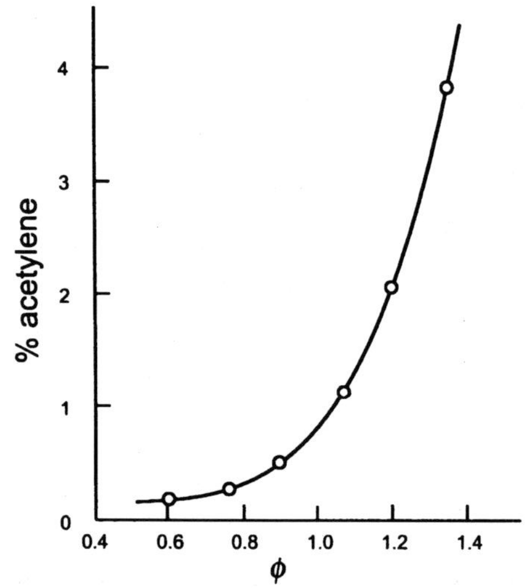

Figure 8-13. Dependence of acetylene formation in CH⁴-0² flames on the equivalence ratio ϕ (After Wagner 1979).

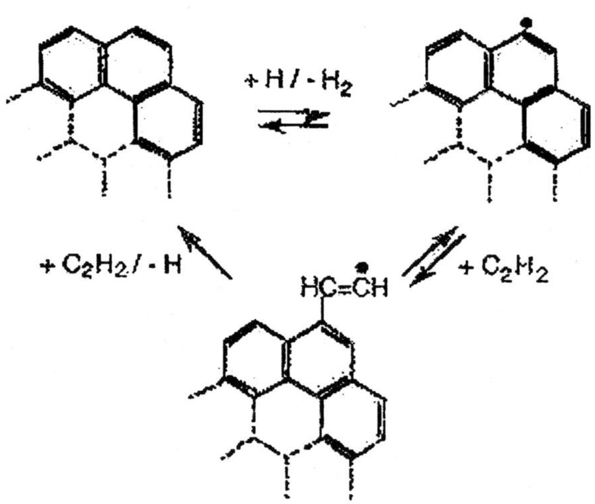

Figure 8-14. Ring growth by radicalic acetylene addition in PAH formation (After Frenklach and Clary 1983, Frenklach and Wang 1991)

Measured and calculated PAH formation also available (See Figure 8-15)

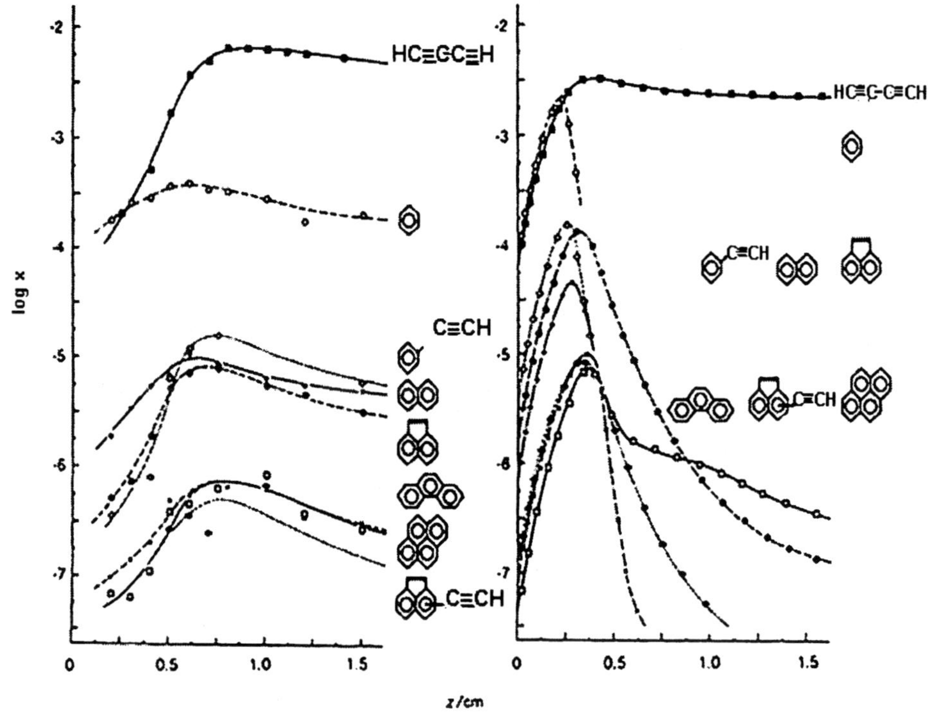

Figure 8-15. Measured (After Bockhorn et al. 1982, left) and calculated (After Frenklach and Warnatz 1987, right) profiles of PAH in a laminar premixed acetylene-oxygen-argon flame at low pressure and fuel rich conditions.

Simonneit has proposed the zigzag addition process for PAH formation to account for the non-bimarkers (Figure 8-16).

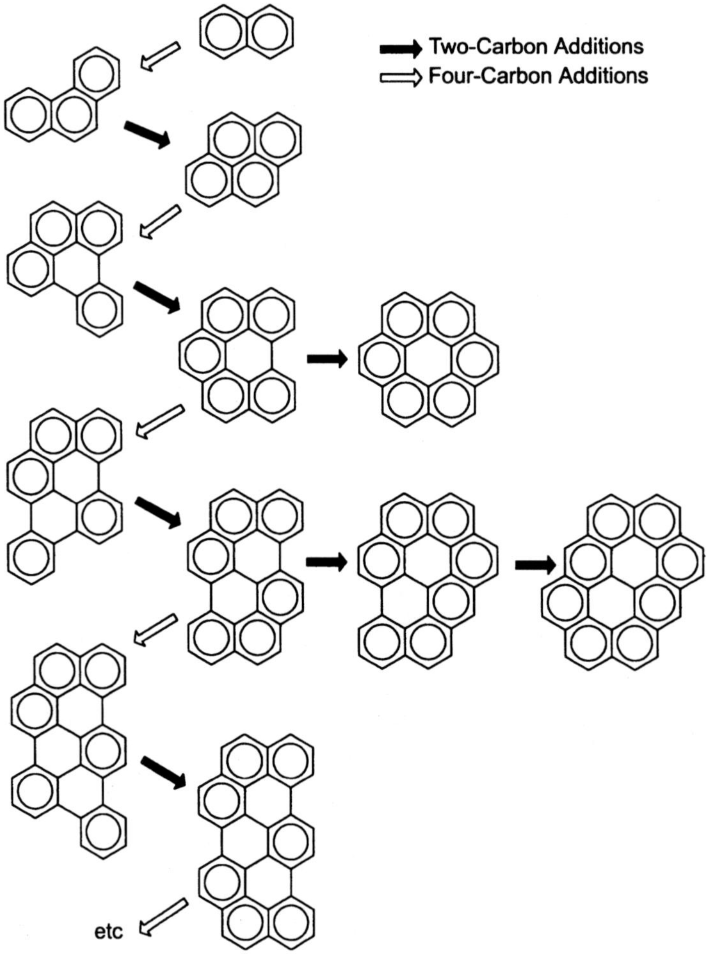

Figure 8-16. The "zigzag addition process" for PAH formation (two carbon additions are acetylene radicals, and four carbon additions are 1,3 butadiene radicals) (After B.R.T Simonneit, 2002).

In the presence of oxygen, soot of articles grow by surface growth through coagulation as shown in Figure 8-17.

$$C_nH_m + kO_2 \rightarrow 2kCO + m/2\,H_2 + (n\text{-}2k)\,C,\ \ n > 2k, \text{ or } C/O > 1$$

(See also Table 8-8)

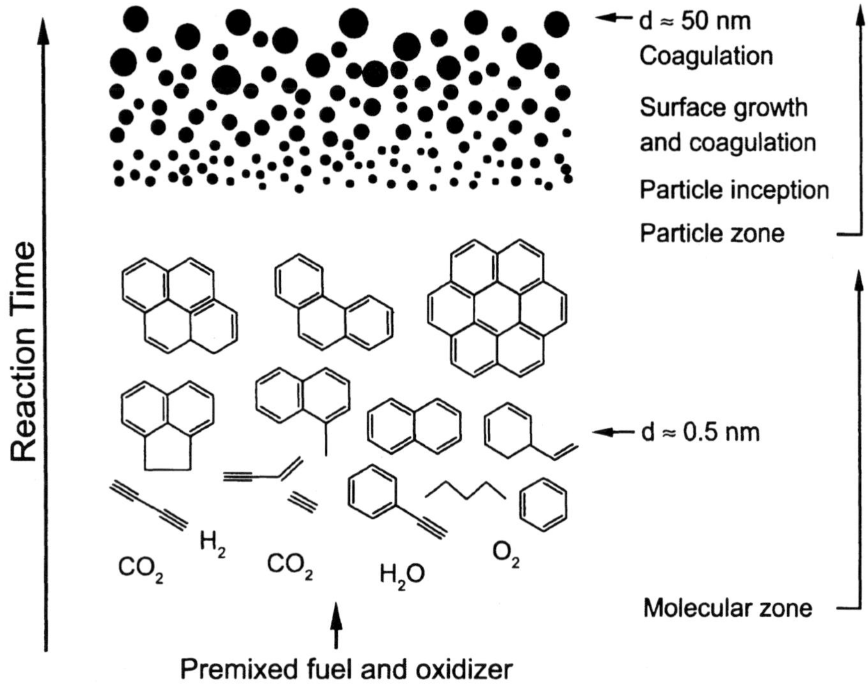

Figure 8-17. Schematic reaction path leading to soot formation in homogenous mixtures or premixed flames (Bockhorn 1994); ≡ is a acetylenic triple bond (≡ = acetylene e.g.)

Define a value fraction f_v, which is the ratio of the soot value V_{soot} and the total volume V_{total}.

$$f_v = V_{soot} / V_{total}$$

Also define the soot particle number density $[n]_{soot}$ which is the ratio of soot particle number, n_{soot} and the total volume,

$$[n]_{soot} = n_{soot} / V_{total}$$

Table 8-8. Measured soot limits in terms of C/O ratios at 1800 K (After Haynes and Wagner 1981)

Combustion System		Bunsen burner	Well stirred reactor	Flat flame 1bar	26mbar
CH_4	$-O_2$	------	------	0.45	------
C_2H_6	-air	0.48	------	0.47	------
C_3H_8	-air	0.47	------	0.53	------
C_2H_4	-air	0.61	0.70	0.60	------
C_2H_4	$-O_2$	------	------	0.71	------
C_4H_8	-air	0.52	0.68	------	------
C_2H_2	-air	0.83	------	------	------
C_2H_2	$-O_2$ (T≈3000 K)	0.95	------	------	0.95
C_6H_6	-air	0.57	0.57	0.65	------
C_6H_6	$-O_2$	------	------	------	0.74
$C_{11}H_{10}$	-air	0.42	0.50	------	------

Now C_{soot} is the concentration of soot. If the soot particles are assumed to be monodispersed, the resulting soot particle size can be found. N_A is the Avogadro's constant,

$$d_{soot} = 3\,(6\,f_v / \pi C_{soot} N_A)^{1/2}$$

Experimentally, small spheres of soot always appeared as grape-like clusters and the amount of soot is dependent on C/O ratio and temperature treatment between 1000 and 2000 K. It has been found that at different pressures the probability for collision of O_2 with soot is different (see Figure 8-19). A detailed simulation of soot formation by oxidation can be found as shown in Figure 8-20. The Schulz-Flory distribution is applied here (see Chemistry for Engineers chapter on Polymer Chemistry).

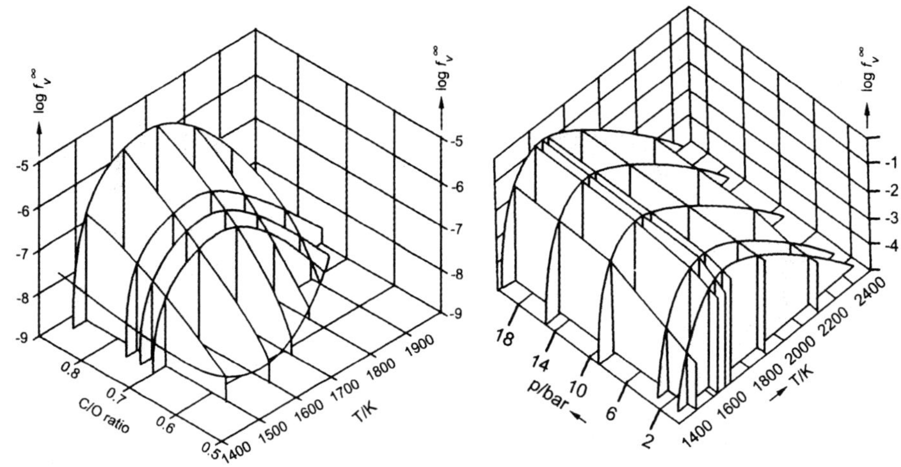

Figure 8-18. Experimental temperature, pressure and C/O dependence of the final soot volume fraction f_v^∞ and soot yield = fraction of C appearing as soot (After Bohm et al. 1989, Jander 1995)

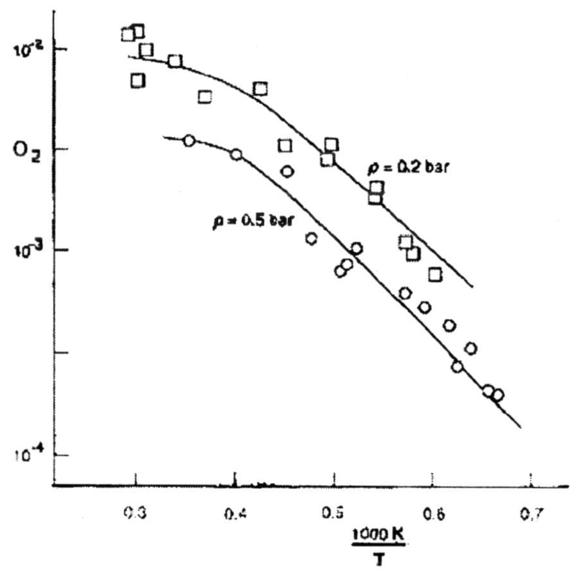

Figure 8-19. Temperature dependence of the probability γ of the reaction per collision of O^2 molecules with soot at different pressures (After Roth et al. 1990, von Gersum and Roth 1992)

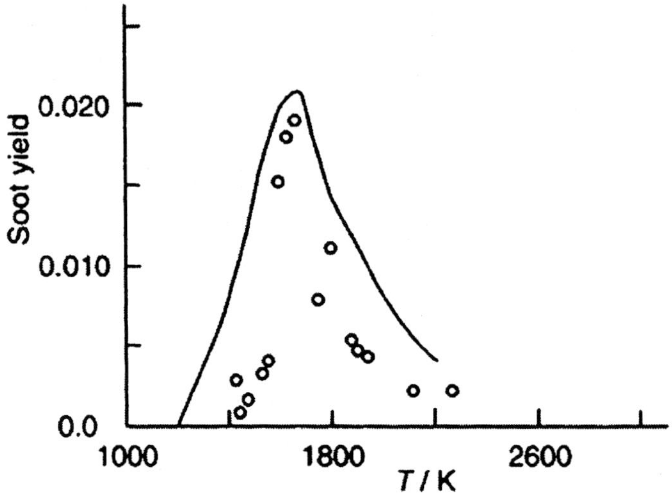

Figure 8-20. Soot yield in a spatially homogenous C_2H_2-O_2-Ar mixture at a C/O ratio of 3.1 and p = 1 bar as function of temperature; points: uncalibrated) measurements (Based on Frenklach et al. 1984), line: simulation (Based on El-Gamal 1995)

8.4 DIOXIN

Both the incineration of industrial waste and the burning of garbage in municipal incinerators will generate chlorinated dioxins or 2,3,7,8-tetrachlorodibenzo-*p*-dioxin (TCDD) (Fig. 8-21). In general, dioxin can include both polychlorinated dibenzo-*p*-dioxins (PCDD) and polychlorinated dibenzo furans (PCDF). TCDD is definitely the most toxic in terms of LD_{50}. Historically, dioxin originated from the use of 2,4-D or 2,4,5-T as herbicides. A wide spread of dioxin has been found in a number of different fish types (Table 8-9). Besides being toxic, it has been proven carcinogenic and teratogenic in animals. How to remediate large amounts of contaminated soil is a serious problem. At present, the supercritical water (SCW) oxidation, which utilizes the properties of water at pressures greater than 218 atmosphere with a temperature over 374°C, is one choice (see Figure 8-22 for a schematic). An alternative method is to use lignin and other peroxidase from fungal metabolites as indicated by Figure 8-23.

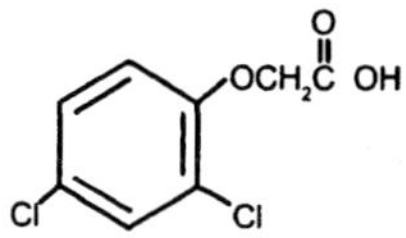

2,4-Dichloro-phenoxyacetic acid
(2,4-D)

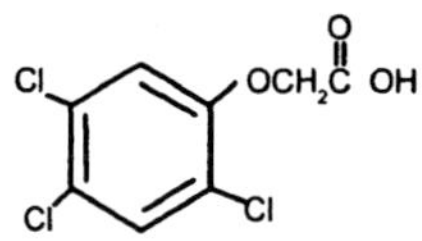

2,4,5-Trichloro-phenoxyacetic adic
(2,4,5-T)

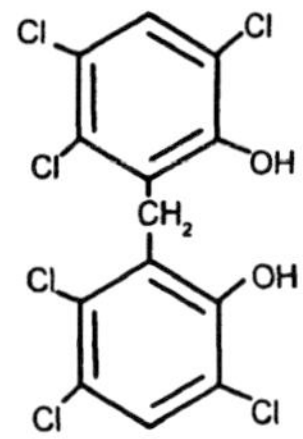

Hexachlorophene

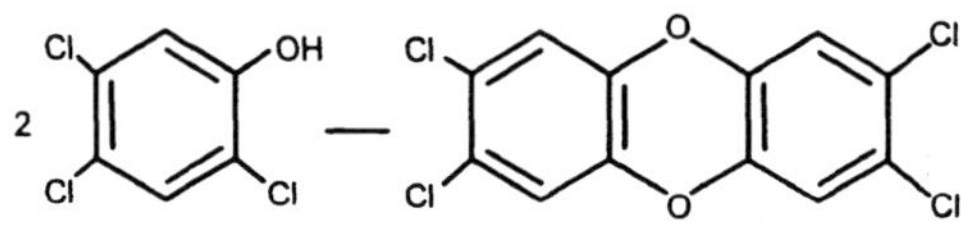

2,3,7,8-Tetrachloro dibenzon-p-dioxin (TCDD)

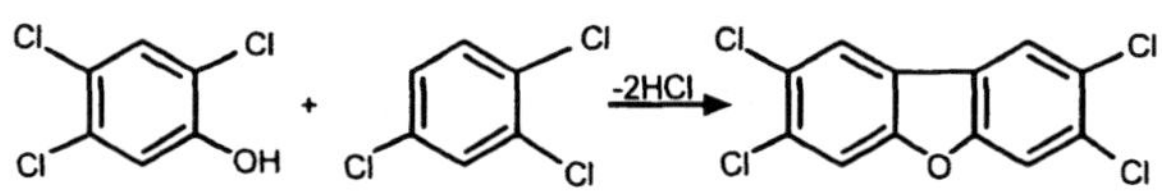

2,3,7,8-Tetrachloro-dibenzon-furan (TCDF)

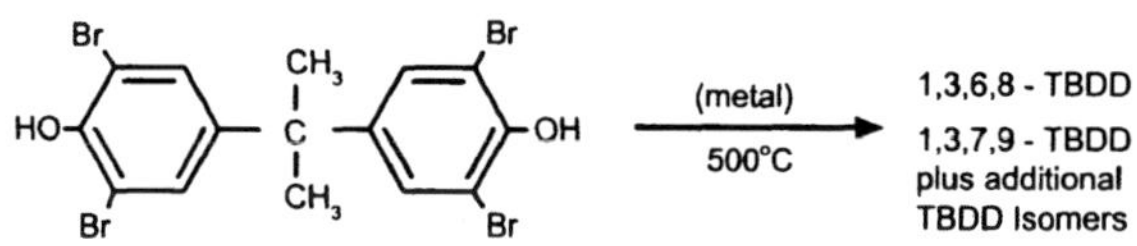

Tetrabromobisphenol-A

Firemaster BP-6

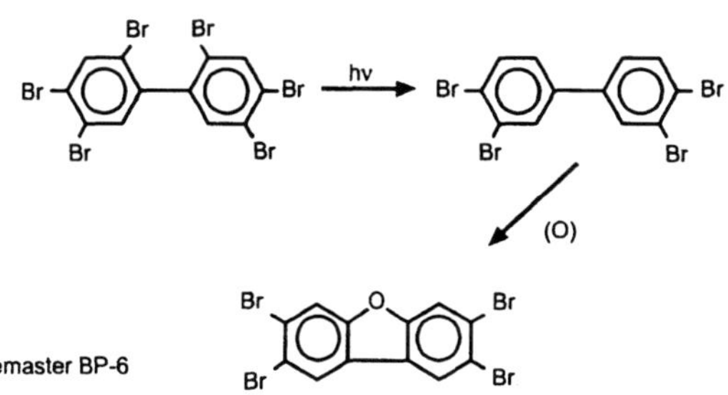

Figure 8-21. Herbicides, defoliants, and forest fire retardants.

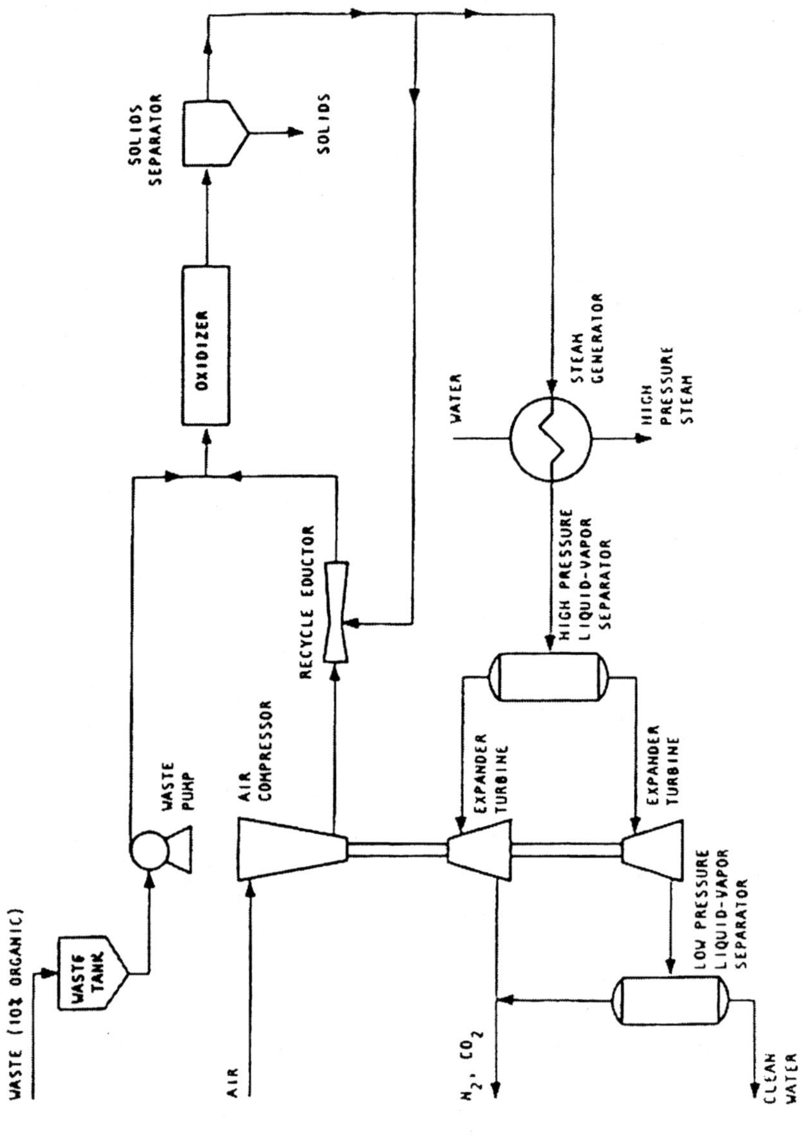

Figure 8-22. Process schematic for supercritical oxidation of an aqueous waste with a heating value of 1750 Btu/lb. (Modell, et al., 1984. Proceedings of the 2nd International Symposium on Operating European Centralized Hazardous Waste Management Facilities, Odense, Denmark).

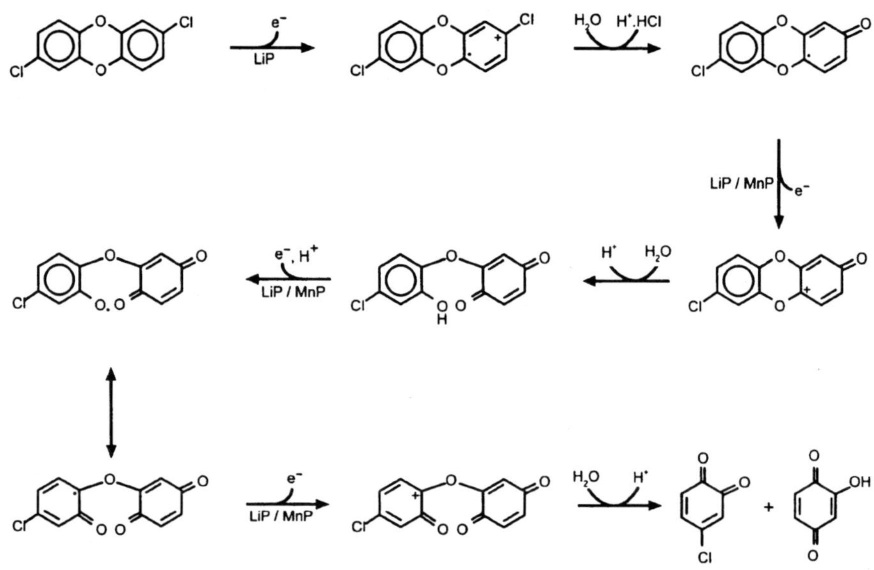

Figure 8-23. Proposed mechanism for the lignin peroxidase-catalyzed dioxin ring cleavage of 2,7-dichlorodibenzo-p-dioxin. (LiP is lignin peroxidase, MnP is manganese peroxidase)

(All contain the polychorophenol component, especially with ortho-chloro phenol as an important constituent; T stands for tetra, P stands for poly, in dioxins).

How are fugitive molecules such as dioxin formed during high temperatures? What is the source of chlorine? One explanation is the discovery of more than one species of the chlorophenols found in the wastewater of coal liquefaction or the aqueous washings of coal liquids. What is the precursor to chlorine? There are no known chloro-containing compounds in coal. The origin of chlorine may be chloride. Under elevated temperatures and in the proper oxidative environment, chloride may be oxidized to chlorine, and the ensuing electrophilic substitution of phenol is well known. The coupling of phenols will generate dioxin (see Fig. 8-21). Bromine analogs of dioxin are known to be toxic. They originate from the fire retardant in polyester laminates, such as tetrabromobisphenol-A, and from fire-fighting agents, such as Firemaster BP-6. A chick embryo assay method based on the quantitation of aryl hydroxylase (AHH) and 8-ethoxyresorufin deethylase (7-ER) is shown in Table 8-10. Dioxin has been located in various parts of an incinerator (Fig. 8-24), and so far the quantities analyzed far exceed the EPA standard of 10 ng/dscm (Fig. 8-25). This figure (10 ng per dry standard cubic meter) is equivalent to about 6 parts per trillion, which approaches the limit of the sensitivity of analytical instruments.

Table 8-9. FDA's analysis results for dioxin residues in fish collected from various locations in 1981-1983.

Type of Fish	No. of Fish Samples	Origin	2,3,7,8-TCDD	
			pg/g[a]	Recovery (%)b
Buffalo Fish	5	Mississippi River, vicinity St. Louis	ND	58-64
Buffalo Fish	2	Mississippi River, Sauget, IL	ND	53,25[c]
Catfish	1	Mississippi River, vicinity St. Louis	10	61
Catfish	1	Mississippi River, vicinity St. Louis	ND	43
Carp	2	Mississippi River, vicinity St. Louis	ND	48, 84[c]
Carp	13	13 Michigan rivers[d]	ND	50-62
Walleye	1	Clinton River, MI	ND	56
Carp	1	Tittabawassee River, MI	93	74
Coho Salmon	1	Detroit River, MI	ND	86
Carp	1	Muskegon Lake, MI	10	84
Coho Salmon	1	Trout Run, PA	ND	95
Carp	1	Lake Pippin, WI	ND	48
Lake Trout	1	Lake Huron	6	81
Whitefish	1	Lake Huron, Thunder Bay	ND	69
Coho Salmon	1	Lake Huron, Tawas Bay	ND	60
Whitefish	4	Lake Michigan	ND	54, 78
Sucker	1	Lake Michigan	ND	--
Coho Salmon	3	Lake Michigan	ND	74-78
Lake Trout	1	Lake Michigan	ND	79
Carp	1	Lake Michigan	ND	54
Whitefish	1	Lake Superior, Whitefish Bay	ND	--
Lake Trout	1	Lake Superior	ND	96
Coho Salmon	3	Lake Erie	ND	55-82
Sucker	1	Lake Erie, West Seneca, NY	ND	56[c]
Sucker	1	Lake Erie, Angola, NY	ND	62[c]
Various Fish[e]	6	Lake Erie, Dunkirk, NY	ND	50-67[c]
Coho Salmon	1	Lake Ontario, Salmon River, NY	ND	88
Coho Salmon	1	Lake Ontario, Salmon River, NY	35	73
Sucker	2	Lake Ontario, Wilson, NY	ND	58, 94
Lake Trout	1	Lake Ontario, Wilson, NY	46	76
Brown Trout	1	Lake Ontario, Wilson, NY	8	73
Rainbow Trout	1	Lake Ontario, Wilson, NY	21	73
White Perch	1	Lake Ontario, Oswego, NY	25	96
Yellow Perch	1	Lake Ontario at Chaumont Bay	ND	--
Butterfish	1	Atlantic Ocean[f]	ND	85
Flounder	1	Atlantic Ocean[f]	ND	73
Hake	3	Atlantic Ocean[f]	ND	83-103
Herring	1	Atlantic Ocean[f]	ND	85

[a]ND, not quantitated or confirmed; if 2,3,7,8-TCDD is present, it is present at a level below about 10pg/g. Values corrected for reagent blank (~3pg/g) and recovery.
[b]Test portions fortified at level of 50-100 pg/g with 2,3,7,8-TCDD standard solution. Each test portion is a composite of from 1 to 10 fish.
[c]Test portions fortified with 1,3,7,8-TCDD recovery surrogate.
[d]Au Sable, Cass, Clinton, Flint, Grand, Huron, Kalamazoo, Muskegon, Pine, Raisin, Shiawassee, St. Clair, and St. Joseph.
[e]White bass, catfish, drum, yellow perch, Chinook salmon, and brown trout.
[f]8-12 mi off Long Beach, NJ, collected in 1984.

Table 8-10. Relative potencies of 2,3,7,8-Tetrabromodibenzo-p-dioxin and 2,3-dichloro-7,8-dibromodibenzo-p-dioxin.

	Dose producing ½ maximal induction (pmol/egg)	Relative potency	
		7-ER[a]	7-ER[b]
2,3,7,8-TCDD	11.1	1.0	—
2,3,-Dichloro-7,8-dibromo DD	7.35	1.51	—
2,3,7,8-TBDD	9.40	1.18	—
		AHH	
2,3,7,8-TCDD	9.70	1.0	1.0
2,3,7,8-Dichloro-7,8-dibromo DD	6.50	1.49	1.1
2,3,7,8-TBDD	9.00	1.08	1.0

[a]B. Ranglingam et al., 1986.

[b]A.S. Kende et al., 1974.

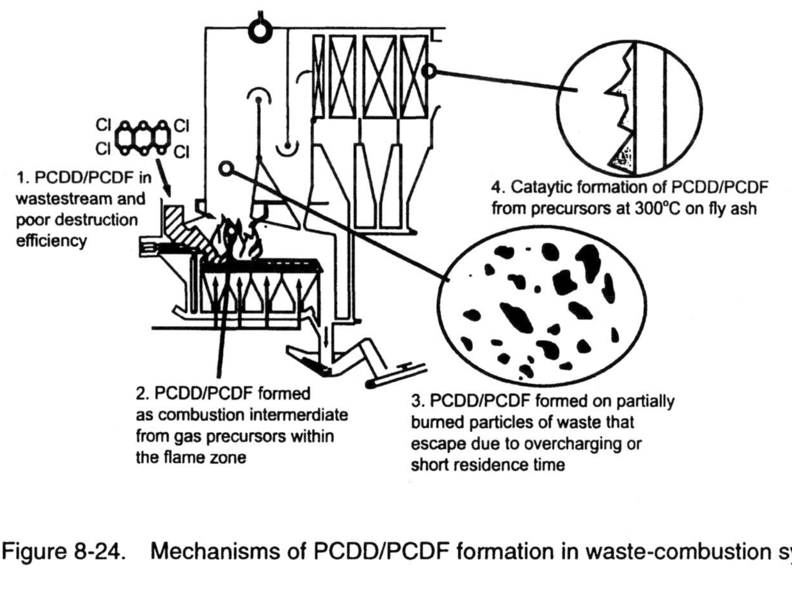

Figure 8-24. Mechanisms of PCDD/PCDF formation in waste-combustion systems.

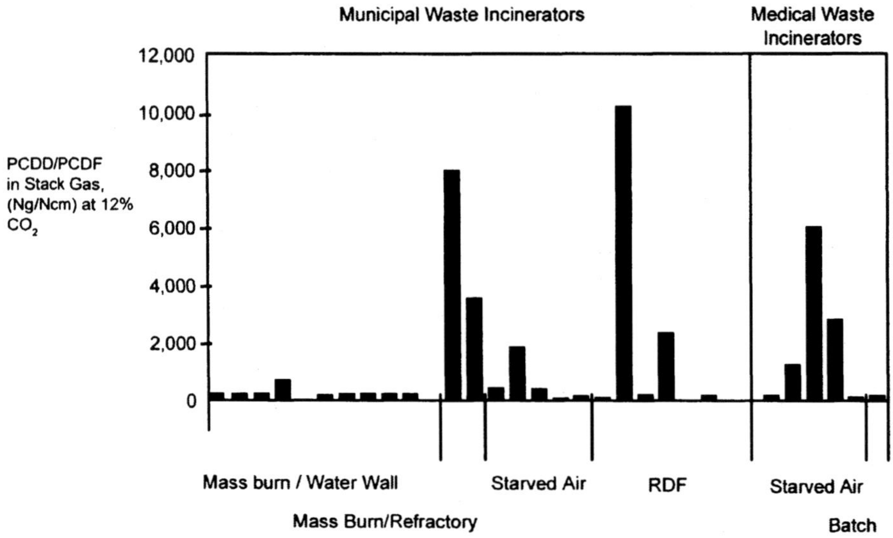

Figure 8-25. Polychlorinated dibenzo(p)dioxin and furan emissions from waste incinerator systems.

8.4.1 Abatement of PCDD/PCDF

Catalytic post-combustion methods (end of the pipe approach) are proven to be effective for flue-gas cleaning of PCDD/PCDF substances to the ppm level. PCDDs/PCDFs are oxidized by the catalyst to form CO_2 at a very high efficieny, reaching 99% at temperatures less than 300°C. Flue gas cleaning, using Selective Catalytic Reduction (SCR) of $H0_x$ by means of NH_3, have been proven to be most effective for MWC plants and fossil (electric) power stations. Platinum Catalysts supported on a Si/Ti carrier (honeycomb or corrugated form) were used in a pilot plant experiment to study the PCDDs/PCDFs destruction ratios (□) versus temperature at various space velocities (SV) (See Fig 8-26).

$$SV = \text{Gas Flow } (m^3/hr) \, / \, \text{Reactor Volume } (m^3)$$

Also, the destruction ratios (□) are directly proportional to the geometric surface area of the catalyst in contact with the flue gases at a constant SV. The artificial velocity (AV) is defined as:

AV = Actual Gas Flow Rate (m^3/hr)/Total Surface Area of Catalyst (m^2) (See Fig. 8-27)

The destruction ratios can be expressed by:

$$\square = K_1(AV) + K_2$$

where $K_1 = -4.19$, $K_2 = 125.57$ at 250°C
$K_1 = -1.928$, $K_2 = 115.21$ at 300°C
$K_1 = -1.526$, $K_2 = 111.24$ at 359°C

Reduction of municiple waste combustion emissions of PCDDs and PCDFs to the acceptable levels published by EPA of oil ng (I-TEF)/ds cm is now possible.

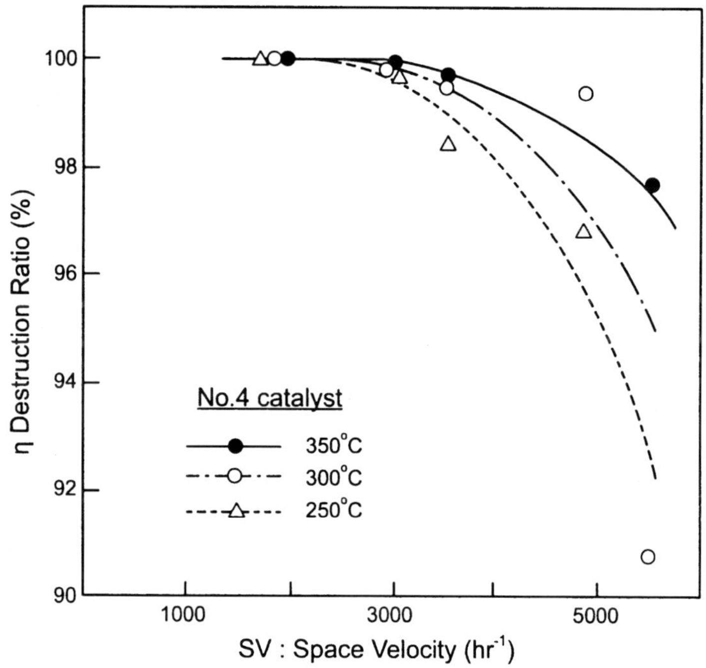

Fig. 8-26 PCDDs destruction ratio versus SV for pilot plant experment by Hiraoka and Takeda, 1991

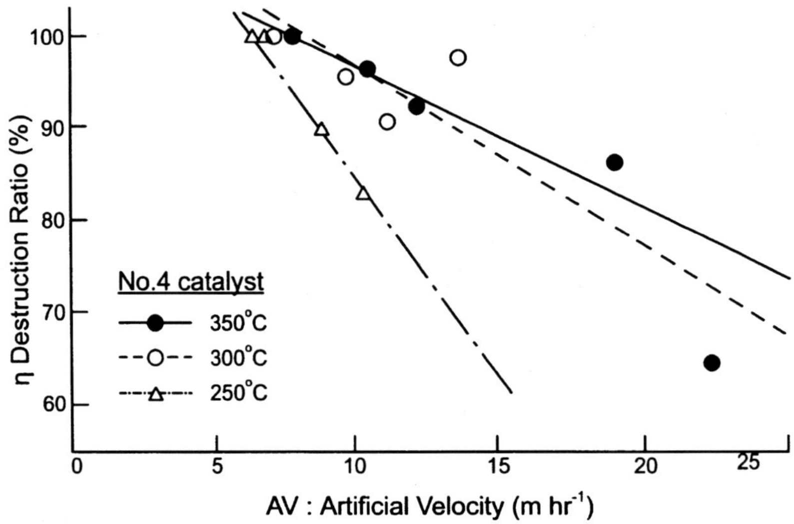

Fig. 8-27 PCDDs destruction ratio versus AV by Hiraoka and Takeda, raq

REFERENCES

8-1. W. Bartok and A. F. Sarofim, *Fossil Fuel Combustion, A Source Book*, Wiley-Interscience, New York, 1991.

8-2. Bockhom, H., *Soot Formation in Combustion*, Springer-Verlag, Berlin, 1994.

8-3. Bohm, H, D. Hesse and H. Jander, B Cuers, J. Pietsdier, H.G Wagner and M. Wiess, The Influence of Pressure and Temperature Soot Formation in Premixed Flames, 22nd Symp. Comb. , The Combustion Institute, Pittsburgh, PA. p. 403 (1989).

8-4. Brunner, C. R., *Handbook of Incineration Systems*, McGraw-Hill, New York, 1991.

8-5. Brunner, C. L., *Incineration System Selection and Design*, Van Nostrand-Reinhold, New York, 1984.

8-6. Bryers, R. W., *Incinerating Municipal and Industrial Waste*, Hemisphere, New York, 1991.

8-7. Buckhorn, A., F. Fetting, H.W Wenz, Investigation of The Formation of High Molecular Hydrocarbons and Soot in Premixed Hydrocarbon-Oxygen Flames, Ber Bunsenges Phy. Chem. 87, 1067 (1983).

8-8. Corey, R. C., *Principles and Practices of Incineration*, Wiley-Interscience, New York, 1969.

8-9. DePaul, F. T. and J. R. Crouder, *Control of Emissions from Municipal Solid Waste Incinerators*, Noyes, Park Ridge, New Jersey, 1989.

8-10. El-Gamal, M., Simulation der Rußkildung in Vorgemischten Verbrennungs System en, Dissertation, University of Stuttgart, 1995.

8-11. Frenklach, M. and J. Warnatz, Detailed Modeling of PAH Profiles in a Soot Low Pressye Acetylene Flame, Comb. Sci. Technology 51, 265 (1987).

8-12. Frenklach, M. and H. Wang, Detailed Modeling of Soot Particle Nucleation and Growth, 23rd Symp. Comb. , The Combustion Institute, Pittsburgh, PA,p.1559 (1991).

8-13. Gardiner, W. C., Jr., *Combustion Chemistry*, Springer-Verlag, New York, 1984.

8-14. Hao, O. J. and K. K. Plinil, Wet air oxidation of hazardous waste, in *Remediation of Hazardous Waste Contaminated Soils* (D. L. Wise and D. J. Trantolo, ed.), Marcel Dekker, New York, 1994, pp. 849–868.

8-15. Haynes, R. S. and H.G Wagner, Soot Formation, Prog. Energy Combustion Sci. 7, 229 (1981).

8-16. Hiraoka, M., and N. Takeda. "Catalytic Destruction of PCDDs/PCDFs in Municipal Solid Waste Flue Gas." *Chemosphere,* Vol. 23, Nos. 8-10, pp 1445-1452, (1991).

8-17. Jones, J. L. and S. B. Redding, *Thermal Conversion of Solid Wastes and Biomass*, American Chemical Society, Washington, DC, 1980.

8-18. Kuryla, W. C. and A. J. Papa, *Flame Retardency of Polymeric Materials*, Vol. 1 to Vol. 5, Marcel Dekker, New York, 1979.

8-19. Lee, C. C. and G. Huffman, Update of thermal destruction technologies, in *Thermal Processes* (H. Freeman, ed.), Technomic, 1990.

8-20. Tsuji, H., A. K. Gupta, T. Hasegawa, M. Katsuki, K. Kishimoto, and M. Morita, *High Temperature Air Combustion: From Energy Conservation to Pollution Reduction*, CRC Press, New York, 2003

8-21. Li, L., P. Chen, and E. F. Gloyna, Generalized kinetic model for wet oxidation of organic compounds, *AIchE J.*, 37(11), 1687-1697 (1991).

8-22. Lin, S. H., S. J. Ho, and C. L. Wu, Kinetic and performance characteristics of wet air oxidation of high concentration wastewater, *Ind. Eng. Chem. Res.*, 35(1), 307-314 (1996).

8-23. Linan, A. and F.A Williams, *Fundamental Aspects of Combustion*, Oxford University Press, New York, 1993.

8-24. Nemerow, N. L. and A. Dasgupta, *Industrial and Hazardous Waste Management*, Van Nostrond-Reinhold, New York, 1991.

8-25. Patterson, D. J. and N. A. Henein, *Emissions from Combustion Engines and Their Control,* Ann Arbor Science, Ann Arbor, Michigan, 1972.

8-26. Schwarzenback, R. P., P. M. Gschwend, and D. M. Imboden, *Environmental Organic Chemistry*, Wiley, New York, 1993.

8-27. Simonneit, B. R. T., *Molecular Indicators (Biomarkers) of Past Life, Anatomical Record*, 268, 186-195 (2002).

8-28. Spellman, Frank R., *Incinerating Biosolids,* Technomic Publishing Company, Inc., Lancaster, Pennsylvania, 1997.

8.29. Tchobanoglons, G., H. Theisen, and R. Eliassen, *Solid Wastes: Engineering Principles and Management Issues*, McGraw-Hill, New York.

8-30. McBean, Edward A., Rovers, Frank A. and Farquhar, Grahame J., *Solid Waste Landfill Engineering and Design*, Prentice Hall PTR, Englewood Cliffs, New Jersey, 1995.

8.31. Wagner, H. G., Soot Formation in Combustion, 17th Symp. Comb., The Combustion Institute, Pittsburgh, PA p.3 (1979).

8.32. Warnatz, J., U.Mass, and R.W Dibble, Combustion, Physical and Chemical Fundamentals, Modeling and Simulation, Experiments, Pollution Formation, 2nd ed; Springer-Verlag, Berlin, 1999.

8.33. Weinstein, N. J. and R. F. Toro, *Thermo Processing of Municipal Solid Waste for Resource and Energy Recovery*, Ann Arbor Science, Ann Arbor, Michigan, 1976.

PROBLEM SET

1. Approximate the specific heat of solid waste and sludge.

2. A particular crude oil is heated to 510 K (458°F) and charged at 10 L/hr (0.01 m^3/hr, or 2.6 gal/hr) to the flash zone of a laboratory distillation tower. The flash zone is at an absolute pressure 110 kPa (16 psi). Determine the percent vaporized and the amounts of the overhead and bottom streams. Assume that the vapor and liquid are in equilibrium. Hint: 30% (by volume) of feed is vaporized. Flash zone temperature is 483 K.

3. A coal has the following ultimate analysis: C = 0.8339, H_2 = 0.0456, O_2 = 0.0505, N_2 = 0.0103, S = 0.0064, ash = 0.0533, total = 1.000. The coal is burned in a steam boiler furnace. Determine the weight of air required for theoretically perfect combustion, the weight of gas formed per pound of coal burned, the volume of flue gas at the boiler exit temperature of 600°F (589K) per pound of coal burned, the air required with 20% excess air and the volume of gas formed with this excess; and the CO_2 percentage in the flue gas on a dry and wet basis.

CHAPTER 9

PYROLYSIS AND CATALYSIS

In contrast with combustion which requires the use of air, thermal decomposition by pyrolysis can occur with limited or no air, by using an inert gas instead. However, both combustion and pyrolysis can use catalysts to enhance the rate of reaction. It is well known that pyrolysis of hydrocarbons and a variety of other compounds can give products that are useful in the production of petrochemical feed stocks. This chapter will first discuss the reaction mechanism by which these products are formed. Next, we will discuss a model system followed by a description of isolation of essential products as well as byproducts. The chapter will end with a very simple introduction to catalysis and catalyst chemistry.

9.1 PYROLYSIS CHEMISTRY

Pyrolysis is often called "destructive distillation" or "retorting." It involves a process in which organic material is heated to a high temperature in a controlled oxygen environment. Unlike combustion in an excess of air, which is highly exothermic and produces primarily heat and carbon dioxide, pyrolysis of organic material is analogous to a distillation process and is endothermic.

In the petroleum or petrochemical industry, cracking is often used instead of pyrolysis. Especially due to the requirement of ethylene production, efforts have been devoted to catalytic cracking, hydrocracking etc. Furthermore, during thermal treatment, organic molecules can form smaller molecules (cracking or bond scission); in the meantime, the larger molecules can also form (polycondensation or polymerization). In the latter case, tar, char, and coke are formed. The consequence is the concern of public health e.g. the diesel particulate matter of $PM_{2..5}$ (particulate matter with aerodynamic diameter smaller than 2.5 mm) or even the Aitken particulate (particulate less than unit micron). The carcinogenic and toxic properties of these waste by-products require a critical understanding of their formation mechanism.

9.1.1 Paraffins Cracking

The following is a review of the chemical reactions of pyrolysis. For paraffins, both dehydration and chain scission can occur:

$$C_nH_{2n+2} = C_nH_{2n} + H_2 \qquad [9\text{-}1]$$

$$C_nH_{2(m+n)+2} = C_nH_{2n} + C_mH_{2m+2} \qquad [9\text{-}2]$$

Usually, the reactions are conducted at high temperature and low pressure. Ideal gas law is valid. The reaction equilibrium constant k_f and the standard free energy change □G° are given by:

$$\Delta G^o = \text{-RTln}\, k_f \qquad [9\text{-}3]$$

$$k_f = \pi p_i^{\nu_i} \qquad [9\text{-}4]$$

where v_i is the chemical stoichiometric coefficient, p_i is partial pressure for hydrocarbonic molecules in the reactant for $v_i < 0$, and molecules in the product for v_i are greater than zero. In the standard state, if partial pressure of all components is equal to 1, then

$$\Delta G^o = \sum_i v_i \Delta G^o_{fi} \qquad [9\text{-}5]$$

This can also be written as:

$$\Delta G^o = \{\sum_i v_i \Delta G^o_{fi}\}_{product} - \{\sum_i v_i \Delta G^o_{fi}\}_{reactant} \qquad [9\text{-}6]$$

Also, the standard free energy of formation is related to the standard heat of formation, ΔH^o_f:

$$\left[\frac{\partial(\Delta G^o_f / T)}{\partial T}\right]_P = -\frac{\Delta H^o_f}{T^2} \qquad [9\text{-}7]$$

Integration of the above equation gives:

$$\Delta G^o_f / T = -\int \frac{\Delta H^o_f}{T^2} dT + C \qquad [9\text{-}8]$$

□G□□° can also be expressed as a polynomial of T:

$$\Delta G^o_f = A + BT + CT^2 + DT^3 \qquad [9\text{-}9]$$

Some of the values of A, B, C, and D are listed in Table 9-1. Some of the free energy - □G□□° at 1000 K for common short drain alkanes are listed in Table 9-2. It is evident that chain scission outweighs the hydrogenation.

From Arrhenius equation,

$$k = A\exp(-T_2 / RT) \qquad [9\text{-}10]$$

After taking logarithm form,

$$\ln k = \ln A + \frac{T_2}{R}\frac{1}{T} \qquad [9\text{-}11]$$

If ln k is plotted vs. T^{1}, then ln A is the intercept and $\frac{T_2}{R}$ is the slope. For n-paraffin pyrolisates and olefin pyrolisates, the results are shown in Figure. 9-1 (A and B).

Taking n-pentane as a standard k_s, a number of different hydrocarbons k_i have been examined. The rate constants increase with the carbon homolog's number (see Figure 9-2).

Table 9-1. Values of coefficients in multinomial

Hydrocarbon	A	B × 10^{-3}	C × 10^{6}	D × 10^{9}	Maximum Deviation
Methane CH_4	-17.018912	12.69051	13.680543	-4.7779211	0.01085
Ethane C_2H_6	-19.002584	31.820685	20.738727	-7.4354358	0.01802
Ethylene C_2H_4	13.221134	6.9630128	12.613755	-4.2571673	0.01456
Acetylene C_2H_2	54.312343	-14.815563	1.1757617	-0.0641951	0.00445
Propane C_3H_8	-23.341913	52.519228	26.004186	-9.4852413	0.02939
Propylene C_3H_6	6.0635378	24.694603	19.737445	-7.0643459	0.01725
n-Butane C_4H_{10}	-28.397933	73.208547	31.090844	-11.414663	0.02948
1-Butene C_4H_8	1.5788555	44.973295	26.100911	-9.5852931	0.03620
n-Pentane C_5H_{12}	-32.884323	93.690343	37.146266	-13.798675	0.04210
1-Pentene C_5H_{10}	-3.1136216	65.643697	31.211192	-11.514629	0.03266

Table 9-2. Values of ΔG^o_{1000} of pyrolysis reactions for alkanes at 1000 K

	Reaction		ΔG^o_{1000} (kJ.mol^{-1})		Reaction		ΔG^o_{1000} (kJ.mol^{-1})
Dehydrogenation	$C_nH_{2n+2} \longleftrightarrow C_nH_{2n} + H_2$			Chain Scission	$C_{m+n}H_{2(m+n)+2} \longleftrightarrow C_nH_{2n} + C_mH_{2m+2}$		
	$CH_4 \longleftrightarrow \frac{1}{2}C_2H_4 + H_2$	(A)	-39.96		$C_3H_8 \longleftrightarrow C_2H_4 + CH_4$	(G)	53.93
	$C_2H_6 \longleftrightarrow C_2H_4 + H_2$	(B)	-8.88		$C_4H_{10} \longleftrightarrow C_3H_6 + CH_4$	(H)	69.04
					$C_4H_{10} \longleftrightarrow C_2H_4 + C_2H_6$	(I)	42.37
	$C_3H_8 \longleftrightarrow C_3H_6 + H_2$	(C)	9.55		$C_5H_{12} \longleftrightarrow C_4H_8 + CH_4$	(J)	69.12
	$C_4H_{10} \longleftrightarrow C_4H_8 + H_2$	(D)	5.96		$C_5H_{12} \longleftrightarrow C_3H_6 + C_2H_6$	(K)	61.17
					$C_5H_{12} \longleftrightarrow C_2H_4 + C_3H_8$	(L)	42.75
	$C_5H_{12} \longleftrightarrow C_5H_{10} + H_2$	(E)	8.08		$C_6H_{14} \longleftrightarrow C_5H_{10} + CH_4$	(M)	70.13
	$C_6H_{14} \longleftrightarrow C_6H_{12} + H_2$	(F)	7.41		$C_6H_{14} \longleftrightarrow C_4H_8 + C_2H_6$	(N)	60.12
					$C_6H_{14} \longleftrightarrow C_3H_6 + C_3H_8$	(O)	60.42
	… …		… …		$C_6H_{14} \longleftrightarrow C_2H_4 + C_4H_{10}$	(P)	45.30
					… …		… …

*For convenience, the ordinal numbers of the reactions are started from (A) in each of the following illustrations so that they have only a local meaning but not suitable for those cases outside the case in question.

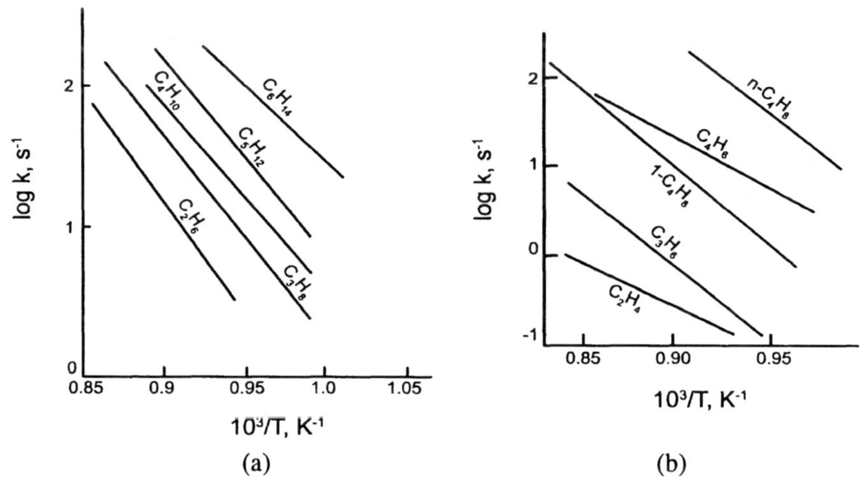

Figure 9-1. (a) Reaction rate constants for n-paraffin pyrolysis versus temperature, (b) Reaction rate constants for olefin pyrolysis versus temperature

In addition, pyrolysis of olefin can have the following qualities:

- Disproportionation e.g. propylene can be:

$$2C_3H_6 = C_2H_4 + C_4H_8$$
$$2C_3H_6 = C_2H_6 + C_4H_6$$
$$2C_3H_6 = C_5H_8 + CH_4$$
$$2C_3H_6 = C_6H_{10} + H_2$$

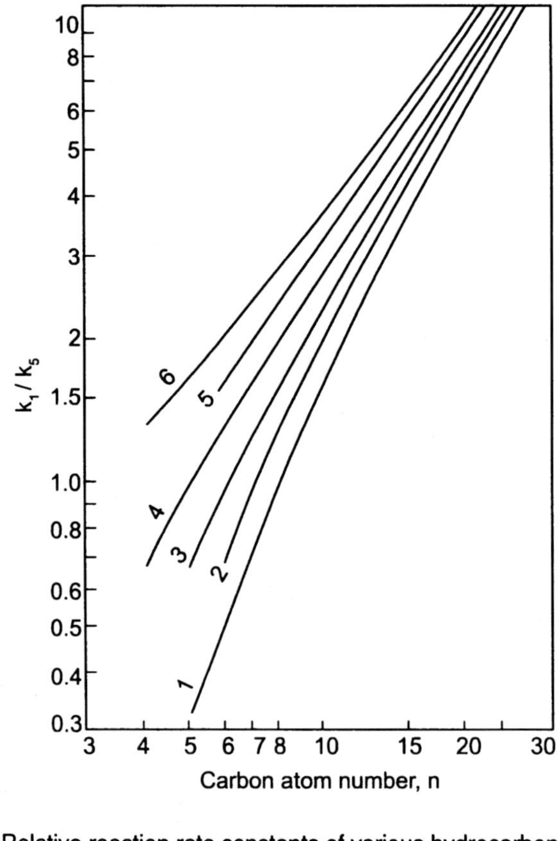

Figure 9-2. Relative reaction rate constants of various hydrocarbons 1-alkyl cyclopentane , 2-alkyl cyclohexane, 3-2-methyl alkane, 4-n-alkane, and 5-dimethyl branched chain alkane in which two methyls are not attached to one carbon atom 6-n-1-alkene.

- Diels-Alder reaction

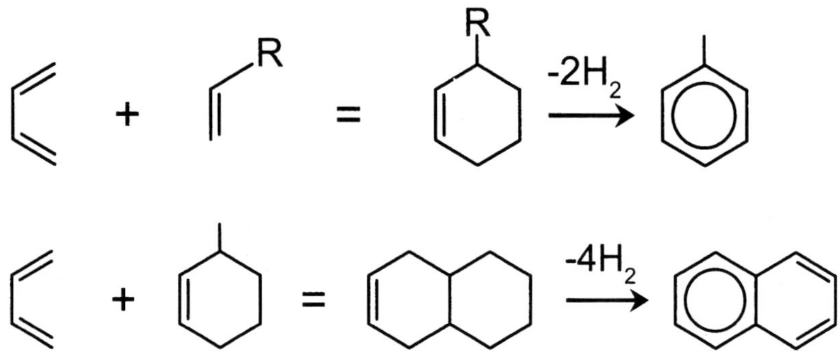

- Aromatization

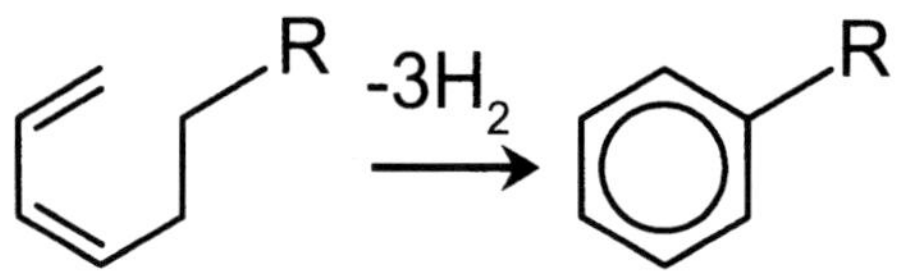

[Example 9.1] Predict the yield of propylene and ethylene if propane is subjected to pyrolysis. Predict the yield of ethylene from pyrolysis of methane.

From Table 9-2, the $-\Delta G^o_{1000}$ values of

$$C_3H_8 \longrightarrow C_2H_4 + CH_4 \text{ , 53.93 KJ/mole}$$

$$C_3H_8 \longrightarrow C_3H_6 + H_2 \quad \text{, 9.55 KJ/mole}$$

Thus,

$$\left(k_{p1000}\right)_{propylene} = \exp(-\Delta G^o / RT)$$

$$= \exp(-9.55/0.001987 \times 1000)$$

$$= 8.18 \times 10^{-3}$$

$$(k_{p1000})_{ethylene} = \exp(-\Delta G^o / RT)$$

$$= \exp(-53.93/0.001987 \times 1000)$$

$$= 1.63 \times 10^{-12}$$

The ethylene-propylene equilibrium constant ratio is thus ethylene/propylene = 1.99×10^8.

Thus, ethylene is predominant.

In the case of the reaction,

$$CH_4 \longrightarrow \frac{1}{2} C_2H_4 + H_2$$

The $-\Delta G^o_{1000}$ value is –39.96, and the equilibrium constant calculated out is 5.42×10^8. Therefore, methane cannot crack to form ethylene.

9.1.2 Naphthenics Pyrolysis

Pyrolysis of naphthenics usually abides by the following rules:

- Alkyl in the side chain is easier to rupture than the hydrocarbon ring. Thus, the yield of ethylene for pyrolysis is of naphthenics with a long side chain is higher than without a side chain.

- Dehydrogenation of napthenics is easier than the ring-opening. Therefore, the dehydrogenation to form aromatics is superior to the ring-opening reactions to form olefins.

- For ring-opening of cyclohexane, the following scheme is obtained:

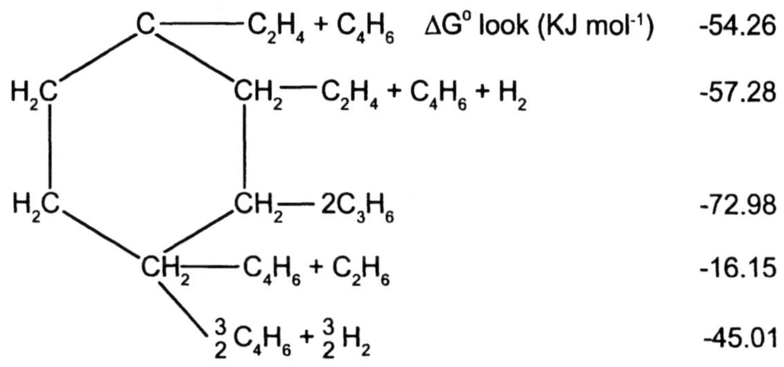

- The 5-membered ring napthenic is more refractory than the 6-membered ring napthanics. The pyrolysis rate of the alkyl substituted cyclopentane is lower than that of the unsubstituted cyclohexane.

- Pyrolysis of alkyl cyclopentane or alkylcyclohexane is lower than that of n-alkane with the same carbon number.

- If the relative content of cyclic to n-alkane in a pyrolysis feedstack is increased, the yield of ethylene will be decreased. On the other hand, the yields of aromatics and butadiene will be increased.

In a typical petroleum feedstock, the pyrolysis products are illustrated as in Fig. 9-3. In refining gas oil, the earlier stage pyrolysis is evolved with the primary reactions. The secondary reactions belong to the pyrolysis of the products as indicated by Fig. 9-4.

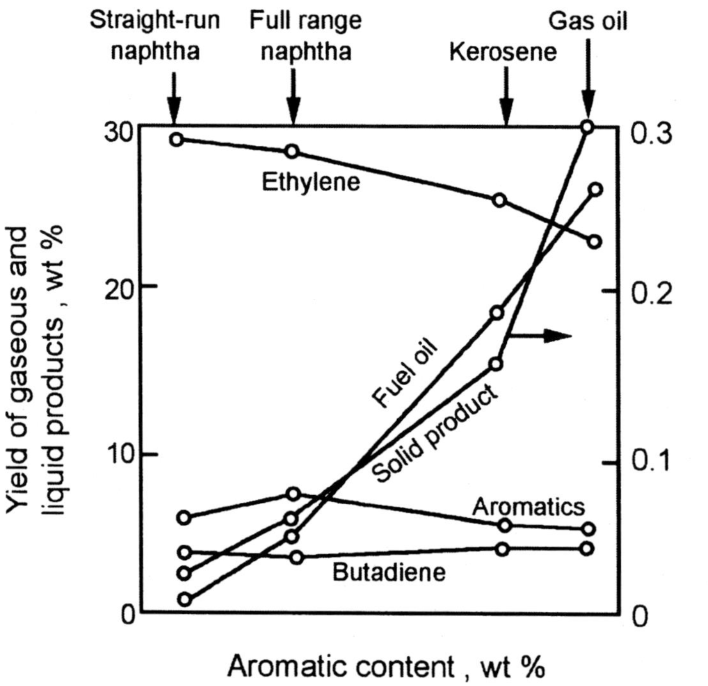

Figure 9-3. Effect of aromatic content on the yields of pyrolysis products. Reaction conditions: outlet temperature, 770 °C; outlet pressure, 0.11 Mpa; residence time, 0.25 s; dilution ratio of steam to hydrocarbon feedstock, 0.8.

Chemical equilibrium is discussed here with ethane pyrolysis. At high temperature, ethane dehydrogenation system is involved in the following five reactions:

$$C_2H_6 \xleftrightarrow{k_{pr}} C_2H_4 + H_2 \qquad [9\text{-}12a]$$

$$C_2H_4 \xleftrightarrow{k_{p2}} C_2H_2 + H_2 \qquad [9\text{-}12b]$$

$$C_2H_2 \underset{k_{f22}}{\overset{k_{p2}}{\rightleftarrows}} 2C + H_2 \qquad [9\text{-}12c]$$

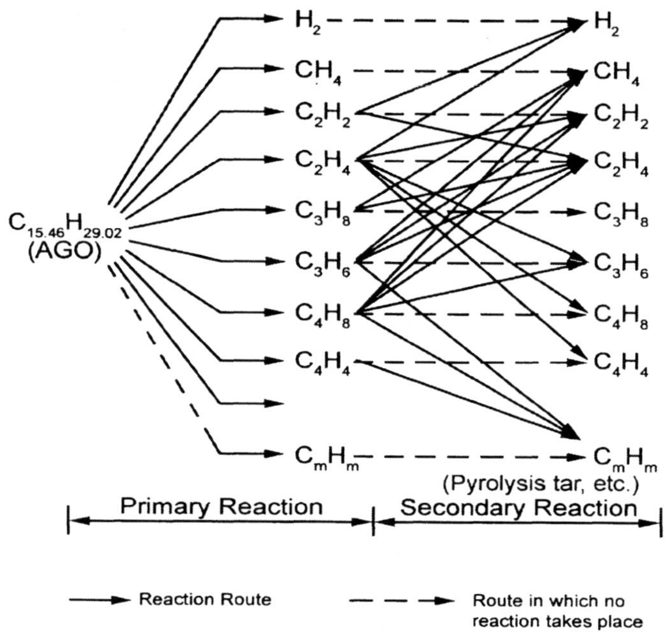

Figure 9-4. Scheme of the primary and secondary reactions for AGO pyrolysis

$$C_2H_4 \underset{k_{f24}}{\overset{k_{p4}}{\rightleftarrows}} 2C + 2H_2 \qquad [9\text{-}12d]$$

$$C_2H_6 \underset{k_{f26}}{\overset{k_{p5}}{\rightleftarrows}} 2C + 3H_2 \qquad [9\text{-}12e]$$

chemical equilibrium constants for dehydrogenation = $k\text{p}_1$, $k\text{p}_2$

chemical equilibrium constants for formation of ethane, ethylene, and acetylene = k_{f26}, k_{f24}, k_{f22}

chemical equilibrium constants of ethane, ethylene, and acetylene = k_{f26}, $k_{f24,}$ k_{p5}

Further,

$$k_{p3} = k_{f22}^{-1}$$

$$k_{p4} = k_{f24}^{-1}$$

$$k_{p5} = k_{f26}^{-1}$$

Expressing the stoichiometric coefficient of the five equations for propane pyrolysis in matrix form,

$$
\begin{array}{c} \\ M = \end{array}
\begin{array}{c}
\begin{array}{ccccc} C_2H_6 & C_2H_9 & C_2H_2 & C & H_2 \end{array} \\
\begin{bmatrix} -1 & 1 & 0 & 0 & 1 \\ 0 & -1 & 1 & 0 & 1 \\ 0 & 0 & -1 & 2 & 1 \\ 0 & -1 & 0 & 2 & 2 \\ -1 & 0 & 0 & 2 & 3 \end{bmatrix}
\begin{matrix} a \\ b \\ c \\ d \\ e \end{matrix}
\end{array}
$$

Transformation of any two rows of this matrix will give the following. Remember that each row represents one reaction.

$$
M = \begin{bmatrix} -1 & 1 & 0 & 0 & 1 \\ 0 & -1 & 1 & 0 & 1 \\ 0 & 0 & -1 & 2 & 1 \\ 0 & -1 & 0 & 2 & 2 \\ -1 & 0 & 0 & 2 & 3 \end{bmatrix}
\begin{matrix} a \\ b \\ c \\ d-b-c \\ e-a-d \end{matrix}
$$

$$
\begin{bmatrix} -1 & 1 & 0 & 0 & 1 \\ 0 & -1 & 1 & 1 & 0 \\ 0 & 0 & 0 & 0 & 0 \\ 0 & -1 & 0 & 2 & 2 \\ 0 & 0 & 0 & 0 & 0 \end{bmatrix}
\begin{matrix} a \\ b \\ c+b-d \\ d \\ e-a-d \end{matrix}
$$

$$
\begin{bmatrix} -1 & 1 & 0 & 0 & 1 \\ 0 & -1 & 1 & 0 & 1 \\ 0 & 0 & 0 & 0 & 0 \\ 0 & 0 & 0 & 0 & 0 \\ -1 & 0 & 0 & 2 & 3 \end{bmatrix}
\begin{matrix} a \\ b \\ c+b-d \\ d+a-e \\ e \end{matrix}
$$

$$\left[\begin{array}{ccccc} -1 & 1 & 0 & 0 & 1 \\ 0 & 0 & 0 & 0 & 0 \\ 0 & 0 & -1 & 2 & 1 \\ 0 & -1 & 0 & 2 & 2 \\ 0 & 0 & 0 & 0 & 0 \end{array}\right]\begin{array}{c} a \\ b+c-d \\ c \\ d \\ e-a-d \end{array}$$

$$\left[\begin{array}{ccccc} -1 & 1 & 0 & 0 & 1 \\ 0 & 0 & 0 & 0 & 0 \\ 0 & 0 & -1 & 2 & 1 \\ 0 & 0 & 0 & 0 & 0 \\ -1 & 0 & 0 & 2 & 3 \end{array}\right]\begin{array}{c} a \\ b+c-d \\ c \\ d+a-c \\ e \end{array}$$

$$\left[\begin{array}{ccccc} 0 & 0 & 0 & 0 & 0 \\ 0 & -1 & 1 & 0 & 1 \\ 0 & 0 & -1 & 2 & 1 \\ 0 & 0 & 0 & 0 & 0 \\ -1 & 0 & 0 & 2 & 3 \end{array}\right]\begin{array}{c} a-e+d \\ b+c-d \\ c \\ d \\ e \end{array}$$

$$\left[\begin{array}{ccccc} 0 & 0 & 0 & 0 & 0 \\ 0 & -1 & 1 & 0 & 1 \\ 0 & 0 & 0 & 0 & 0 \\ 0 & -1 & 0 & 2 & 2 \\ -1 & 0 & 0 & 2 & 3 \end{array}\right]\begin{array}{c} a-c+d \\ b \\ c+b-d \\ d \\ e \end{array}$$

$$\left[\begin{array}{ccccc} 0 & 0 & 0 & 0 & 0 \\ 0 & 0 & 0 & 0 & 0 \\ 0 & 0 & -1 & 2 & 1 \\ 0 & -1 & 0 & 2 & 2 \\ -1 & 0 & 0 & 2 & 3 \end{array}\right]\begin{array}{c} a-e+d \\ b+c-d \\ c \\ d \\ e \end{array}$$

This study lends to the conclusion that there are three independent reactions, with the exception of reactions a, d, and e and the reactions b to d. In other words, in any set of three independent reactions, there must be reactions a or e and reactions b or c.

Calculations of chemical equilibrium compositions are as follows for the equations (a), (b), and (c). From chemical thermodynamics,

$$K_p = \underset{i}{\pi} N_i^{*} v_i \left(\frac{p}{\underset{i}{\Sigma N_i^{*}}} \right) \Delta v_i$$

where N_i^* is the number of equilibrium moles of the compound in gas phase and p is the total pressure of the system. For equations (a) to (c),

$$\square N_i^{*}=1$$

$$N_{26}^{*} + N_{24}^{*} + N_{22}^{*} + N_{02}^{*} = 1$$

$$Kp_1 = N_{29}^{*} N_{02}^{*} / N_{26}^{*}$$

$$Kp_2 = N_{22}^{*} N_{02}^{*} / N_{24}^{*}$$

$$Kp_3 = N_{02}^{*} / N_{22}^{*}$$

By solving the above non-linear equations, we have:

$$N_{01} = \left| \left\{ \frac{K_{p1}K_{p2}K_{p3}}{2} + \frac{K_{p1}^2 K_{p2}(HK_{p2})}{6} - \frac{K_{p1}^2}{27} + \left[\left(\frac{K_{p1}K_{p2}(1+K_{p3})}{3} - \frac{K_{p1}^2}{9} \right)^3 + \left(-\frac{K_{p1}K_{p2}K_{p3}}{2} - \frac{K_{p2}^2 K_{p2}(1+K_{p3})}{6} + \frac{K_{p1}^3}{27} \right)^2 \right]^{1/2} \right\}^{1/3} + \left\{ \frac{K_{p1}K_{p2}K_{p3}}{2} + \frac{K_{p1}^2 K_{p2}(1+K_{p2})}{6} - \frac{K_{p1}^3}{27} - \left[\left(\frac{K_{p1}K_{p2}(1+K_{p2})}{3} - \frac{K_{p1}^2}{9} \right)^3 + \left(-\frac{K_{p1}K_{p2}K_{p3}}{2} - \frac{K_{p1}^2 K_{p2}(1+K_{p3})}{6} + \frac{K_{p1}^3}{27} \right)^2 \right]^{1/2} \right\}^{1/3} - \frac{K_{p1}}{3} \right|$$

[9-13]

Also,

$$N_{22}^* = N_{02}^*/K_{p2}$$
$$N_{24}^* = N_{22}^*N_{02}^*/K_{P2}$$
$$N_{26}^* = N_{24}^*N_{02}^*/K_{p2}$$

Thus, moles of H_2, C_2H_2, C_2H_2, C_2H_4, and C_2H_6 are obtained. Different computations of the synthesis gas composition in comparison with the experimental results have been made. Recently different models for catalytic reforming have been worked out by Klein and coworkers. Autothermal reformery (ATR) becomes useful to the fuel cell application.

9.1.3 Pyrolysis of Aromatics and Others

Some aromatic hydrocarbons have been studied before soot formation set in. For example, in the case of 1- or 2-methyl naphthalene, recombination of methyl radicals forms ethane. The radical assisted isomerization can be important (See Fig. 9-5)

Pyrolysis of heterocyclics are more complex, such as with the pyrolysis of cellulose, in which a main portion of the solid waste can go through many intermediate types (Fig. 9-6 and 7)

Pyrolysis of polymers can never yield quantitatively the monomers as it is made. This explains why the recovery of virgin rubber is very difficult. Also, the recycling and treatment of plastic waste, such as polyvinyl chloride and polyvinyline chloride, can be troublesome. In many cases, the treatment with free radical generator or quencher such as sulfur or asphaltene will be helpful. Modern rubber recycling has been dependent on an extruder equipped with huge ultrasound power input.

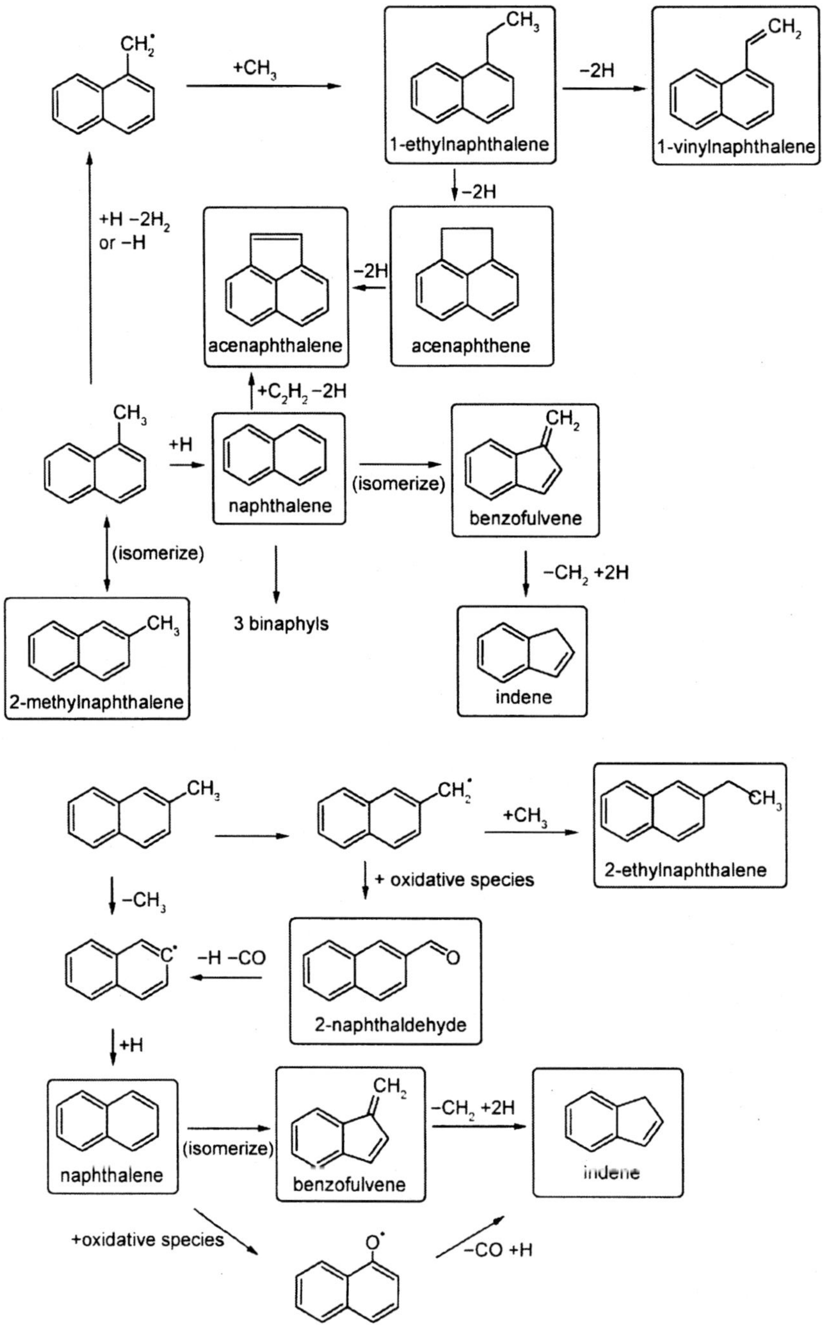

Figure 9-5. Hypothesized pathways of 1-methylnaphthalene pyrolysis (top panel) and 2-methylnaphthalene oxidation (bottom panel)

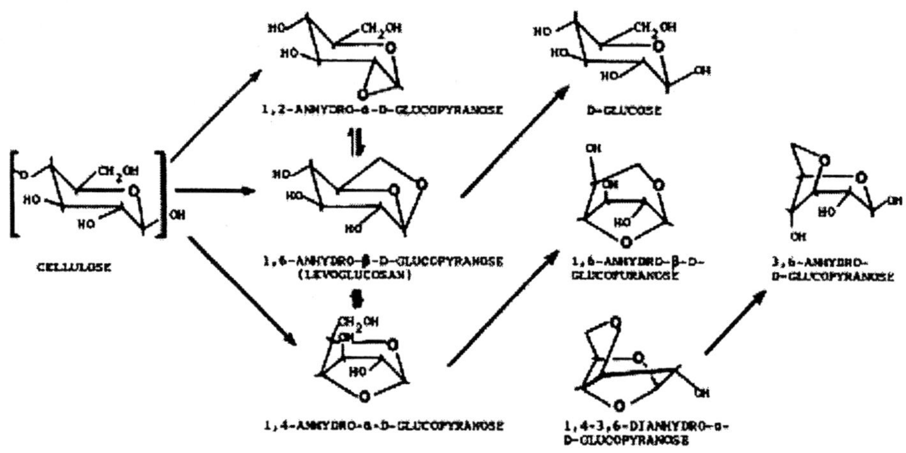

Figure 9-6. Pyrolysis of cellulose: initial reactions

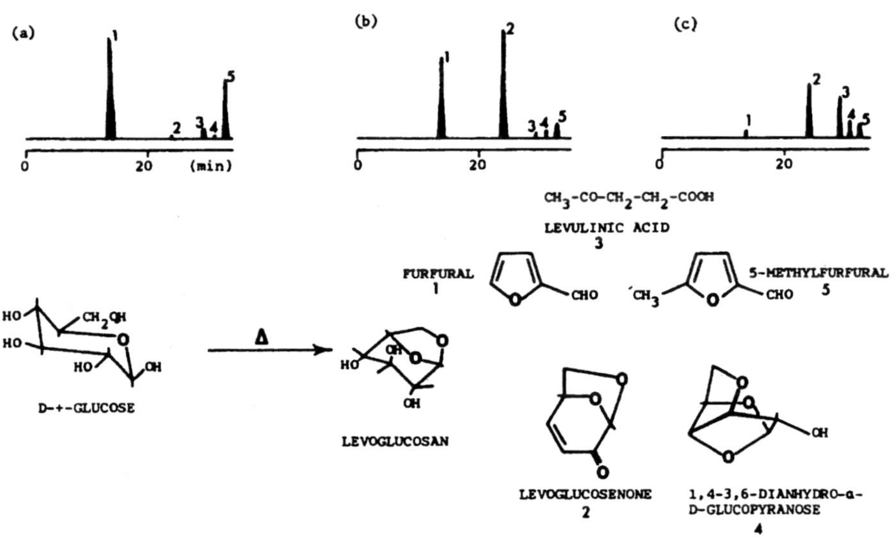

Figure 9-7. Curie-point pyrolysis of D-glucose, cellulose, and levoglucosan: (a) glucose, (b) cellulose, and (c) levoglucosan

9.2 Model For Pyrolysis

9.2.1 Gas-Tar-Char system

The high temperatures (900° – 2000°F) and lack of oxygen result in a chemical breakdown of the waste organic materials into three component streams: (a) a gas consisting of primarily hydrogen, methane, carbon monoxide, and carbon dioxide, (b) a "tar" and "oil" that is liquid at room temperature and includes organic chemicals such as acetic acid, acetone, and methanol, and (c) a "char" consisting of almost pure carbon plus any inerts (glass, metals) and mineral salts that enter the process unit. Residence time, temperature, and pressure can be controlled in the pyrolysis reactor to produce various product combinations. Most complex organic molecules upon pyrolysis will yield a tar, often referred to as bitumen, and oil and gas will evolve upon further heating. Both tar and oil are soluble; they are often referred to as the liquid portion. The residue will be char, which often is referred to as carbon or "coke." The so-called kerogen pyrolysis is a model for the pyrolysis of solid waste or sludge. The chemical compositions of kerogen, coal, and solid waste are summarized in Table 9-7.

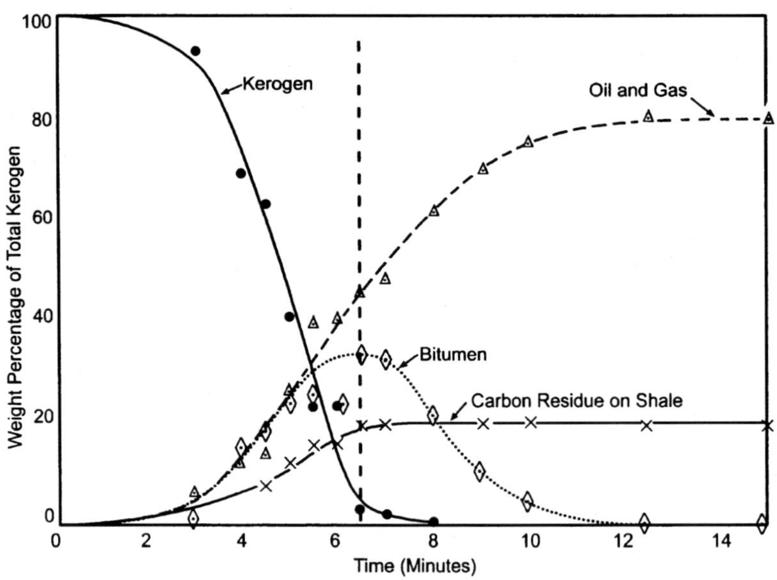

Figure 9-8. Rates of kerogen disappearance and product appearance during heating of Green River oil shale at 900°C, 1650°F. (After Allred, 1966)

A model kerogen pyrolysis curve is illustrated in Figure 9-8, where experimental curves can be estimated via computation. Many kinetic schemes can fit the experimental curves. One of the schemes involves two first-order reactions and three second-order reactions. For example, assuming that

$$A_1 = \text{kerogen (MSW)}$$

$$A_2 = \text{bitumen (tar)}$$

$$A_3 = \text{oil and gas}$$

$$A_4 = \text{carbon and coke (char)}$$

$$A_1 \xrightarrow{k_1} A_2$$

$$A_2 \xrightarrow{k_2} A_3$$

$$A_1 + A_2 \xrightarrow{k_3} A_2 + A_2$$

$$A_1 + A_2 \xrightarrow{k_4} A_3 + A_2$$

$$A_1 + A_2 \xrightarrow{k_5} A_4 + A_2 \qquad [9\text{-}14]$$

Let $x_i(i)$ = concentrations of substance, A_i, at time, t, and

$$1 \le i \le 4$$

and

$$\sum x_i(t) = 1 \quad \text{for} \quad t \ge 0 \qquad [9\text{-}15]$$

The initial conditions are:

$$x_1(0) = 1$$

$$x_2(0) = x_3(0) = x_4(0) = 0$$

Write out the kinetic equations of $\dot{x}_1(t)$ with k_j rate constants, j = 1...5 as function:

$$\dot{x}_1 = -k_1 x_1 - (k_3 + k_4 + k_5) x_1 x_2$$

$$\dot{x}_2 = k_1 x_1 - k_2 x_2 + k_3 x_1 x_2$$

$$\dot{x}_3 = k_2 x_2 + k_4 x_1 x_2$$

$$\dot{x}_4 = k_5 x_1 x_2 \qquad [9\text{-}16]$$

and $\dot{x}_1(t)$ is the derivative, with respect to time, of concentration of A_i at time t, which is also denoted by $t_i(t)$, the rate of change. These differential equations can be expressed in matrix notation:

$$f_i = \begin{pmatrix} \dot{x}_1 \\ \dot{x}_2 \\ \dot{x}_3 \\ \dot{x}_4 \end{pmatrix} = \begin{pmatrix} -x_1 & 0 & -x_1x_2 & -x_1x_2 & -x_1x_2 \\ x_1 & -x_2 & x_1x_2 & 0 & 0 \\ 0 & x_2 & 0 & x_1x_2 & 0 \\ 0 & 0 & 0 & 0 & x_1x_2 \end{pmatrix} \begin{pmatrix} k_1 \\ k_2 \\ k_3 \\ k_4 \\ k_5 \end{pmatrix} \qquad [9\text{-}17]$$

At a particular value of time, say t_j, the above matrix equation has the form

$$x'(t_j) = X[x(t_j)]K \qquad [9\text{-}18]$$

After numerical integration and a least-square procedure, the specific rate constants (vector K) can be obtained by solving the matrix equation. Numerical solution of matrix equations by the Runge-Kutta-Grill method will yield the dashed lines shown in Figure 9-8.

Furthermore, k_j also can be expressed in terms of the Arrhenius equation as:

$$k_j = K_{j0} \exp\left(\frac{-E_j}{RT}\right) \quad [9\text{-}19]$$

where k_{jo} (frequency factor) and E_j (activation energy) are as listed in Table 9-3.

Table 9-3. Activation Energies and Frequency Factors

Reaction[a]	Activation Energy, E, (kcal/g·mol)	Exponent of Frequency Factor (e^{k_o})
$A_1 \rightarrow A_2$	20.3	8.86
$A_2 \rightarrow A_3$	37.4	24.25
$A_1 + A_2 \rightarrow A_2 + A_2$	33.8	23.67
$A_1 + A_2 \rightarrow A_3 + A_2$	28.2	18.75
$A_1 + A_2 \rightarrow A_4 + A_2$	31.0	20.70

A_1 = Kerogen; A_2 = pyrolytic bitumen; A_3 = oil and gas; A_4 = carbon residue.

[a] Temperature range of reactions is 425–475°

9.2.2 Optimization

Optimization of pyrolysis also can be made based on Pontryagin's maximum principle. For example, if we want to maximize the producing tar, x_2, by using the temperature, T, as the control variable at a given heating residual time, from t_0 to t_f, then the performance function can be written:

$$J(T) = \int_{t_f}^{t_0} f_0(x, T, t)dt \qquad [9\text{-}20]$$

as in this case, we choose $f_0 = -x_2$, or

$$J(T) = \int_{t_0}^{t_f} -\dot{x}_2 dt = x_2(t_f) + 2(t_0) \qquad [9\text{-}21]$$

The Hamiltonian function with optimum temperature profile $T(t)$ of our problem is:

$$H(x, p, T, t) = f_2(x, T, t) - p^T f_I \qquad [9\text{-}22]$$

where p is a transposed form of auxiliary vector p_i. After substitution,

$$H(x, p, T, t) = -\dot{x}_2 = \begin{pmatrix} p_1 & p_2 & p_3 & p_4 \end{pmatrix} \begin{pmatrix} f_1 \\ f_2 \\ f_3 \\ f_4 \end{pmatrix} \qquad [9\text{-}23]$$

$$= -k_1 x_1 (1 - p_1 + p_2) + k_2 x_2 (1 + p_1 x_1 + p_2)$$

$$+ k_3 x_1 x_2 (-1 - p_2) + k_4 x_1 x_2 p_1 + k_5 x_1 x_2 p_1$$

or

$$H(x,p,T,t) = -k_{10}\left[\exp\left(\frac{-E_1}{RT}\right)\right](x_1)(1 - p_1 + p_2)$$

$$+ k_{20}\left[\exp\left(\frac{-E_2}{RT}\right)\right](x_2)(1 + p_1 x_1 + p_2)$$

$$+ k_{30}\left[\exp\left(\frac{-E_3}{RT}\right)\right](x_1 x_2)(-1 - p_2) \qquad [9\text{-}24]$$

$$+ k_{40}\left[\exp\left(\frac{-E_4}{RT}\right)\right](x_1 x_2 p_1)$$

$$+ k_{50}\left[\exp\left(\frac{-E_5}{RT}\right)\right](x_1 x_2 p_1)$$

and the auxiliary variable,

$$\dot{p}_i = -\frac{\partial H}{\partial x_i}$$

where $i = 1,2$ and

$$\dot{p}_i = 0 \qquad [9\text{-}25]$$

where $i = 3,4$.

These computations can help to approximate and predict temperatures, heating residence time, and thermal requirement. Often minimizing or maximizing a given fraction from the pyrolyzed product becomes essential for practical applications. The optional temperature profile to obtain the maximum of bitumen is illustrated in Figure 9-9.

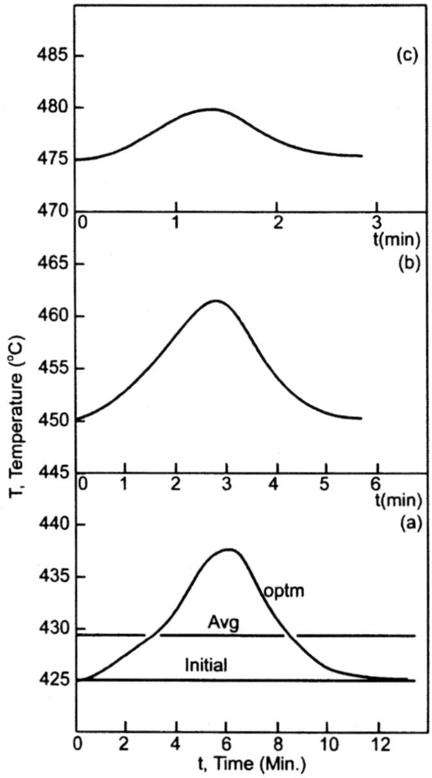

Figure 9-9. Optimal temperature profile for pyrolitic bitumen maximization. Initial heating temperatures were approximately: (a) 425°C, (b) 450°C, and (c) 475°C. (From C. S. Wen and T. F. Yen, Optimization of oil shale pyrolysis, Chem. Eng. Sci., 32, 346-349, 1977.)

9.3 CATALYSIS AND CATALYST CHEMISTRY

9.3.1 Transition state and Stabilization

Every chemical reaction is required by the reactants to proceed through a series of different states until the products are formed. In molecular terms, the intermediate states have changed in bond length, angles, conformations, etc. Often the gradual transition from reactants to products is simplified by assuming that the reactants are in equilibrium with a single **transition state**, and that the products are produced by the breakdown of the transition state. The overall rate of the reaction can be expressed in terms of the equilibrium constants characterizing the equilibrium between the reactants and the transition state and the rate of the breakdown of the transition state into products. For example, the rate constant, k, for reactions in liquid solutions is:

$$k = \frac{RT}{Nh}\exp\left(\frac{-\Delta G^0}{RT}\right) = \frac{RT}{Nh}\exp\left(\frac{\Delta S^0}{R}\right)\exp\left(\frac{-\Delta H^0}{RT}\right) \qquad [9\text{-}26]$$

where ΔG^0 ($= \Delta H^0 - T\Delta S^0$) is the standard free energy change for the conversion of reactants to transition state (usually called the free energy of activation) and ΔH^0 and ΔS^0 are the standard enthalpy change and entropy change, respectively.

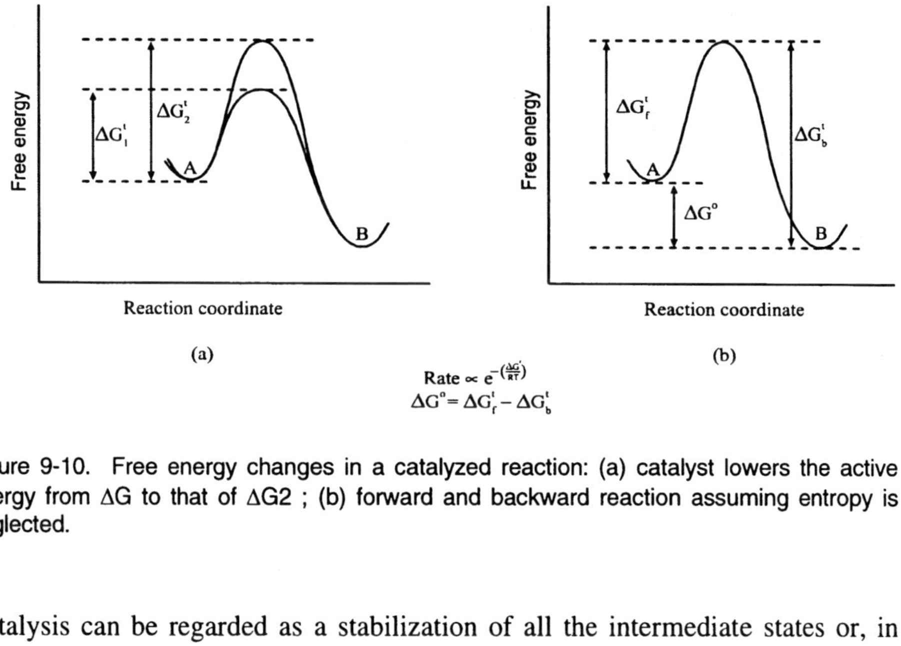

Figure 9-10. Free energy changes in a catalyzed reaction: (a) catalyst lowers the active energy from ΔG to that of ΔG2 ; (b) forward and backward reaction assuming entropy is neglected.

Catalysis can be regarded as a stabilization of all the intermediate states or, in terms of transition state, as a stabilization of the transition state with respect to the reactants. In other words, ΔG^0 is decreased, as indicated in Figure 9-10, where the activation energy of the catalyzed reaction is lower than that of the uncatalyzed reaction. In other words, catalysts increase the rate of chemical reaction but are not consumed during the reaction. A simple example of an S_N2 reaction will require an activated complex:

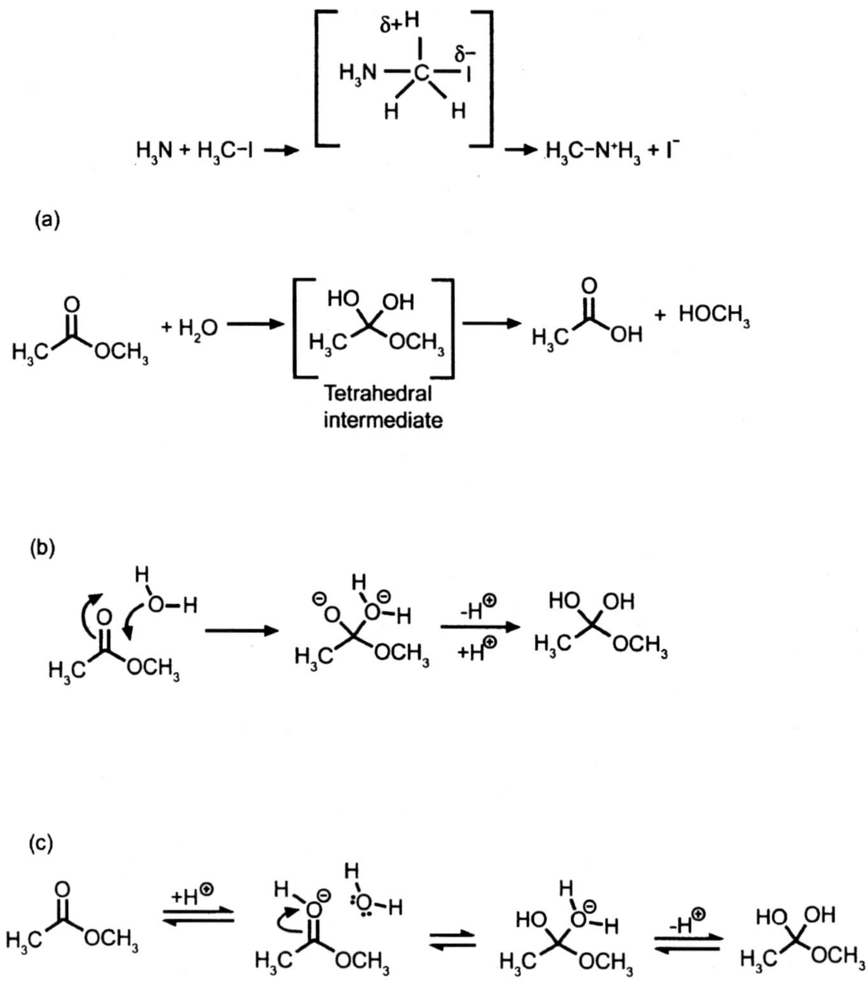

Figure 9-11. Hydrolysis of ester: (a) and (b) are uncatalyzed reactions and (c) is an acid catalyzed reaction.

This transition state can arrange the reactants into a form suitable for the desired bimolecular reaction. The catalyst in this case is water (a protic solvent) rather than the traditional ethyl acetate (an aprotic solvent). Water can lower the activation energy.

9.3.2 Hydrogen Transfer

For multistep reactions, it is important for the intermediates to avoid charge separations. Usually the hydrolysis of esters requires a tetrahedral intermediate, as shown in Figure 9-11(a). Under uncatalyzed (neutral) conditions, nucleophilic attack by water on the ester is slow. The intermediate zwitterion (Fig. 9-11(b)) contains one negatively charged and one positively charged oxygen atom. Deprotonation and reprotonation would produce the tetrahedral intermediate (Fig. 9-11(b)). In the catalyzed (trace acid) addition of water to the ester, the carbonyl oxygen is first protonated in the reactant. When water attacks, no further charge is created; a cationic reactant is converted into a cationic intermediate with a charge on the protonated carbonyl oxygen. Deprotonation of this intermediate produces the same diol intermediate as in Figure 9-11(c).

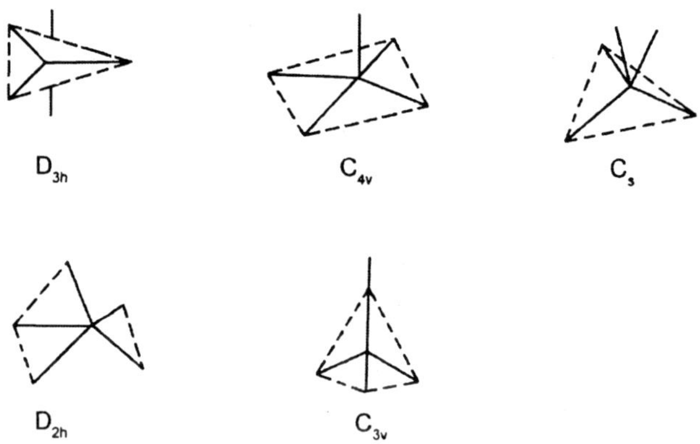

Figure 9-12. Suggested configuration of the methonium ion. [From G.A Olah, J.Am. Chem. Soc., 94, 808 (1972)].

Carbon atoms that are associated with charges of the cationic type are formed as **carbocations**. Their existence in superacid media has been proven by NMR in Olah's investigation and are pertinent to petroleum refining and cracking. Carbocations such as CH_3^+ are referred to as **carbenium** ions, and those such as CH_5^+ are referred to as **carbonium** ions. Carbenium ions are generally planar (sp^2 hybridization), and carbonium ions have quite a number of configurations (Fig. 9-12). Protonation of saturated hydrocarbons, especially the alkanes, is possible. For example, protonation of propane results in three structures bearing the minimum energy (Fig. 9-12). For other high alkanes, this principle also applies (also see Fig. 9-13). Structures A and D result in isomerization, and structures C, F, and G lead to cracking with olefins

$$C_7H_{17}^+ \rightarrow C_3H_8 + C_4H_9^+$$
$$C_4H_9^+ \rightarrow CH_2{=}CH_2 + CH_2CH_3^+$$

In this manner, branching, alkylation, and hydrogen transfer occur, resulting in today's high octane fuel.

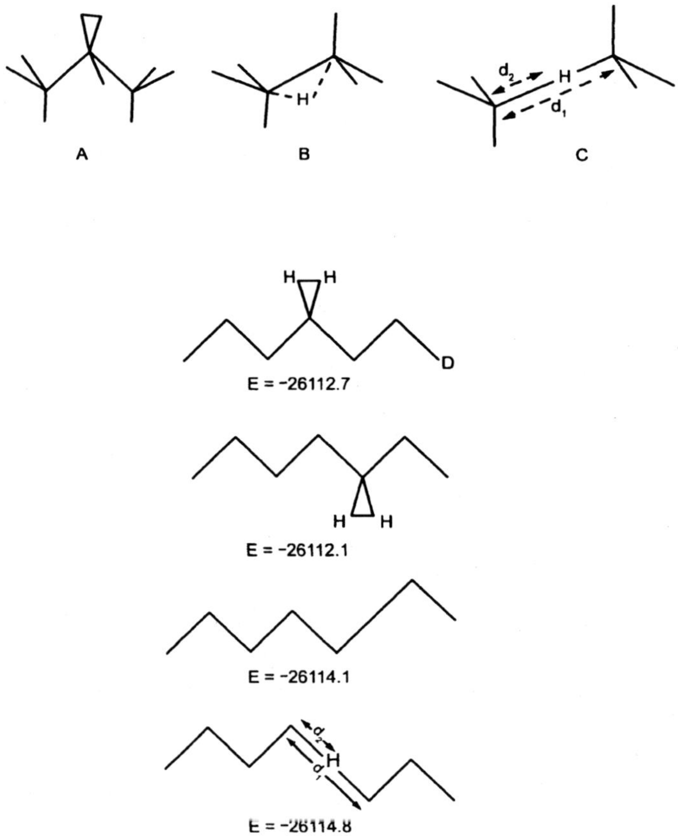

Figure 9-13. Protonated alkanes: A to C are propanes; D to G are *n*-heptanes. A, strong H-H complex; B, nonlinear bridge; C, linear bridge; D and E like A, results in isomerization; F and G like C, results in cracking.

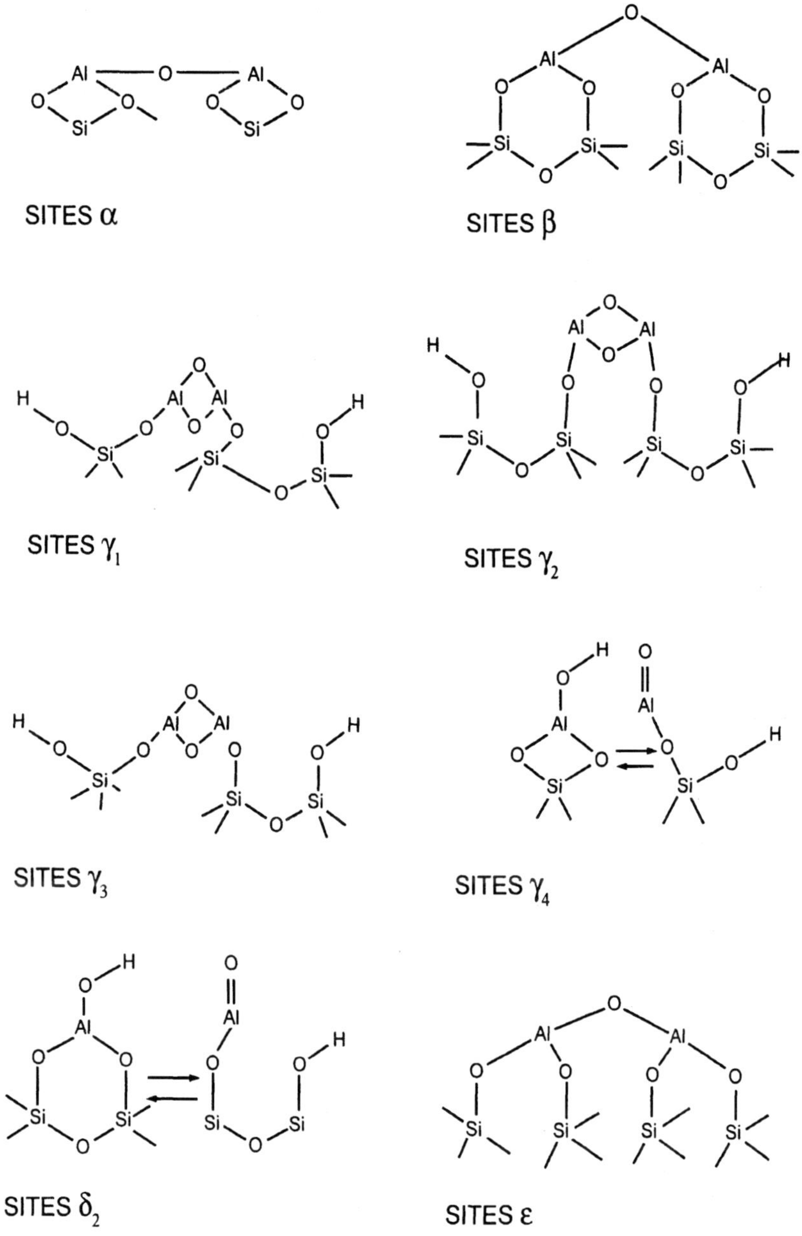

Figure 9-14. Possible sites on silica alumina. [From J. B. Peri, *J. Catalysis*, 41, 227 (1976).]

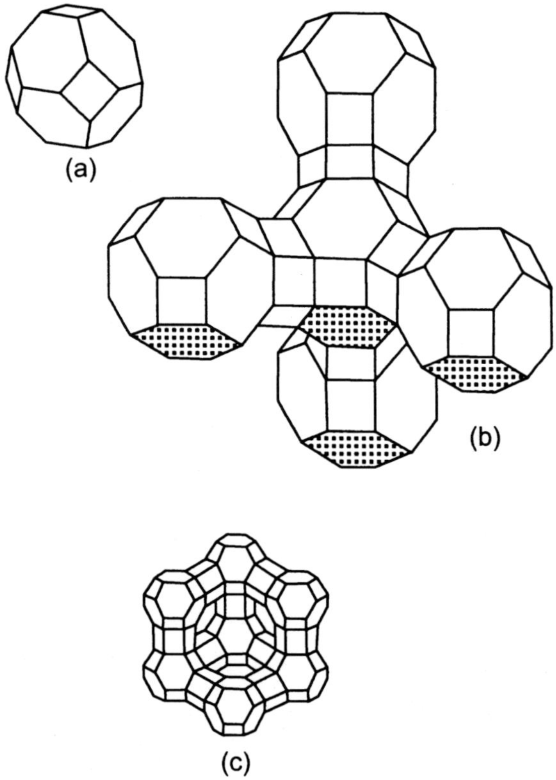

Figure 9-15. Building blocks of faujasite-type zeolites: (*a*) sodalite cage; (*b*) 4 sodalite cages arranged in tetrahedral configuration around the sodalite cage; (*c*) hexagonal prisons.

Amorphous silica aluminas serves as a proton transfer source. Actually the aluminum to silica site exists in a variety of bonding groups, as shown in Peri's classes (Fig. 9-14). For crystalline aluminosilicate catalysts, the zeolites are most widely used. The crystal structure belongs to the truncated octahedron known as a sodalite cage (see Fig. 9-15). The large pore network within the space of the crystal is accessible to a variety of molecules present in petroleum. The negative charges on the tetrahedrally coordinated aluminas are balanced by the cations present in the vicinity of each aluminum atom. The existence of exchangeable cations is the key to the acidity of good cracking catalysts. The catalyst contains both the Brönsted site as well as the Lewis site, the latter being the consequence of the first.

9.3.3 Transition Metal Catalysis and Solubility Parameters

It is well known that platinum and palladium catalyze the hydrogenation of olefins (alkenes) and that thorium and silver catalyze oxygenation processes. Platinum hydrogenation may be used here as an example. Usually a transition metal is required for the π-bonding of the d-orbitals to unsaturated double bonds. At first, hydrogen molecules are absorbed by the metal surfaces in which the σ bonds in hydrogens are homolytically cleaved to be replaced by Pt-H bonds. As indicated in Figure 9-16, hydrogen atoms migrate freely in the platinum network. As hydrogen atoms approach the π-bond region, a five-membered transition state is reached. The energy consumed in the cleavage of the Pt-H bond and the C-C π bond is compensated for by the Pt-C and C-H bonds. This is followed by a deinsertion step to force the hydrogen to bond to the carbon and complete the hydrogenation.

A critical requirement to complete the catalytical process is the free mobility of hydrogenation in the transition metal or the alloy. Of course the crucial step is the solubility of hydrogen atoms in the transition metals or intermetallic (alloys). According to the Hildebrard-Scatchard theory, if the solubility parameters of the two components, A and B, are equal in a system, then A and B will be mutually soluble or miscible in each other. The solubility parameter is defined as the positive square root of the molecular cohesive energy density, which is the molecular cohesive energy, E per unit volume (V = molecular volume).

$$\delta = C^{\frac{1}{2}} = \left(\frac{-E}{V}\right)^{\frac{1}{2}} \qquad [9\text{-}27]$$

The Gibbs free energy of mixing, ΔG_m, must be negative when A and B are soluble or

$$\Delta G_m = \Delta H_m - T\Delta S$$

Since ΔS is always positive, ΔH_m determines dissolution. Dissolution will take place only when the heat of mixing is small,

$$\Delta H_m = V_m(\delta_A - \delta_B)^2 \phi_A \phi_B \qquad [9\text{-}28]$$

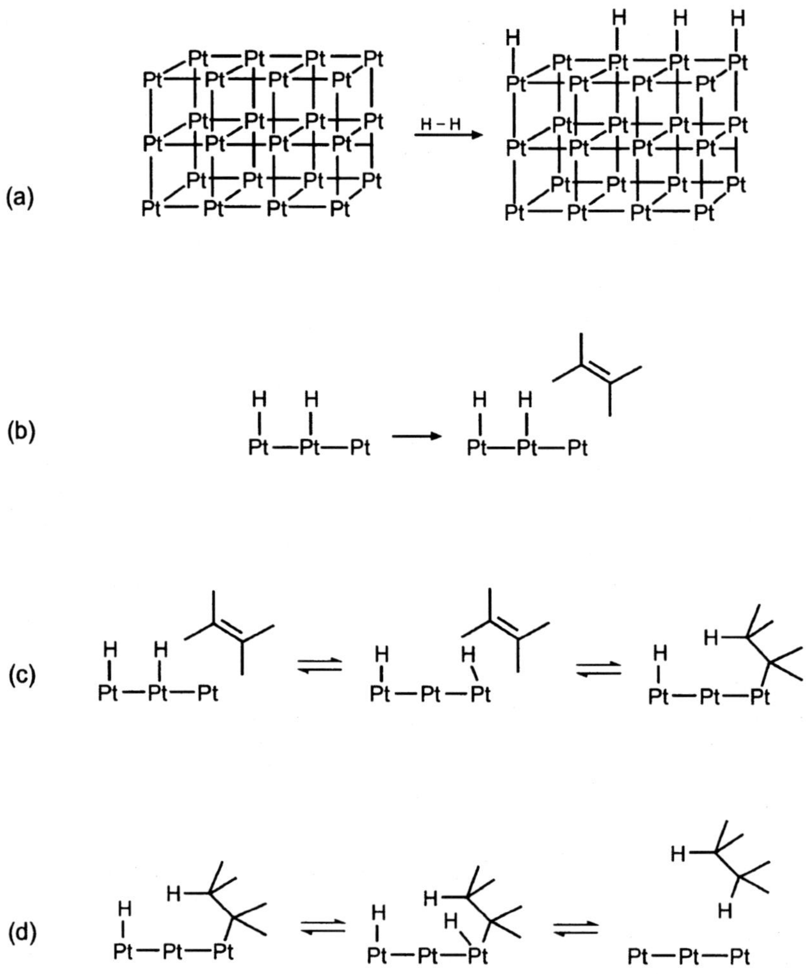

Figure 9-16. Catalytic hydrogenation mechanism: (*a*) H atom is compatible with Pt network; (*b*) formation of π-bond with olefins; (*c*) more H to the olefinic carbon; and (*d*) displacement of H by Pt.

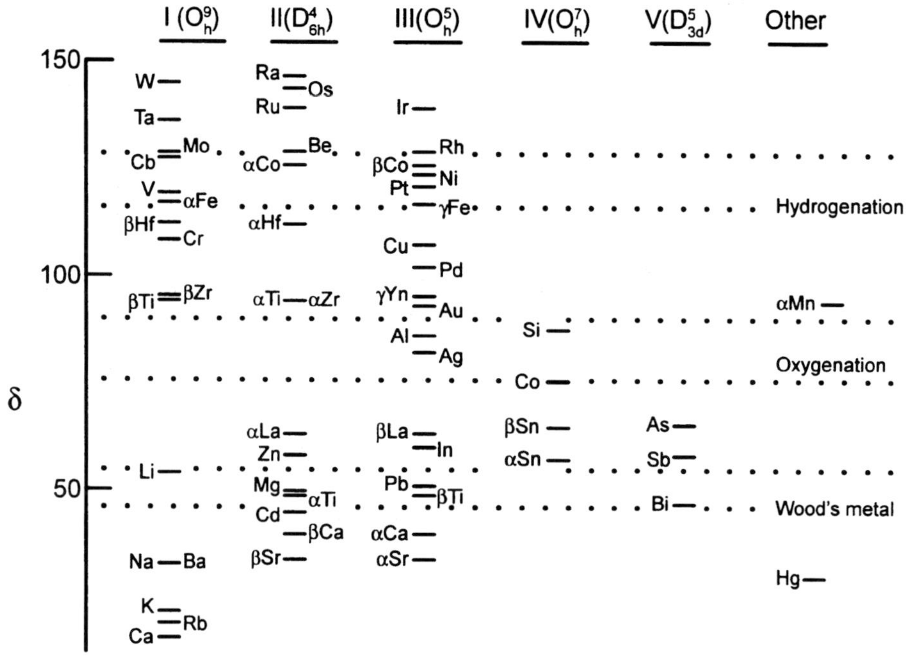

Figure 9-17. Solubility parameters of metals grouped into crystal classes. Classes are as follows: I, body-centered cubic; II, hexagonal close packed; III, face-centered cubic; IV, diamond structure; and V, sheet structure.

where V_m is molar volume and ϕ_A ϕ_B is the molar volume fraction. When δ_A is close to δ_B, the heat of mixing will be either small or zero, and thus the Gibbs free energy change will be negative. Usually the dimensions of δ are $cal^{1/2}$ $cm^{-3/2}$ or hidebrands, which are equivalent to $MPa^{1/2}$. To convert from hidebrands to $MPa^{1/2}$, multiply by $(4.187)^{1/2} = 2.046$. To apply this principle to hydrogenation using a transition metal, such as Pt, it is evident from Table 9-4 that the solubility parameter of the H atom and Pt atom are, respectively, 254 and 253 $MPa^{1/2}$. Also, for oxygenation and oxidation, the metals close to 171 $MPa^{1/2}$ should be suitable. It can also be seen from Figure 9-17 that the metals can work as either reduction or oxygen catalysts. From the same diagram one can infer the compatibility of two or more metals to successfully become metallic solutions, Wood's metal, which contains Sn, Bi, Pb, and Cd and has a melting point below 80°C (see the bands in Fig. 9-17).

There are many cases of transition metal catalysis. One famous example is the Ziegler-Natta catalysts used in conjugated diene polymerization. Often this involves a bimetallic anionic insertion mechanism using both Ti halides and Al alkyls for α-olefins (see Fig.9-18). Much enzyme catalysis also involves trace metals. For enzyme catalysis, the entropy advances are the result of two factors. First, many functional groups are simultaneously present, so intramolecular catalysis via transitional intermediates can occur. Second, a restriction of substrate conformation and orientation occurs when the substrate is bound to the enzyme. The rate of anhydride formation is known for a series of compounds in which rotational freedom is severely restricted with a 10^7 order of magnitude (Table 9-5).

Table 9-4. Solubility Parameters of Hydrogen, Oxygen, and Related Metals for Catalysts

Atom	δ (MPa$^{1/2}$)	Atom	δ (MPa$^{1/2}$)
H	254	O	171
Pt	249	Th	170
V	248	Ag	167
Ni	255	Ge	166
Co	253	Al	180
Mo	265	Lu	153
Ta	267	Si	194
Ru	259	Mn	196

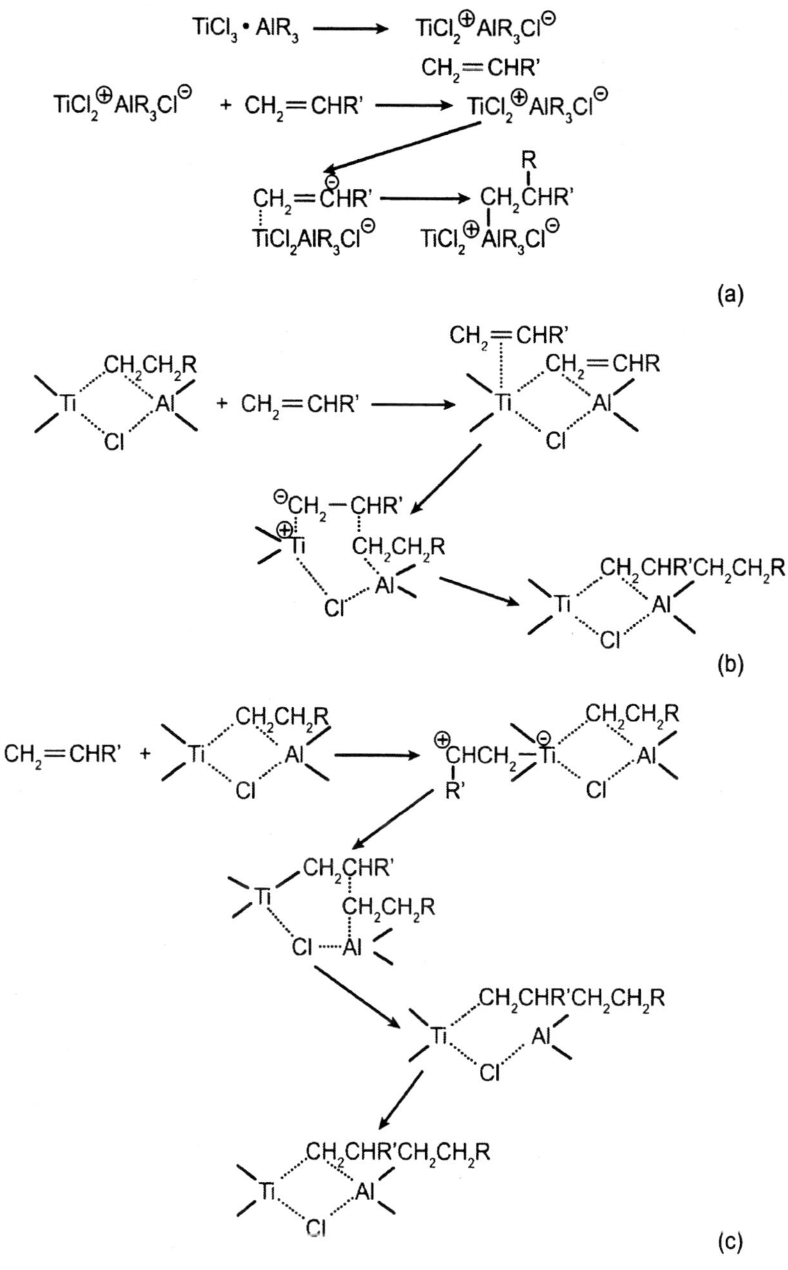

Figure 9-18. Mechanisms for bimetallic catalysis of polymerization of α-olefins. (After R.W. Lenz, Synthetic High Polymers, Wiley, New York, 1967.)

Table 9-5. Effects of Rotational Restriction on Reaction Rates

Reactant	Relative Rate of Anhydride Formation
$CH_3COO^- + CH_3COOC_6H_5Br$ (*p*)	1.0
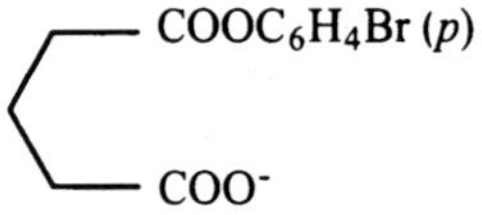	~ 10^3 M
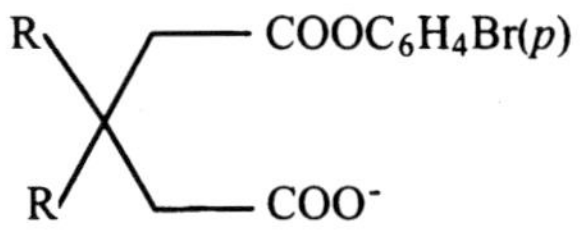	3 × 10^3 – 10^6 M
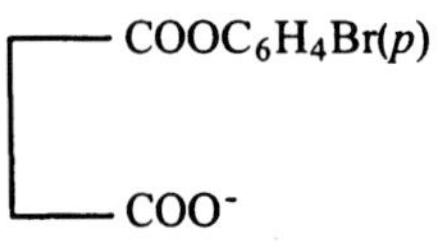	2 × 10^5 M
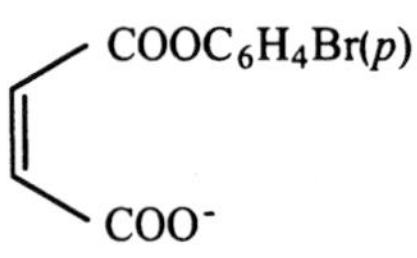	10^7 M
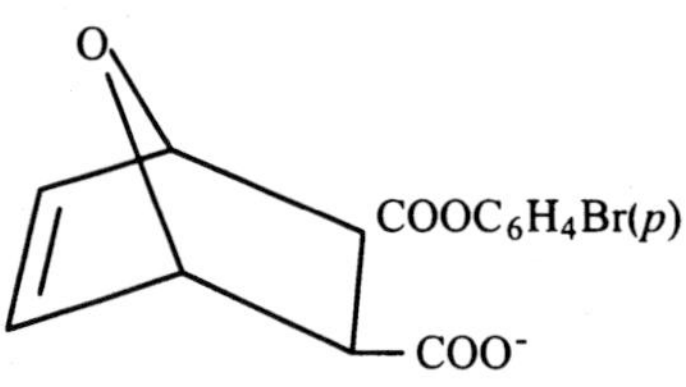	5 × 10^7 M

Source: Modified from T.C. Bruice, *Ann. Rev. Biochem*, 45, 331, (1976)

As an answer to see the tremendous increase in the efficiency of the biocatalyst, this could be due to the orientation of the rotational restriction. In Table 9-5 the increase of reaction rate of 10^7 fold may be due to the **turn over frequency** (TOF).

For the discussion on TOF please refer to Chapter 8, T. F Yen, *Chemistry for Engineers*, Imperial College Press.

REFERENCES

9-1. Berlin, A.A., I.A. Novakov, N.A Khalturinskiy. and G.E. Zaikov, *Chemical Physics of Pyrolysis, Combustion and Oxidation*, Nova Science Publishers, Inc., New York, 2005.

9-2. Irwin, J.W., *Analytical Pyrolysis: A Comprehensive Guide*, Chromatographic Science Series, Vol. 22, Marcel Dekker, Inc., New York, 1982.

9-3. Jan, M. R., J. Shok, and Z. Hussain, *Investigation of a New Method for the Cracking of Polyethylene, American Lab.,* Oct. 53-55 (2000).

9-4. Yang, J. and M. Lu, *Thermal Growth and Decomposition of Methyl napthalenes, Environmental Science Technical* 39, 3077-3082 (2005).

9-5. Zou , R., *Fundamentals of Pyrolysis in Petrochemistry and Technology,* Citic Publishing House, Beijing, China, 2000.

9-6. Wen, C.S., T.F. Yen, *Optimization of Oil Shale Pyrolysis*, Chemical Engineering Science, 32(3) 396-349(1977).

9.7. Weinberg, V. and T.F. Yen, *Solubility Parameters in Coal and Coal Liquefaction Product*, Fuel 59(5), 287-289(1980).

PROBLEM SET

1. Predict the remainder of organics in a coal field after kerogen has been decomposed, in particular the ratio of gas (oil has been evaporated) and carbon (char) in a Cambrian coal field. Use the chart in Figure 9-8 and assume temperature to be 900°C in a trade-off with the geological ages.

2. Predict the composition of products of the propane pyrolysis.

CHAPTER 10

PHOTOLYSIS

No one doubts the eminent importance of photolysis related to our environments, photochemical reactions are essential for the biosphere, since visible and ultraviolet solar radiation is necessary for all life on earth. For example, the source for the atmospheric oxygen comes from the photosynthesis in the biosphere and the photo-dissociation of water in the upper atmosphere. Here we first introduce principle of photometry and photochemistry, the transfer of energy in a molecule including excitation. Then we discussed the term symbols which are used in photochemical smog in troposphere and in chemical species formation in the upper atmosphere (thermosphere/ionosphere). Followed by the direct and sensitized photolysis the applications of photolysis are exemplified, for example, the optical brighteners, the photocatalytic process, the photogalvanic cell.

10.1 PHOTOMETRY AND PHOTOCHEMISTRY

First, let us become familiar with some of the energetic units. Consider an isotropic point source that radiates equally in all directions in space. The flux, Φ, from the source is expressed in joules per second, i.e., in watts. The intensity, I, of this isotropic point source is defined by:

$$I = d\Phi/d\Omega \qquad [10\text{-}1]$$

which equals the flux radiated per unit solid angle, Ω. The units are watts per steradian ($W \cdot sr^{-1}$).

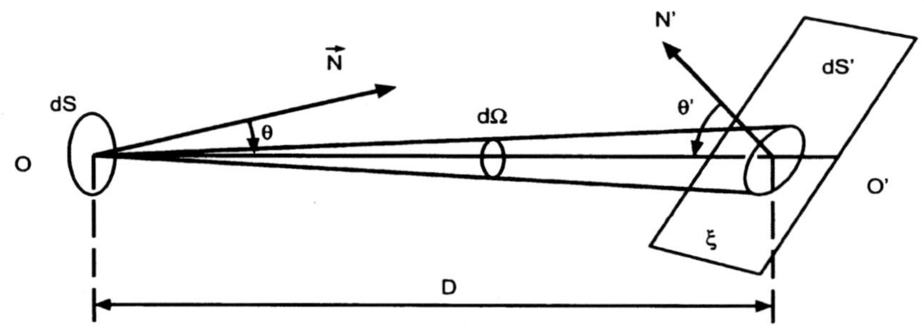

Figure 10-1. Luminance of a source.

Now if a source of a small area, dS, whose normal N deviates from the observation line by the angle OO′ θ, as shown in Figure 10-1, then the luminance L of the source can be expressed as:

$$d^2\Phi = L \cdot dS \cdot \cos\theta d\Omega \qquad [10\text{-}2]$$

The unit of luminance is watts per square meter per steradian ($W\ m^{-2}\ sr^{-1}$). Similarly, the inadiance, E, at the surface dS' is defined as:

$$E = d\Phi/dS' \qquad [10\text{-}3]$$

for which units are watts per square meter ($W\ m^{-2}$). Another term used is radiance, sometimes referred to as emittance, or exitance, R, of radiating surface, while dS represents the total power emitted per unit in all possible directions.

$$R = \int \Omega L \cos \theta d\Omega = d\Phi / dS \qquad [10\text{-}4]$$

for which units are watts per square meter (W m^{-2}).

All four units mentioned above can have the spectra units. For example, spectral flux:

$$\Phi_\lambda = d\Phi / d\lambda\ (\lambda) \qquad [10\text{-}5]$$

$$I_\lambda = dI / d\lambda\ (\lambda) \qquad [10\text{-}6]$$

Table 10-1. Different Systems of Units and Their Relations

Nature	Visual Units	Symbol	Dimensional Relations	Energetic Units
Flux	Lumen	ϕ		W.
Intensity	Candela	I	1 cd = 1 lm sr^{-1}	W. sr^{-1}
Luminance	Candela m^{-2}	L	1 cd m^{-2} = 1 lm m^{-2} sr^{-2}	W. m^{-2} sr^{-1}
Irradiance	Lux	E	1 lx = 1 lm m^{-2}	W. m^{-2}

The energy units and the equivalent visual units of the four equations are summarized in Table 10-1. The irradiance, E, from a surface dS' situated at a range of D from the source is:

$$E = d\Phi / dS' = Id\Omega / dS' = I \cos \theta' / D^2 \qquad [10\text{-}7]$$

The total flux emitted by a source in a hemisphere is given by:

$$d\Phi = \pi L\ dS \qquad [10\text{-}8]$$

If a surface dS' receives a flux $d\Phi$, the surface scatters a flux $d\Phi'$, which is a fraction of the incident flux. A term called the reflection, ρ, of the surface is:

$$\rho = d\Phi' / d \qquad [10\text{-}9]$$

Light is transmitted both in wave- and particle-like manner. If the wavelength (between two consecutive maxima) is λ, then the relation to the frequency, ν, is as follows:

$$\lambda = \frac{c}{\nu} \qquad [10\text{-}10]$$

where c is the speed of light or 3×10^8 ms^{-1}. For a particle-like consideration, light can be quantitized and emitted, transmitted, and absorbed in photons or quanta. The energy E of a quantum is given by:

$$E = h\nu = h\, c/\lambda \qquad [10\text{-}11]$$

where h is the Planck constant, 6.63×10^{-34} Js. On a molar basis, the unit of light is the einstein, which is the equivalent of N (6.02×10^{23} = 1 mol) of photons or quanta.

$$E = Nhc/\lambda = \frac{1.196 \times 10^5}{\lambda}\, kJ/einstein \qquad [10\text{-}12]$$

The energy of UV or visible light is in the same order of magnitude as that of the usual covalent bonds of single, divalent or triple type (Table 10-2), consequently such bonds can be cleaved or broken (dissociated) in the case of light absorption.

When a photon approaches to a molecule, interaction between the radiation and the bonds of the molecule takes place. The absorption of radiation will be effective to produce photochemical change.

Table 10-2. Typical Energies for Some Single Bonds and the Approximate Wavelengths of Light Corresponding to This Energy[a]

Bond	Bond Energy E^b (kJ mol^{-1})	Wavelength λ (nm)
O-H	465	257
H-H	436	274
C-H	415	288
N-H	390	307
C-O	360	332
C-C	348	344
C-Cl	339	353
Cl-Cl	243	492
Br-Br	193	620
O-O	146	820

[a]Compare Eq.10-12

[b]Values from Ref. 10-5

[Example 10-1] What wavelength would give sufficient energy to dissociate oxygen molecules?

$$O_2 \rightarrow 2O$$

The dissociation energy is 5.1 eV which is equivalent to 492 kJ/mole (118 kcal/mole)

$$\lambda = \frac{hc}{E}$$

$$= \frac{6.63\times10^{-27}\,erg\ \sec/\,molecule\,(3\times10^{10}\,cm/\sec)}{118\,kcal\,/\,mole}$$

$$\times\frac{6.02\times10^{23}\,molecule/\,mole}{4.184\times10^{10}\,erg/\,kcal}$$

$$= 2.43 \times 10^{-5} \, cm$$
$$= 243 \; nm$$

Compare this value with all the single bond dissociated energy listed in Table 10-2, it requires UV side of light to achieve any useful efficiency.

The bond energy of molecules can be calculated in the UV-visible range to indicate the same order of magnitude. In this manner, the bonds of the molecules are susceptible to cleavage, forming the basis of photolysis.

If light is absorbed in a solution containing certain compounds, two empirical laws are applicable: **Lambert's law** states that the fraction of radiation absorbed by solution is independent of the intensity of the original radiation, while **Beer's law** concludes that the amount of radiation absorbed by the solution is proportional to the number of molecules absorbing the radiation. Using these ideas,

$$I(\lambda) = I_0(\lambda)\; 10^{-[\alpha(\lambda) + \varepsilon(\lambda)C]l} \qquad [10\text{-}13]$$

where $I_0(\lambda)$ is the incident light intensity, $I(\lambda)$ is light intensity given by the solution, C is molar concentration (mol/ L), l is the path length of light expressed in cm, $\alpha(\lambda)$ is the decadic attenuation coefficient in cm^{-1} of the medium, and $\varepsilon(\lambda)$ is the molar extinction coefficient in L mol^{-1} cm^{-1} of the compound in solution at wavelength λ. Absorbance is:

$$A = \log \frac{I_0(\lambda)}{I(\lambda)} = [\alpha(\lambda) + \varepsilon(\lambda)C]l \qquad [10\text{-}14]$$

Usually in a spectrophotometer,

$$A = \varepsilon(\lambda)\, Cl \qquad [10\text{-}15]$$

and transmittance is the reverse of A:

$$T = A^{-1} \qquad [10\text{-}16]$$

10.1.1 Photochemical Excitation

When a molecule is photochemically excited during the light-absorption process, the excited molecules could undergo a number of chemical reactions or physical processes(Figure 10-2).

$$AB^* \rightarrow A + B, \text{ dissociation}$$

$$AB^* + E \rightarrow AE + B, \text{ direct reaction}$$

$$AB^* \rightarrow BA, \text{ spontaneous isomerization}$$

$$AB^* \rightarrow AB + h\nu, \text{ luminescence}$$

$$AB^* \rightarrow AB^{\#}, \text{ intramolecular energy transfer}$$

$$AB^* + CD \rightarrow AB + CD^{\ddagger}, \text{ intermolecular energy transfer}$$

$$AB^* + M \rightarrow AB, \text{ physical quenching}$$

The symbols *, #, and ‡ are used only to illustrate the pressure of electronic excitation and not a difference of states. If we consider the vibration levels including the ground state, the excited molecules may cascade down to another level of the ground state by giving off energy in small increments of heat to the environment. This process is referred to as **internal conversion** (IC). Alternatively, the excited molecule may go to another excited state. This is termed **intersystem crossing** (ISC). Afterwards, the excited molecule will settle at the low level of the ground state by giving off all energy at once as light. This process called **fluorescence** can be generalized as a luminescent process. Usually, the drop-off is direct. If the drop-off proceeds to another excited state by ISC, the process is called **phosphorescence**. These luminescent processes can be shown in a Jablonski diagram, as in Figure 10-3. Furthermore, an excited molecule can transfer excess energy to another molecule in the environment. This process is called **photosensitization**. When a compound transfers its energy to another molecule, it is called a photosensitizer. The compound that receives the energy is called the quencher. Usually the quantum yield $\Phi_i(\lambda)$ is defined as the ratio of the number of molecules by the process, *I,* (a number of physical and chemical processes have been shown) to the total number of photons (at wavelength λ) absorbed by the system. Thus,

$$\Phi_f + \Phi_p + \Phi_{isc} + \Phi_{ic} = 1 \qquad [10\text{-}17]$$

If a molecule exhibits both florescence and phosphorescence, the phosphorescent state T_1 is excited by ISC from S_1, and

$$S_0 = h\nu \rightarrow S_1\text{, excitation}$$

$$S_1 + M \xrightarrow{k_1} S_0 + \text{M, fluorescence quenching}$$

$$S_1 \xrightarrow{A'} S_0 + h\nu'\text{, fluorescence}$$

$$S_1 \xrightarrow{k_2} T_1\text{, ISC}$$

$$S_1 \xrightarrow{k_3} S_0\text{, IC}$$

$$T_1 + M \xrightarrow{k_4} S_0 + M\text{, phosphorescence quenching}$$

$$T_1 \xrightarrow{A''} S_0 + h\nu''\text{, phosphorescence}$$

$$T_1 \xrightarrow{k_5} S_0\text{, ISC}$$

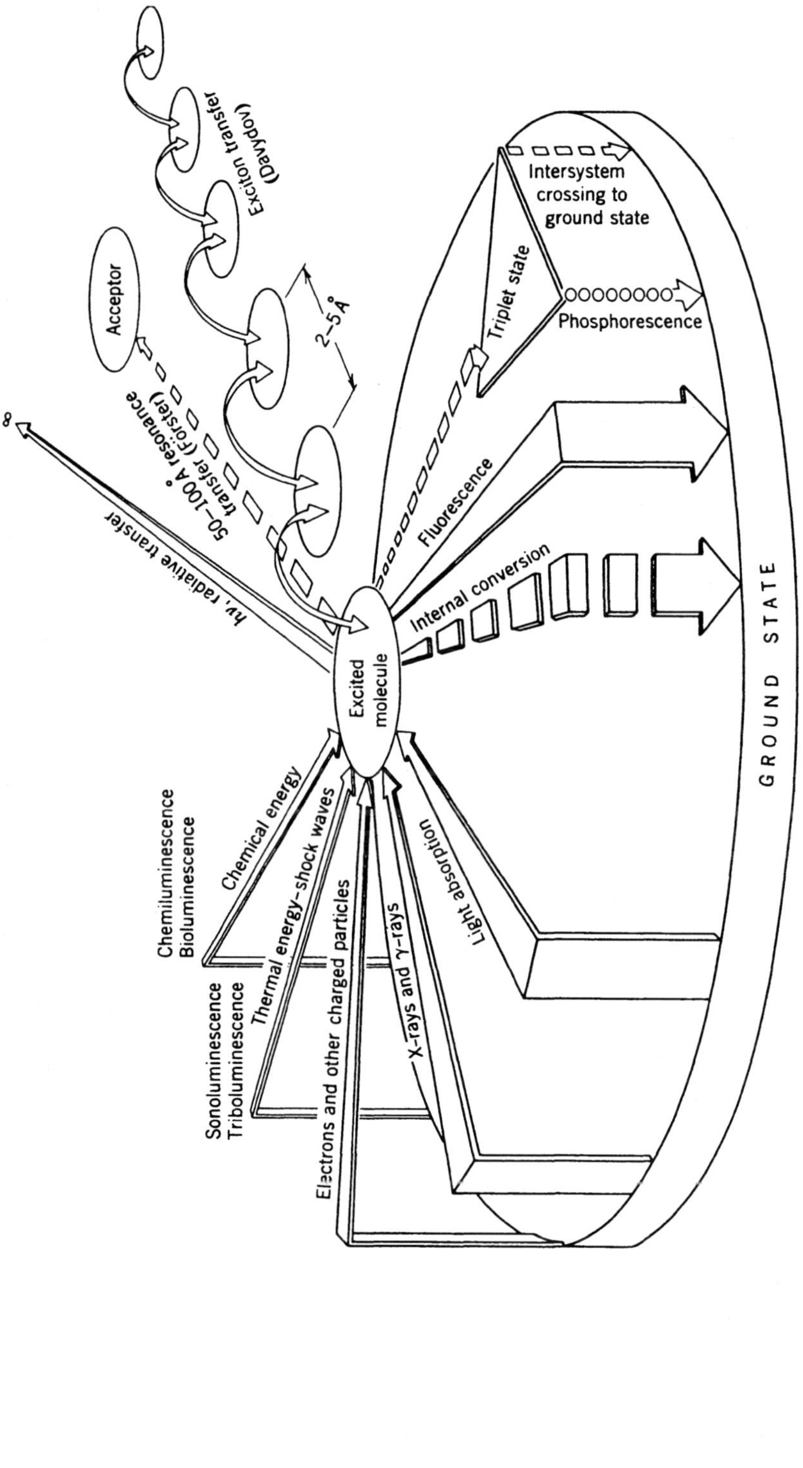

Figure 10-2. The life of an excited molecule.

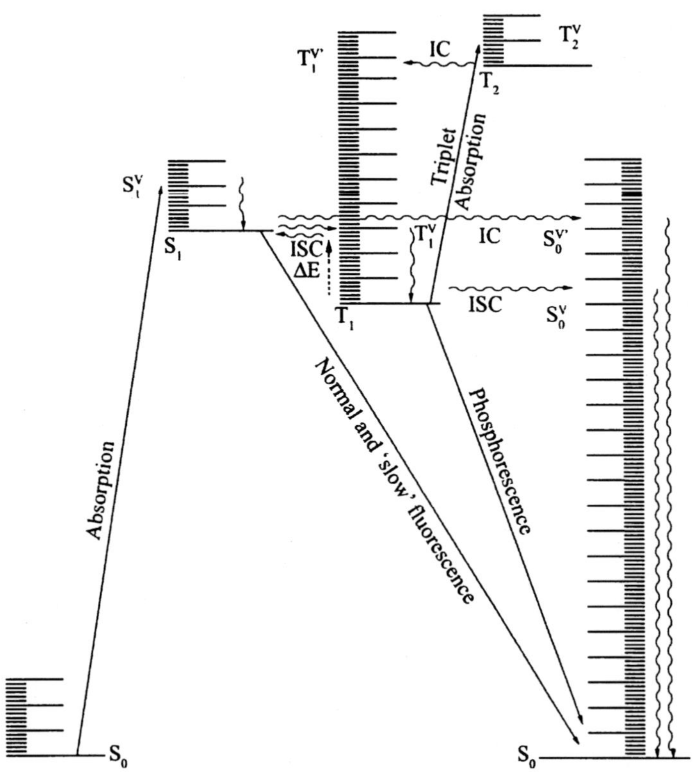

Figure 10-3. Jablonski diagram showing absorption and the emission processes of fluorescence and phosphorescence.

[Example 10-2] An aqueous quinine sulfate solution can be quenched by the addition of chloride ion. Evaluate the quenching rate constant if data of relative emission intensity and the corresponding chloride ion concentration are known.

Using a simple excitation and deactivation scheme:

$$X + h\nu_{\text{abs}} \rightarrow X^*, \text{ absorption } k_q$$

$$X^* + M \rightarrow X + M, \text{ quenching rate} = k^q\,[X^*]\,[M]$$

$$X^* \rightarrow X + h\nu_{\text{emi}}, \text{ emission rate} = A[X^*]$$

Solution of the steady-state equations for $[X^*]$ with $d[X^*]/dt = 0$ leads to:

$$I_{emi} = A[X^*] = \frac{AI_{abs}}{A + k_q[M]}$$

$$I_{\text{abs}} = k_q\,[X^*]\,[M] + A\,[X^*]$$

The above equation can be taken as a Stern-Vomer equation:

$$I_{emi}^{-1} = I_{abs}^{-1}\left(1 + \frac{k_q}{A}[M]\right)$$

where $[M]$ is the Cl^- concentration and A is the Einstein coefficient for spontaneous emission. If one plots the relative emission intensity to $[Cl^-]$, a straight line will yield a slope as in k_q/A and an intercept as I_{abs}. The A factor also can be evaluated (if not known) from the B factor by using the ν^3 relationship.

$$A = \frac{\delta\pi h\nu^3}{c^3}B$$

which can be related to Planck's law. The B factor can be easily measured from absorption band measurement,

$$B = \int x d\nu$$

As in the example, the generalized kinetic scheme can also be solved. Thus, according to the steady-state equation for $[S_1]$ and $[T_1]$, the relation can be:

$$\phi_p = \frac{A'}{(A' + k_2 + k_3 + k_1[M])} \qquad [10\text{-}18]$$

and

$$\phi_p = \phi_f \frac{A''}{A'} \left(\frac{k_2}{A'' + k_5 + k_4[M]} \right) \qquad [10\text{-}19]$$

Many of the rate constants can be evaluated in ways similar to the previous example.

10.2 SPECTROSCOPIC STATES

10.2.1 Atom Designation

The state of an electron in the hydrogen atom or hydrogen-like ion (e.g., He^+ and Li^{2+}, each containing only one electron) can be specified by three quantum numbers, n, l, and m_l. The principal quantum number n, with a value of 1, 2, 3..., defines the shell of the atom occupied by the electrons. Therefore, the total energy of the single bond electron (neglecting relativistic effects) is specified by n. The azimuthal orbital angular momentum quantum number can have a value of 0, 1, 2... (n-1). It determines the total angular momentum of the electron, which is quantified and can be represented by a vector having a magnitude of $\hbar\,[l\,(l + 1)]^{1/2}$, where $\hbar = h / 2\pi$. The states (or atomic orbitals) with l = 0, 1, 2, 3... are designated by s, p, d, f... orbitals so that the electron with lowest electronic state of the hydrogen atom (n = 1, l = 0) is called a 1s electron. The quantum number m_l is called magnetic quantum number and can have the values $-l$, $-(l - 1)$...0...$(l - 1)$, l. It determines the allowed components of the angular momentum in a definite z-direction, which are also quantified and can only have the value of $m_l\hbar$.

In addition to the three quantum numbers described above, in order to explain the fine structure i atomic species, another factor, electron spin, must also be considered. In addition to orbiting the nucleus, an electron beyond the angular momentum has an intrinsic magnetic moment as if (classically) it were spinning about its own axis. Associated with this moment is the fourth quantum number, M_s, such that the total spin angular momentum vector has a magnitude of $\hbar\,[s\,(s + 1)]^{1/2}$. However, s can only have a single value of ½; the spin angular momentum for the z-component is quantified with allowed values of magnitudes $m_s\hbar$, where $m_s = \pm 1/2$. The electron spin and magnetic moment interact so that l and s are coupled to give a total electronic angular momentum magnitude of $\hbar j(j + 1)]^{1/2}$, where $j = l \pm ½$.

With polyelectronic atoms, complexities arise as a result of electrostatic interactions among the electrons as well as between the electrons and nucleus. In this case, only approximate solutions by quantum mechanical treatment are possible. Energy of the state is a function of principal quantum number, n, and the total angular momentum of the electrons. The number of electrons in a given orbital is governed by the Pauli exclusion principle, which states that no two electrons can exist in the same quantum state. Therefore, an atomic orbital is characterized by a specific set of *n l,* and m_l values only if their spins are opposite, i.e., m_s = +1/2, m_s = -1/2. Also, the Aufbau principle states that the orbitals are filled in the order of increasing energy. According to Hund's rule, if two or more orbitals have the same energy (i.e. they are degenerate), then the electrons go into different orbitals as much as possible, thus ending electron-electron repulsion by keeping them apart with spin parallel in the same direction rather than opposed.

For light elements with a charge number of less than 40, the orbital angular momentum of all electrons are strongly coupled to give stable results and orbital angular momentum. This is designated by the quantum number, L. Similarly, the spin angular momentum can be combined to yield a quantum number S. It follows in the z-direction.

$$m_l = \square m_L$$
$$m_s = \square m_s$$

Allowed values of m_l are –1, - (L-1)... Allowed values of m_s are –s, -(s-1)...0, (s-1), s, or $|M_s| \leq S$. It is assumed that completely filled shells and subshells have a zero net resultant orbital and spin angular J. Possible values of J are:

J = L + S, L + S – 1, L + S – Z...□L – S + 1□,□L – S□

where □L – S□ = L – S if L ≥ S or □L – S□ = S – L if S ≥ L.

In this manner, J can only have positive values. Therefore, the complete term symbol can be represented by $n^{2s+1}L_j$.

In this term symbol, the quantity (2s + 1) is termed the multiplicity of the atom, or the singlet, doublet, triplet, etc. for values of 1, 2, 3... respectively. It indicates the total number of possible spin orientations. Often, the values of n and J are omitted from the term symbol, partially if the energy differences among the J states are small. Similar to hydrogen or hydrogen-like atoms, the atomic states of L = 0, 1, 2, 3... are designated by S, P, D, F...

As an example for the term symbol, oxygen atoms with eight eletrons have an electronic configuration of $1s^2 2s^2 2p^4$. The $1s^2$ and $2s^2$ shells are completely filled, hence only the four 2p ($l = 1$) electrons need to be considered. The state of the lowest energy, called the ground state, is most stable. According to Hund's rule:

- the state is of maximum multiplicity
- for a given multiplicity, the state is of minimum J if the partially-filled subshell is less or half occupied
- for a given L and S, the state is of maximum J if the partially filled subshell is more than half occupied.

As shown below, the ground state of oxygen given by the four electrons occupying the three 2p orbitals:

The quantum numbers of the four electrons can be listed as:

electrons	n	l	m_l	m_s
1	2	1	+1	+1/2
2	2	1	+1	-1/2
3	2	1	0	+1/2
4	2	1	-1	+1/2

For the above configuration of electron filling,

$$m_s = \Sigma m_s = ½ - ½ \_ ½ + ½ = 1$$

This is consistant with the Pauli exclusive principle, which concerns placement of four electrons in three orbitals. Furthermore,

$$m_l = \Sigma m_l = 1 + 1 + 0 - 1 = 1$$

still,

$$|M_L| \leq L$$
$$L = 1$$

Hence, multiplicity = 2S + 1 = 3. The quantity of S and L are maximum allowed values, so this state is of the lowest energy or can be written as:

2^3P

a triplet P state. According to Hund's rule, if the subshell is more than half-filled, a maximum J gives the most stable state. This way the complete term symbol may be written as:

2^3P_2

Also, the *n* and J can often be omitted, so ^3P is used.

Electron excitation may result in a change of J, transition to another orbital within the subshell, or to a high energy subshell. For example, two possible excited states of an oxygen atom within the same subshell (same *n* and *l*) are as follows:

in which

M_l +1,+1, 0, 0
M_s +½, -½, +½, -½

and

m_l +1, +1, -1, -1
m_s +½, -½, +½, -½

The resulting term symbols of both excited states are:

2^1S_2

and

2^1S_0

The former of these terms has maximum J. Therefore, 2^1D_2 is more stable. In conclusion, the stability of electronic states of oxygen atoms is:

$^3P_2 > 2^1D_2 > 2^1S_0$

The singlet 2^1S_0 is the least stable. The production of the singlet state of an oxygen atom requires the shortest UV light (see Environmental Chemistry, Global Cycles.)

10.2.2 Molecule Designation

Just like with atoms, it is possible to designate a set of quantum numbers for diatomic molecules. In a manner similar to polyelectronic atoms, the resultant orbital angular momentum along the internuclear axis can be a quantum number □. It can have values of 0, 1, 2...L, where L is the quantum number for the resultant orbital angular momentum for all the electrons in the molecule. Thus, □ are designated respectively as □, π, □... states (analagous to s, p, d... for the hydrogen atom and to S, P, D... for polyelectric atoms). Similarly, the total molecular spin quantum number is S, and the component of S along the intermolecular axis is with possible values of S, S – 1...0...-(S-1), -S. The sum of three true quantum numbers is □ = | □ + □ | (analogous to J in atoms). Hence, the total term symbol for a specific diatomic or linear polyatomic electronic state is:

$$^{(2S+1)}\square_{\square}$$

- If inversion of all electrons through a center of symmetry of the molecule leads to no change of sign of the electronic wave function, the state is gerude (g) or even. Otherwise, if a change of sign results, then the state is ungerude (u) or odd. The symbols G and U are included in the molecular term symbol as a second subscript to □.
- If reflection in a plane of symmetry passing through the nuclei centering the intermolecular axis of symmetry leads to no change of sign in the electronic state, the state is thus positive, +. Otherwise, the state is negative, -. The symbol +, - is written as a second subscript to □ (This element of symmetry applies to □ = 0 or □ states.

An example is given using molecular oxygen. The electronic configuration for ground-state of O_2 with 16 electrons is:

$$(15T_g)^2\,(15T_u^*)^2\,(25T_q)^2\,(25T_u^*)^2\,(2P_xT_g)^2\,(2P_y\,\pi_u)^2\,(2P_z\,\pi_u)^2$$
$$(2P_y\,\pi_g^*)^1\,(2P_z\,\pi_g^*)^1$$

The parentheses and their contents designate the molecular orbitals. The asterisk * refers to antibonding orbitals (the density of electrons is less than that of free atoms). The superscript indicates the number of electrons in the molecular orbital. For the last two orbitals, $(2P_y\ \pi_g^*)^1$ and $(2P_z\ \pi_g^*)^1$, there are two unpaired electrons according to Hund's rule of maximum multiplicity. The total spin is 1, and the ground state is a triplet. The quantum number □ can be z or 0. If □ = 2, both electrons would have the same □ = ± value, which is forbidden by the Pauli exclusion principle for electrons in equivalent orbitals with the same species. Thus, only □ = 0 or □ state is allowed for S = 1. The following Figure indicates the wave functions for the two electrons in the two orbitals.

Figure 10-4 indicates the interchange though the center of symmetry does not lead to a change of sign. Thus, g is the subscript. Whenever the interchange does not occur through the center, the plane does not change sign. Thus, the subscript is -. In conclusion, the term symbol for this normal stable state (ground state) of O_2 is:

$$^3\square_g^-$$

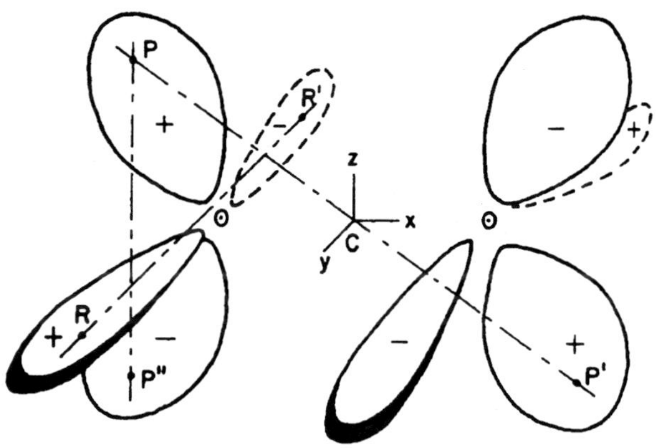

Figure 10-4. Schematic drawing of the $(2P_{y\,g}{}^*)$ and $(2P_{z\,g}{}^*)$ orbitals of O_2. The $(2P_{y\,g}{}^*)$ orbital is in the x – y plane, and the $(2P_{z\,g}{}^*)$ orbital is in the x – z plane. Inversion through the center of symmetry C from P to P' does not change sign, whereas inversion through a plane of symmetry passing through the two nuclei such as the x – y pane (from P to P") or the x – z pane (from R to R') does lead to a change in sign.

[Example 10-3] What are the species of N? Of O^{+}? Write out the term symbols for them, and indicate their orders of stability. What are the aurora spectrum lines of N?

For both, the electronic structures are $2s^2 2p^3$. There are three electrons distributed in three p-orbitals. Ideally, each electron occupies one orbital or:

(↑) (↑) (↑)

M_l +1 0 -1 L = 0

M_s +½ +½ +½ S = 3/2

Therefore, the resulting term symbol is:

$2^4S_{3/2}$

Another Arrangement is:

(↑↓) (↑) ( )

M_l +1,+1 0 L = 2

M_s +½,-½ +½ S = ½

Here, J = L – S, since L ≥ S. Hence, the term symbol is:

$2^2D_{3/2}$

An excited arrangement is:

(↑↓) ( ) (↑)

M_l +1,+1 -1 L = 1

-½, -½ +½ S = ½

Here, J = L + S since L ≤ S. Hence, the term symbol is:

$2^2P_{3/2}$

Accordingly,

$2^4S_{3/2} > 2^2D_{3/2} > 2^2P_{3/2}$

The aurora spectra lines of N at 3466A are from $^2P_{3/2}$ □ $^4S_{3/2}$ via $^2D_{3/2}$.

10.3 DIRECT AND SENSITIZED PHOTOLYSIS

Many valuable contributions to this subject have been made by Swiss researchers such as Schwarzenbach, Zepp, Haog, and Zafirriou. Photolysis can be direct or indirect (sensitized) through another molecule in the vicinity. First, light can be attenuated in natural water bodies. Applying the Lambert-Beer law,

$$W(Z_{mix},\lambda)=W(\lambda)\,10^{-\alpha_D(\lambda)z_{mix}} \quad [10\text{-}20]$$

where $W(\lambda)$ is incident light intensity at the surface in einstein cm^{-2}s^{-1}, Z_{mix} is the average depth of the mixed water body in cm, which can be obtained by $z_{mix} = V/A$, $\alpha_D(\lambda)$ is commonly referred to as diffuse attenuation coefficient in cm^{-1}, and $W(Z_{mix}, \lambda)$ is the intensity at depth Z_{mix} in einstein cm^{-2}s^{-1}. Thus,

$$\alpha_D(\lambda)=\frac{1}{Z_{mix}}\log\frac{W(\lambda)}{W(Z_{mix},\lambda)} \quad [10\text{-}21]$$

The rate of light absorption by water body per unit volume, $I'(\lambda)$, is:

$$I'(\lambda)=W(\lambda)-W(z_{mix},\lambda)\frac{A}{V}=W(\lambda)\left[1-10^{-\alpha_D(\lambda)z_{mix}}\right]z_{mix}^{-1} \quad [10\text{-}22]$$

For a given pollutant in water exhibiting a molar extraction coefficient (λ), the attenuation coefficient $\alpha(\lambda)$ is changed to the sum of $\alpha(\lambda) + \varepsilon(\lambda)C$, where C is the molar concentration of the pollutant. The small fraction of light $F(\lambda)$ absorbed by the pollutant is:

$$F(\lambda)=\frac{\varepsilon(\lambda)C}{\alpha(\lambda))+\varepsilon(\lambda)C}\approx\frac{\varepsilon(\lambda)C}{\alpha(\lambda)} \quad [10\text{-}23]$$

because $\varepsilon(\lambda) << \alpha(\lambda)$. Hence, the rate of light absorption by the pollutant per unit volume, $I(\lambda)$, is:

$$I(\lambda)=I'(\lambda)F(\lambda)=\frac{W(\lambda)\varepsilon(\lambda)\left[1-10^{-\alpha_D(\lambda)Z_{mix}}\right]C}{Z_{mix}\,\alpha(\lambda)}=k_a(\lambda)C \quad [10\text{-}24]$$

$k_a(\lambda)$ is referred to as the specific rate of light absorption of a given pollutant in a given system and has the units:

$$k_a(\lambda)=\frac{\text{einstein cm}^{-2}\text{ sec}^{-1}\text{ cm}^{-1}\text{ mol}^{-1}\text{ L}}{\text{cm}\cdot\text{cm}^{-1}} \quad [10\text{-}25]$$
$$=\text{einstein cm}^{-3}\text{ L mol}^{-1}\text{ sec}^{-1}=10^{3}\text{ einstein mol}^{-1}\text{ sec}^{-1}$$

The units for $I(\lambda)$ are einstein $L^{-1}s^{-1}$. A number of $k_a(\lambda)$ values for pollutants at near surface and well-mixed water bodies have been determined. Furthermore, the rate of transformation of a given pollutant under photolysis is:

$$-\left(\frac{dc}{dt}\right)_{p,\lambda}=\phi_r(\lambda)I(\lambda)=\phi_r(\lambda)k_a(\lambda)C=k_p(\lambda)C \quad [10\text{-}26]$$

where $k_p(\lambda)$ is the direct photolysis first-order rate constant at wavelength λ with unit of reciprocal time. The remaining qualities have been discussed previously.

It should be emphasized that the rate of photolysis depends on both k_a, which can be evaluated by a computer-aided method and the quantum yield, ϕ_r which can be determined experimentally by chemical actinometry using an adequate reference compound. The reaction quantum yield values of some selected polyaromatic hydrocarbons (PHA) and other pollutants in aqueous median are listed in Table 10-3.

When xenobiotics are present in natural water systems, the dissolved organic matters (DOM) such as humic material, which is known to carry large numbers of chromophores, will be light absorbers to sensitize or indirectly photolyse the pollutant molecules. It is claimed that the presence of an organic carbon content of DOM of 4 mg CL^{-1} (ca. 30 μM) would excite 270 times per day or more than 10 times per hour. The pathways of indirected photolysis are illustrated in Figure 10-5. These can be achieved by photooxidation by oxygen from a ground triplet state to the first excited singlet state. The singlet oxygen, $^1O_2$, concentrations have been evaluated in water from a number of Swiss rivers and lakes and were proportional to the DOC concentration in these waters (see Fig. 10-6). The determination of singlet oxygen concentration was based on the use of a probe compound, furfuryl alcohol, which undertakes a **Diels-Alder type of reaction**:

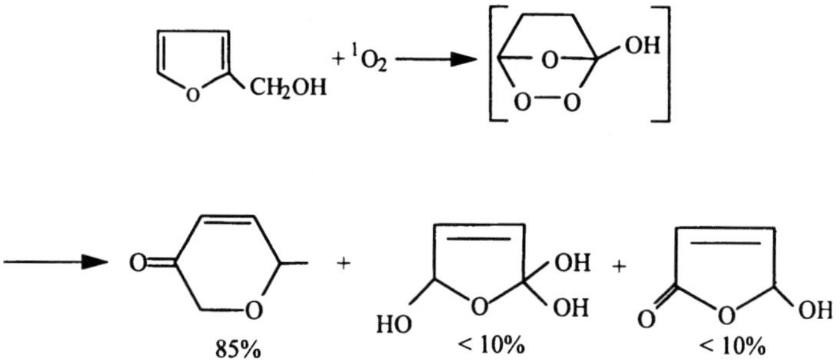

Other photoreactants in natural waters include the hydroxyl radical, HO, and the peroxyl radical, ROO.

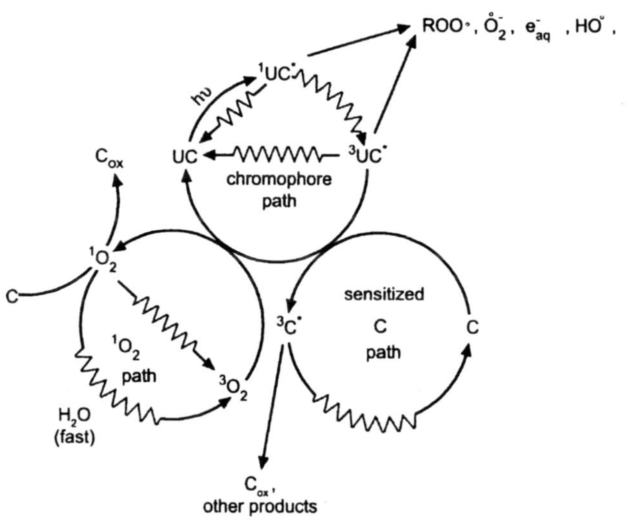

Figure 10-5. Pathways for indirect photolysis of organic chemical, C. UC refers to unknown chromophores. Wavy arrows symbolize radiationless transition. (Source: Zafiriou et al., 1984)

Table 10-3. Direct Photolysis Reaction Quantum Yields of Some Selected Organic Pollutants in Aqueous Solution

Compound	Wavelength[a] (nm)	Solvent Other Than Water (pH)	Reaction Quantum Yield (Φr)[b]
Naphthalene	313		1.5×10^{-2}
Phenanthrene	313		1.0×10^{-2}
Anthracene	313		3.0×10^{-3}
1,2-Benzanthracene	313, 366, sun	1% AN	3.2×10^{-3}
Benzo(*a*)pyrene	313, 366, sun	1–20% AN	8.9×10^{-4}
3,4-Dichloroaniline	313, polychr.	pH 7–10	4.4×10^{-2}
3,5-Dichloroaniline	313, sun	pH 4–10	5.2×10^{-2}
Pentachlorophenolate	314, polychr.	pH 8–10	1.3×10^{-2}
4-Nitrophenol	313, polychr.	pH 2–4	1.1×10^{-4}
4-Nitrophenolate	365, polychr.	pH 9–10	8.1×10^{-6}
Nitrobenzene	313		2.9×10^{-5}
4-Nitrotoluene	366		5.2×10^{-3}
2,4-Dinitrotoluene	313		2.0×10^{-3}
2,4,6-Trinitrotoluene	313, 366, sun		2.1×10^{-3}

[a] Wavelength or wavelength range at which ϕ_r had been determined; sun = sunlight, polychr. = polychromatic artificial light (> 290 nm). AN = acetonitrile.

[b] Moles of compound converted per moles of photons absorbed.

Source: R.P. Schwarzenbach, P.M. Gachwend and D.M. Imboden, 1993.

10.4 APPLIED PHTOLYSIS

10.4.1 Optical Brighteners

Colored fluorescent dyes and pigments are used for the brilliant "day-glo" colors used in textile dyes. They are also used for special fluorescent substances such as optical bleaches in "whiter-than-white" domestic detergent powders. The principle behind this process is that substances absorb ultraviolet light and radiate within the visible region so that the washed (white) textiles apparently reflect more light.

Requirements of optical brighteners are quite stringent. They must satisfy the following:

1. Brighteners must not absorb at all visible ranges, since this would lead to coloration of the fabric, but they must absorb strongly in the near-ultraviolet range where there is still residual intensity available from natural or artificial sources.
2. The fluorescence must lie in the short wavelength and of the visible spectrum. Otherwise, the fluorescence would give an undesirable yellow color to the fabric.
3. The fluorescent compound must be photochemically stable, and it must not enable the degradation or oxidation of the fibers.
4. The compound must be soluble or dispersible in an aqueous detergent solution but must be strongly absorbed by the textile fibers at an appreciable concentration during washing and usage.

Mechanistically, the uptake of optical brighteners by cellulosic fibers is different from that by synthetic fibers. In cottons, the fibers are hydrophilic and swell in water to allow the brightener molecules to adhere to the amorphous pores (15–30 Å in diameter). On the other hand, for synthetic fibers such as polyamides or polyesters, smaller brightener molecules either penetrate the canals between the fiber polymer chains or graft to the polar groups at the polymer backbone. Figure 10-7 shows the brightener molecules used for cotton fibers and synthetic fibers, respectively. The toxic effects of brighteners on human health are not known.

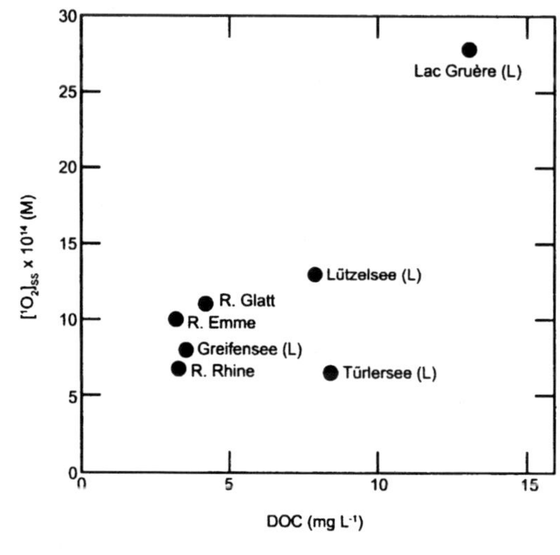

Figure 10-6. Observed $[^1O_2]_{ss}$ in water samples from some Swiss rivers (R) and lakes (L) as a function of the dissolved organic carbon (DOC) concentrations of these waters. The results apply for noontime light density on a clear summer day at 47.5°N. (Source: Hagg and Hoigne, 1986)

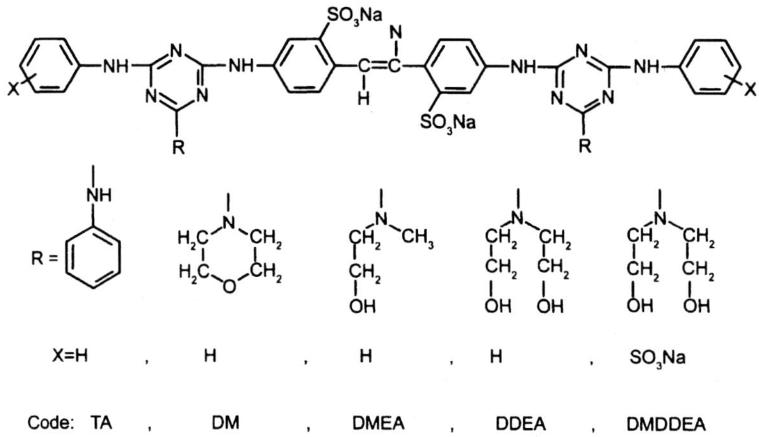

(a)

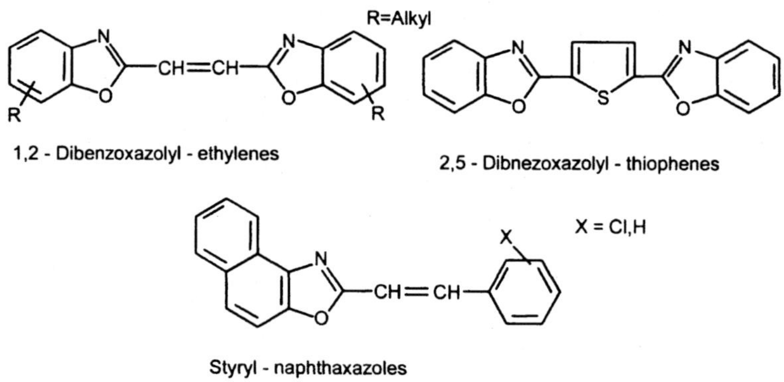

(b)

Figure 10-7. (a) Formulas of optical brighteners (*bis*-triazinyl-4-4′-diamino-stilbene-2-2′-disulfonic acid derivatives) suitable for cellulosic (e.g., cotton) fibers. (b) Brighteners suitable for polyamide (e.g., nylon) and polyester (e.g., Terylene) fibers.

10.4.2 Photocatalytic Process

The most common process is the use of photocatalyst TiO_2 and ultra violet light to decompose organic contaminants in water. When photons of sufficient energy are absorbed on the catalyst surface, electrons and holes are produced which form very powerful oxidizing agents: peroxide and hydroxyl radicals. These radicals have been found to completely mineralize a large number of organic compounds (zenobiotics) including solvents, PCBs, dioxins, pesticides, dyes, etc. resulting in only carbon dioxide, water, and dilute acids or ammonia as the products.

$$C_xH_yO_zN_nCl_w + O_2 \xrightarrow[TiO_2]{solar-UV} CO_2 + H_2O + HCl + NH_3$$

Photocatalytical oxidation process involving the use of semiconductors has gained considerable attention in the treatment of hazardous organic wastes. Upon irradiation, a semiconductor generates electron/hole pairs with free electrons produced in the nearby empty conduction bands (cb) and positive holes remaining in the valence band (vb). The holes, acting as strong oxidizing agent, migrates to the semiconductor's surface and react with the organic contaminant. A typical reaction scheme of a n-type semiconductors such as TiO_2 is illustrated in Figure10-8.

The life-time of an electron/hole separation process ranges from nanoseconds to a few hours depending on ambient condition. The recombination of electron/hole pair can take place between energy bands or on the surface, often in this manner the photocatalytical reaction is retarded. In order to impede the recombination process, conducting metals such as noble metals can be incorporated into the semiconductor. In this manner the electron transfer can be facilitated and the life-time of electron/hole separation prolonged.

Usually 10% platinized TiO_2 or TiO_2 with 5% platinum deposition is used. The UV source is a 500 watt xenon lamp. The photocatalytic reactor can be a falling-film / heliostat system or a glass-tube/trough system. A parabolic trough reactor is illustrated in Figure 10-9.

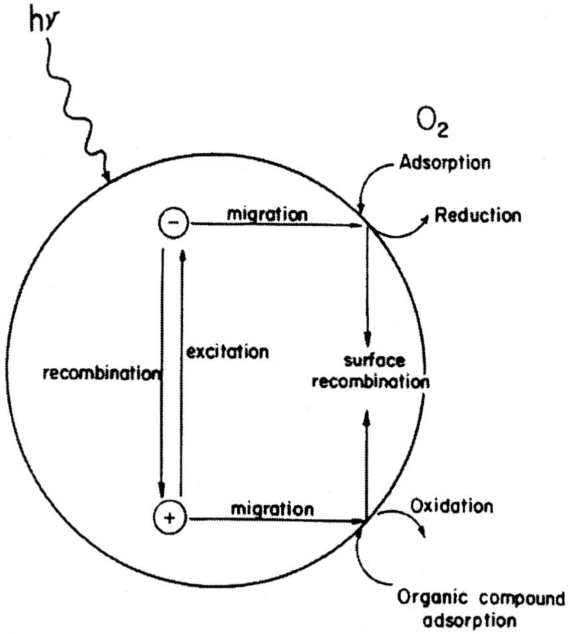

Figure 10-8. Schematic Redox Reactions of an n-type semiconductor

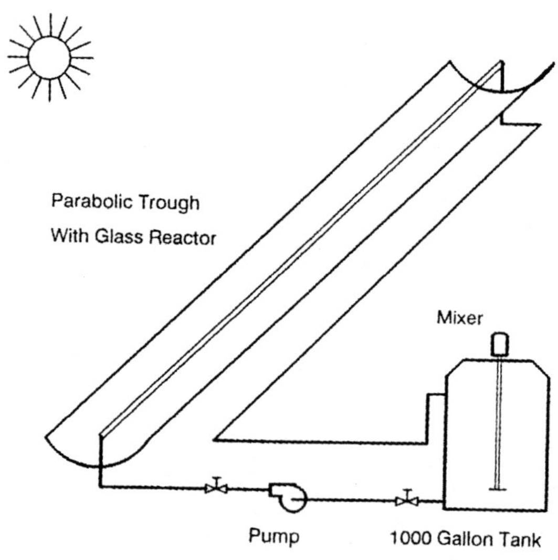

Figure 10-9. Schematic of parabolic trough system used for solar detoxification studies.

C.P. Huang has studied the treatment of phenol in wastewater. Average concentration of phenol in petroleum and coke industry wastewater ranges from 10-100 mg/L. For decomposition and mineralization of phenol the following mechanism has been purposed:

$$TiO_2 + \rightarrow e_{cb}^- + h_{\square b}^+ \qquad [10\text{-}27]$$

(excitation; electron/hole pair production)

$$e_{cb}^- \rightarrow e_{tr}^- \qquad [10\text{-}28]$$

(electron migration to the surface)

$$h_{\square b}^+ \rightarrow h_{tr}^+ \qquad [10\text{-}29]$$

(hole migration to the surface)

$$H_2O \rightarrow H^+ + OH^- \qquad [10\text{-}30]$$

(water dissociation)

$$O_2 + e_{tr}^- \rightarrow \square O_2^- \qquad [10\text{-}31]$$

$$H^+ + \square O_2^- \rightarrow \square HO_2 \qquad [10\text{-}32]$$

$$\square HO_2 \rightarrow \tfrac{1}{2} O_2 + \square OH \qquad [10\text{-}33]$$

$$2\,OH^- + 2\,h_{tr}^+ \rightarrow 2\,\square OH \qquad [10\text{-}34]$$

$$\square O_2^- + e_{tr}^- \rightarrow \square O_2^{\,2-} \qquad [10\text{-}35]$$

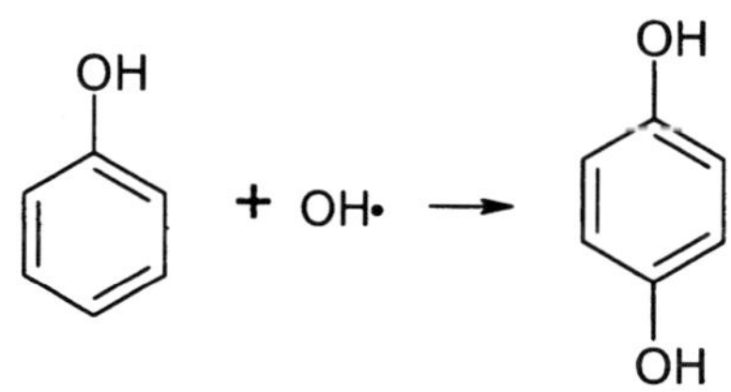

[10-36]

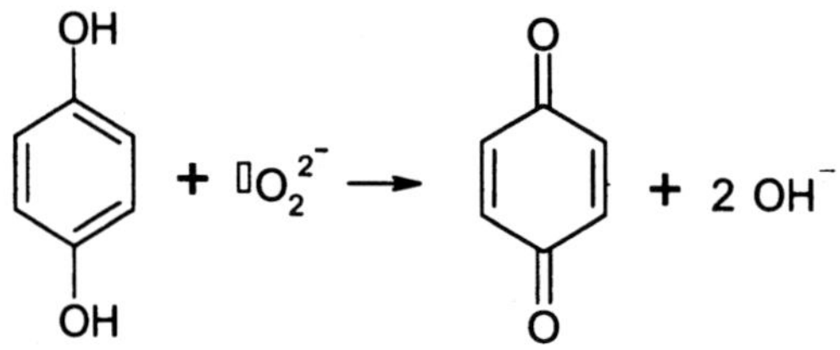

[10-37]

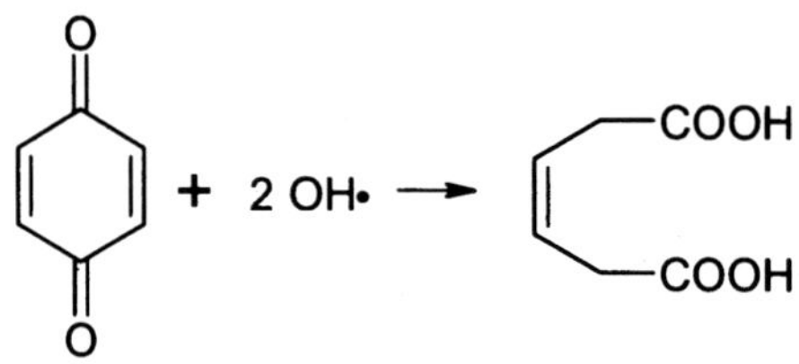

[10-38]

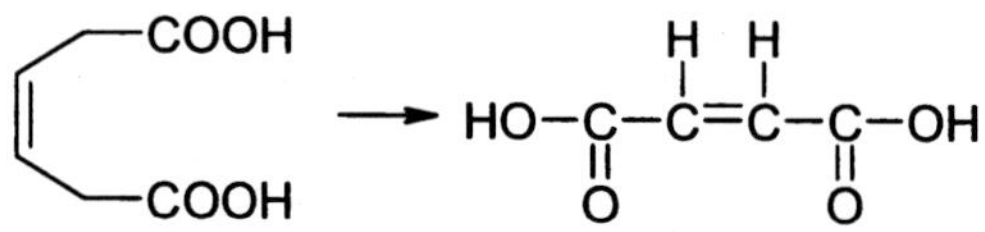

[10-39]

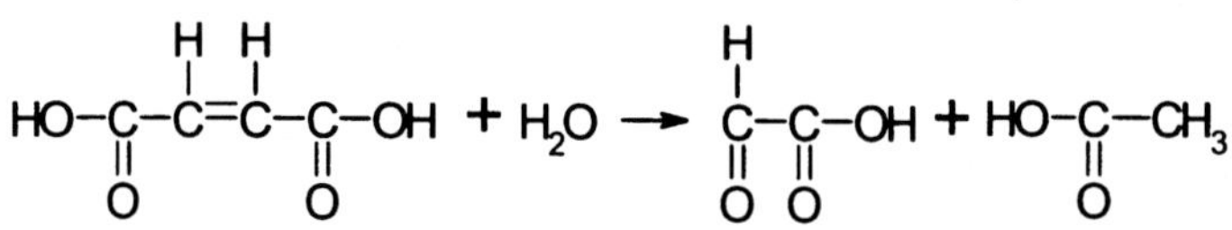

[10-40]

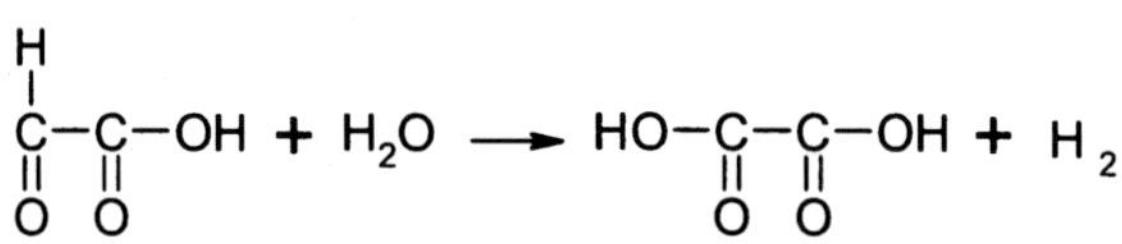

[10-41]

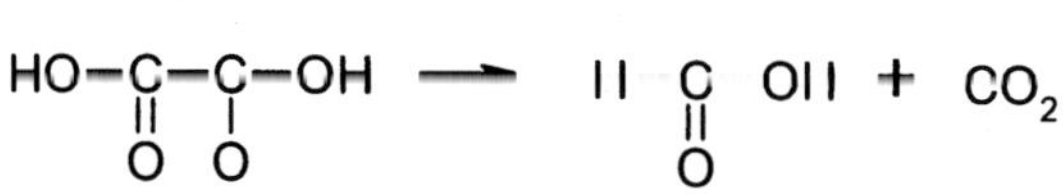

[10-42]

$$\mathrm{H{-}C({=}O){-}OH + 1/2\ O_2 \longrightarrow CO_2 + H_2O} \quad [10\text{-}43]$$

$$\mathrm{HO{-}C({=}O){-}CH_3 + 2\,O_2 \longrightarrow 2\,CO_2 + 2\ H_2O} \quad [10\text{-}44]$$

Photocatalytic oxidation can be used for a large variety of naturally-occurring materials such as cellulose, lignin, wood, algae lignite even dead animals and their excrement.

10.4.3 Photogalvanic System

Photogalvanic cells are photoelectrochemical devices in which a photochemical reaction takes place in the homogeneous liquid phase, producing species which react at the electrodes. One photogalvanic cell that illustrate this system is the iron-thionine cell by Rabinowitch. Initially the purple dye thionine (in 0.1 N H_2SO_4) is produced via the thionine triplet excited state

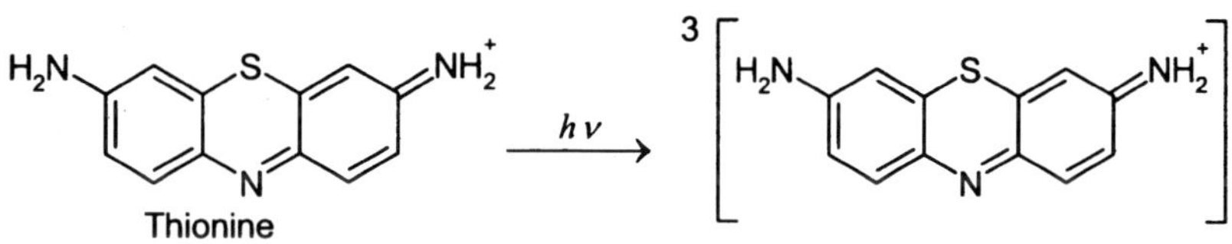

[10-45]

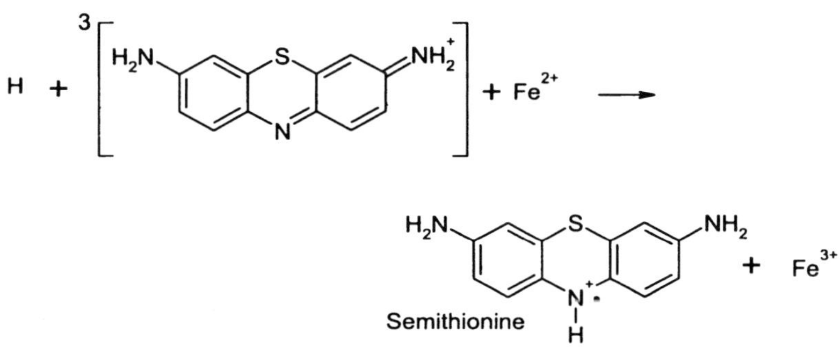

[10-46]

to end up to a radical ion, semithionine. The back reaction between the ferric ion and the half-reduced radical ion (semithionine) is then photogenerated spontaneously.

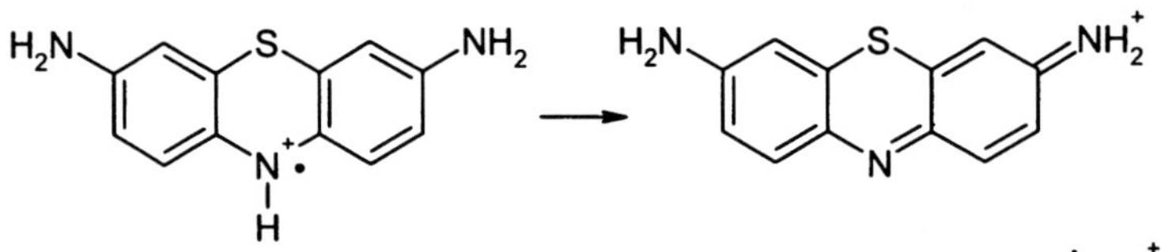

$$+ H^{+} + e \qquad [10\text{-}47]$$

and

$$Fe^{3+} + e \rightarrow Fe^{2+} \qquad [10\text{-}48]$$

If two inert electrodes are placed into a thionine-ferrous ion solution and only one of the electrodes is irradiated, concentration gradients involving the oxidized and reduced forms of thionine and iron are established. Between the two electrodes, an electrode potential of approximately 0.4 V is found. The two half-cell reactions are Equations 10-47 and 10-48. In general

$$D \xrightarrow{h\nu} D^{*} \qquad [10\text{-}49]$$

$$D^{*} + M^{n+} \rightarrow D^{-} + M^{(n+1)+} \qquad [10\text{-}50]$$

and

$$D^{-} + M^{(n+1)+} \rightarrow D + M^{n+} \qquad [10\text{-}51]$$

also

$$D^{-} \rightarrow D + e \text{ (anode)} \qquad [10\text{-}52]$$

$$M^{(n+1)+} + e \rightarrow M^{n+} \text{ (cathode)} \qquad [10\text{-}53]$$

Other photoredox couple such as vanadium (III) bis (2, 2'-bipyridyl) chloride and Fe (III). Similarly in solution

$$[V(bpy)_2Cl_2]^{+} \xrightarrow{h\nu} [V(bpy)_2Cl_2]^{*+} \quad [10\text{-}54]$$

$$[V(bpy)_2Cl_2]^{*+} + Fe^{3+} \rightarrow [V(bpy)_2Cl_2]^{2+} + Fe^{2+} \quad [10\text{-}55]$$

And at the electrodes

illuminated (cathode) $[V(bpy)_2Cl_2]^{2+} \xrightarrow{+e} [V(bpy)_2Cl_2]^{+}$

[10-56]

dark (anode) $Fe^{2+} \rightarrow Fe^{3+} + e$

Also $Ru(bpy)_3{}^{2+} - Fe^{3+}$ is a good photoredox couple. (Figure 10-10)

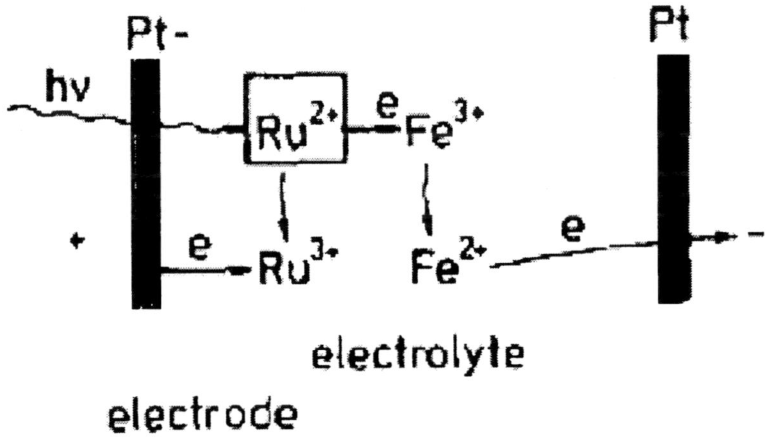

Figure 10-10. A Photogalvanic Cell (After Memming, 1990)

Semiconductor can be formed with pn junction of Group 5 and Group 7 elements as seen in Table 10-4 , since they contain the valence of III and V. At Group 12 and Group 8 many also do the same trick, examples are Cd, Se, Zn, etc. Even tin antimonide with trace indium impurities can be used for sulfur filters based on galvanic effort.

Table 10-4. Elements which can form pn gap energy for semiconductor

II	III	IV	V	VI
	5 **B**	6 **C**	7 **N**	7 **O**
	13 **Al**	14 **Si**	15 **P**	16 **S**
30 **Zn**	31 **Ga**	32 **Ge**	33 **As**	34 **Se**
48 **Cd**	49 **In**	50 **Sn**	51 **Sb**	52 **Te**
80 **Hg**	81 **Tl**	82 **Pb**	83 **Bi**	84 **Po**

Regardless of pn type or a Schottky semiconductor-metal type, the following scheme is applicable

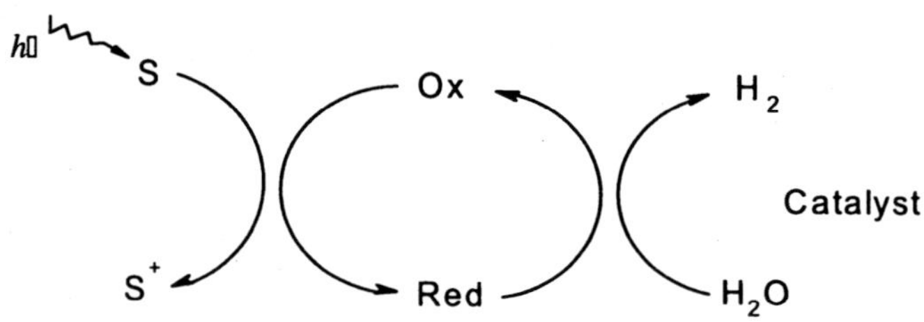

[10-57]

where S is light adsorbing redox system

$$S + h\nu \rightarrow S^* \qquad [10\text{-}58]$$

$$S^* + H_2O \rightarrow S^+ + H + OH^- \qquad [10\text{-}59]$$

$$S + \tfrac{1}{2} H_2O \rightarrow S + H^+ + \tfrac{1}{4} O_2 \qquad [10\text{-}60]$$

in which

$$S = \text{acridine orange or } Ru(bipy)_3{}^{2+}$$

$$Ox/Red = \text{methylviologen } {}^{2+}/_{1+}$$

or

$$= Eu\ {}^{3+}/_{2+}$$

$$Cat = PtO_2$$

In order to avoid the back reaction to produce S^+ and Red from Equation 10-61 which is

$$S^+ + Red \xrightarrow{dark} S + Ox \qquad [10\text{-}61]$$

another electron donor, DH, such as triethanol-amine or cysteine, is added which can react with S^+. Thus

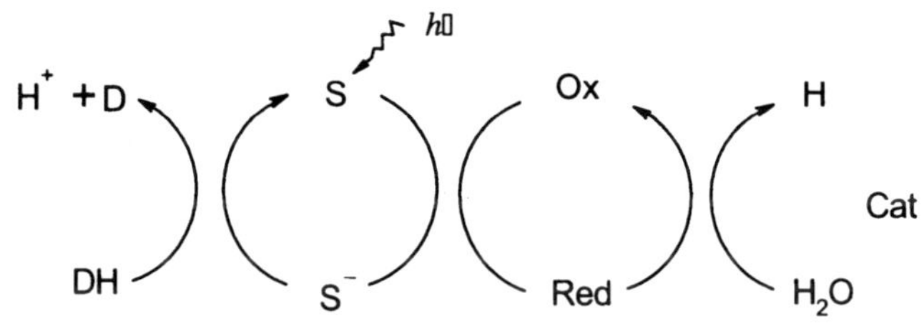

micellar phase can be used for electron transfer process. If a dye D can be dissolved in micelle together with other organic molecule A, then

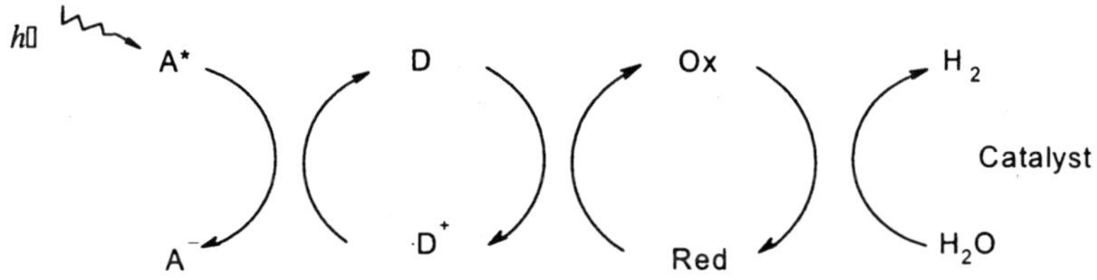

DA is in micellar phase which forms the Helmholtz layer

$$D^* + A \xrightarrow{h\nu} D^+ + A^-$$

Recently the photogalvanic effects in heteropolar nanotubes have been studied. The study of the $B_xC_yN_z$ nanotubes have produced a new class of photovoltaic effects at the molecular level by being noncentrosymmetric (NCS) and polar. The photoexcitation across the band gap with polarized light will produce the so called electric shift current due to the asymmetric generation of that carriers. The shift current can be understood from the BN forms of figure 10-11.

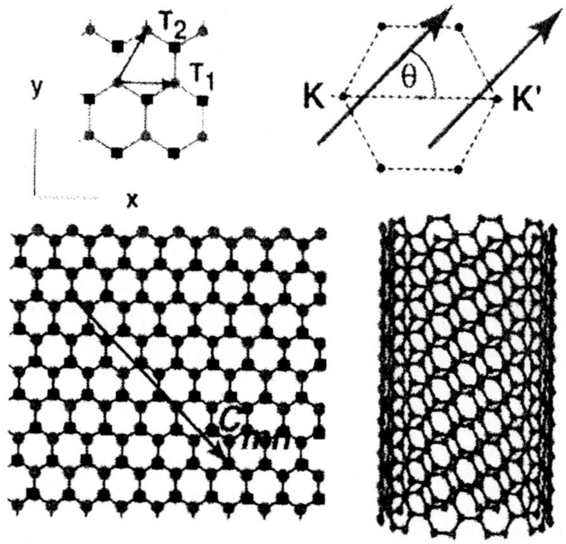

Figure 10-11. Planar BN forms a honeycomb lattice with B and N occupying alternate sites (upper left). A BN tube is formed by wrapping the sheet through the translation vector C_{mn} (lower level) quantizing the transverse crystal momenta. The upper right-hand panel shows two representative lines of allowed momenta for a tube with a nonzero chiral index ν = mod (m-n, 3) and chiral angle . The lower right panel shows the structure for the nonchiral armchair (10,10) wrapping of the BN sheet. (After P. Kr l et al.,2000)

REFERENCES

10-1. Ramamurthy, V. and Kirk S. Schanze, *Molelcular and Supramolecular Photochemistry*, Volume 2: *Organic and Inorganic Photochemistry.*

10-2. Ramamurthy, V. and Kirk S. Schanze, *Molelcular and Supramolecular Photochemistry*, Volume 6: *Organic, Physical, and Materials Photochemistry*, Marcel Dekker, Inc., New York, 2000.

10-3. Tseng, J. and C.P. Huang, mechanistic Aspects of the Photocatalytic Oxidation of Phenol in Aqueous Solutions, *ACS Symposium Series*, **422**, 12-39 (1990)

10-4. Pacheco, J.E. and J.T. Holmes, Falling Film and Glass-Tube Solar Photocatalytic Reactors for Treating Contaminated Water, *ACS Symposium Series*, **422**, 12-39 (1990)

10-5. Schwarzenback, R. P., P. M. Gschwend, and D. M. Imboden, *Environmental Organic Chemistry*, Wiley, New York, 1993.

10-6. Memming , R., *Electricchimica Acta*, **25**, 77-88 (1980)

10-7. Aliwi, S. M., S. A. Naman, and I. K. Al-Daghstani, *Solar Cells*, **18**, 85-91 (1986).

10-8. Bailey, R. A., H. M. Clark, J. P. Ferris, S. Kranse and R. C. Stung, *Chemistry of the Environment*, Academic Press, New York, 1978.

10-9. Rubinowitch, E. J., *Chem.Phy*, **8,** 551 (1940)

10-10. Lenz, R. W., Organic Chemistry of Synthetic High Polymers, Interscience, New York, 1967.

PROBLEM SET

1. Assuming atmospheric oxygen is produced from photodissociation of water only, what amount of oxygen had accumulated by the start of Silurian era (~ 4.2×10^8 years)?

2. What is the amount of singlet oxygen in the atmosphere, if the source is from the collision of O ($^1D_2$) based on the following?

$$O(^1D_2) + O_2(^3\Sigma_g^-) \rightarrow O(^3P_2) + O_2(^1\Sigma_g^+)$$

$$\downarrow O_2(^3\Sigma_g^+)$$

$$O_2(^1A_g)$$

$$k = 6 \times 10^{-11}\ \text{cm}^3 / \text{molecule-sec}$$

CHAPTER 11

OXIDATION AND REDUCTION

Redox chemistry controls the exogenic cycle. It deals with numerous species dissolved in natural waters. For example, metal sulfide can be dissolved by oxygenation, but manganous and ferrous species can be easily precipitated by oxygenation. The examination of several biological systems reveals that life is a redox process, as evidenced by such mechanisms as photosynthesis, assimilation, metabolism, enzymatic reactivity, biodegradation, energetic coupling, and stoichiometric coupling. Redox is the basis of many associated phenomena, such as combustion, corrosion, and disinfection.

A good portion of this chapter is devoted to the principles of redox chemistry. The first part explains electrode potential and electron activity. Predominance area diagrams for redox systems are discussed. Other applications, such as the removal of metals from waters, advanced oxidation processes, and denitrification, follow.

A recent technological achievement, the one-electron reduction in soil environment and the zero-valent metal for contaminant remediation in groundwater, will be highlighted. Also, oxidative filtration using native minerals as surface-induced catalytic media in China is mentioned.

11.1 Redox Reaction and Its Stoichiometry

The process in which a substance loses or donates an electron is termed **oxidation**, whereas the process in which a substance gains or accepts an electron is termed **reduction**. For example, Zn is an electron donor,

$$Zn = Zn^{2+} + 2\ e^-, \quad \text{oxidation}$$

and Cu^{2+} is an electron acceptor,

$$Cu^{2+} + 2\ e^- = Cu, \quad \text{reduction}$$

These reactions are called half-reactions. They can be coupled to form a **redox reaction**,

$$Zn + Cu^{2+} = Cu + Zn^{2+}$$

The half oxidation reaction and half reduction reaction must be coupled in such a way that no free electrons are available. Usually an electron donor that is oxidized to a higher state is a reducing agent for the electron acceptor, and an electron acceptor that is reduced to a lower state is an oxidizing agent for the electron donor.

Since the electron generated (oxidation) must be balanced by the electron consumed (reduction), stoichiometry is applied to equate the redox reactions. First, the reactants and products of both half-reactions are identified. Next, the oxidation states of the responsible atoms in both reactants and products of both half-reactions are determined. Then the oxygen is balanced with H_2O, the hydrogen is balanced with H^+, and the oxidation state difference is balanced (from reactants to products) with electrons (e has a charge of −1; sometimes e and e^- are used indiscriminately). Finally, the electrons are balanced in both half-reactions to achieve an overall redox reaction. Empirical rules used to establish the oxidation states of compounds are as follows:

1. The sum of the oxidation states of atoms in a molecule is zero; for ions this equals the formal charge.

2. Oxidation states for H and O are known as +1 and -2, respectively, and are used as a basis to evaluate states of other atoms, including N, C, S, etc.
3. For multiple linked atoms, each atom is zero, e.g., N for N_2 is zero, S for S_8 is zero. Therefore, the empirical results read as follows: NO_2^-, N = +3; NH_4^+, N = −3; NO_3^-, N = +5. If N and S are used for evaluation, and N = −3, S = −2, then for HCN, C = +2; SCN, C = +5; for S, H_2S, S = −2; SO_3^{2-}, S = +4; SO_4^{2-}, S = +6; $S_2O_3^{2-}$, S = +2; $S_4O_6^{2-}$, S = +2.5; $S_2O_6^{2-}$, S = +5. For C compounds, HCO_3^-, C = +4; HCOOH, C = +2; $C_6H_{12}O_6$, C = 0; CH_3OH, C = −2; CH_4, C = −4; C_6H_5COOH, C = −2/7, etc. In general, for $A_aO_oH_h^n$, the oxidation state for A is $(2o - h + n)/a$.

[Example 11-1] Balance the equation for the oxidation of phenol with permanganate in wastewater for separation of precipitates.

Reactant: C_6H_5OH (−2/3), $KMnO_4$(+7)

Products: CO_2 (+4), MnO_2 (+4)

Oxidation half-reactions:

$$C\,(-2/3) = C\,(+4) + 14/3\ e$$

$$1/6\ C_6H_5OH = C\,(-2/3) + 1/6\ H_2O + 2/3\ H^+$$

$$2O\,(-2) + C\,(+4) = CO_2$$

$$2\ H_2O = 2O\,(-2) + 4\ H^+$$

$$1/6\ C_6H_5OH + 11/6\ H_2O = 14/3\ H^+ + CO_2 + 14/3\ e$$

or

$$C_6H_5OH + 11\ H_2O = 6\ CO_2 + 28\ H^+ + 28\ e \qquad (1)$$

Reduction half-reactions:

$$Mn\ (+7) + 3\ e = Mn\ (+4)$$

Table 11-1. Equilibrium Constants of Redox Processes Pertinent in Aquatic Conditions (25°C)

Reaction	$p\varepsilon^0$ (≡ log *K*)	$p\varepsilon^0$ (W)
(1) ½ O_2 + H^+ (W) + *e* = ½ H_2O	+20.75	+13.75
(2) 1/5 NO_3^- + 6/5 H^+ (W) + *e* = 1/10 N_2(g) + 3/5 H_2O	+21.05	+12.65
(3) ½ MnO_2(s) + ½ HCO_3^-(10^{-3}) + 3/2 H^+ (W) + *e* = ½ $MnCO_3$(s) + H_2O	—	+8.5[a]
(4) ½ NO_3^- + H^+ (W) + *e* = ½ NO_2 + ½ H_2O	+14.15	+7.15
(5) 1/8 NO_3^- + 5/4 H^+ (W) + *e* = 1/8 NH_4^+ + 3/8 H_2O	+14.90	+6.15
(6) 1/6 NO_2^- + 4/3 H^+ (W) + *e* = 1/6 NH_4^+ + 1/3 H_2O	+15.14	+5.82
(7) ½ CH_3OH + H^+ (W) + *e* = ½ CH_4(g) + ½ H_2O	+9.88	+2.88
(8) ¼ CH_2O + H^+ (W) + *e* = ¼ CH_4(g) + ¼ H_2O	+6.94	−0.06
(9) FeOOH(s) + HCO_3^-(10^{-3}) + 2H^+ (W) + *e* = $FeCO_3$(s) + 2H_2O	—	−1.67[a]
(10) 1/2CH_2O + H^+ (W) + *e* = ½ CH_3OH	+3.99	−3.01
(11) 1/6 SO_4^{2-} + 4/3 H^+ (W) + *e* = 1/6 S(s) + 2/3 H_2O	+6.03	−3.30
(12) 1/8 SO_4^{2-} + 5/4 H^+ (W) + *e* = 1/8 H_2S(g) + ½ H_2O	+5.75	−3.50
(13) 1/8 SO_4^{2-} + 9/8 H^+ (W) + *e* = 1/8 HS^- + ½ H_2O	+4.13	−3.75
(14) ½ S(s) + H^+ (W) + *e* = ½ H_2S(g)	+2.89	−4.11
(15) 1/8 CO_2(g) +H^+ (W) + *e* = 1/8 CH_4(g) + ¼ H_2O	+2.87	−4.13
(16) 1/6 N_2(g) + 4/3 H^+ (W) + *e* = 1/3 NH_4^+	+4.68	−4.68
(17) ½ ($NADP^+$) + ½ H^+ (W) + *e* = ½ (NADPH)	−2.0	−5.5[a]
(18) H+ (W) + *e* = ½ H_2(g)	0.0	−7.00
(19) Oxidized ferredoxin + *e* = reduced ferredoxin	−7.1	−7.1
(20) ¼ CO_2(g) + H^+ (W) + *e* = 1/24 (glucose) = ¼ H_2O	−0.20	−7.20
(21) ½ $HCOO^-$ + 3/2 H^+ (W) + e = ½ CH_2O + ½ H_2O	+2.82	−7.68
(22) ¼ CO_2(g) + H^+ (W) + *e* = ¼ CH_2O + ¼ H_2O	−1.20	−8.20
(23) ½ CO_2(g) + ½ H^+ (W) + *e* = ½ $HCOO^-$	−4.83	−8.73

[a] These data correspond to (HCO_3^-) = 10^{-3} M rather than unity and so are not exactly $p\varepsilon^0$ (W); they represent typical aquatic conditions more nearly than $p\varepsilon^0$ (W) values do.

$$MnO_4^- = Mn\ (+7) + 4\ O\ (-2)$$

$$Mn\ (+4) + 2\ O\ (-2) = MnO_2$$

$$4\ H^+ + 2\ O\ (-2) = 2\ H_2O$$

$$MnO_4^- + 4\ H^+ + 3e = MnO_2 + 2\ H_2O$$

3(1) + 28(2),

$$3\ C_6H_5OH + 28\ MnO_4^- + 28\ H^+ = 28\ MnO_2 + 23\ H_2O + 18\ CO_2$$

Some common redox reactions with their electrode potentials are listed in Table 11-1.

11.2 Electrode Potential and Electron Activity

The Zn-Cu system discussed in the previous section is called the Daniel cell when the zinc electrode in a zinc chloride solution is connected to a copper electrode in a cupric chloride solution. Electrons will flow from the Zn-electrode (anode) to the copper electrode (cathode) to be detected by a voltmeter (Fig. 11-1). The half-reaction will occur in the half-cell. The potential of the cell or the electromotive force of the cell at 1 atm and 25°C is called E^0, the electrode potential of the cell.

$$E^0 = E^0_{ox} + E^0_{red} \qquad [11\text{-}1]$$

E^0_{ox} and E^0_{red} are the half-reaction potentials for oxidation and reduction. E^0 is the electrochemical version of ΔG related to free energy by:

$$\Delta G^0 = -\,nFE^0$$

$$\Delta G = -\,nFE \qquad [11\text{-}2]$$

where n is the number of electrons involved and F is the Faraday = 23.06 kcal/volt-equivalent (96,500 coulombs/equivalent).

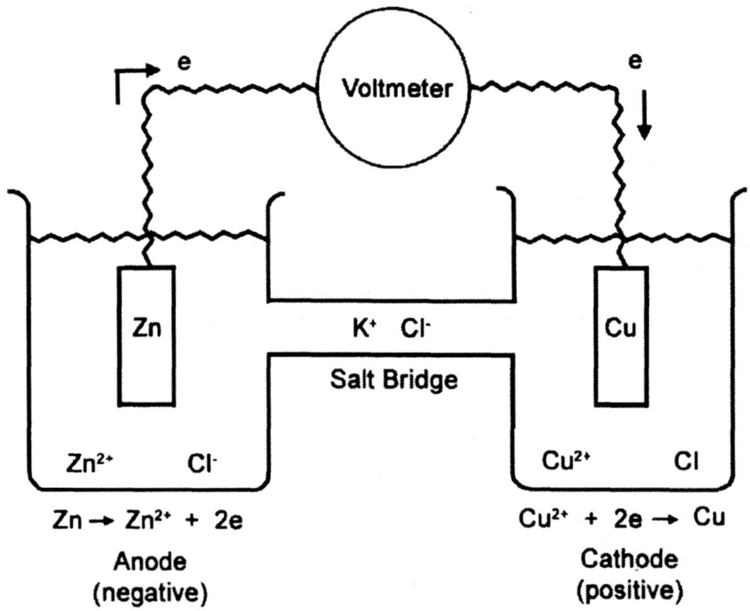

Figure 11-1. Daniel cell.

In this convention, the reference hydrogen electrode, or the reduction of hydrogen ion to hydrogen gas, is zero. The free energy governs whether the redox reaction will proceed or not, using the equation:

$$\Delta G = \Delta G^0 + RT\ln Q \qquad [11\text{-}3]$$

Dividing by $-nF$, we obtain from Equation [11-2],

$$E = E^0 - \left(\frac{RT}{nF}\right) \ln Q \qquad [11\text{-}4]$$

Often $E = E^0 - (0.059/n) \log Q$, which is the Nernst equation. At equilibrium, $E = 0$, $Q = K$, it can be shown that:

$$E^0 = \frac{RT}{nF} \ln K \qquad [11\text{-}5]$$

For example, from Table 12-1 it can be seen from cell potentials that:

	$\Delta G^0 = -nFE^0$
$Zn = Zn^{2+} + 2\,e$	$-2F\,(+0.76)$
$2\,e + Cu^{2+} = Cu$	$-2F(+0.34)$
$Zn + Cu^{2+} = Zn^{2+} + Cu$	$-2F(+1.10)$

$$E^0 = 1.1 = \frac{0.059}{2} \log K$$

and

$$K = 10^{37.2}$$

As shown in Figure 11-2, a scale is established for oxidative or reductive capacity, where the capacities of H_2 are used as a standard. The availability of electrons in a solution is termed electron activity, and like pH, pε is defined as:

$$p\varepsilon = -\log\,[e] \qquad [11\text{-}6]$$

If we have a half-reaction of a redox reaction involving a higher oxidation state (high val) and a lower oxidation state (low val), we can conveniently write the forms with the balance of electrons:

$$\text{high val} + ne = \text{low val}$$

and

$$K = \frac{[\text{low val}]}{[\text{high val}][e]^n}$$

or

$$[e] = \left(\frac{1}{K}\frac{[\text{low val}]}{[\text{high val}]}\right)^{\frac{1}{n}} \qquad [11\text{-}7]$$

$$\begin{aligned} p\varepsilon &= -\log[e] = \frac{1}{n}\left(\log K - \log\frac{[\text{low val}]}{[\text{high val}]}\right) \\ &= \frac{1}{n}\left(-pK + \log\frac{[\text{high val}]}{[\text{low val}]}\right) = p\varepsilon^0 + \frac{1}{n}\log\frac{[\text{high val}]}{[\text{low val}]} \end{aligned} \qquad [11\text{-}8]$$

For example, for the $Fe^{3+} - Fe^{2+}$ couple,

$$Fe^{3+} + e = Fe^{2+}$$

$$K = \frac{(Fe^{2+})}{(Fe^{3+})(e)}$$

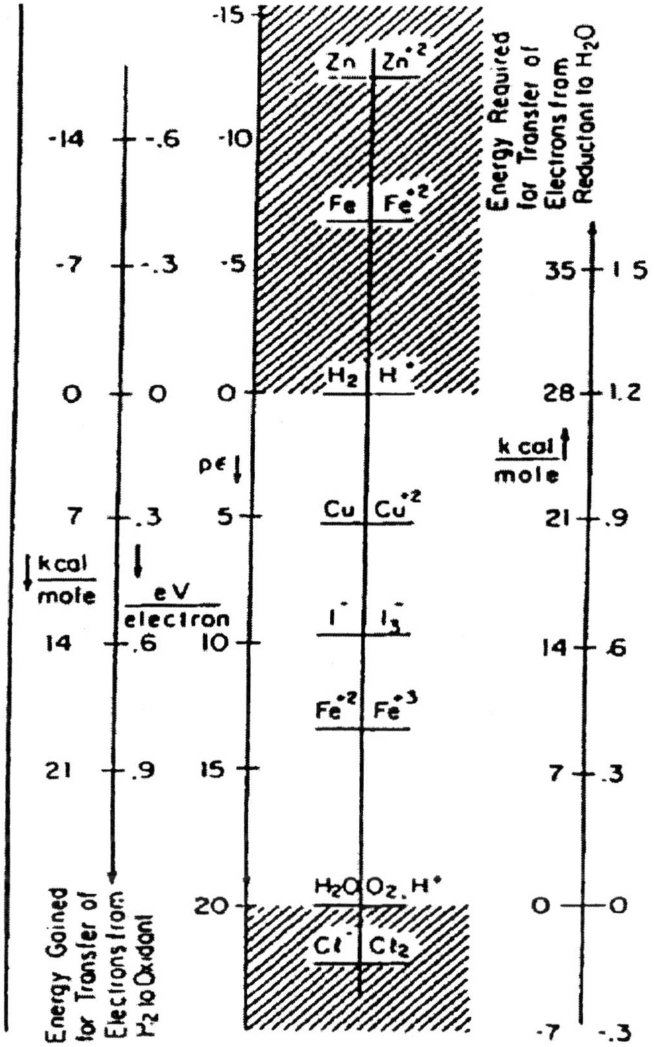

Figure 11-2. Electron free-energy levels.

then

$$\log K = \log \frac{(\mathrm{Fe}^{2+})}{(\mathrm{Fe}^{3+})} - \log e$$

or

$$p\varepsilon = p\varepsilon^0 - \log \frac{(\mathrm{Fe}^{2+})}{(\mathrm{Fe}^{3+})} \qquad [11\text{-}9]$$

when $(\mathrm{Fe}^{2+}) = (\mathrm{Fe}^{3+})$, then

$$p\varepsilon = p\varepsilon^0 = \log K$$

Hence,

$$p\varepsilon^0 = \log K$$

if we define

$$p\varepsilon^0 = \frac{\log K}{n} = -\frac{pK}{n} \qquad [11\text{-}10]$$

If we take the $p\varepsilon^0$ equation [11-10] and the E^0 equation [11-5], we can equate them by ln k and log k, which introduces 2.3, and thus we can derive the following:

$$p\varepsilon^0 = \frac{FE^0}{2.3RT} \ (= 16.9E^0 \text{ at } 25°\text{C})$$

$$p\varepsilon = \frac{FE}{2.3RT} \ (= 16.9E \text{ at } 25°\text{C}) \qquad [11\text{-}11]$$

or if we want to correlate this expression with free energy, then

$$p\varepsilon^0 = -\frac{\Delta G}{2.3nRT}$$

or

$$p\varepsilon = -\frac{\Delta G}{2.3nRT} \qquad [11\text{-}12]$$

From the above discussion for a generalized reduction reaction,

$$\sum n_i A_i + ne = 0,\ K$$

Where A_i is the participating species and n_i is the numerical coefficient positive for reactants and negative for products. Thus,

$$p\varepsilon = \frac{1}{n}\log K + \frac{\log[\prod [A_i]^{n_i}]}{n} \qquad [11\text{-}13]$$

The following three examples involve determining the equation of $p\varepsilon$ and its value.

[Example 11-2] Determine the electron activity of three waters: (a) water containing an acid solution of 10^{-5} M of Fe^{3+} and 10^{-3} M of Fe^{2+}; (b) water in equilibrium with atmosphere at pH 7.5; and (c) water with 10^{-5} dissolved manganese salt in equilibrium with $\gamma MnO_2(s)$ at pH 8.

a) $Fe^{3+} + e^- = Fe^{2+},\ K = \dfrac{[Fe^{2+}]}{[Fe^{3+}][e]} = 10^{13.01}$

Thus,

$$p\varepsilon = 1301 + \log\frac{[Fe^{3+}]}{[Fe^{2+}]} = 11.01$$

b) This water at pH 7.5 in equilibrium with the atmosphere can be expressed as:

$$1/2\ O_2 + 2\ H^+ + 2e = H_2O\ (l),\ K = 10^{41.55}$$

and

$$K = \frac{1}{P_{O_2}^{\frac{1}{2}} [H^+]^2 [e]^2}$$

$$p\varepsilon = 41.55/2 + 1/2 \log\{ P_{O_2}^{\frac{1}{2}} [H^+]^2\} = 12.94$$

c) The water at pH 8 containing 10^{-5} M Mn^{2+} in equilibrium with $\gamma MnO_2(s)$:

$$\gamma MnO_2\ (s) + 4H^+ + 2e = Mn^{2+} + 2H_2O \quad \log K = 40.84$$

$$K = \frac{[Mn^{2+}]}{[H^+]^4 [e]^2}$$

$$p\varepsilon = 20.42 + 1/2 \log \{[H^+]^4/[Mn^{2+}]\} = 6.92$$

11.2.1 *pε*-pH Predominance Area Diagrams

Many oxidation-reduction schemes are illustrated as electron activity as a function of pH. A good example will explain the usefulness of this approach. Boundaries between two species are usually drawn in such a manner that the concentrations of the two species are equal on the line. Let us examine a common situation — 0.01 M chlorine dissolved in water, given that

$$e + H^+ + HOCl = 1/2\ Cl_2 + H_2O, \quad p\varepsilon^0 = 27.0$$

$$2\ e + Cl_2 = 2\ Cl^-, \qquad p\varepsilon^0 = 23.5$$

$$HOCl = H^+ + Cl^-, \qquad pK_a = 7.5$$

$$C_{T,Cl} = 10^{-2}\ M = 2\ (Cl_2) + (HOCl) + (OCl^-) + (Cl^-)$$

All of this information should be used to derive the lines shown in Figure 11-3.

In many of these diagrams, the first approach is to establish the water stability lines, i.e., oxygen and hydrogen in their extremes. Thus,

$$4\ e + 4\ H^+ + O_2 = 2\ H_2O,\ \ E^0 = 1.23$$

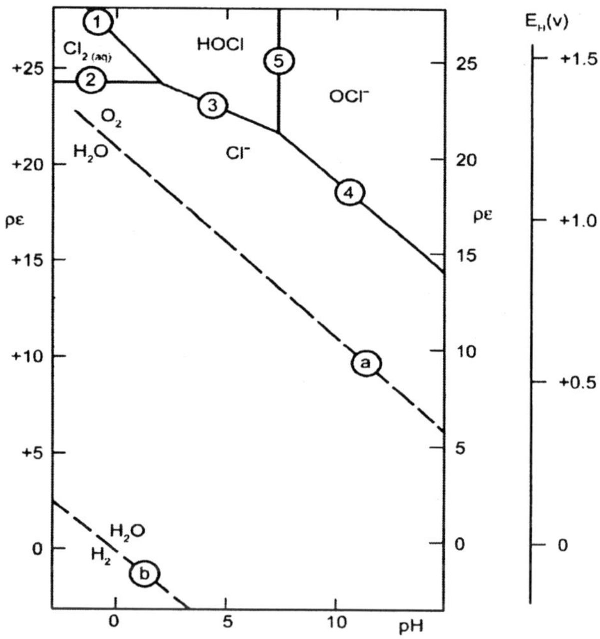

Figure 11-3. pε versus pH diagram for the chlorine system. In dilute solutions $Cl_2(aq)$ exists only at low pH. Cl_2, OCl^-, and HOCl are all unstable or metastable in water.

$$p\varepsilon = p\varepsilon^0 = \frac{1}{n}\log\frac{1}{P_{O_2}\left[H^+\right]^4}$$

$$= (1.23)(16.9) + \frac{1}{4}\log P_{O_2} - pH = 20.6 - pH \qquad \text{(line a)}$$

Since P_{O_2} = 0.21 atm., which is line a in Figure 11-3,

$$2\,H_2O + 2\,e^- = H_2 + 2\,OH^-,\ E^0 = -\,0.83$$

$$p\varepsilon = p\varepsilon^0 - \frac{1}{2}\log P_{H_2}\,[OH]^2 = (16.9)(0.83) - 1/2\ P_{H_2} + pOH$$

Since P_{H_2} = 1 the above can be written as:

$$p\varepsilon = -14 + pOH$$

or

$$p\varepsilon = -\,pH \qquad \text{(line b)}$$

Line (b) can be found in Figure 11-3. The remaining five lines are as follows:

Line 1 — relationship between Cl_2 and HOCl
Line 2 — relationship between Cl_2 and Cl^-
Line 3 — relationship between HOCl and Cl^-
Line 4 — relationship between OCl^- and Cl^-
Line 5 — relationship between HOCl and OCl^-

Line 1. From the equation

$$p\varepsilon = p\varepsilon^0 - \log\frac{[Cl_2]^{\frac{1}{2}}}{[HOCl][H^+]}$$

or

$$p\varepsilon = 27.0 - pH - \log\frac{[Cl_2]^{\frac{1}{2}}}{[HOCl]}$$

Since this ratio is a function of $C_{T,Cl}$ in this region 2 (Cl_2) + (HOCl) >> (OCl^-) + (Cl^-) or $C_{T,Cl} \approx 2\,(Cl_2) + (HOCl) = 10^{-2}$, at Line 1, ($Cl_2$) = (HOCl)

$$3\,(Cl_2) = 10^{-2},\ (Cl_2) = 3.3 \times 10^{-3}$$

or

$$\frac{[\mathrm{Cl}_2]^{\frac{1}{2}}}{[\mathrm{HOCl}]}=\frac{[\mathrm{Cl}_2]^{\frac{1}{2}}}{C_{T,\mathrm{Cl}}-2[\mathrm{Cl}_2]}=\frac{\left(3.3\times10^{-3}\right)^{\frac{1}{2}}}{3.3\times10^{-3}}=5.5$$

Therefore,

$$p\varepsilon = 27.0 - \mathrm{pH} - \log 5.5 = 26.3 - \mathrm{pH} \qquad \text{(line 1)}$$

Line 2. From the equation:

$$p\varepsilon \;=\; 23.5-\frac{1}{2}\log\frac{[\mathrm{Cl}^-]^2}{[\mathrm{Cl}_2]}$$

In this region,

$$2[\mathrm{Cl}_2] + [\mathrm{Cl}^-] >> [\mathrm{HOCl}] + [\mathrm{OCl}^-]$$

or

$$C_{T,\mathrm{Cl}} = 10^{-4} \cong 2[\mathrm{Cl}_2] + [\mathrm{Cl}^-]$$

at

$$[\mathrm{Cl}_2] = [\mathrm{Cl}^-],\ [\mathrm{Cl}_2] = 3.3 \times 10^{-3}$$

$$p\varepsilon = 23.5 - 1/2\log(3.3 \times 10^{-3}) = 24.2 \qquad \text{(line 2)}$$

Line 3. As before,

$$\mathrm{HOCl} + \mathrm{H}^+ + \mathrm{e}^- = 1/2\ \mathrm{Cl}_2 + \mathrm{H_2O} \qquad E^0 = 1.60$$

$$e + 1/2\ \mathrm{Cl}_2 = \mathrm{Cl}^- \qquad E^0 = 1.39$$

$$\mathrm{HOCl} + \mathrm{H}^+ + 2\mathrm{e} = \mathrm{Cl}^- + \mathrm{H_2O} \qquad E^0 = 2.99/2 = 1.50$$

$$p\varepsilon = 25.4 - 1/2 \log \frac{[\mathrm{Cl^-}]}{[\mathrm{HOCl}][\mathrm{H^+}]}$$

$$= 25.4 - 1/2\ \mathrm{pH} - 1/2 \log \frac{[\mathrm{Cl^-}]}{[\mathrm{HOCl}]}$$

$$[\mathrm{Cl^-}] = [\mathrm{HOCl}],\ p\varepsilon = 25.4 - 1/2\ \mathrm{pH} \qquad \text{(line 3)}$$

Line 4. Similarly,

	ΔG^0
$HOCl + H^+ + 2e = Cl^- + H_2O$	$-2F\,(1.50) = -69.2$
$OCl^- + H^+ = HOCl$	$-RT \ln 10^{7.5} = -10.2$
$OCl^- + 2H^+ + 2e = Cl^- + H_2O$	-79.4

Therefore,

$$p\varepsilon = (16.9)\left(\frac{79.4}{2F}\right) - \frac{1}{2}\log\left(\frac{[\mathrm{Cl^-}]}{[\mathrm{OCl^-}][\mathrm{H^+}]^2}\right)$$

$$= 29.1 - \mathrm{pH} - \frac{1}{2}\log\frac{[\mathrm{Cl^-}]}{[\mathrm{OCl^-}]}$$

at

$$[\mathrm{Cl^-}] = [\mathrm{OCl^-}],\ p\varepsilon = 29.1 - \mathrm{pH} \qquad \text{(line 4)}$$

Line 5. When [HOCl⁻] = [OCl⁻]

$$\log\frac{[\mathrm{OCl^-}]}{[\mathrm{HOCl^-}]}=0$$

$$\mathrm{pH}=\mathrm{p}K_a=7.5 \qquad \text{(line 5)}$$

which is a vertical line.

It can be concluded from the above that lines 1 and 2 are concentration dependent.

11.3 Removing Iron and Manganese

Traditional air oxidation in water treatment is used for the removal of divalent forms of iron and manganese for public health purposes. These ions are frequently present in ground waters with low oxygen content. The oxygenation of ferrous ion is known to be catalyzed.

$$2\,\mathrm{Fe^{2+}} + 1/2\,\mathrm{O_2} + 5\,\mathrm{H_2O} = 2\,\mathrm{Fe(OH)_3} + 4\,\mathrm{H^+}$$

The kinetics of the oxidation are second order with respect to pH:

$$\frac{-d[\mathrm{Fe(II)}]}{dt} = k\,[\mathrm{Fe(II)}]\,[\mathrm{OH^-}]^2\,P_{\mathrm{O_2}}$$

Cations such as $Cu^{2+}$and Co^{2+} and anions such as HPO_4^{2-} and $H_3SiO_4^-$ are known to be active catalysts. Copper sulfate is also a good catalyst. Water-bearing species can often enhance oxygenation. For bivalent manganese, autocatalysis plays an important role in oxygenation, and a multiple-step mechanism has been proposed:

$$\mathrm{Mn\,(II)} + \mathrm{O_2} \xrightarrow{\text{(slow)}} \mathrm{MnO_2(s)}$$

$$\mathrm{Mn\,(II)} + \mathrm{MnO_2(s)} \xrightarrow{\text{(fast)}} \mathrm{Mn(II)\,MnO_2(s)}$$

$$\mathrm{Mn(II)\,MnO_2(s)} + \mathrm{O_2} \xrightarrow{\text{(very slow)}} 2\,\mathrm{MnO_2}$$

This actually involves a heterogeneous catalysis with a sorption process. Ninety percent conversion was completed within 50–80 minutes at pH 9.3. Aeration equipment was used with a tray tower over which water was allowed to run.

Actual chlorine oxidation is faster with air oxidation.

$$Fe^{2+} + 1/2\ Cl_2 + 3\ H_2O = Fe(OH)_3 + Cl^- + 3\ H^+$$

$$Mn^{2+} + Cl_2 + 2\ H_2O = MnO_2 + 2\ Cl^- + 4\ H^+$$

In some cases potassium permanganate is used as an oxidant.

$$3\ Fe^{2+} + MnO_4^- + 7\ H_2O = 3\ Fe(OH)_3 + MnO_2 + 5\ H^+$$

$$3\ Mn^{2+} + 2\ MnO_4^- + 2\ H_2O = 5\ MnO_2 + 4\ H^+$$

This step is a very efficient separation if a coagulant is added.

In China, Li and Liu adopted the aeration contact oxidation process using native available minerals as filter media to remove iron and manganese from groundwater with success. Data from this process are shown in Table 11-2. In general, ferrous ions diffused to the surface of the filter media are the controlling factor for the rate of iron removal. The diffusion rate for traditional groundwater filters may be expressed as:

$$-\frac{dC}{dt} = \frac{DS}{\sigma}(C - C') \approx \frac{DS}{\sigma}C \qquad [11\text{-}14]$$

where C is the ferrous concentration in water, C' is the ferrous concentration on the surface of the filter, σ is the thickness of the boundary layer, D is the diffusion constant, and S is the external surface of the filter per unit volume.

$$S = 6\phi(1-\varepsilon)d_m \qquad [11\text{-}15]$$

where ϕ is the shape factor, ε is porosity, and d_m is media grain diameter. Let us define β as the contact catalytic activity coefficient of the filter and L as the thickness of the filter bed:

$$-\frac{dC}{dL} = \beta C \qquad [11\text{-}16]$$

Table 11-2. Iron and Manganese Removal

	Iron (mg/L)		Manganese (mg/L)		pH		D.O. (mg/L)	
	Raw Water	Effluent	Raw Water	Effluent	Raw Water	Aeration	Raw Water	Aeration
Wujiang Machinery Factory	3.6	0.10	1.0	Nil	7.0	7.2	3.3	7.9
Hanjiang Machinery Factory	3.0	0.08	0.28	Nil	7.0	7.7	2.2	7.0
Guangxi Medical Institute	5.0	0.02	1.4	Nil	7.1	7.4	2.8	5.4
Acheng Polyster Fiber Factory	1.5	0.02	1.4	Nil	7.22	7.5	3.0	8.8
Weijian Machinery Factory	1.65	0.03	1.3	0.03	7.1	7.6	—	9.4
Shangrao Army Barracks	3.0	0.02	0.36	Nil	6.5	6.9	—	3.8
Danling Water Plant	9.15	0.00	0.4	Nil	6.7	7.1	—	—

Therefore,

$$\beta = \frac{6D\phi\varepsilon(1-\varepsilon)}{d_m U} \qquad [11\text{-}17]$$

where $t = \varepsilon L/u$ and u is the filtration rate.

When flow is laminar, σ is a constant and β is inversely proportional to the filtration rate. On the other hand, when flow is turbulent, the thickness of the boundary layer, σ, is inversely proportional to the filtration rate. Where a is a proportionality coefficient,

$$\sigma = \frac{a}{u} \qquad [11\text{-}18]$$

then

$$\beta = \frac{6D\phi\varepsilon(1-\varepsilon)}{ad_m} \qquad [11\text{-}19]$$

When the flow is in the transitional region between laminar and turbulent, β is inversely proportional to the pth power of the overflow rate:

$$\beta = \frac{6D\phi\varepsilon(1-\varepsilon)}{bd_m U^p} \qquad [11\text{-}20]$$

where b is a proportionality coefficient and $0 < p < 1$. Usually the flow condition can be determined from the Reynolds number,

$$\mathrm{Re} = \frac{\rho du}{6\mu\phi(1-\varepsilon)}$$

where Re < 2, the flow is laminar.

11.4 REDOX REACTION MECHANISM

Since **free radicals** are involved in reduction and oxidation reactions, a brief review is made here. Either an atom or a functional group bearing an unpaired electron (mostly from broken bonds) can be a free radical. Two types of free radicals are known. One is a stable radical, stabilized by large resonating π-electrons, in which the unpaired electron can freely mix with other electrons in the π-system. The other type is a reactive free radical found in certain locations of an atom. Free radicals bearing this odd electron have the tendency to gain an electron to be paired. The odd electron of a free radical can exhibit an electron spin resonance signal at $g \cong 2$ under a magnetic field. Radicals are not neutral: radical ions, such as the amine cation radical or naphthalene anion radical may be formed. Radical reactions usually occur in a chain sequence and consist of steps, such as initiation, propagation, and termination. **Initiation** can be photolysis of chlorine or thermal homolysis of peroxide:

$$\mathrm{ROOR} \xrightarrow{\Delta} 2\mathrm{RO}\cdot \quad (\text{or } 2\mathrm{RO})$$

(There are two ways of expressing radicals, one with a dot, the other without the dot. They are used interchangeably, as in the following reaction.)

$$\mathrm{Cl_2} \rightarrow 2\mathrm{Cl}\cdot$$

The reaction of peroxide with a transition metal is as follows:

$$\mathrm{ROOR} + M^{(n-1)+} \rightarrow M^{n+} + \mathrm{RO}\cdot + \mathrm{RO}^-$$

The principal **propagation** reactions will involve hydrogen abstraction, addition, fragmentation, and rearrangement. Often a hydrogen donor is involved for the hydrogen abstraction:

$$\mathrm{R} + \mathrm{SH} \rightarrow \mathrm{RH} + \mathrm{S}$$

Also, a radical can be added to a double bond

$$\mathrm{R_3C}\cdot + {=}\mathrm{C}{=}\mathrm{C}{=} \rightarrow {=}\mathrm{C(CR_3)}-\dot{\mathrm{C}}{=}$$

Fragmentation is the reverse of addition, usually through β-scission, and a radical is split

$$\mathrm{R_3}\dot{\mathrm{C}}\mathrm{O} \rightarrow \mathrm{R}\cdot + \mathrm{R_2CO}$$

Table 11-3 is a summary of the reactions of hydroxyl radicals with a number of organic compounds. **Termination** of a free radical is usually accomplished either by coupling with another reactive radical or by dimerization (coupling) itself.

$$\mathrm{R}\cdot + \cdot\mathrm{OH} \rightarrow \mathrm{ROH}$$
$$2\mathrm{R_2CH}\cdot \rightarrow \mathrm{R_2CH}-\mathrm{CHR_2}$$

Coupling also can lead to disproportionation

$$2(\mathrm{CH_3})_2\mathrm{CH} \rightarrow \mathrm{CH_3CH_2CH_3} + \mathrm{CH_3}=\mathrm{CH_2}$$

where a hydrogen atom is exchanged.

Table 11-3. Absolute Rate Constants for Reactions of Hydroxyl Radicals in Aqueous Solution at Room Temperature

Reactant	$k \times 10^{-8}$ ($M^{-1}sec^{-1}$)	Type of Reaction
Acetic acid	0.14	Abstraction
Acetone	0.6	Abstraction
Aniline	50	Addition
Benzene	35	Addition
2-Bromoethanol	4.6	Abstraction
t-Butanol	3	Abstraction
Chloroform	0.08	Abstraction
Dioxane	20	Abstraction
Ethanol	11	Abstraction
Ethylene	41	Addition
Hydroquinone	120	Addition
Methanol	5	Abstraction
Phenol	~70	Addition
2-Propanol	13	Abstraction
Toluene	30	Addition

Metal ion–**catalyzed decomposition** of hydrogen peroxide in aqueous media usually can yield a hydroxyl radical as a powerful oxidant for a number of organic compounds. For example,

$$Fe^{2+} + H_2O_2 \xrightarrow{k_1} Fe^{3+} + \cdot OH + OH^-; \; k_1 = 55\ M^{-1}sec^{-1}$$

$$Fe^{3+} + H_2O_2 \xrightarrow{k_2} Fe^{2+} + H^+ + HO_2$$

$$H_2O_2 + \cdot OH \xrightarrow{k_3} H_2O + HO_2; \; k_3 = 4.5 \times 10^7\ M^{-1}sec^{-1}$$

$$Fe^{2+} + \cdot OH \xrightarrow{k_4} Fe^{3+} + OH^-, \; k_4 = 310 \times 10^8\ M^{-1}sec^{-1}$$

$$RH + \cdot OH \rightarrow H_2O + R$$

The first equation is called the Fenton reaction, and the Fe^{2+} ion can be regenerated in a cyclic fashion. Instead of water, an organic solvent can be used if the metal can be solubilized as a chelate. Di-*t*-butyl peroxide can be used in 2-propanol with Fe^{2+} or Cu^{+} or Co^{2+} to generate the *t*-butyl radical:

$$Fe^{2+} + (Me_3CO\text{-})_2 \rightarrow Me_3CO\cdot + Me_3CO^- + Fe^{3+}$$

The *t*-butyl radical on the α-H of 2-propanol will yield a ketyl radical:

$$Me_3CO^- + Me_2CHOH \rightarrow Me_2\dot{C}OH + Me_3COH$$

The ketyl radical will enhance the decomposition of di-*t*-butyl peroxide.

$$Me_2\dot{C}OH + (Me_3CO-)_2 \rightarrow Me_2C{=}O + Me_3COH + Me_3CO\cdot$$

In this manner, we will have a constant radical concentration for ready use.

Reductive dehalogenation of halogenated aliphatics always results in an intermediate carbon-centered radical:

$$\equiv C-C-X + e \rightarrow \equiv C-C\cdot + X^-$$

Subsequent reaction of this radical intermediate will involve hydrogenolysis — abstraction of a hydrogen atom from a donor

$$\equiv C\text{–}C\cdot \xrightarrow{H\cdot} \equiv C\text{–}CH$$

or via dehydrohalogenation by loss of a hydrogen atom to form a double bond

$$=C\text{–}C\cdot \xrightarrow{-H\cdot} =C{=}C=$$

or a radical complexing by pairing

$$2 \equiv C\text{–}C\cdot \rightarrow \equiv C\text{–}C\text{–}C\text{–}C\equiv$$

or vicinal dehalogenation

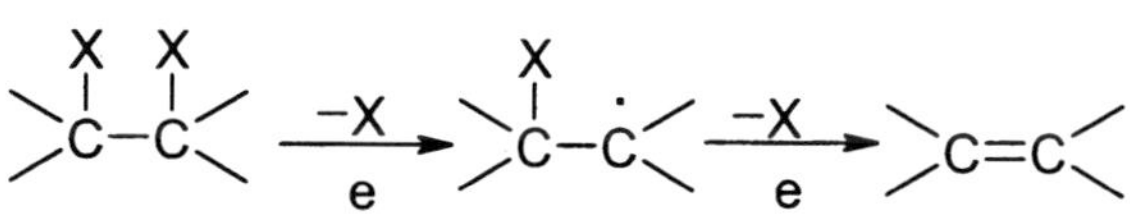

to form a double bond.

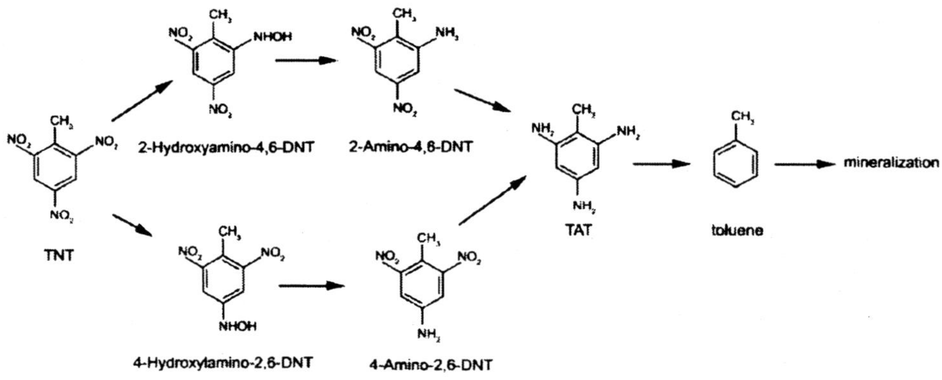

Figure 11-4. Proposed pathway of TNT under anaerobic conditions.

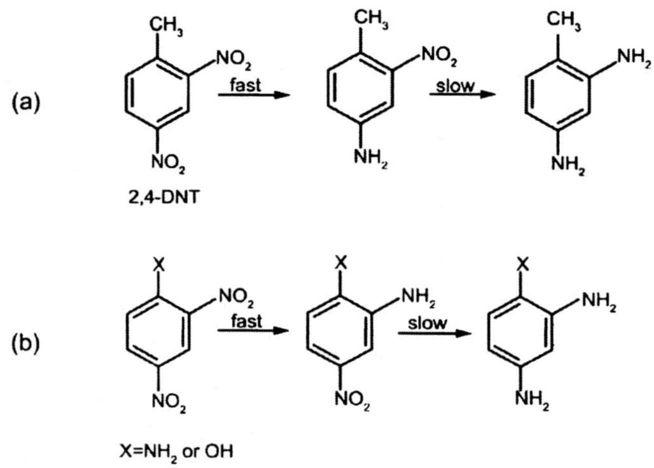

Figure 11-5. The reduction of 2,4-dinitrobenzene derivatives depends on the 1-position substituent: (*a*) methyl — *para* reduction; (*b*) amino or hydroxyl — *ortho* reduction.

Reduction in **nitroaromatics**, such as dinitro-herbicides or dinitro-munitions, is interesting. For TNT reduction, product distribution follows the positions and the availability of particular positions substituted in the ring (Fig. 11-4). The product of reduction from 2,4-dinitrotoluene is different from that of 2,4-dinitroaniline or 2,4-dinitrophenol (see Fig. 11-5). Normally, the 1-position is directed by electron-releasing groups, such as CH_3, NH_2, or OH, and the *para* position can be the subsequent site for reduction, as the *ortho* position is sterically hindered. However, in electron-releasing groups and superconjugation of the methyl roup, the π-electron will force the *ortho*-position substituent out of the way so that the *ortho* position is more easily reduced than the *para* position. This type of reaction is termed **regioselectivity** (see also Fig. 11-6). Reduction of ***N*-nitrosoamine** can occur either at the N-O bond or at the N-N bond. It is difficult to tell which is the case since the end product is the same (Fig. 11-7). In this instance again, the substituent in the anine function has more influence. For example, for *N*-nitroso diphenylamine, which has two aromatic groups, the principal route is N-O cleavage, and in the latter case extensive conjugation helps to stabilize the N-N linkage. We emphasize here again that reductive potential cannot always predict the reaction rate. Structural types are extremely important and can be obtained from the basic reaction mechanism in organic chemistry. The reductive products of **anthraquinone dyes** are a good example. Due to the large accumulation of the residues from heavy use in the textile industry, the hydrogen-bonded anthrone is more persistent in nature than its non–hydrogen-bonded counterpart (Fig. 11-8).

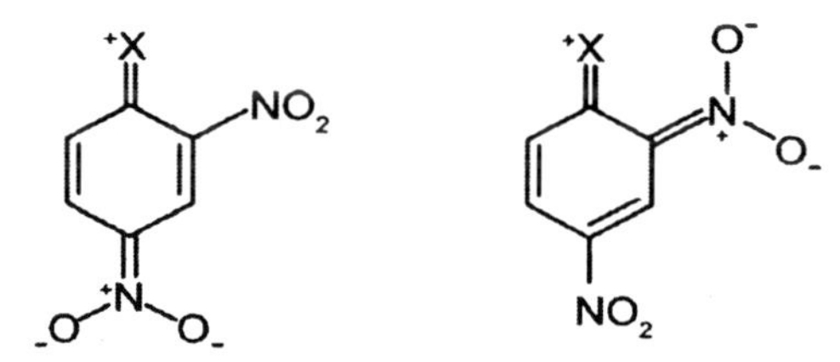

more favorable (a)

less favorable (b)

Figure 11-6. Regioselectivity: (*a*) electron-donating group will activate the *para* position; (*b*) in some cases, either by steric hindrance or by hydrogen-bridge formation, the *ortho* position becomes activated.

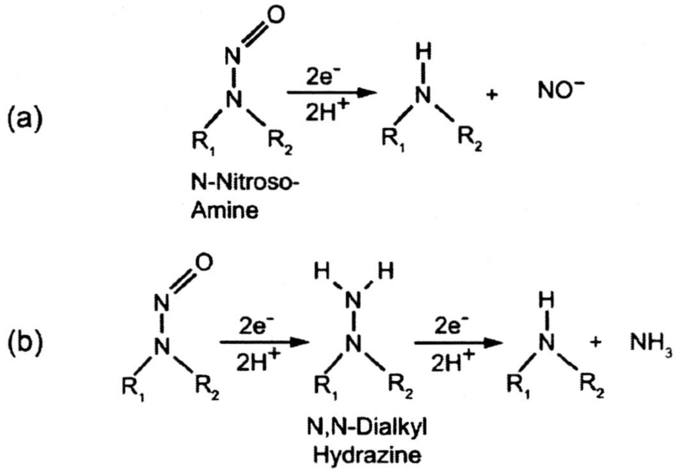

Figure 11-7. Reductivity of *N*-nitrosoamine: (*a*) N-O breaking; (*b*) N-N breaking.

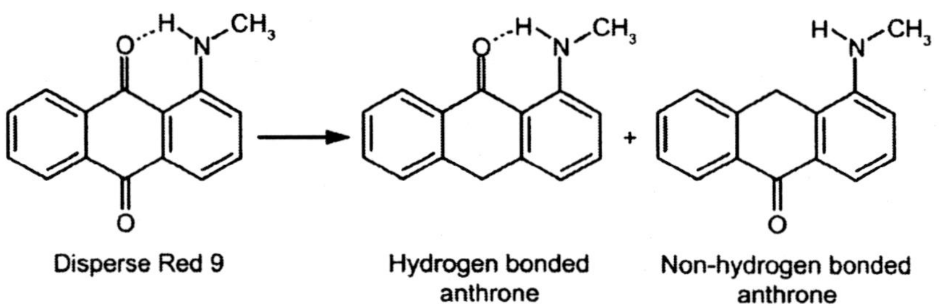

Figure 11-8. Anthraquinone dyes. In the reducing environment hydrogen-bonded anthrone is very stable compared to the non–hydrogen-bonded isomer.

In Chapter 8 we described the mediation of combustion by the free radical chain mechanism. This information pertains to the termination (or protection) of the free radical reaction by the use of a stable radical. The function of an antioxidant, such as butylated hyroxytoluene (BHT), vitamin E, or ascorbic acid (vitamin C), is to intercept the peroxy radicals or to interrupt any autocatalytic chain reactions.

The Redox mechanism also depends on the environment in which the reactants reside. The solid surface area in the presence of organometallics (mediators), in raindrops or aerosols, or in the presence of hydrogen donors will have an influence about the reactivity. The following sections address the particular areas of geochemistry and photochemistry.

11.4.1 Geochemical Redox Processes in Soil Environment

Without a doubt, the redox process is essential in soil. The O_2-H_2O pair is the redox pair controlling reactions in the aerated solution, so the reaeration of the anoxic soils drives reduced species such as Fe^{2+} toward the oxidized state. As shown in Figure 11-9, large discrepancies exist between the theoretical and measured *Eh* values as indicated by O_2-H_2O, NO_{3-}, -N_2, and MnO_2-Mn^{2+}. This may reflect spatial variability in aeration of soil aggregates, causing slow diffusion in micropores. Moreover, a redox reaction carried out at the mineral surface is important. Most silicate clays and oxides are insulators and semiconductors. As such, they possess a band gap, *Ef*, and this energy is analogous to the redox potential, *Eh*. For example, the solid is able to oxidize xenobiotics adsorbed from the solution and electrons flowing into the solid will cause the Fermi energy to rise. Reduction of the solid continues until the Fermi energy rises enough to balance the redox potential of the solution.

As pointed out by McBride, many soil minerals such as Fe oxide behave as insulators. Others such as Mn oxides are often semiconductors or conductors in which electron flow through the conduction band is possible. For example, Mn oxides can oxidize NO_2^- to NO_3^- without release of Mn^{2+} to the solution. The electrons accepted by the oxide from NO_2^- are delocalized in the entire solid so that no Mn^{2+} is released. In solids, the crystal field of a metal ion is likely to alter its reduction potential. For example, Cu^{2+} reduction by Fe^{2+} is not favorable, yet structurally Fe^{2+} (in the octahedral sheet of 2:1 silicate minerals) is a much more potent oxidant than aqueous Fe^{2+}. This is explained by the high stability of Fe^{3+} relative to Fe^{2+} in octahedral sites of silicates and oxides. It is thought that the Mn oxides of mixed oxidation state of Fe^{3+}-bearing layered silicates have more potent power as oxidants than dissolved O_2, but this cannot be gauged by classical electrochemical concepts. Mineral surfaces may attract the substrate molecules by some other forms of bonding so that the probability of electron transfer is enhanced.

The following are examples of this concept:

1. Benzidine oxidation by birnessite (a Mn oxide) and smectite
2. Phenol oxidation by Mn and Fe oxides
3. Cr^{3+} oxidation by Mn oxides

Many metal-catalyzed reactions have to go through the ion radical for redox system. Figure 11-10 shows the scheme for benzidine oxidation. Another example is the oxidative degradation of aromatic amines and phenols by Al^{3+}, as shown in Figure 11-11, which also includes the intermediate species of radical ion.

Most adsorption characteristics of pesticides can be explained based on the nature of the functional groups of the pesticide molecules and the soil components, such as clays, Fe and Al oxides, and humus. These components can be basic compounds, such as S-triazine herbicides (Fig. 11-12), acidic pesticides (Table 11-4), and ampholytic substance such as glyphosate (Roundup):

$$\mathrm{COOC-CH_2NHCH_2-P(=O)(-OH)_2}$$

The mixed acid-base nature of a molecule makes it easy to form a chelate with a surface metal ion.

11.4.2 One Electron Transfer Process

On the other hand, the reduction process in soils often involves a one-electron transfer process. The reducing agents, such as Fe^{2+}, sulfide, and phenols, that accumulate in soils under anaerobic conditions may react with xenobiotics, for example,

$$>\mathrm{C-C(-X)}< + \mathrm{Fe^{2+}} \xrightarrow{\text{slow}} >\mathrm{C-C}\cdot< + \mathrm{Fe^{3+}} + \mathrm{X^-}$$

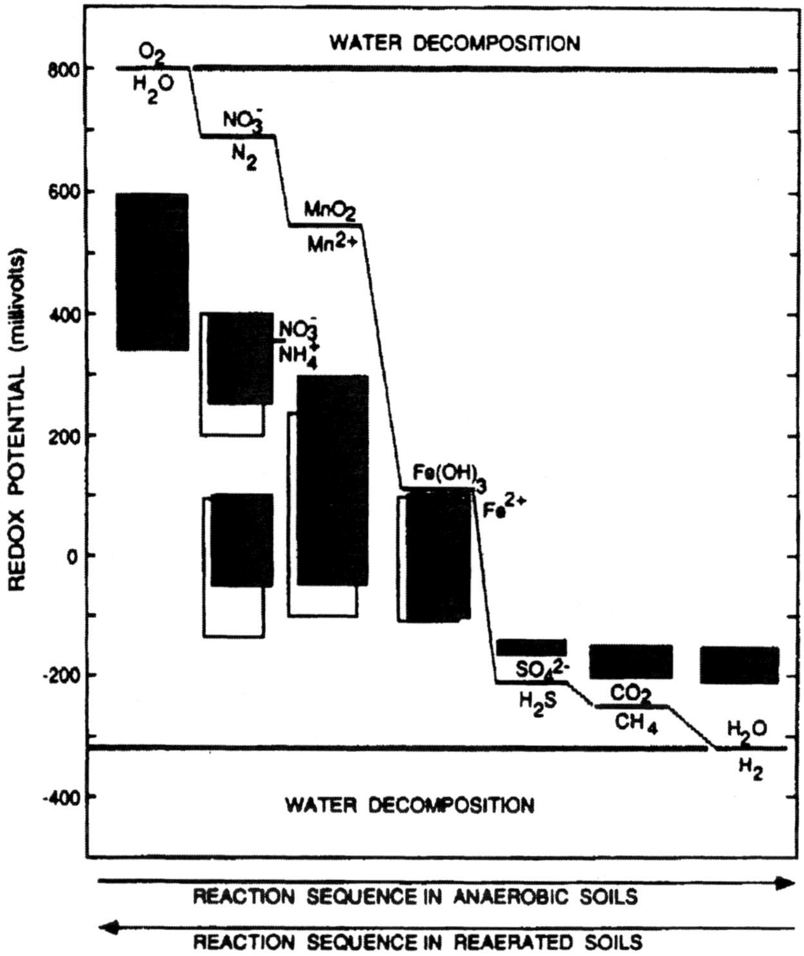

Figure 11-9. The reduction and oxidation sequence in soil solution at pH 7. Theoretical potentials are indicated by solid lines, assuming equal activities of reduced and oxidized species unless otherwise noted (the pressure of H_2 is arbitrarily set at 10^{-3} atm). Measured ranges of soil potentials over which the indicated species react (change concentration) during soil reduction and oxidation are specified by boxes (shaded for reduction, open for oxidation, black for initial appearance of the reduced form during reduction). [Data, in part, from W. H. Patrick and A. Jugsujinda, Sequential reduction and oxidation of inorganic nitrogen, manganese, and iron in flooded soil, *Soil Sci. Soc. Am. J.*, 56,1071-1073 (1992).]

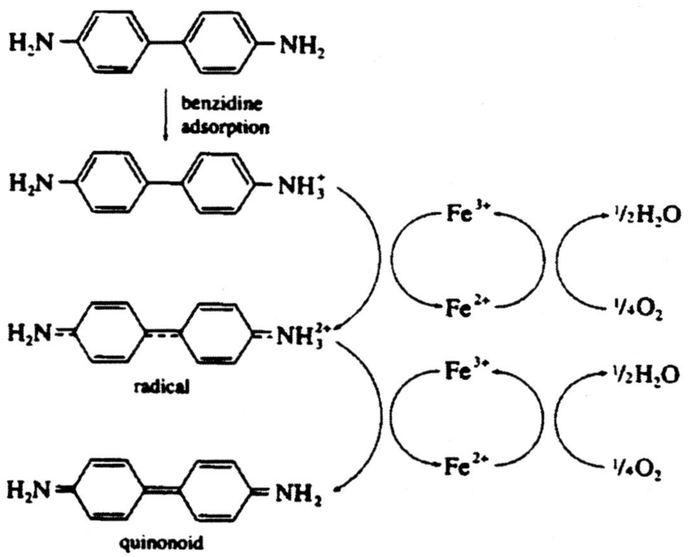

Figure 11-10. Benzidine oxidation by electron transfer.

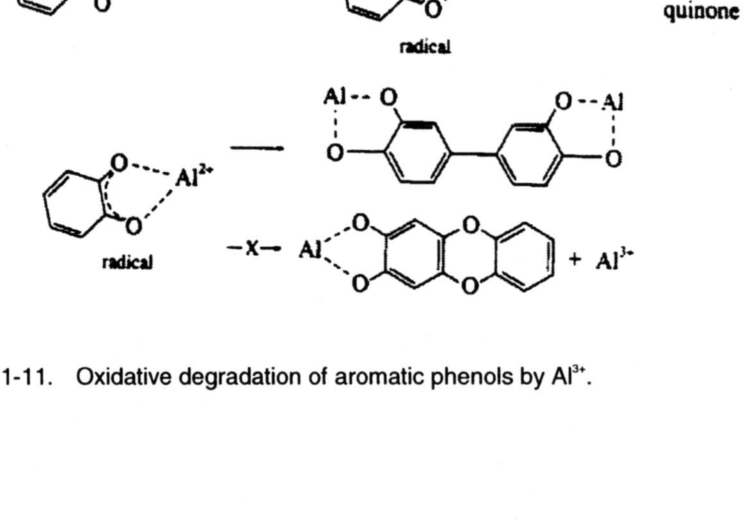

Figure 11-11. Oxidative degradation of aromatic phenols by Al^{3+}.

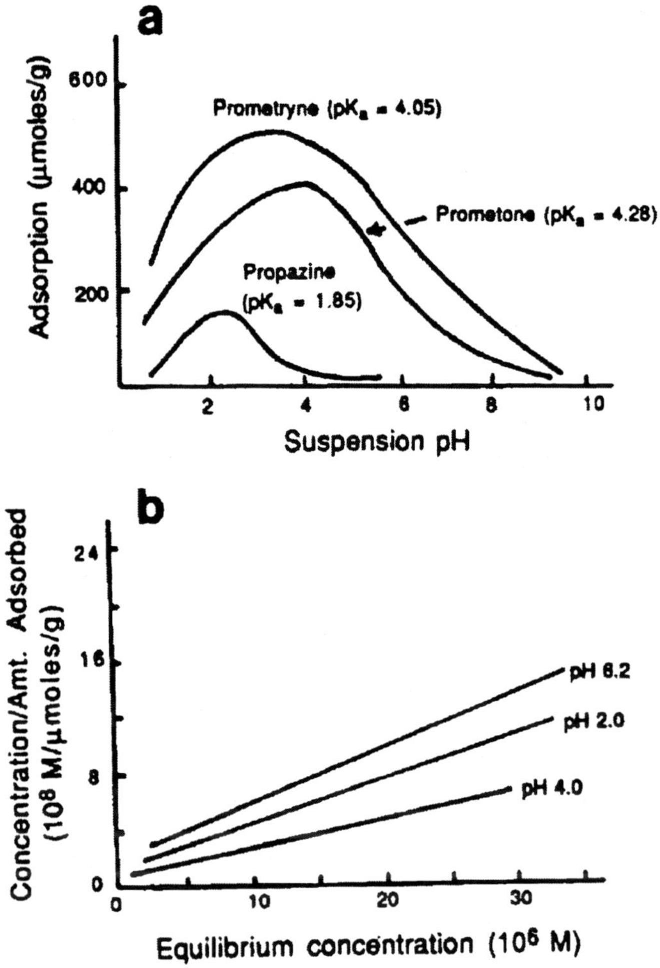

Figure 11-12. Adsorption of s-triazines by Na^{+}-montmorillonite as (*a*) a function of pH; (*b*) a function of s-triazine (prometone) concentration (Adapted from J.B. Weber, "Mechanisms of Adsorption of S-Triazines by Clay Colloids and Factors Affecting Plant Availability," *The Triazine Herbicides*, Springer-Verlag, New York, 1970).

Table 11-4. Structure and Properties of Several Pesticides

	Dinoseb	Picloram	2,4-D	Dicamba
Structure	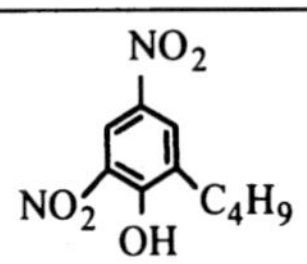	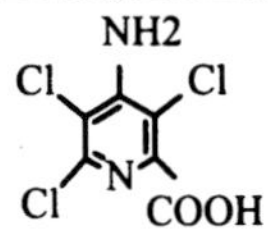	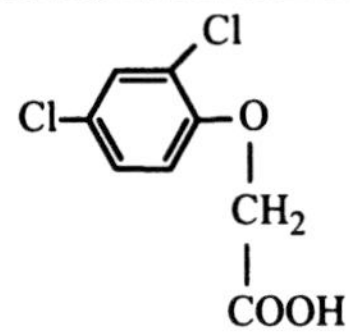	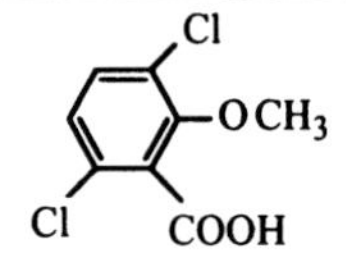
pK_a	4.40	3.6	2.80	1.93
Water solubility (ppm)	45	430	650	4500

Source: S.B. Weed and J.B. Weber, Pesticide-organic matter interactions, *Pesticides in Soil and Water* (W.D. Guenzi, ed.), Soil Science Society of America, Madison, Wisconsin, 1974.

The dehalogenated aliphatic compound is formed by the radical to interact with H^+.

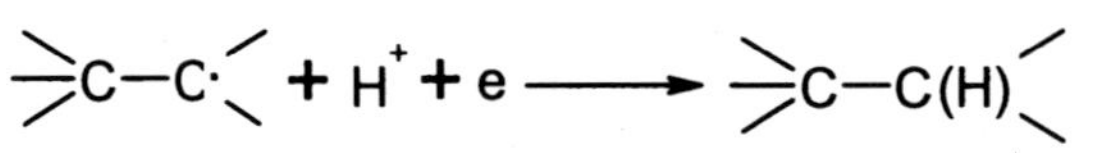

or, if another halogen is bonded to adjacent carbon, an alkene can be formed:

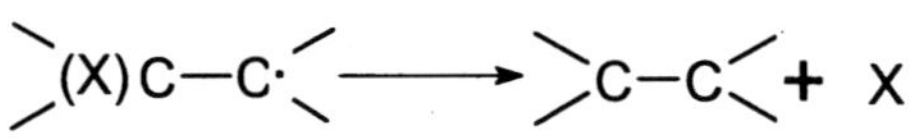

11.4.3 Zero-valent Metal Remediation

Similarly, recent development of zero-valent metals used for contaminant mediation involves a two-electron transfer reaction. For example, in the case of iron, Fe^0, a known reducing agent like Zn or Sn can reduce oxygen, water, or an alkyl chloride, respectively:

$$2Fe^0 + O_2 + 2H_2O \rightarrow 2Fe^{2+} + 4OH^-$$

$$Fe^0 + 2H_2O \rightarrow Fe^{2+} + H_2 + 2OH^-$$

$$Fe^0 + R + H^+ \rightarrow Fe^{2+} + RH + Cl$$

All the above can be related to the corrosion of iron, causing precipitates such as siderite, goethite, and green rust 1, respectively:

$$HCO_3^- + Fe^{2+} \rightarrow FeCO_3\,(S) + H^+$$

$$4Fe^{2+} + O_2 + 6H_2O \rightarrow 4\alpha\text{–}Fe\ OOH + 8H^+$$

$$7Fe(OH)_2 + Fe^{2+} + 2Cl^- + 0{\cdot}5O_2 + (2n+1)\ H_2O \rightarrow 2[3Fe(OH)_2\ Fe(OH)_2Cl{\cdot}nH_2O]$$

Regardless of the corrosion byproducts, it has been found that groundwater containing tricholoroethylene (TCE), percholoroethylene (PCE), dichloroethylene (*cis* and *trans*), chloroform, PCBs, and carbon tetrachloride can be greatly reduced, as seen in Figure 11-13. This reduction process is also useful for the in situ remediation of chromium or technetium. Cr (VI) can be quantatively reduced to Cr (III), since Cr (III) forms precipitate with Fe (III).

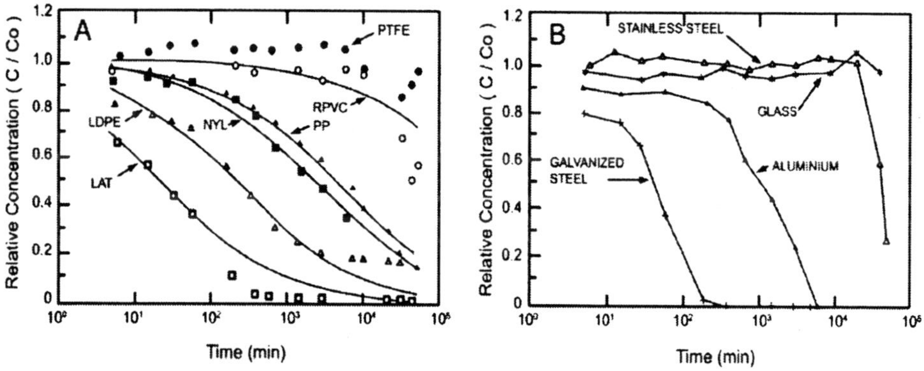

Figure 11-13. Decline in aqueous bromoform concentration over time: (A) in contact with polymer materials; (B) in contact with metal materials. (Adapted from Reynolds et al., 1990.)

$$xCr(III) + (1\text{-}x)\,Fe(III) + 3H_2O \rightarrow (Cr_xFe_{1-x})(OH)_3 + 3H^+$$

For a number of metals, such as mercury, the soil environment provides a number of transformations with different intermediates as shown in Figure 11-14. The soil environment provides major pathways for many metals to be deposited in various forms in the lithosphere as well as the pedosphere.

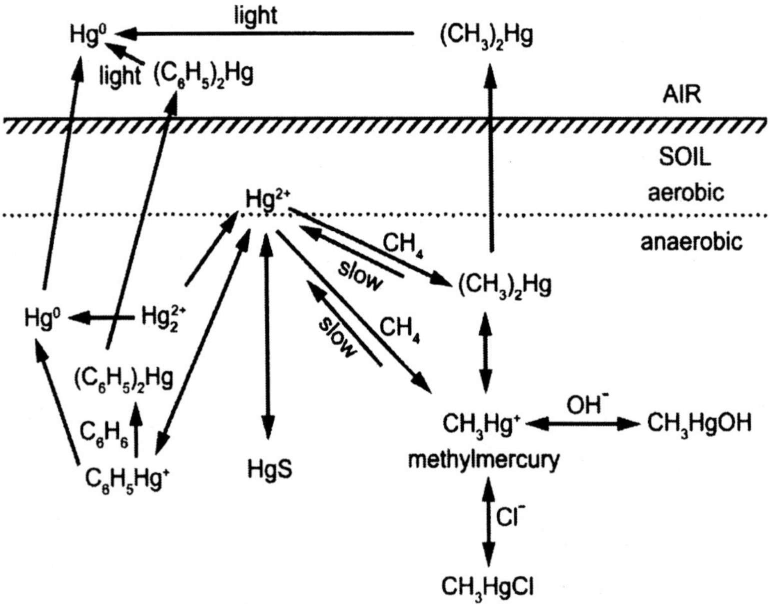

Figure 11-14. Biological and chemical transformations of mercury in the soil. Broken arrows denote the loss of volatile forms of Hg to the atmosphere or the air-filled pores of the soil. (Modified from B. E. Davies, Trace element pollution, Applied Soil Trace Elements, Wiley, New York, 1980.)

11.4.4 Miscellaneous Photochemical Cycles

We have introduced some basic information on photochemistry in Chapter 10. We also learned how to use the term symbol for atoms. For example, for an O atom, stability increases from $^1S_0$ to $^1D_2$ to $^3P_2$. Usually, $O(^3P_2)$ is the ground state, and $O(^1D_2)$ and $O(S_0)$ are the excited states. For an oxygen molecule, the ground state is $O_2(^3\Sigma_g^-)$ (which is a triplet), a higher state $O_2(^1\Delta_g)$ and $O_2(^1\Sigma_g^+)$ (which are singlets), and the excited state $O_2(^3\Sigma_u^-)$ (all can be seen in Fig. 11-15). Ordinary O_2 can be photodissociated to atoms at a wavelength of 243 nm (corresponding to bond energy of O–O, which is 5.1 eV).

$$O_2(^3\Sigma_g^-) \xrightarrow{h\nu} O_2(^3\Sigma_u^+) \rightarrow 2O(^3P_2)$$

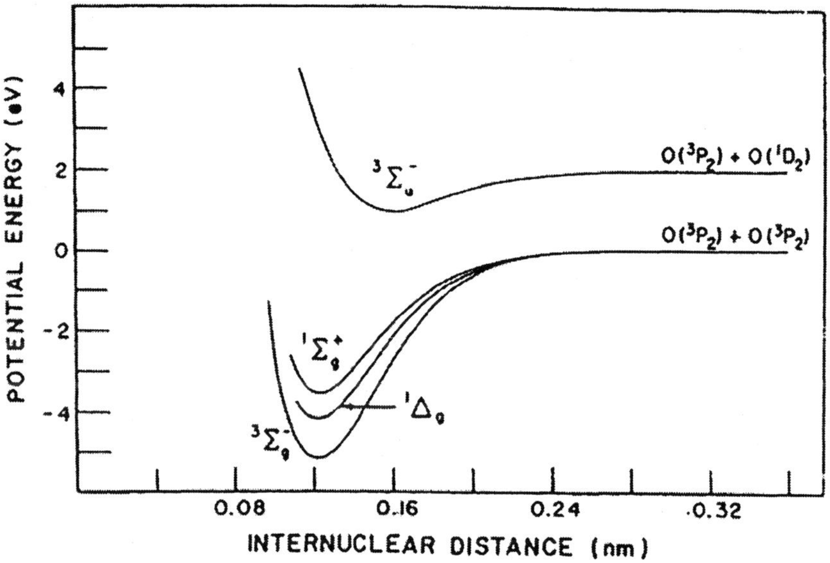

Figure 11-15. Potential energy curves from O_2 (Curves drawn from data of F. R. Gilmore, J. Quant. Spectrosc. Radiat. Transfer 5, 369, 1965)

The excited electronic states of O_2 are involved in the smog cycle: ozone photolysis, ore $O_2(^1\Delta_g)$ "singlet oxygen" 22.5 kcal, and $O_2(^1\Sigma_g^+)$, 37.5 kcal above the ground state, respectively. Singlet oxygen is sometimes represented as $^1O_2$, but is more correctly $O_2(\Delta_g)$, a very reactive species. It is spin-paired and very easily interacts with the π-system (e.g., olefins). In the atmosphere singlet, oxygen originates from excited oxygen atoms via energy transfer

$$O(^1D_2) + O_2(^3\Sigma_g^-) \rightarrow O(^3P_3) + O_2(^1\Sigma_g^+) \xrightarrow{M} O_2(^1\Delta_g)$$

The lifetime of $^1\Delta_g$ is 60 minutes, compared to that of $^1\Sigma_g^+$, which is 12 seconds. $^1\Sigma_g^+$ is easily prepared in the laboratory as follows:

$$Cl_2 + H_2O_2 \rightarrow 2HCl + O_2(^1\Delta_g)$$

When anthracene reacts with singlet oxygen, resulting in the endoperoxide bridge at the 9,10-position, the reaction is of the Diels-Alder type (4 + 2 cycloaddition). Many natural peroxides can be formed from Diels-Alder reversible reactions with singlet oxygen, for example, ascridol can be formed from α-terpinene (Fig. 11-16). The formation of the ozone in the atmosphere occurs as follows:

$$O(^3P_2) + O_2(^3\Sigma_g^-) + M \rightarrow O_3 + M$$

$$O_3 \xrightarrow{h\nu} O(^1D_2) + O_2(^1\Delta_g)$$

Therefore, ozone is a powerful oxidant. Dissociation of water is:

$$H_2O \rightarrow H_2 + O(^3P_2) + O(^1D_2)$$

and

$$H_2O \rightarrow H(^2S_{1/2}) + HO(^2\Pi) + HO(^2\Sigma^+)$$

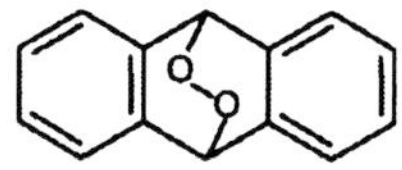

(a)

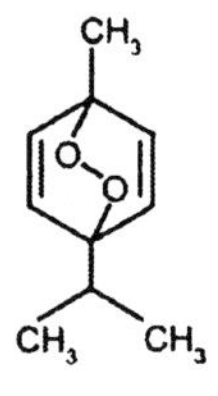

(b)

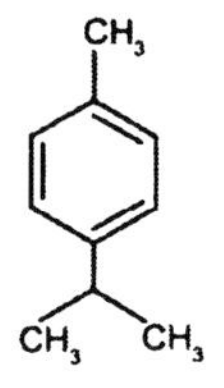

(c)

Figure 11-16. (*a*) Anthracene endoperoxide; (*b*) ascaridole, a natural peroxide; (*c*) 2-terpinene, the diene precusor of (b).

The application of photochemistry to the above reaction is not practical for the water split off. The longest possible wavelength for the direct water split off is 245 nm (116 kcal), which is below the stratosphere ozone layer and not available in the biosphere. In order to combat the fact that most free radicals cannot be stored, a photo oxidation is required in which intermediate free radicals are limited. For example, cerous ions change to ceric ions in aqueous perchloric acid solution by photooxidation:

$$Ce^{3+} + H_2O \xrightarrow{h\nu} CeOH^{3+} + H$$

Here the hydrogen atom is a strong oxidizing agent:

$$H + Ce^{3+} + H_2O \rightarrow CeOH^{3+} + H_2$$

Thus one photon absorbed is equivalent to the two molecules of water reduced and the oxidation of two cerous ions. Ceric ions are known to liberate oxygen from aqueous systems:

$$2CeOH^{3+} \rightarrow H_2O + 2Ce^{3+} + \frac{1}{2}O_2$$

Therefore, **water splitting** is possible.

$$H_2O \xrightarrow[Ce^{3+}]{h\nu} H_2 + \frac{1}{2}O_2$$

The d-orbitals of transition metals are ready to donate to O_2 and form a dioxygen adduct:

$$M^{2+} + O_2 \rightarrow (MO_2)^{2+}$$

This adduct will attract additional metal ions and form a bionuclear complex

$$(MO_2)^{2+} + M^{2+} \rightarrow (MO_2M)^{4+}$$

Consequently, the O-O bond will cleave to form a pair of ferryl or manganyl ions:

$$(MOOM)^{4+} \rightarrow 2(MO)^{2+}$$

In this way a metal ion with higher valence is formed by the resonance forms:

$$M^{2+}O^0 = M^{3+}O^- = M^{4+}O^{2-}$$

or

$$M^{3+}O^- + H^+ \rightarrow M^{3+} + HO$$

or

$$(MO)^{2+} \rightarrow M^{3+} + O_2^-$$

The superoxide radical can form hydrogen peroxide at low pH:

$$2O_2^- + 2H^+ \rightarrow O_2 + H_2O_2$$

Additional H_2O_2 can also disrupt the proportion of hydroperoxyl radical:

$$O_2^- + H^+ \rightarrow HO_2$$

$$HO_2^- \rightarrow O_2 + H_2O_2$$

When hydrogen peroxide encounters a M^{2+} ion, a further electron is transferred and a hydroxyl radical is formed:

$$M^{2+} + H_2O_2 + H^+ \rightarrow M^{3+} + H_2O + HO$$

Since Fe^{2+} is a trace mineral, under ambient pH conditions M^{3+} ions can be reduced back to M^{2+} by the action of sunlight.

$$M^{3+}OH^- + h\nu \rightarrow M^{2+} + HO$$

The hydroxyl ion can produce more radicals by reacting with oxygen. A photochemical cycle is set up in the aqueous phase, which is analogous to the gas-phase photolysis of NO_2 forming O_3 (and NO), which in turn produce the OH radical found in the smog reaction.

11.5 Advanced Oxidation Process

The advanced oxidation process (AOP) involves the generation of hydroxyl radicals in sufficient quantity to affect water purification, according to the definition given by Glaze, who has made numerous contributions in this area. Various combinations of ozone (O_3), hydrogen peroxide (H_2O_2), and ultraviolet radiation (UV) are used to generate the highly reactive intermediate hydroxyl radical by-product, which is currently thought to be responsible for the high oxidation potential of AOP. Ozone has the highest thermodynamic oxidation potential (2.07 V) among the common oxidants and has been used for more than eight decades in water treatment (see Table 11-5). However, its strong oxidizing strength appears to be very selective in practice, and its reaction rates for the oxidation of many water contaminants are very slow. In combination with H_2O_2, UV, or at high pH, hydroxyl radicals are generated. The rate constants for the reaction of OH radicals with organic species are commonly in the range of 10^9–10^{10} M^{-1} sec^{-1}, which is higher than any other oxidant currently known.

Table 11-5. Standard Potentials of Selected Chemical Disinfectants

Compound	Formula	Potential (V)	Assumed Reactions
Chlorine	Cl_2	1.36	$Cl_2 + 2e^- = 2Cl^-$
Bromine	Br_2	1.09	$Br_2 + 2e^- = 2Br^-$
Iodine	I_2	0.54	$I_2 + 2e^- = 2I^-$
Ozone	O_3	2.07	$O_3 + 2e^- + 2H^+ = O_2 + H_2O$
Chlorine dioxide	ClO_2	1.91[a]	$ClO_2 + 5e^- + 2H_2O = Cl^- + 4OH^-$
		0.95[b]	$ClO_2 + e^- = ClO_2^-$

[a] Complete reaction, traditionally used to describe chlorine dioxide.

[b] Reaction often occurs in water.

Four types of AOP are currently being restudied for applications to wastewater treatment, including the treatment of contaminated groundwater: ozone at high pH (O_3–pH); ozone with hydrogen peroxide (O_3–H_2O_2); ozone with ultraviolet radiation (O_3–UV); and hydrogen peroxide with ultraviolet radiation (H_2O_2–UV).

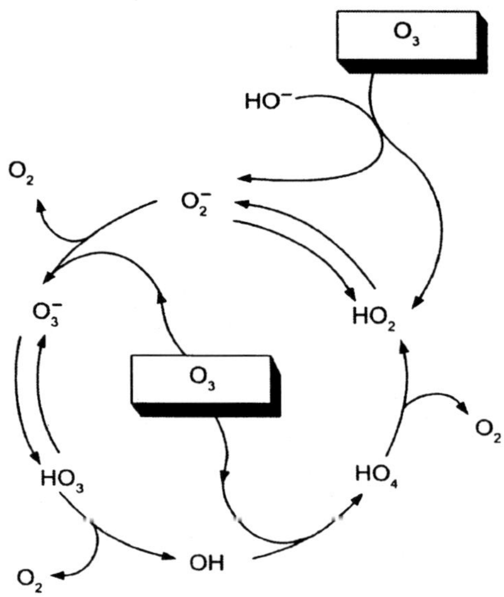

Figure 11-17. Scheme showing the principal species in the decomposition of ozone in pure water initiated by hydroxide ions. [After W. H. Glaze, J. W. Kang, and D. H. Chapin, Ozone Sci Technol., 9(4), 335 (1987).]

11.5.1 Ozonation at Various pH Values

The basic mechanism of ozone decomposition in water involves a complex chain reaction initiated by a variety of substances, including hydroxide, hydroperoxide, and humic substances. A series of single-electron and atom-transfer processes occurs as the ozone decomposes, producing OH radicals as intermediates in the reaction (Fig. 11-17). Hundreds of ozone molecules can degrade in a single initiation step, thereby accounting for the short half-life of ozone in distilled water at pH 7. The cycle can be broken by radical-radical coupling processes and, in the presence of bicarbonate, carbonate, and humic substances, by a trapping of the hydroxyl radical. Bicarbonate ions will trap hydroxyl radicals with such efficiency that the half-life of ozone in distilled water at pH 7 will increase from 10^3 to 10^4 seconds in 0.2 mM HCO_3^-. Bicarbonate ions or humic substances can interfere with the efficiency of ozone oxidation of micromolar concentrations of organic contaminants through competition and scavenging of OH radicals. In natural waters, ozone may react with an organic substance directly (e.g., by reaction of the O_3 molecule with the carbon-carbon bonds of TCE or PCE) or indirectly by reaction of the substance with the OH intermediate from O_3 decomposition. Therefore, when ozone is used at high pH as an oxidant, reaction with another substance occurs by direct and OH pathways.

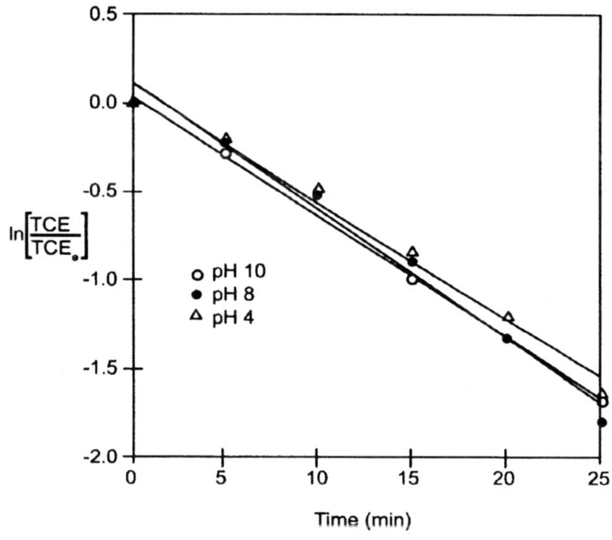

Figure 11-18. Effect of pH on the ozonation of trichloroethylene (TCE) in North Hollywood well No. 26 water. Bicarbonate alkalinity: 200 mg/L; TOC 1.1 mg/L; pH adjusted upward with sodium hydroxide, downward with hydrochloric acid; 70 L CSTR with ozone generated from pure oxygen. [After W. H. Glaze, J. W. Kang, and D. H. Chapin, *Ozone Sci Technol.*, 9(4), 335 (1987).]

The variation in effectiveness of ozone as an oxidant at high pH depends upon the composition of the water being treated. In highly alkaline waters, with characteristics similar to those of some groundwaters, the increased oxidation activity of OH at increased pH is canceled out by the scavenging effects of bicarbonate. At pH values greater than 10.3, carbonate becomes the predominant form of alkalinity and has an affinity for OH, and unless softening of the water precedes the treatment by ozonation, little benefit will be derived by increasing the pH between 4 and 10 (Fig. 11-18). Increasing the pH above 10 may cause no net increase in oxidation.

11.5.2 Ozonation with Ultraviolet Radiation

Decomposition of O_3 in wet air, initiated by UV rays, yields two hydroxyl radicals after an intermediate oxygen atom-forming stage (in ^1D electronic states; for term symbols please see Chapter 8).

$$O_3 + h\nu \rightarrow O\,(^1D) + O_2$$

$$O\,(^1D) + H_2O \rightarrow 2(OH)$$

In water, hydrogen peroxide can form when OH radicals are formed and do not escape from the solvent cage:

$$O_3 + h\nu + H_2O \rightarrow H_2O_2$$

For substances that do not absorb ultraviolet radiation, the O_3/UV system (illustrated in Fig. 11-19) and the O_3/H_2O_2 system are the same, with hydrogen peroxide formed in situ in the case of O_3/UV. However, for compounds that strongly absorb UV, oxidation can occur by direct means, including molecular ozone or UV radiation, or by indirect means, based on the reaction with OH radicals formed as intermediates (Fig. 11-20).

In order to observe the effects of O_3 dosage at a constant UV dose on the pseudo first-order rate constant for the degradation of compounds that absorb UV to varying degrees, lab experiments have been performed using PCE and TCE as substrates. The results, illustrated in Figures 11-21 and 11-22, respectively, indicate that O_3/UV is an oxidation and a photolysis mechanism at the same time. For a substrate like PCE, which can absorb UV rays, large fluxes in UV radiation will accelerate the destruction of the substrate. For photolytically labile substances, such as some pesticides, the direct photolysis constant is so large that little can be gained from using ozone. For substances that do not absorb UV, such as TCE, O_3/UV is impractical, and H_2O_2 should be added directly to the ozonation process. Factors that determine the effectiveness of O_3/UV include intensity and wavelength of UV radiation, ratio of O_3/UV doses, rate of ozonation, concentration and type of substance to be removed, concentration of bicarbonate, and other radical traps.

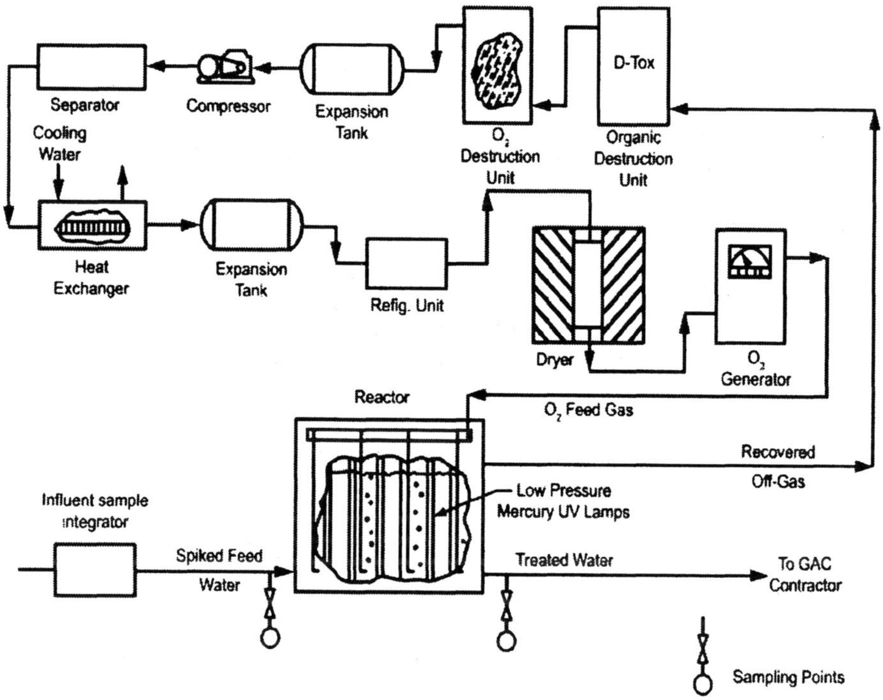

Figure 11-19. Site layout and experimental set-up of UV/ozone oxidation system with oil-gas recycle. (After T. K. Underbrink et al., Berlin International Ozone Conference, 1989).

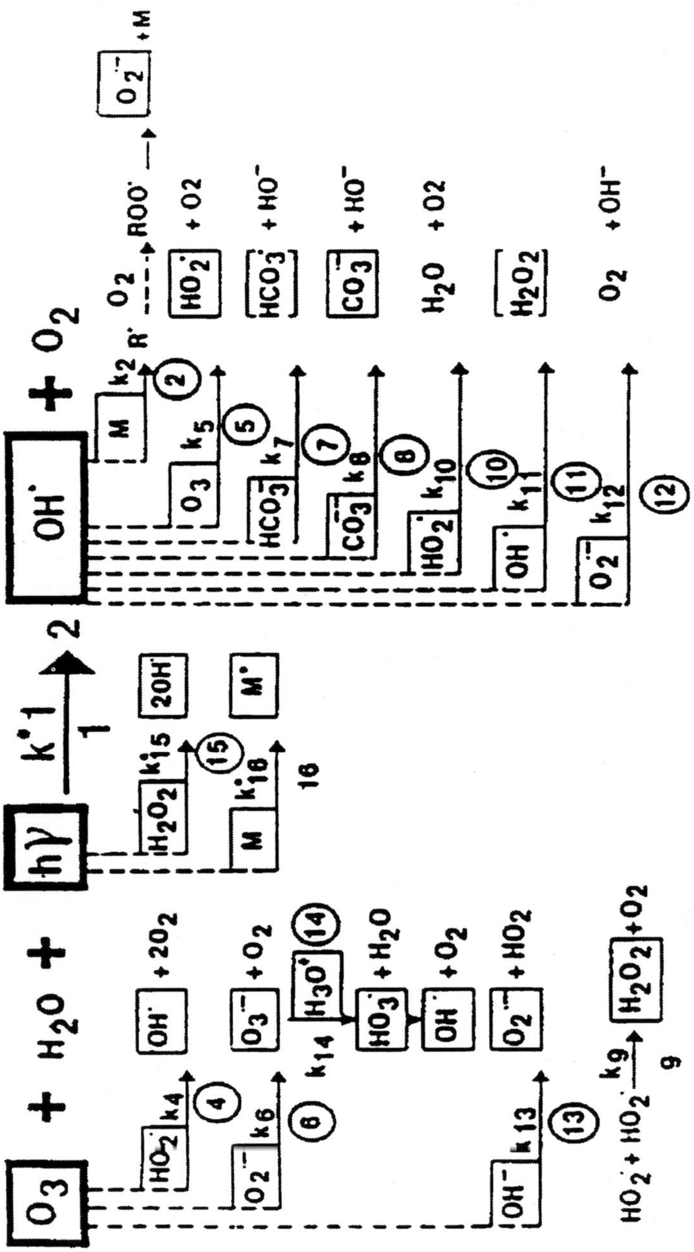

Figure 11-20. O3/UV system: chemical reaction diagram (After Paillard H., R.Brunet, and M.Dore, Ozone Sci. Technol., 9(4), 3911 (1987).

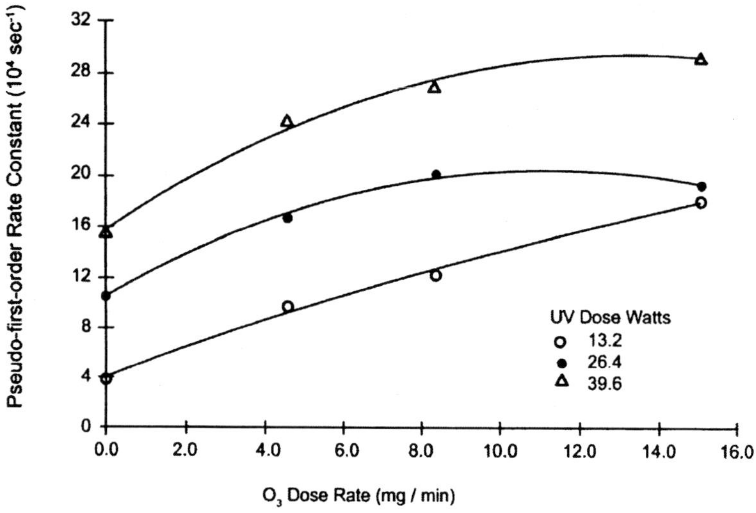

Figure 11-21. Effect of ozone and UV dose rates on the photolytic ozonation of perchloroethylene (PCE) in North Hollywood Well No. 14 water. Rate constants (s-1) multiplied by 104; UV dose in watts at 254 nm calibrated by chemical actinometry. [After W. H. Glaze, J. W. Kang, and D. H. Chapin, *Ozone Sci Technol.*, 9(4), 335 (1987).]

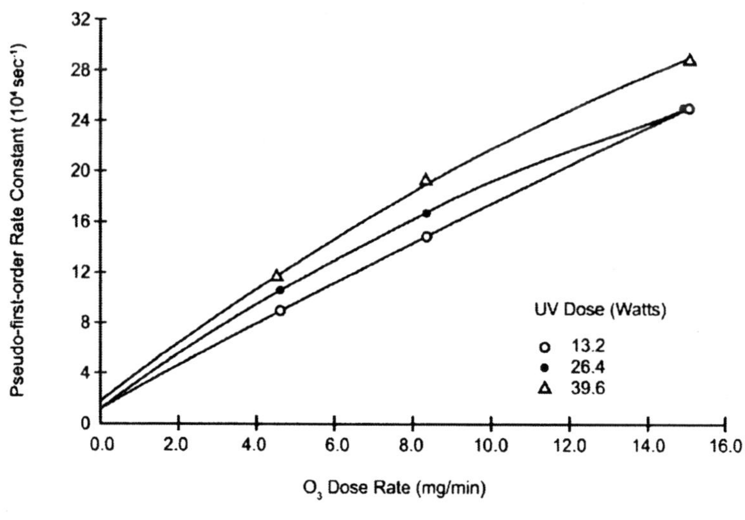

Figure 11-22. Effect of ozone and UV dose rates on the photolytic ozonation of trichloroethylene (TCE) in North Hollywood well water. Rate constants multiplied by 104; UV dose in watts at 254 nm calibrated by chemical actinometry. [After W. H. Glaze, J. W. Kang, and D. H. Chapin, *Ozone Sci Technol.*, 9(4), 335 (1987).]

A pilot-scale study was conducted in 1987 by James M. Montgomery Consulting Engineers (JMM, now Montgomery-Waterson) to evaluate the feasibility of applying two commercially available O_3/UV with PCE, TCE, and carbon tetrachloride (CTC). Results of the pilot tests indicated that the maximum volatile organic compound (VOC) removals attainable are as follows: TCE 99%, PCE 99%, and CTC 80%. An investigation of the applicability of O_3/UV treatment on pesticide-contaminated groundwater and wastewater was conducted by Ultrox in 1987. Pilot-scale results of the treatment of organic phosphate pesticides and organic chloride pesticides under specified operating conditions are listed in Tables 11-6 and 11-7, respectively. The results of the organic phosphate pesticide treatment indicated that Mevinphos was more resistant to oxidation by O_3/UV oxidation than Dimethoate and Disulfoton even at longer residence times, higher ozone dosages, and longer exposure time to UV light. Results of organic chloride pesticides treated by O_3/UV indicated that these compounds will oxidize rapidly with a lower retention time and produce lower analytical results, such as those listed in Table 11-8, which indicate that DBCP was reduced from a concentration of 0.034 mg/L to less than one part per trillion in 60 minutes. Chloroform required a contact time of 90 minutes to reduce the concentration from 750 mg/L to a nondetectable concentration.

Table 11-6. Test No. 1, 30 UV Lamps On, Using a Mixture of Mevinphos, Dimethoate, and Disulfoton, Using Continuous Flow

Operating Conditions			Analysis					
Water Flow Rate (gpm)	Air Flow Rate (scfm)	Ozone Conc. (wt%)	Dichloropropene (ppb)		Endosulfan (ppb)		Parathion (ppb)	
			Inf.	Eff.	Inf.	Eff.	Inf.	Eff.
1	3	2.7	0.200	ND[a]	0.043	ND[a]	2720	ND[a]
0.75	3	2.5	0.200	ND[a]	0.043	ND[a]	2720	ND[a]
0.50	3	2.5	0.200	ND[a]	0.043	ND[a]	2720	ND[a]
Recycle Runs								
1	3	2.6			Not Run			
1	3	2.6	NR	ND[a]	NR	ND[a]	NR	ND[a]

Table 11-7. Test No. 2, 30 UV Lamps On, Using a Mixture of Dichloropropene, Endosulfan, and Parathion, Using Continuous Flow

Operating Conditions			Analysis					
Water Flow Rate (gpm)	Air Flow Rate (scfm)	Ozone Conc. (wt%)	Mevinphos (ppb)		Dimethoate (ppb)		Disulfoton (ppb)	
			Inf.	Eff.	Inf.	Eff.	Inf.	Eff.
1	3	2.3	4200	45	3900	88	5500	38
0.75	3	2.3	3800	61	6600	28	7800	11
0.50	3	2.3	3800	8	6600	ND[a]	7800	ND
Recycle Runs								
1	3	2.3			Not Run			
1.5	3	2.3	NR	ND[a]	NR	ND[a]	NR	ND[a]

[a] Analysis conducted by Montgomery Laboratories.

NR = Not run; ND = nondectable.

Table 11-8. Test No. 4, 30 UV Lamps On, Well Water No. 31A Batch Test (60 gal.)

Reaction Time (min)	Air Flow Rate (scfm)	Ozone Concentration (wt.%)	DBCP (ppb)	Chloroform (ppb)
0	3	2.0	0.034	750
30	3	2.0	0.001	80
60	3	2.0	ND	16
90	3	2.0	ND	ND

ND = Nondetectable.

11.5.3 Ozonation–Peroxidation

The decomposition cycle of O_3 can be initiated by the conjugate base of hydrogen peroxide, the hydroperoxide ion, $HO_2\cdot$:

$$HO_2\cdot + O_3 \xrightarrow{k_1} HO_2 + O_3\cdot$$

Both of the products, ozonide ion, $O_3\cdot$, and hydroperoxide radical, HO_2, can form hydroxyl radicals (Table 11-9). Therefore, 2 OH radicals are formed from each initiation step. When ozone and peroxide are fed into the experimental reactor simultaneously, residuals are generally below 0.1 mg/L, with ozone utilization greater than 90%.

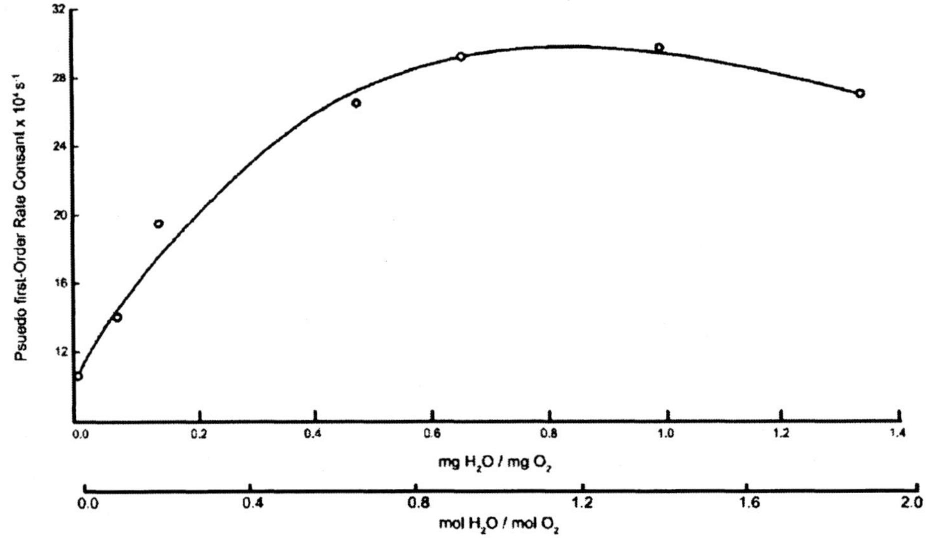

Figure 11-23. Effect of hydrogen peroxide on the pseudo-first-order rate constants of oxidation of TCE. (After W. H. Glaze, J. AWWA, May, 1988.)

Table 11-9. Principal Reactions in the Ozone–Hydrogen Peroxide Process

Reaction		Constant
$H_2O_2 = HO_2^- + H^+$		$K_{PER} = 1.6 \times 10^{-12}$ M
$HO_2^- + O_3 \rightarrow O_3^- + HO_2$	(1)	$k_1 = 2.8 \times 10^6\ M^{-1}s^{-1}$
$HO_2 = H^+ + O_2^-$		$K_{HO2} = 1.6 \times 10^{-5}$ M
$O_2^- + O_3 \rightarrow O_3^- + O_2$	(2)	$k_2 = 1.6 \times 10^9\ M^{-1}s^{-1}$
$O_3^- + H^+ \rightarrow HO_3$	(3)	$k_3 = 5.2 \times 10^{10}\ M^{-1}s^{-1}$
$HO_3 \rightarrow OH + O_2$	(4)	$k_4 = 1.1 \times 10^5\ s^{-1}$
OH + TCE $\rightarrow$ Products (?)	(5a)	$k_{5a} = 4.0 \times 10^9\ M^{-1}s^{-1}$
OH + PCE $\rightarrow$ Products (?)	(5b)	$k_{5b} = 2.3 \times 10^9\ M^{-1}s^{-1}$
$OH^- + O_3 \rightarrow O_2^- + HO_2$	(6)	$k_6 = 70\ M^{-1}s^{-1}$
$OH + HO_2^- \rightarrow OH^- + HO_2$	(7)	$k_7 = 7.5 \times 10^9\ M^{-1}s^{-1}$
$OH + H_2O_2 \rightarrow H_2O + HO_2$	(8)	$k_8 = 2.7 \times 10^7\ M^{-1}s^{-1}$
OH + Si $\rightarrow$ Products (?)	(9)	
$OH + HCO_3^- \rightarrow OH^- + HCO_3$	(10)	$k_{10} = 1.5 \times 10^7\ M^{-1}s^{-1}$
$OH + CO_3^{2-} \rightarrow OH^- + CO_3^-$	(11)	$k_{11} = 4.2 \times 10^8\ M^{-1}s^{-1}$
$HCO_3^- = H^+ + CO_3^{2-}$	(12)	$k_{HCO3} = 5.0 \times 10^{-11}$ M

ΣS_i is the sum of the various matrix substrates that quench hydroxyl radicals.

While H_2O_2 is a promoter of O_3 decomposition and hydroxyl radical formation, at higher H_2O_2:O_3 ratios it can act as an OH ratio on rate constants for TCE degradation, illustrating this phenomenon (Fig. 11-23). The hyperbolic shape of the curve indicates that at a molar ratio of H_2O_2 to O_3 of approximately 1.0 for a groundwater sample containing 300 mg/L alkalinity, the optimum rate of TCE destruction is attained; however, at higher concentrations, H_2O_2 acts as an OH trap, and the rate of TCE destruction declines.

$$H_2O_2 + OH \xrightarrow{k_1} O_2\cdot + H_2O + H^+; \quad k_1 = 2.7 \times 10^7 M^{-1}sec^{-1}$$

$$HO_2\cdot + OH \xrightarrow{k_2} H_2O + O_2\cdot; \quad k_2 = 7.5 \times 10^9\ M^{-1}sec^{-1}$$

Superoxide, $O_2\cdot$, is a promoter of O_3 decomposition and is formed by the preceding reactions; therefore, it is not clear why H_2O_2 inhibits decomposition of TCE at high H_2O_2 concentrations.

Pilot scale studies of H_2O_2/O_3 treatment of hard groundwater contaminated with PCE and TCE have been conducted to examine the feasibility of such a method for groundwater remediation. The study consisted of two series of experiments to determine optimum H_2O_2:O_3 ratios, O_3 dosages, and contact times and to investigate mass-transfer enhancement from liquid phase reaction. For PCE and TCE destruction, the optimum H_2O_2-to-O_3 dose ratio by weight was estimated to be approximately 0.5. However, several factors may influence the optimum ratio, and it is therefore dependent on the type of reactor and the nature of the water to be treated (see Table 11-10).

When the H_2O_2:O_3 ratio was maintained at 0.5 by weight, the O_3 dosage was varied from 4 to 10.9 mg/L, and the contact time was varied from 2.7 to 15 minutes. The highest PCE and TCE removal efficiency was observed to be independent of contact time and only dependent on O_3 dosage. Therefore, the rate of removal of PCE and TCE was correlated to the rate of mass transfer of O_3 from the gas phase to the liquid phase.

According to the penetration theory of mass transfer, when the dissolving gas reacts in the liquid phase, the transfer from the gas phase to the liquid phase is enhanced. The presence of H_2O_2 in the liquid phase has been experimentally proven to significantly increase the transfer efficiency of O_3 from the gas phase to the liquid phase.

Table 11-10. Test Procedures and Analytical Results from Ultrox® Demonstration at South Gate Groundwater Supply Site[a]

Test No.	O_3 Dosage	H_2O_2 Dosage	UV Lamp Complement	PCE Analysis Inf. (ppb)	PCE Analysis Eff.	TCE Analysis Inf. (ppb)	TCE Analysis Eff.	Percent Removal PCE	Percent Removal TCE
1	Med.	None	Full	14	4	3	2.4	71.4	20
2	High	None	Full	20	0.2	ND	ND	99	—
3	High	Med.	None	17	0.4	—	—	97.65	—
4	0	High	None	17	14	—	—	17.6	—
5	0	Med.	One-third	17	0.1	—	—	99.4	—
6	0	0	Full	17.5	4	0.4	0.5	77.14	—
7	Med.	Med.	None	18	0.2	0.5	0.4	93	20
8	0	Med.	One-fourth	16	13	1.3	1.3	18.75	—
9	0	Med.	One-half	16	1.3	1.2	0.3	92	75
10	Low	Low	None	18	2.2	—	—	87.7	—
11	Med.	0	None	—	0.32	—	—	—	—
12	Low	0	None	18	4.4	—	—	75.5	—

[a] Analysis conducted by Ultrox's Analytical Laboratory. All test run at 50 gpm.

ND = Nondetectable.

11.5.4 UV – Peroxidation

Cleavage of hydrogen peroxide is, in principle, the most direct method for the generation of hydroxyl radicals. Photolysis of H_2O_2 results in a quantum yield of two OH radicals per quantum of radiation absorbed:

$$H_2O_2 + h\nu \rightarrow OH\cdot + OH\cdot$$

However, due to the very low molar extinction coefficient of H_2O_2, a very high concentration of H_2O_2 is required for the photochemical reaction. At 254 nm, the molar extinction coefficient is 19.6 $M^{-1}sec^{-1}$ for H_2O_2 as compared to 3,300 $M^{-1}sec^{-1}$ for O_3.

When H_2O_2 was added to CSTR at a rate of 10 mg/min and was photolyzed with three 13-watt mercury arc lamps, TCE was destroyed at a reasonable rate; however, residual H_2O_2 accumulated to unacceptable levels, as plotted in Figure 11-24. A conclusion was reached: UV/H_2O_2 currently is not a viable method for water purification except in special cases.

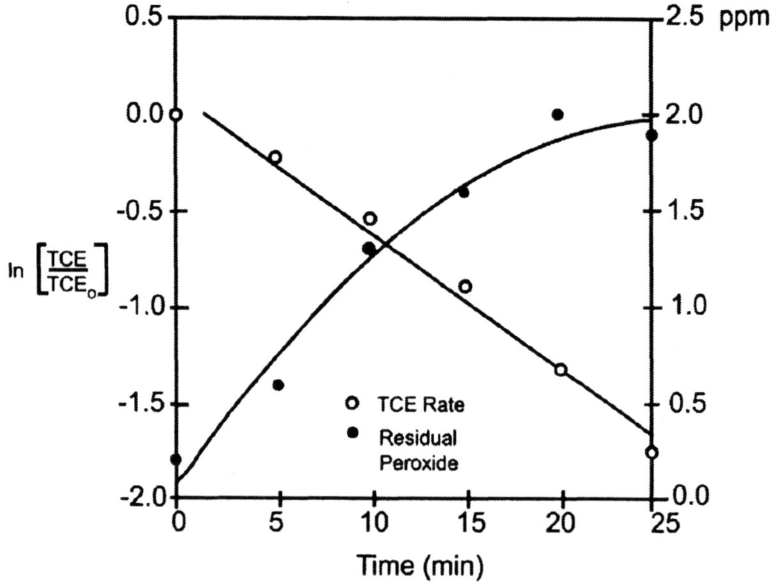

Figure 11-24. Pseudo-first-order plot for destruction of TCE by H2O2/UV process in North Hollywood well No. 14 water. Reaction carried out in 70-L CSTR with feed rate of 10 mg H_2O_2 per minute. Residual peroxide determined by horseradish peroxidase fluorimetry. [After W. H. Glaze, J. W. Kang, and D. H. Chapin, Ozone Sci Technol., 9(4), 335 (1987).]

REFERENCES

11-1. Cooper, P. W., *Explosive Engineering*, VCH, New York, 1996.

11-2. Faust, B., Photochemistry of clouds, fogs, and aerosols, *Environ. Sci. Technol.*, 28, 217A–222A (1994).

11-3. Larson, R. A. and E. J. Weber, *Reaction Mechanisms in Environmental Organic Chemistry*, Lewis, Boca Raton, Florida, 1994.

11-4. Lingane, J. J., *Electromagnetic Chemistry*, 2nd ed., Interscience, New York, 1966.

11-5. McBride, M. B., *Environmental Chemistry of Soils*, Oxford University Press, New York, 1994.

11-6. Wilhemi, A. R. and P. V. Knopp, Wet air oxidation—an alternative to incineration, *Chem. Eng. Prog.*, 46, 46–52 (1979).

11-7. Nardi, L., Use of chromotography in the study of the oxidative decomposition of spent organic solvents from reprocessing plant, *J. Chromatogr.*, 463, 81 (1989).

PROBLEM SET

1. Calculate the equilibrium composition of an acid solution (pH = 2) containing total concentration of iron (Fe_T = 10^{-4}M), this solution is at equilibrium with atmosphere at 25°C.

2. Wastewater containing significant amounts of lead are generated during the production of lead-acid batteries. Lead is normally removed from wastewater by precipitating it in the form of a slightly soluble salt. One method that is commonly used is the addition of trisodium phosphate to form lead phosphate. Lead phosphate is very insoluble under acid conditions, and at pH 3.5 the solubility of lead is approximately 0.15 mg/l. The chemical reaction that describes this process is

$$3Pb + 2Na_3PO_4{\cdot}12H_2O \rightarrow Pb_3(PO_4)_2 + 6Na^+ + 24H_2O$$

Because of the extremely low solubility of lead phosphate it can be assumed that the forward reaction goes to complete. For this situation, how many pounds (dry weight) of lead phosphate sludge are formed per million gallons of wastewater treated, if excess trisodium phosphate is added and the wastewater initially contains 20 mg/l of Pb^{2+}?

3. Construct the *Eh* vs. pH diagram for the Pb-H_2O system under standard conditions.

$$PbO_{(s)} + 2H^+_{(aq)} \rightarrow Pb^{2+}_{(aq)} + H_2O_{(l)} \qquad (K_a)eq.=10^{12.7}$$

$$Pb^{2+}_{(aq)} + 2e^- \rightarrow Pb_{(s)} \qquad E^0= -0.126$$

$$PbO_{(s)} + 4H^+_{(aq)} + 2e^- \rightarrow 2H_2O_{(l)} + Pb^{2+}_{(aq)} \qquad - E^0=+1.455$$

$$PbO_{(s)} + 2H_2O_{(l)} \rightarrow Pb(OH)_3^-{}_{(aq)} + H^+_{(aq)} \qquad (K_a)_{eq.}=10^{-15.4}$$

CHAPTER 12

CORROSION

Corrosion is defined as the destruction or deterioration of a material as it reacts with its aggressive environment. The environment consists largely of moist air, rural and urban polluted air, municipal and industrial wastewater, seawater, deep formation brine, marine drainage water, soils, leachates from solid waste dumps, etc. Such aggressive environments are corrosive to some degree. It is difficult to cite a nonaggressive environment for background studies. The annual cost of corrosion in the United States is estimated to be about $30 billion. Globally, the cost is indeed more expensive. A large percentage of the annual production of steel is used for replacement only.

Corrosion is important in every engineering discipline. Wastewater corrosion, for example, has caused the following problems: (a) a loss in the hydraulic carrying capacity of pipes and fittings (In New England alone, losses of one half of such capacity were observed for the past 30 years.), (b) possible structural failures after heavy rains in southern California, (c) debased quality of water transported, (d) exhaustion of any chlorine residue by corrosion products adsorbed by organic substances, and (e) enhancement of red water problems.

We will begin with an introduction to electrochemistry, followed by the mechanism and cause of corrosion. General principles of corrosion such as the Pourbaix diagram, the Tafel slopes, the corrosion current density and potential, the Evans diagram, and the control technology of corrosion will be discussed.

12.1 ELECTROCHEMISTRY

Let us go back to Chapter 11 on the oxidation and reduction of the Zn-Cu system. Both e and e^- can be used to indicate electron.

$$Zn = Zn^{2+} + 2e$$

$$Cu^{2+} + 2e = Cu$$

Table 12-1. Standard Electrode Potentials (25°C)[a]

Red = Ox + $n^o e^-$	E_ϕ^0 (V vs. H_2 electrode)
$2Cr^{3+} + 7H_2O = Cr_2O_7^{2-} + 14H^+ + 6e^-$	1.33
$2H_2O = O_2 + 4H^+ + 4e^-$	1.23
$Cu = Cu^{2+} + 2e^-$	0.34
$H_2 = 2H^+ + 2e^-$	0.00
$Pb = Pb^{2+} + 2e^-$	–0.13
$Sn = Sn^{2+} + 2e^-$	–0.14
$Ni = Ni^{2+} + 2e^-$	–0.24
$Cd = Cd^{2+} + 2e^-$	–0.40
$Fe = Fe^{2+} + 2e^-$	–0.44
$Zn = Zn^{2+} + 2e^-$	–0.76
$Al = Al^{3+} + 3e^-$	–1.67
$Mg = Mg^{2+} + 2e^-$	–2.37

[a] Solvent: water; unit activity: 1 molal.

Source: IUPAC Convention.

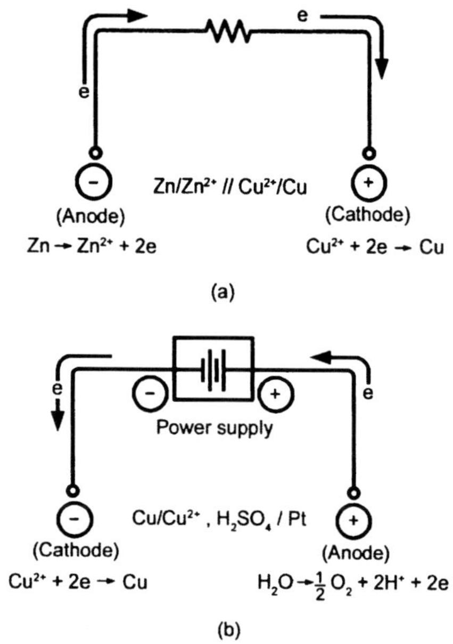

Figure 12-1. Typical electrochemical cells: (a) galvanic; (b) electrolytic.

These are the first type of electrochemical cells, called galvanic cells (see Fig. 12-1). The reaction occurs spontaneously at the electrode when no dissimilar metals of the galvanic series (Table 12-1) are connected externally by a conductor (Fig. 12-1(a)). The primary cell (nonrechargeable) such as the Leclanche Zn–MnO_2 cell, the secondary (rechargeable) cell such as Pb–PbO_2, or the fuel cell H_2-O_2 belong to this class. The second type of electrochemical cell falls into the category of electrolytic cell in which the metals have to be imposed by expending electrical energy (Fig. 12-1(b)). Examples of this type would be electrolytic synthesis (e.g., production of chlorine), electrorefining (e.g., Cu), electroplating (e.g., Ag), or recharge of the storage cell. An electrode on the oxidation side is known as an **anode** and on the reduction side as a **cathode**. The terms anodic oxidation and cathodic reduction are often used. Usually the potential that causes electrons to flow from solution to metal is termed **anodic current**, while the potential causing electrons to flow from metal to solution is called **cathodic current**. Note that the potential of (+) and (-) signs do not always differentiate the electrodes as anode or cathode.

$$Zn + Cu^{2+} = Zn^{2+} + Cu$$

A notation of this cell is:

$$Zn/Zn^{2+} , Cu^{2+}/Cu$$

A slash represents a phase boundary, and a comma or two slashes separates two components in the same phase. Double slashes signify that the junction potential either is small or can be eliminated by a salt bridge. Typical electrochemical cells are illustrated in Fig. 12-1. In the first case, the cell can be written as:

$$Cu/Cu^{2+}, H_2SO_4/Pt \quad \text{(a)}$$

In the second case,

$$Zn/Zn^{2+} // Cu^{2+}/Cu \quad \text{(b)}$$

If a half-reaction serves as a **reference electrode**, it is termed the **standard hydrogen electrode (SHE)** or **normal hydrogen electrode (NHE)**, which contains the components at unit activity.

$$Pt/H_2\ (a{=}1)/H^+\ (a{=}1, \text{aqueous})$$

Potentials of electrodes are often measured with respect to a reference electrode other than the NHE. The most common reference is the **saturated calomel electrode (SCE)**, which is:

$$Hg/HgCl/KCl\ (\text{saturated in water})$$

and has a potential of 0.242 V versus NHE. The Makereth oxygen electrode analytically measures dissolved oxygen, employing a $KHCO_3$ electrolyte partitioned by an oxygen-permeable membrane with:

$$Pb/Pb^{2+} , OH^-/O_2/Ag$$

The electrode potential is a direct indication of which electrode will be dissolved (or corrode, thus behaving as an anode) and which electrode will be protected (thus behaving as a cathode). Usually the magnitude of the value will determine which metal will be more anodic. Therefore, a series called the **electromotive series**, or **galvanic series**, with electrode potentials is listed in

Table 12-1. The electrode potentials are values in volts with reference to the hydrogen electrode, $\boldsymbol{E}^0$=0. The standard electropotential designated by the IUPAC convention has an opposite sign,

$$E_{\phi}{}^{0} = -E^{0} \qquad [12\text{-}1]$$

[Example 12-1] Evaluate the ion concentration of a Sn-Pb cell and calculate its equilibrium constant.

$$Sn/Sn^{2+},\ Pb^{2+}/Pb$$

$$E^0 = E^0{}_{Sn,\,Sn^{2+}} + E^0{}_{Pb^{2+},\,Pb} = 0.140 + (-0.126) = 0.014\ \text{V}$$

$$\Delta G^0 = -nFE^0 = \frac{-(2)(96{,}500)(0.014)}{4184} = -643\ \text{cal}$$

$$\Delta G^0 = -RT\ln K = -(1.987)(298)(2.3)\log K = -643$$

$$K = \frac{a_{Sn^{2+}}}{a_{Pb^{2+}}} = 2.96$$

If we approximate activity by concentration, a state of equilibrium is reached when the tin ion concentration is 2.96 times greater than that of the lead ion. If the tin ion concentration is lower than the lead ion concentration, the electrochemical reaction does not proceed.

Electrolysis is usually defined as the process of electron flow from electrode to solution or from solution to electrode. Since time is involved, there will be a current, *i*, where *i* represents the number of electrons interacting with the electrode or the number of coulombs of electric charge flowing per second. The current in electrolysis can be expressed as:

$$i\,(\text{ampere}) = \frac{dQ\,(\text{coulombs})}{dt\,(\text{secs})}$$

From Faraday's law,

$$\frac{Q\,(\text{coulombs})}{nF\,(\text{coulombs/mole})} = N\,(\text{moles electrolyzed}) \qquad [12\text{-}3]$$

and the electrochemical rate, *v*, is:

$$v\,(\text{moles/sec}) = \frac{dN}{dt} = \frac{i}{nF}$$

because the rate depends on mass transfer and other surface effects in addition to the usual kinetic variables. Often the electrochemical rate is expressed as:

$$v\,(\text{moles sec}^{-1}\,\text{cm}^{-2}) = \frac{i}{nFA} = \frac{j}{nF} \qquad [12\text{-}5]$$

where j is i/A, the current density (A/cm^2).

The corrosion rate for different elements expressed in lb lost/amp-year is included in Table 12-2. In normal cases, the corrosion rate can be expressed in mdd ($mg/dm^2/day$) or $oz/ft^2/day$. In some cases, assuming the surface is even, ipy (in/year) or mpy (mils/yr) may be used for:

$$\text{mpy} = 534\ W/DAT \qquad [12\text{-}6]$$

where W is the weight loss in mg, D is the density of the specimen in g/cm^3, A is the area in mm^2, and T is the exposure time in hours. Corrosion rate of iron coupled to copper can be located in Table 12-3.

Table 12-2. Nomograph for mpy, ipy, ipm, and mdd

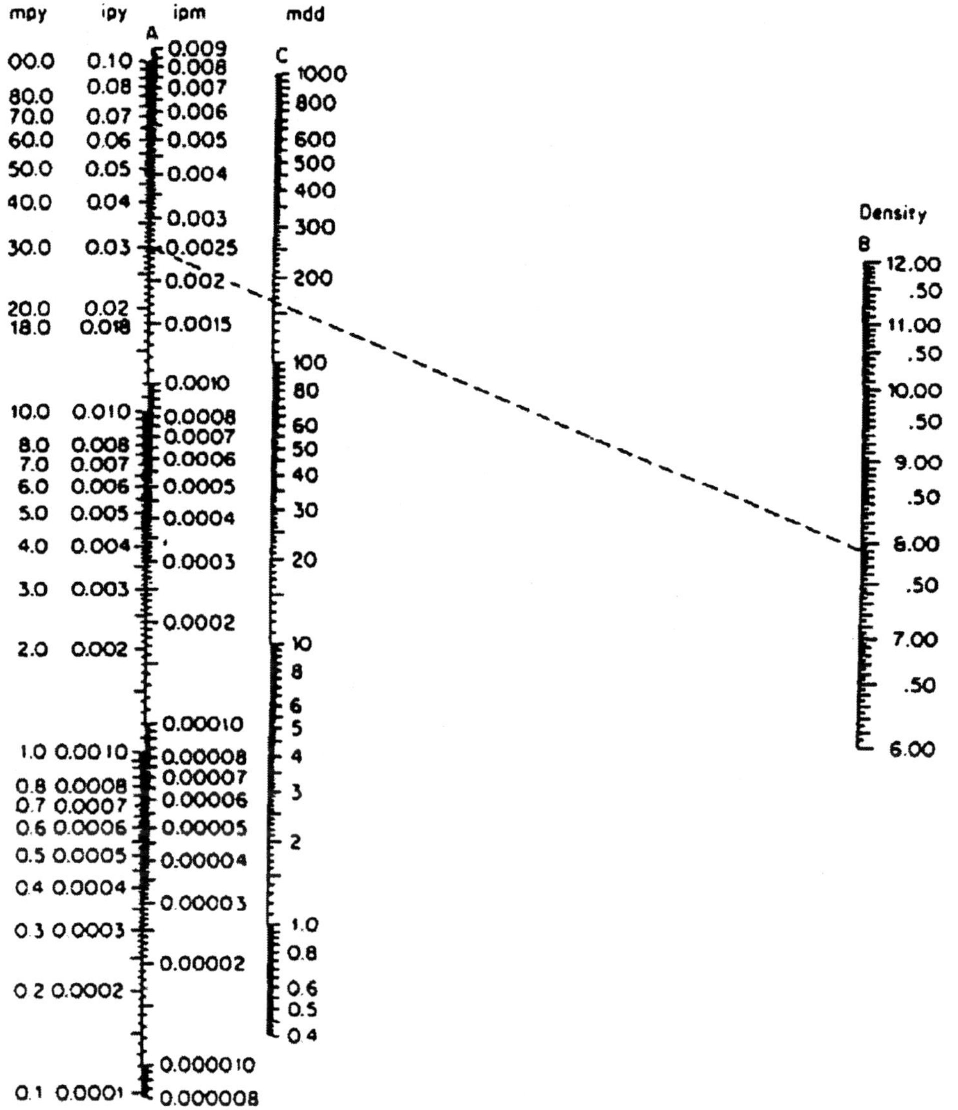

Table 12-3. Corrosion of Iron Coupled to Copper in Aerated, Agitated 3% NaCl (86°F)

Area Ratio, Cu/Fe	Corrosion Rate (mg Fe/day·cm^2 of couple)
1.01	9
2.97	10
5.16	9
8.35	7
11.6	6
18.5	4

12.2 Mechanisms of Corrosion

The factors and mechanisms that cause materials to corrode are numerous and complex. Often the elements required for an electrochemical cell are not obvious. For example, there is no marked difference between anodes and cathodes; the external circuit or the connection between the anode and cathode is hidden. Even the electrolyte or the conducting solution is included in another medium such as soil or stale air. It is difficult to visualize why a metal joint or a bolt remaining in moist air or a pipe buried in marshy and sandy patches of soil begins and continues to corrode. Some of the important mechanisms are discussed here.

12.2.1 The Impressed-Current Cell

In the laboratory, if a voltage is imposed across two electrodes made of the same metal immersed in an electrolyte, an electrochemical cell is established, as shown in Figure 12-2. The metal will dissolve, and electrons will be released at the anode.

$$M^0 = M^{n+} + ne^-$$

At the cathode, hydrogen ions in the solution accept the electrons and yield atomic hydrogen.

$$nH^+ + ne^- = nH^0$$

This atomic hydrogen will form a film on the surface of the cathode. However, if the pH is lowered to less than 4, the molecular hydrogen will then form water.

$$2H^0 + 1/2\ O_2 = H_2O$$

As the hydrogen ion is reduced to atomic hydrogen, the hydroxyl ion that remains after hydrolysis of water can combine with the metal ion released from the anode to form insoluble hydroxides.

$$nOH^- + M^{n+} = M(OH)_n \downarrow$$

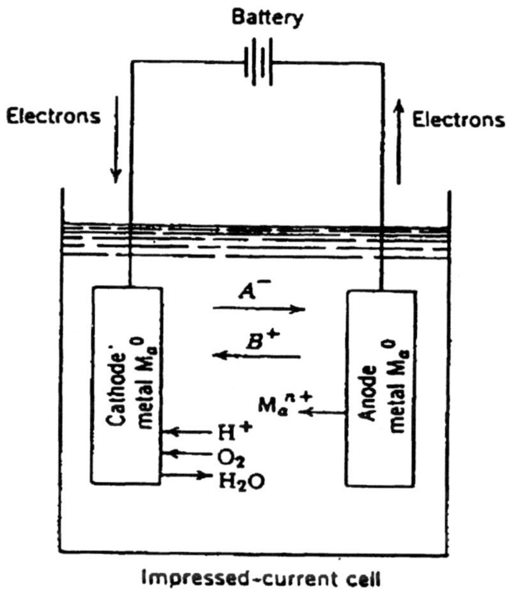

Figure 12-2. First type of electrolytic cell related to corrosion (BA = electrolyte) (same metal).

In actual cases, many buried metal objects corrode at the surface, largely affected by the stray current at the subsurface (see Fig. 12-3). For example, cast-iron water mains become extremely corroded because they inadvertently serve as return conductors for some power supply. Portions of the pipe receiving the current will act as the cathode, whereas places at which current leaves will serve as the anode. Corrosion in this case is external, in contrast to other cases where it is internal. A simple corrosion test is to employ a coupon in a gel contained in a petri dish and cross-wrapped with $K_3Fe(CN)_6$ and phthalocyanine, where the anodic reaction is indicated by white, blue, or red (depending on the species of metal ion) and the cathodic reaction is indicated by the alkalimetry of the indicator.

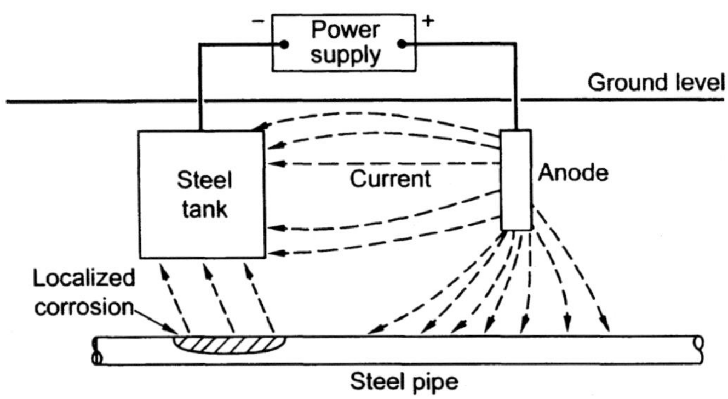

Figure 12-3. Stray currents resulting from cathodic protection.

12.2.2 The Galvanic Cell

If two electrodes of different metals are immersed in a solution of electrolyte and linked by a solid connector, one of the two metals will ionize in a solution state. The electrons will flow through the connector to the other electrode and reduce the hydrogen ions to hydrogen atoms. Such a system, depicted in Figure 12-4, is a **galvanic cell**. Usually, a galvanized pipe is a zinc-coated steel pipe. A galvanized tube is linked by a copper pipe through the common practice of grounding by a neutral wire according to electrical code. Even though the grounding is done in a separate residence, the external circuit can be completed through the neighborhood electricity wiring network. The galvanized end of the pipe will start to corrode even in a different house or building, because the ion transport can be completed through the underground soil to the copper pipe end (in another dwelling). The galvanic series, which usually indicates the order of decreasing tendency to go into solution, is similar to the electromotive force (EMF) series shown in Table 12-1, where the electrode of cell potential is the major factor in determining the occurrence of corrosion. Historically, galvanic corrosion is recorded in a report to a British admiral from the HMS Alarmstrat in 1763. It was found that the iron nails or staples used for fastening the bottom copper protective shield of the ship were completely dissolved in only a single trip, whereas the other iron parts not in contact with copper remained unaffected;

common examples of such corrosion occur in pipe fittings, soldering, and metal impurities in parts of pipes. A metal different from the metal of the main pipe barrel will generate galvanic cells. As the pipe barrel carries water or aqueous wastes acting as electrolytes, the internal surfaces of the pipe will form tubercles, which are insoluble metallic hydroxides that form as a result of corrosion.

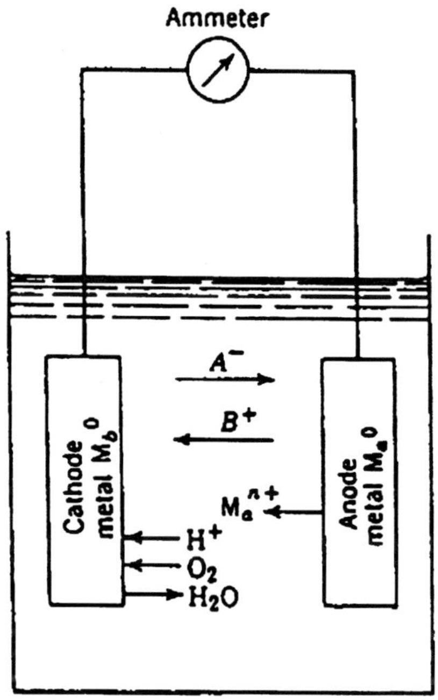

Figure 12-4. Second type of electrolytic cell related to corrosion (BA = electrolyte) (different metals).

12.2.3 Differential Aeration Cells

When two electrodes of the same metal are connected by a solid conductor and immersed in electrolytes, the electrode containing the larger quantity of dissolved oxygen (DO) will become cathodic with respect to the other electrode, which contains less DO. The flow of current in this case (Fig. 12-5) will depend on the removal of polarizing atomic hydrogen film from the electrode that has access to the most oxygen. The electrode that has poor oxygen access will be forced to behave as the anode. Certain areas of the internal surfaces of the metal may have only a limited amount of DO, and consequently they will behave anodically.

Some of these protected areas are pits, corrosion products, barnacle attachments, mill scale, and bacterial slimes. These places certainly are short of DO. In contrast, places where the DO is plentiful are cathodic. A good example would be a metal bolt within two metals (all metals could be the same type) as follows. The metal close to the bolt region has a low DO, whereas the portion of metal plate exposed to the air will have a high DO. Often, corrosion will start on the place underneath the adhesion on a metal surface. Since the hydrogen ion concentration depends on the pH value of the water, variations in pH will have the same effect as DO. Water with a high buffering capacity should be less corrosive according to this reasoning.

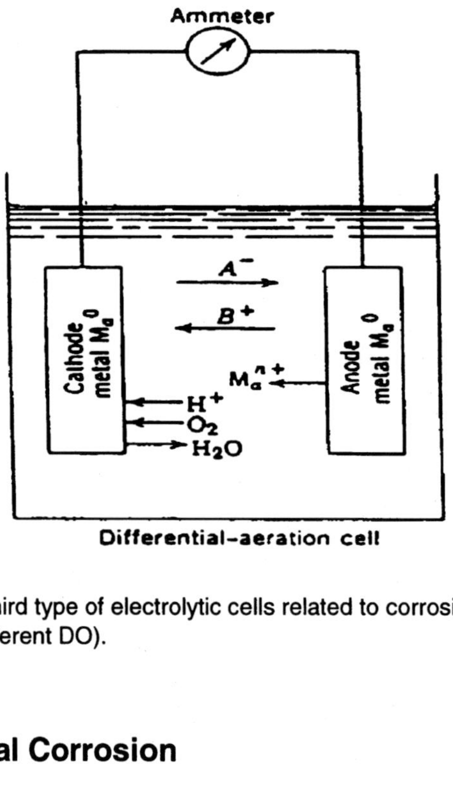

Figure 12-5. Third type of electrolytic cells related to corrosion (BA = electrolyte) (same metal, different DO).

12.2.4 Microbial Corrosion

There are four general types of microbial corrosion: (a) corrosion by sulfate-reducing bacteria; (b) corrosion by acid producers; (c) corrosion due to heavy microbial growth and deposits; and (d) corrosion associated with the breakdown of protective coatings.

Most sulfate reducers are in the genus *Desulfovibrio*, specifically *D. desulfuricans* and *D. vulgaris*. *D. salexigens* is a halophile, which can be found in salt-water habitats. There is also a genus *Desulfotomaculum*; *D. nigrificans* is a thermophilic species that grows at an optimum temperature of 55°C. Although *D. desulfuricans* is an anaerobe, it contains cytochrome c_3. Energy is obtained by the oxidation of molecular hydrogen, and hydrogenase is active in most cultures. The characteristic hydrogen utilization may be a consequence of an endogenous dissimilatory reaction involving the phospho-clastic breakdown of pyruvate, in which hydrogen oxidation is coupled to sulfate or sulfite reduction:

$$SO_4^{2-} + 8H \rightarrow 4H_2O + S^{2-}$$

It is possible for bacteria to remove atomic hydrogen from polarized cathodes, oxidizing it to protons and electrons and utilizing it for reduction. The hydrogen donors used are usually salts of a number of carboxylic acids — lactic, pyruvic, malic, and fumaric — and the end products of growth are acetate and CO_2:

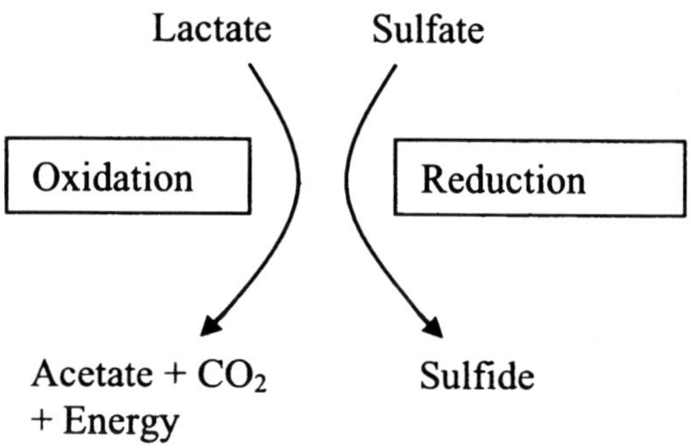

Depolarization of the cathode in the case of a corroding iron pipe that has been buried (Fig. 12-6) occurs as follows:

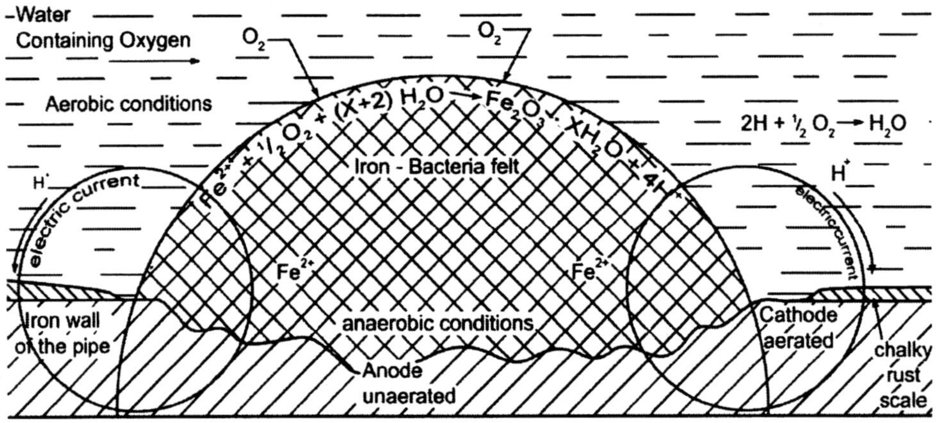

Figure 12-6. Corrosion of buried iron pipe.

Anodic reaction	$4Fe = 4Fe^{2+} + 8e$
Electrolytic dissociation of water	$8H_2O = 8H^+ + 8OH^-$
Cathodic reaction	$8H^+ + 8e = 8H$
Cathode depolarization by bacteria	$SO_4^{2-} + 8H = S^{2-} + 4H_2O$
Corrosion product	$Fe^{2+} + S^{2-} = FeS$
	$3Fe^{2+} + 6OH^- = 3Fe(OH)_2$
Overall reaction	$4Fe + SO_4^{2-} + 4H_2O = 3Fe(OH)_2 + FeS + 2OH^-$

Recent findings indicate that in the presence of excess ferrous ions, the formation of protective sulfide films is prevented. The corrosion rate is high and independent of growth rate. In contrast to the above, anodic depolarization has also been suggested. Corrosion is due to anodic stimulation by sulfide produced by bacteria utilizing electron donors, and the consequent failure in anode potential would increase the EMF for hydrogen production at the cathode. In this connection, the formation of local concentration cells also appears to be a possible factor controlling anaerobic microbial corrosion.

According to Miller's diagram of corrosion in Figure 12-7, a situation favorable to corrosion can be induced by microbes acting either on the fluids contained in metal pipes and vessels or on the protective coatings. In general, sulfur oxidizers can produce sulfuric acid, and the species *Thiobacillus ferrooxidans* can also oxidize ferrous iron. The free acid that is produced can cause serious corrosion of underground metal. It is known that sulfur oxidizers remain viable in a solution of 10–12% H_2SO_4.

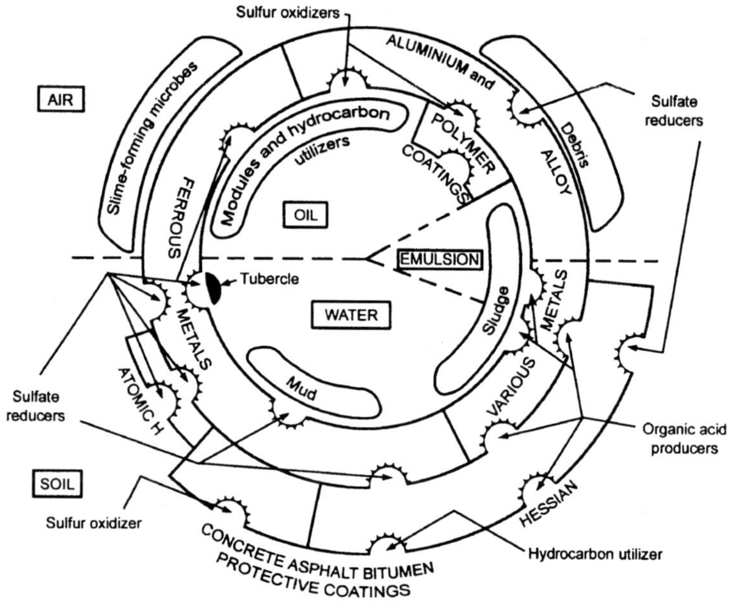

Figure 12-7. Miller's diagram of corrosion.

$$2FeS_2 + 7O_2 + 2H_2O \rightarrow 2FeSO_4 + 2H_2SO_4$$

$$4FeSO_4 + O_2 + 2H_2SO_4 \rightarrow 2Fe_2(SO_4)_3 + 2H_2O$$

$$FeS_2 + Fe_2(SO_4)_3 \rightarrow 3FeSO_4 + 2S$$

$$2S + 3O_2 + 2H_2O \rightarrow 2H_2SO_4$$

Another class of microbes forms large colonial masses or tufts on the substrate on which they grow. These microbes, including algae, fungi, protozoa, and bacteria, are always potential agents of corrosion. Concentration cells can be established from the oxygen and solute gradients present in the ambient water of the end products. These massive growths caused by iron bacteria, together with the ferric hydroxide they produce, can form extensive deposits such as tubercles inside the pipe. Iron bacteria, such as the filamentous forms *Sphaerotilus, Crenothrix,* and the stalk-forming unicellar bacterium *Gallionella*, all can cause this thick deposit in heat exchangers and pipework.

From Miller's diagram one can deduce that bacteria that act on the coating of metals can be anchored on the interface of the coating-metal. For example, cellulose-decomposing microbes damage hessian wraps. After decomposition, fatty acids such as acetic and butyric acid are under anaerobic or partially anaerobic conditions. These acids will further attack the surfaces and cause pitting.

12.2.5 Miscellaneous Mechanisms

There are different forms of corrosion besides those of uniform attachment, for example, crevice corrosion, pitting intergranular corrosion, selective leaching, erosion corrosion, stress corrosion and hydrogen damage. Each of these may have its own mechanisms, many of which are interrelated. The start of pitting is autocatalytic due to a point defect, as shown in Figure 12-8; chloride ions from sodium chloride may aid in the dissolution process and enhance the corrosion. For example, the city of San Francisco does not use aluminum light poles west of Twin Peaks due to the saline content of the ocean water. Stress corrosion may be due to the strain's rupture at the passive film at the tip and the anodic dissolution occurring at the crack tip. Thousands of years ago, Alexander the Great's army discovered that when they repaired the chain bridge over the Euphrates, they had to dispose of older segments of chain because of stress-induced corrosion failure.

Pit growth kinetics can be expressed as

$$i = kt^b \qquad [12\text{-}7]$$

where i is current intensity, t is pit development time, and k is a constant. If the number of pits N per unit area of the species is constant, then $b = 2$ or $b = 3$ (Figure 12-9). A relation can be found:

$$i \sim d^2hN/t \qquad [12\text{-}8]$$

where d is pit diameter and h is depth. If studied with regard to pit geometry and pit numbers, then:

$$d \sim t^{b(d)}$$

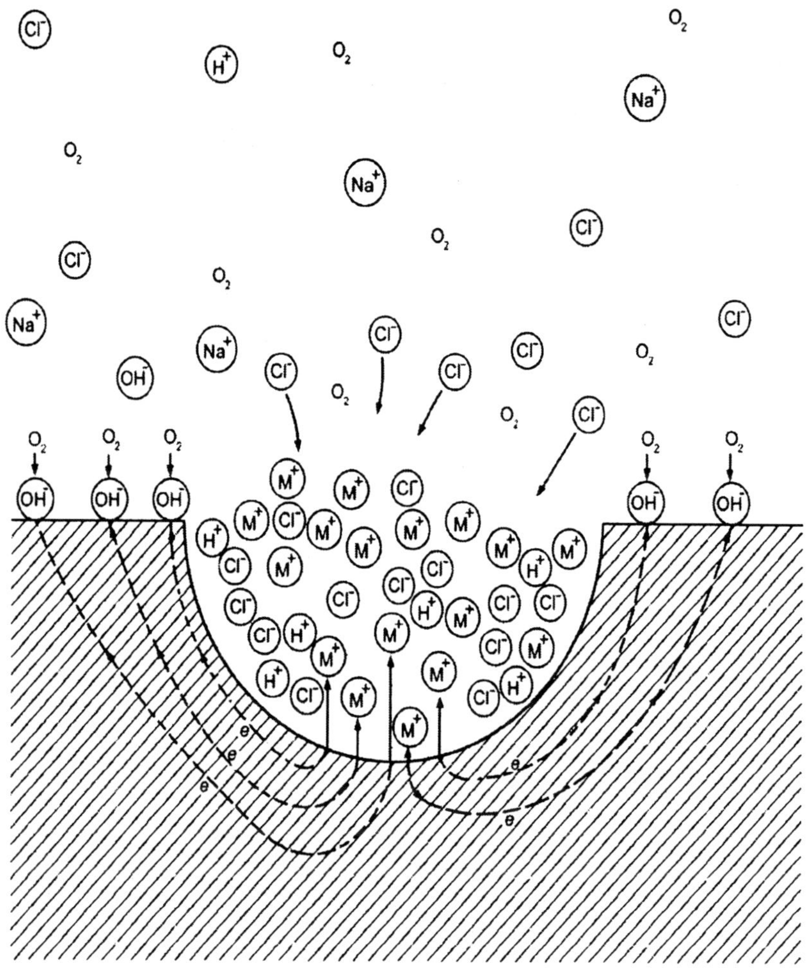

Figure 12-8. Autocatalytic processes occurring in a corrosion pit.

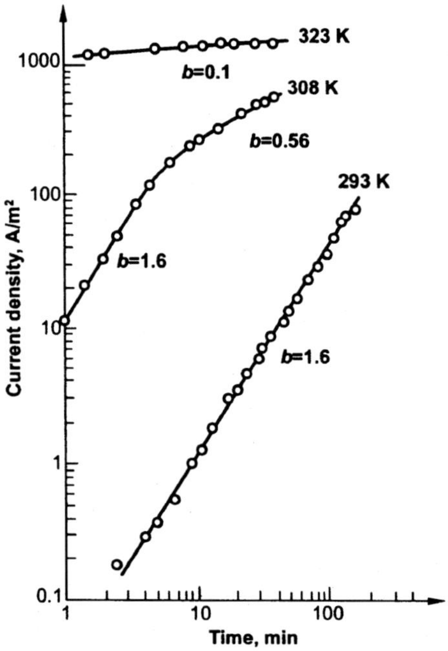

Figure 12-9. Current density versus pit growth time for a 17Cr12Ni2MoTi type steel in 0.5 *M* NaCl + 0.05 *M* H_2SO_4 at 293, 308, and 323 K; E = 880 mV_{NHE}.

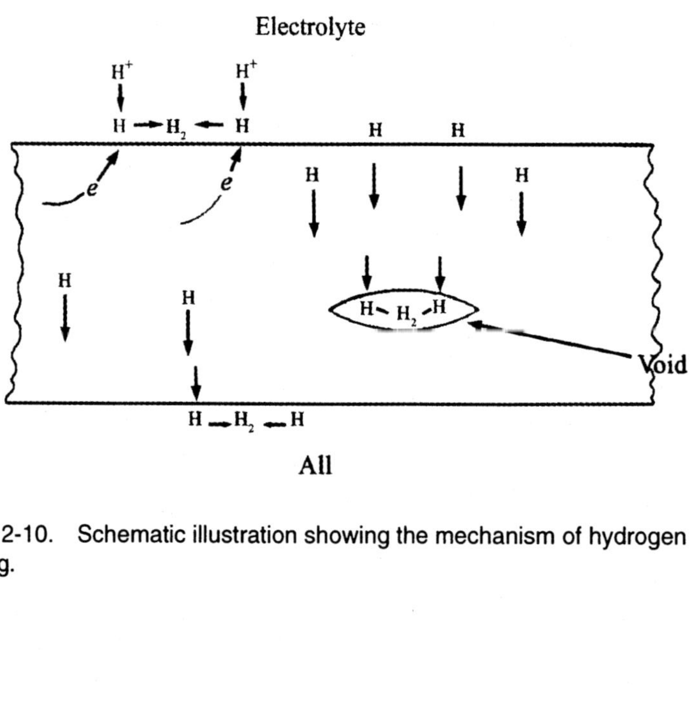

Figure 12-10. Schematic illustration showing the mechanism of hydrogen blistering.

$$h \sim t^{b(h)}$$

$$N \sim t^{b(N)} \qquad [12\text{-}9]$$

where $b(d)$, $b(h)$, and $b(N)$ are values of exponents of time dependence of the diameter, depth, and number of pits, and the relation is

$$b = 2b(d) + b(h) + b(N) - 1 \qquad [12\text{-}10]$$

Table 12-4. Values of the Exponents b, b_d, and b_h for the Experiments Shown in Figure 12-9

	Experimental			Calculated
Temperature, K	b_d	b_h	b	$b = 2b_d + b_h + b_N - 1$
293	0.91	0.75	1.6	1.6
308	0.51	0.50	0.6	0.6
323	0.29	0.35	0.1	0.1

This has been verified experimentally (Table 12-4). Impurities in the grain boundaries often lead to corrosion of alloys. Hydrogen blistering and hydrogen embrittlement may be caused by atomic hydrogen moving inward to the center of the metal as a result of either hydride formation (e.g., martensite in type 304 stainless steel) or a hydrogen gas pocket (hydrogen blistering) as indicated by Figure 12-10. In the petroleum industry, polysulfide ion is used as an inhibitor for storage tanks and hydrogenation equipment.

12.3 Kinetics of Electrode and Transport Control

When current flows between two electrodes, new potential may arise at the electrode surfaces. This type of change of voltage by electrolysis is termed **polarization**. The relation of current versus potential is shown in Figure 12-11, where an ideal polarizable electrode has a greater dependence on potential change. The ideal nonpolarizable electrode does not have this dependence. Current change varies with potential, but there is also a range of potential, which does not alter with the change of current. This is called passivity, and the starting potential value is termed the Flade potential, when $dE_{\Phi}/d_i = 0$. A passive current, i_{pass}, exists when the current is a constant. Different regions of passive, transpassive, transition, and active currents are illustrated in Figure 12-12 when E_{Φ} is plotted versus log i.

For a given metal, a composition map can be constructed of different values of E_{Φ} (or $-E^0$) at different pH values. The diagram of E_{Φ} versus pH is called a **Pourbaix diagram**. Pourbaix diagrams for four metals are shown in Figure 12-13. One can easily determine whether the environment for iron at a location (pH, E_{Φ}) can corrode, make passive, or make immune. These three choices can easily be read off from the diagram of a given metal; e.g., iron and some examples are tabulated for the end-result prediction:

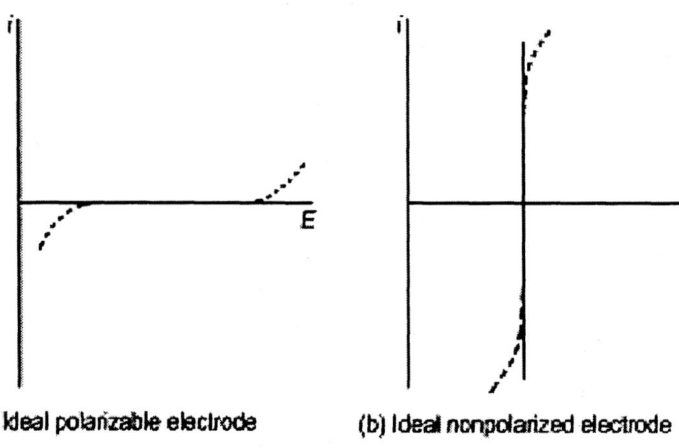

Figure 12-11. Current potential curves for ideal (a) polarized and (b) nonpolarizable electrodes. Broken lines show behavior of actual electrodes that approach the ideal behavior over limited current or potential ranges.

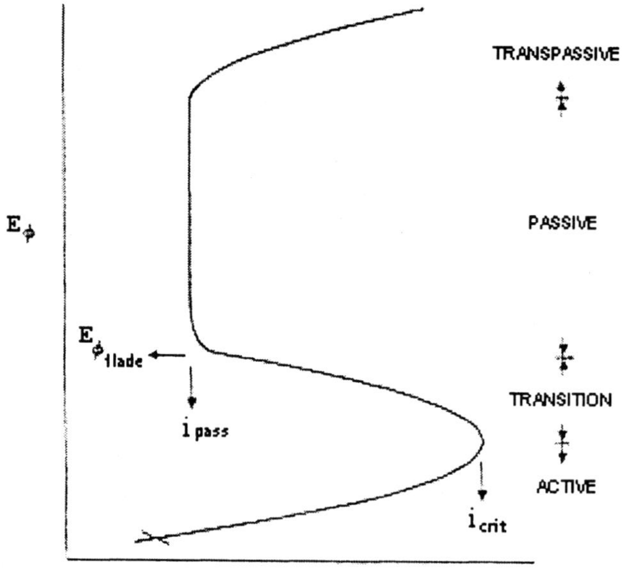

Figure 12-12. Passivity curve.

Table 12-5. Results from Pourbaix diagram

Environment	pH	E_Φ	End result
Distilled H_2O	6.7	–0.480	Corrode
NaCl (1 g/L)	6.8	–0.450	Corrode
K_2CrO_4(1 g/L)	7.3	+0.240	Passive
Tap water	7.2	–0.465	Corrode
Tap water (Fe coupled to Cu)	7.2	–0.450	Fe corrode, Cu immune
Tap water (Fe coupled to Mg)	7.2	–0.930	Fe immune, Mg corrode
Strong base	14	–0.9	Corrode
Strong acid	2	–0.9	Immune

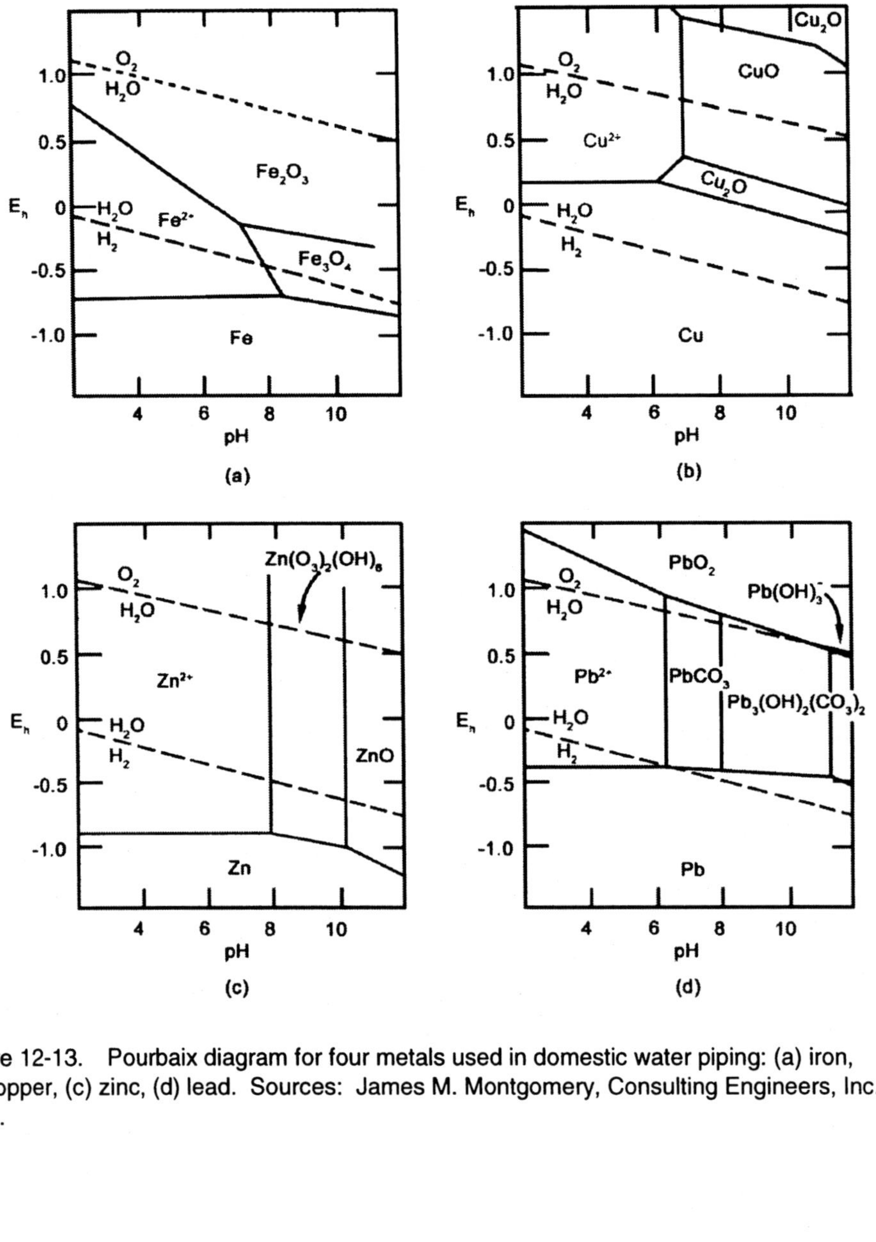

Figure 12-13. Pourbaix diagram for four metals used in domestic water piping: (a) iron, (b) copper, (c) zinc, (d) lead. Sources: James M. Montgomery, Consulting Engineers, Inc. 1985.

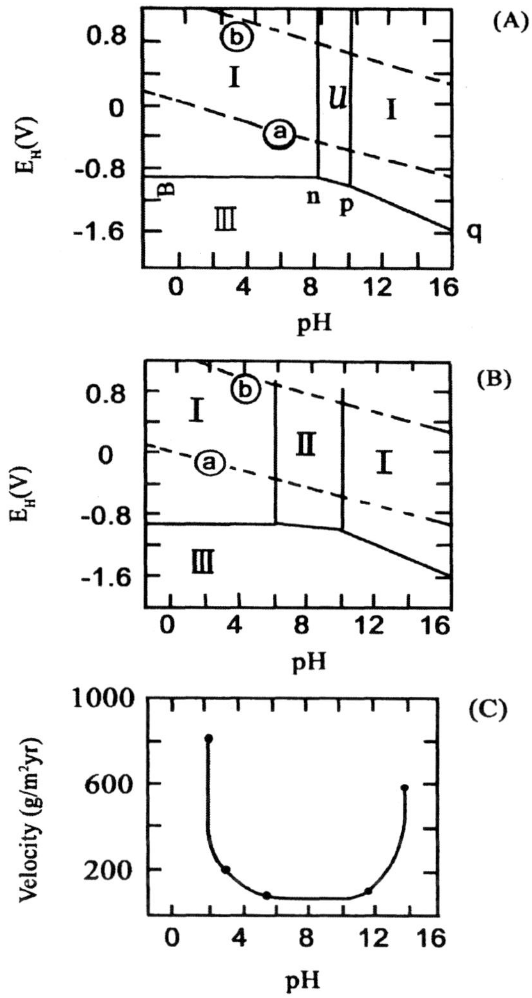

Figure 12-14. Zn - H_2O system (I) corrosion region; (II) passivation region; (III) immunity region.

Figure 12-14 shows the Zn-H_2O system, for which the three important regions of corrosion (I), passivity (II), and immunity (III) are identified. Also, the corrosion rate is given.

The extent of polarization can be measured using overpotential, η, and

$$\eta = E_{\varphi} - E_{\varphi_{eq}} \qquad [12\text{-}11]$$

where $E_{\phi eq}$ represents the equilibrium state of the potential. For example,

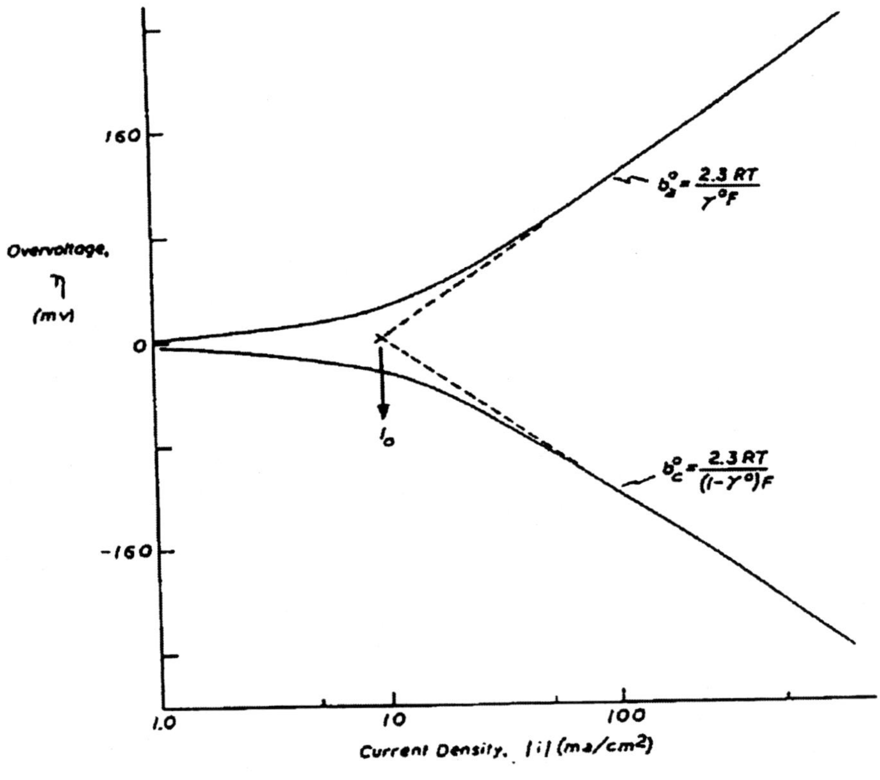

Figure 12-15. Schematic representation of anodic and cathodic polarization curves (the solid line represent the "experimental data"; the dashed line represents the extrapolations of the respective Tafel lines).

$$\eta = E_{\varphi} - E_{\varphi_{corr}} \qquad [12\text{-}12]$$

where $E_{\phi corr}$ is corrosion potential. In terms of current, the perturbation of equilibrium by the anodic(+) or cathodic(-) side (if charge transfer) is considered. Using the polarization (η), the current can be described by the Butler-Volmer equation:

$$i = i_0 \left[e^{\gamma f \eta} - e^{-(1-\gamma) f \eta} \right] \qquad [12\text{-}13]$$

where γ is the charge transfer coefficient and $f = F/RT$. We can express it as:

$$i = i_0 \, e^{\gamma f \eta} \left(1 - e^{-f\eta} \right) \qquad [12\text{-}14]$$

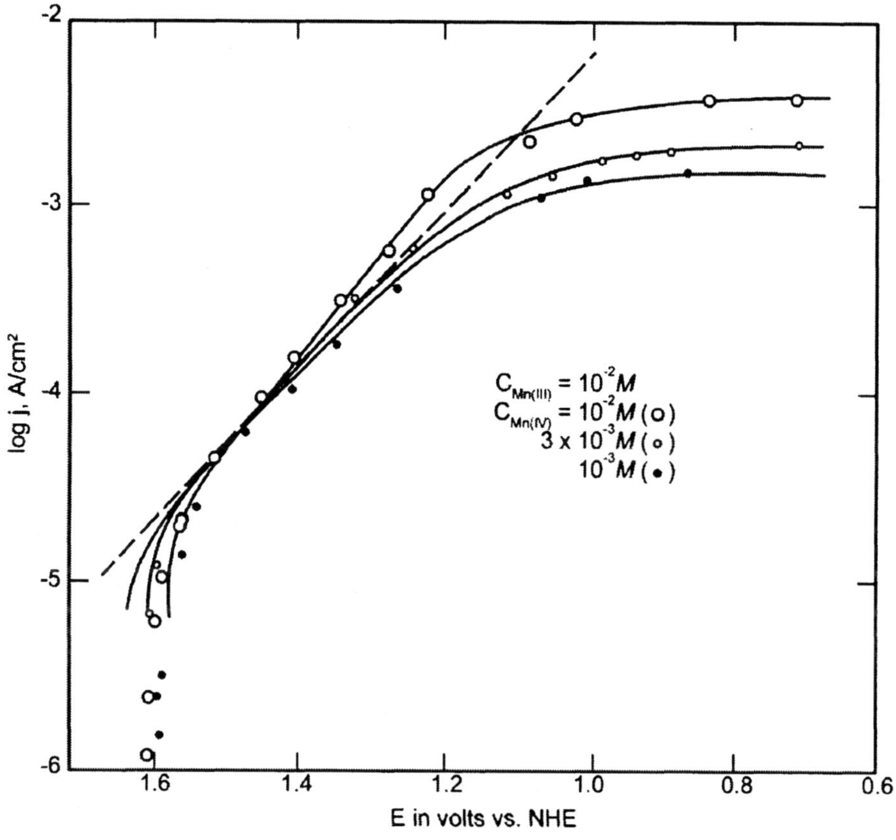

Figure 12-16. Experimental Tafel plots for the reduction of Mn(IV) to Mn(III) at Pt in 7.5 M H2SO4 at 298 K. The dashed line corresponds tp $\alpha = 0.24$. [From K. J. Vetter and G. Manecke, Z. Physik. Chem. (Leipzig), 195, 337 (1950)].

which is

$$\log \frac{i}{\left(1-e^{-f\eta}\right)} = \log i_o + \frac{\gamma F \eta}{2.3RT} \qquad [12\text{-}15]$$

Therefore, when η is plotted versus log i as shown in Figure 12-15, two slopes are obtained. The **Tafel slope** is defined as b^0, and accordingly,

$$b^0 = \left| \frac{\partial \eta}{\partial \log i} \right|$$

$$b_a^0 = \frac{2.3RT}{\gamma F} \qquad [12\text{-}16]$$

$$b_c^0 = \frac{2.3RT}{(1-\gamma)F}$$

where subscripts a and c are anodic and cathodic, respectively. Values of exchange current density and Tafel slope values are shown in Table 12-5. Exponential Tafel slopes of $Mn^{4+} \rightarrow Mn^{3+}$ are illustrated in Figure 12-16. They can be used to approximate the corrosion rate. Evidently for Figure 12-15 the overvoltage η can be approximated from Tafel slope, b^0, and the current density, i:

$$\eta = a + b \log i$$

and

$$b = b^0$$

Common a and b values for a number of metals in both acidic and alkaline aqueous solutions are shown in Table 12-6.

Table 12-6. Exchange Current Densities and Tafel Slopes

Reaction	Electrode	i_0(A/cm^2)[a]	Tafel Slope (mV)[a]
$2H^+ + 2e^- \rightarrow H_2$	Platinum	10^{-3}	120
	Iron	10^{-6}	120
	Nickel	10^{-7}	120
	Copper	10^{-7}	120
	Cadmium	10^{-7}	120
	Carbon	10^{-8}	120
	Zinc	10^{-8}	120
	Tin	10^{-9}	100
	Aluminum	10^{-10}	100
	Lead	10^{-11}	130
	Mercury	10^{-11}	120
$Fe \rightarrow Fe^{2+} + 2e^-$	Iron	10^{-8}	40
$Ni \rightarrow Ni^{2+} + 2e^-$	Nickel	10^{-9}	60
$Cu \rightarrow Cu^{2+} + 2e^-$	Copper	10^{-5}	40
$Zn \rightarrow Zn^{2+} + 2e^-$	Zinc	10^{-5}	40

[a] Approximate values.

Table 12-7. Tafel Constants of *a* and *b* (in A·cm^{-2}, at 293°K) for Some Metals in Presence of Hydrogen Overpotential

	Acidic		Alkaline	
Metals	*a*/V	*b*/V	*a*/V	*b*/V
Ag	0.95	0.10	0.73	0.12
Al	1.00	0.10	0.64	0.14
Au	0.40	0.12	—	—
Ca	1.40	0.12	1.05	0.16
Co	0.62	0.14	0.60	0.14
Cu	0.87	0.12	0.96	0.12
Fe	0.70	0.12	0.76	0.11
Hg	1.41	0.114	1.54	0.11
Mn	0.8	0.10	0.90	0.12
Ni	0.63	0.11	0.65	0.10
Pb	1.56	0.11	1.36	0.25
Pd	0.24	0.03	0.53	0.13
Pt	0.10	0.03	0.31	0.10
Sn	1.20	0.13	1.28	0.23
W	0.43	0.10	—	—
Zn	1.24	0.12	1.20	0.12

[Example 12-2] Compute the electron potential of an iron electrode if electrolysis produces 100 mg·cm^{-2} of hydrogen every hour in an aqueous KOH solution of $a(\mathrm{OH}) = 1$.

For $H^+ + e \rightarrow 1/2\ H_2$, the average potential is:

$$E_\phi = -\frac{RT}{F}\ln\frac{1}{a(\mathrm{H}^+)} = \frac{RT}{F}\ln\frac{10^{-14}}{a(\mathrm{OH}^-)} = \frac{8.314\times 293}{96500}\ln 10^{-14} = -0.814\ \mathrm{V}$$

In order to estimate the overpotential, find the current density:

$$j = i/A = \frac{Q/t}{A} = \frac{Q/A}{t}$$

or

$$\frac{Q}{A}=\frac{nF}{A}=\frac{\left(100\times10^{-3}\right)}{1}\left(96500C\ \text{cm}^{-2}\right)=9650\ C\ \text{cm}^{-2}$$

Thus,

$$j=\frac{9650}{\left(60\times60\right)}\text{amp cm}^{-2}=2.68\ \text{amp cm}^{-2}$$

From Tafel slopes and a values listed in Table 12-6,

$$\eta = a + b\log i = 0.76\ \text{V} + (0.11\times\log 2.68)\text{V} = 0.81\ \text{V}$$

or

$$E_{\Phi\text{eq}} = E_{\Phi} - \eta = (-0.814 - 0.81)\ \text{V} = -1.62\text{V}$$

If the polarization resistance constant, B_1, can be experienced as:

$$B=\frac{b_a\cdot b_c}{2.3\left(b_a+b_c\right)} \qquad [12\text{-}17]$$

or

$$B=\frac{b_a}{2.3}\ \left(if\ b_c\rightarrow\infty\right)$$

then

$$\left(\frac{\Delta E}{\Delta i}\right)_{\Delta E\rightarrow0}=R_p=\frac{B}{i_{corr.}} \qquad [12\text{-}18]$$

where R_p is polarization resistance:

$$B=i_{corr.}R_p$$

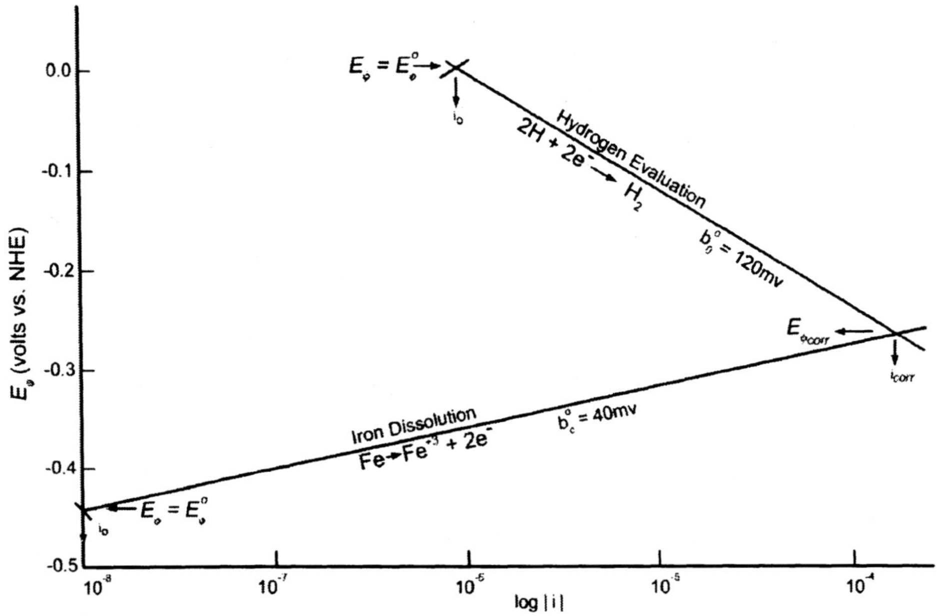

Figure 12-17. Construction of the mixed potential diagram for iron in acid solution (Evans diagram).

$$E - E_{corr.} = \log i_{corr.} \pm b \log i \qquad [12\text{-}19]$$

and log $i_{corr.}$ can be estimated from Tafel slopes.

For mixed potential (see Fig. 12-17), usually i_{corr} (corrosion current density) and $E_{\Phi corr}$ (corrosion potential) are important. Let us enlarge the scope to explain more thoroughly the two lines in Figure 12-17. The construction of E_Φ versus log i or i curves belongs to what is known as an **Evans-type diagram**. In order for corrosion to occur, there must be two events. First, there is metal dissolution, which is generalized as

$$M \rightarrow M^{n+} + ne \qquad \text{(electron sink)}$$

followed by the electronation reaction, which simply means the electron acceptors in the electrolyte remove the electrons produced.

$$xA + mH^+ + ne \rightarrow yD + zH_2O \qquad \text{(electron source)}$$

Usually, the electron source area is cathodic, and the electron sink area is anodic.

In Figure 12-17, the metal dissolution is illustrated by iron and the electronation is illustrated by the hydronium ion (proton) as electron acceptor. The potentials for metal dissolution and electronation reaction are, respectively, $E^0_{\Phi,m}$ and $E^0_{\Phi,so}$. The corresponding current densities are $i_{o,m}$ and $i_{o,so}$ respectively. Also, the corresponding Tafel slopes are b^0_a and b^0_c , respectively, for the two lines. One can easily locate the intersection of the two lines as i_{corr}, $E^0_{\Phi,corr}$. It is possible to derive an equation.

$$i_{corr} = \left(i_{0,m}\, i_{0,s_0}\right)^{\frac{1}{2}} \exp\left[\frac{F\left(E^0_{\phi,s_0} - E^0_{\phi,m}\right)}{4RT}\right] \qquad [12\text{-}20]$$

The physical significance of this equation is clear — sodium erodes in aqueous solution so vigorously that considerable heat is evolved, sparks fly, and the produced hydrogen explodes. Since $E^0_{\Phi,Na} = -2.71V$ (highly negative), $E^0_{\Phi,so} - E^0_{\Phi,so}$ is a very large positive value. Thus, the corrosion current is very high. On the other hand, $E^0_{\Phi,Au} = +1.50$ V (highly positive); by the same reasoning, gold cannot corrode, a primary reason for its use as a monetary standard.

Another concern is if polarization is:

$$\eta = E_{\phi_i} - E_{\phi_{corr}}$$

and

$$i = i_{corr}\left(e^{\frac{\eta}{\beta_a}} - e^{\frac{\eta}{\beta_c}}\right) \qquad [12\text{-}21]$$

where β and b^0 are Tafel slopes, the difference being that β is a ln system while b^0 is a log system. For small polarization, $\eta \rightarrow 0$, and

$$i = i_{corr}\left[\eta\left(\frac{1}{\beta_a} + \frac{1}{\beta_c}\right)\right]_{\eta \rightarrow 0} \qquad [12\text{-}22]$$

The Foxadic or polarization resistance can be obtained from the Tafel slope

$$R_f = \left[\frac{\eta}{i}\right]_{\eta \to 0} = \frac{\beta_a \beta_c}{(\beta_a + \beta_c)}\left(\frac{1}{i_{corr}}\right) = \left(\frac{2.3 b_a^0 b_c^0}{b_a^0 + b_c^0}\right)\left(\frac{1}{i_{corr}}\right) \qquad [12\text{-}23]$$

Thus, there is the addition of partial potentials for the mixed potentials

$$\sum_j i_{a_j}\left(E_{\phi_i}\right) = i_a$$

and

$$\sum_j i_{c_j}\left(E_{\phi_i}\right) = i_c \qquad [12\text{-}24]$$

For galvanic corrosion, Stern developed the following:

$$i\left(E_{\phi_i}\right) = \sum_j f_j i_j\left(E_{\phi_i}\right) \qquad [12\text{-}25]$$

where f_j is the fraction of j metals: for two metals A and B, f_A and f_B are the fractions. With some assumptions the final equations derived are:

$$E_{\phi_{corr}} = E_\phi = \frac{b_a^0 b_c^0}{b_a^0 + b_c^0}\left[\log\left(\frac{f_A i_{0cA} + f_B i_{0cB}}{f_A i_{0aA}}\right) + \frac{E_{\phi oc}}{b_c^0} + \frac{E_{\phi oa}}{b_a^0}\right] \qquad [12\text{-}26]$$

$$\log i_{corr} = \frac{1}{b_a^0 + b_c^0}\left[b_c^0 \log\left(f_A\, i_{ocA} + f_B\, i_{oaB}\right) + b_a^0 \log\left(f_A\, i_{oaA}\right) - E_{\phi_{oa}} - E_{\phi_{oc}}\right] \qquad [12\text{-}27]$$

These two equations are used to predict the current density as well as the potentials for galvanic corrosion.

Often the corrosion rate is required for a particular distribution system during water transportation. A simple assumption can be made that the corrosion rate is proportional to the oxygen transfer rate in the flow pipe. The current density is then governed by the mass transport rate constant, K_f:

$$\bar{j} = nF K_f C \qquad [12\text{-}28]$$

where K_fC is expressed in moles/sec/cm^2 and K_f in cm/sec. For transport of liquid, the following dimensionless groups can be used, namely the Sherwood number, Reynolds number, and Schmidt number:

$$\mathrm{Sh} = \frac{K_f L}{D_l}$$

$$\mathrm{Re} = \frac{uL}{\nu} \qquad [12\text{-}29]$$

$$\mathrm{Sc} = \frac{\nu}{D_l}$$

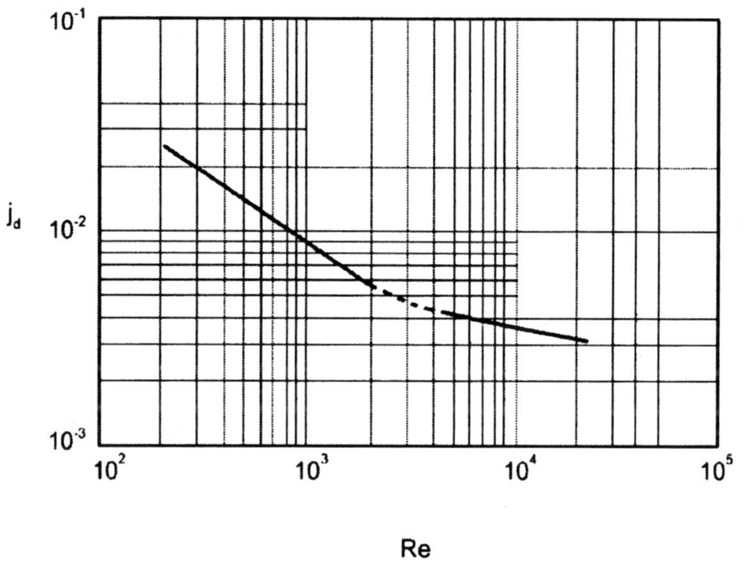

Figure 12-18. Colburn j_d factor for the reduction of oxygen. (From Ref. 12-10.)

where L is characteristic length, D_l is diffusivity (cm^2/sec), u is fluid velocity (cm/sec), and ν is kinematic viscosity. In general K_f can be approximated from:

$$j_d = \left(\frac{K_f}{\nu}\right)(\mathrm{Sc})^{\frac{2}{3}} \qquad [12\text{-}30]$$

or actually determined from the Chilton-Colburn correlation:

$$j_d = \frac{\text{Sh}}{\text{ReSc}^{\frac{1}{3}}} = 0.023\,\text{Re}^{-0.2} \qquad [12\text{-}30a]$$

and j_d can be read off from a correlation group if the Reynolds number is known, as shown in Figure 12-18 (for Calbum and other factors, see Appendix C).

Equation [12-28] can be used to approximate corrosion rate, since it is the current density of the electrode.

$$i_{corr} = nFK_f\,C \qquad [12\text{-}28a]$$

where i_{corr} is expressed in A/cm^2 and K_f in cm/sec. C is expressed as moles/cm^3 of the solution, if the electron-source side of the half-cell is:

$$O_2 + 2H_2O + 4e \rightarrow 4OH^-$$

[Example 12-3] Water is used as the heat exchanger for a reactor as shown in Figure 12-19. The radius of the reactor is 9 inches, and the outer diameter of the tube containing the water is 12.5 inches with a ¼-inch wall. If air is saturated in the water, the velocity of water reaches 6.9 cm/sec, the kinematic viscosity of water is 9.37 × 10^{-3} cm^2/sec, and the mean diffusion coefficient is 1.2 × 10^{-5} cm/sec. Estimate the corrosion rate (after Weber).

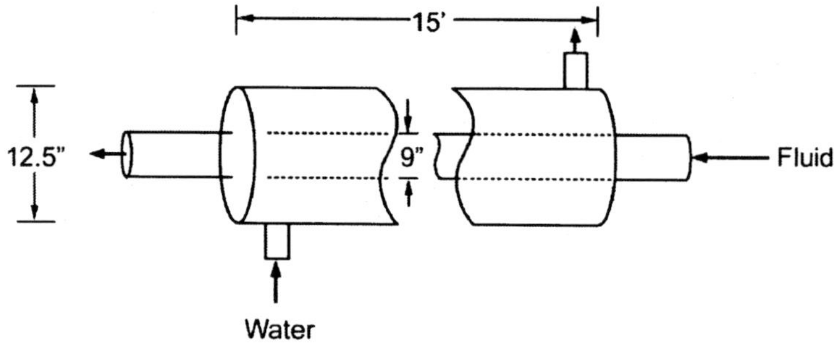

Figure 12-19. Shell and tube heat exchanger. Fluid properties: mean velocity = 6.9 cm/sec; mean kinematic viscosity = 9.37 × 10^{-3} cm/sec. Oxygen property: mean diffusion coefficient = 1.2 × 10^{-5} cm^2/sec.

$$\text{Re} = \frac{vL}{u} = \frac{2.54\left[12.5 - \left(9 + 2 \times \frac{1}{4}\right)\right]6.9}{9.37 \times 10^{-3}} = 5610$$

$$j_d = 0.023\,\text{Re}^{-0.2} = 4.0 \times 10^{-3}$$

$$\text{Sc} = \frac{v}{D_l} = \frac{9.37 \times 10^{-3}}{1.2 \times 10^{-5}} = 392$$

Thus,

$$K_f = \frac{vj_d}{(\text{Sc})^{\frac{2}{3}}} = \frac{(4 \times 10^{-3})(6.9)}{(392)^{\frac{2}{3}}} = 3.25 \times 10^{-4}\ \text{cm/sec}$$

From Equation [12-28],

$$i_{corr} = 4(9.65 \times 10^4)(3.25 \times 10^{-4})(2.39 \times 10^{-7}\ \text{moles}\ O_2/\text{cm}^3\ H_2O\ \text{at}\ 40^\circ C)$$

$$= 3.02 \times 10^{-5}\ \text{A/cm}^2$$

where $n = 4$, since $O_2 + 2H_2O + 4e = 4OH^-$.

12.4 Corrosion Control Technology

A number of methods are available for particular corrosion problems. Some currently utilized methods are summarized in the following sections.

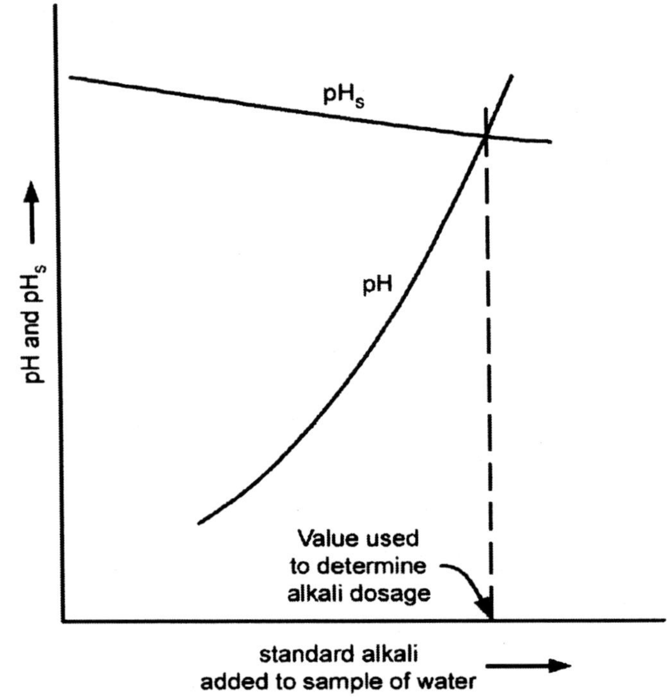

Figure 12-20. Illustration of use of laboratory data to determine the alkali dosage required to achieve stabilization and destabilization for a Langelier index equal to approximately zero.

12.4.1 Protective Coating

The simplest approach is to adjust the Langelier index of the water to a slightly positive value, as discussed in Chapter 3. Adjustment is usually made by adding alkali (generally lime) CaOH or soda ash (Na_2CO_3) in order to change the pH as shown in Figure 12-20. The quantity of alkali at the intersection represents the dosage required for stability. A small excess quantity of alkali ensures the deposit as a thin layer of $CaCO_3$ coating.

For metal surfaces such as iron, different organic and inorganic coatings have been developed. For example, durichlor (a family of Mo) has been developed for coating iron vessels containing sulfuric acid, since some $MoCl_4$ is very inert and insoluble. Organic coatings such as rosin amine, pentachlorophenol, etc. (see Fig. 12-21), used for protecting pipeline metal parts and vehicles, often act as free radical shunts to intercept the derived electrons from corrosion. For ships entering port, derusting washings must be conducted before a new coating is applied. The derusting washing usually contains a mixture of citric acid (complexing agents) and triethanolamine (radical shunts) at different ratios. (See Chapter 3 on masking and demasking.)

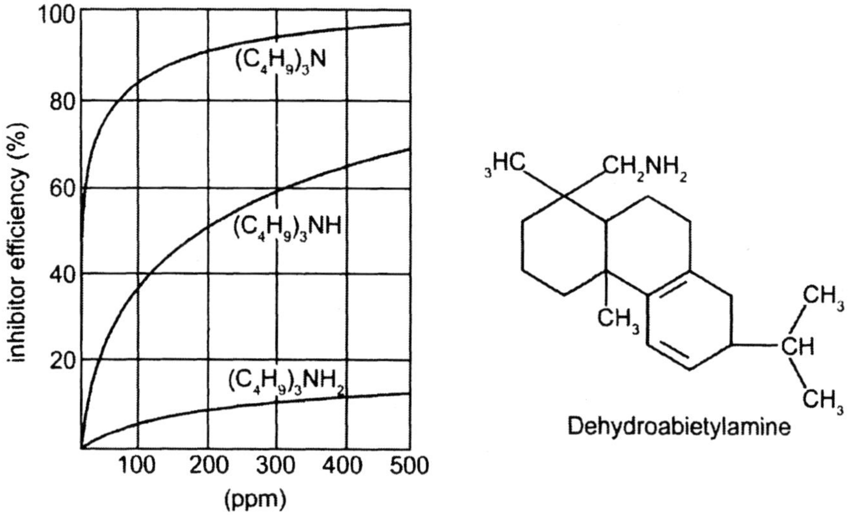

Figure 12-21. A comparison of the effectiveness of primary, secondary, and tertiary butyl amines. The amine inhibitor is derived from rosin.

12.4.2 Deoxygenation

As discussed with regard to differential aeration cells, dissolved oxygen (DO) plays an important role in metallic corrosion. Oxygen in water reacts with atomic hydrogen films in water, thereby depolarizing cathodic regions and accelerating corrosion. Vacuum deaeration has been used for the deoxygenation of water.

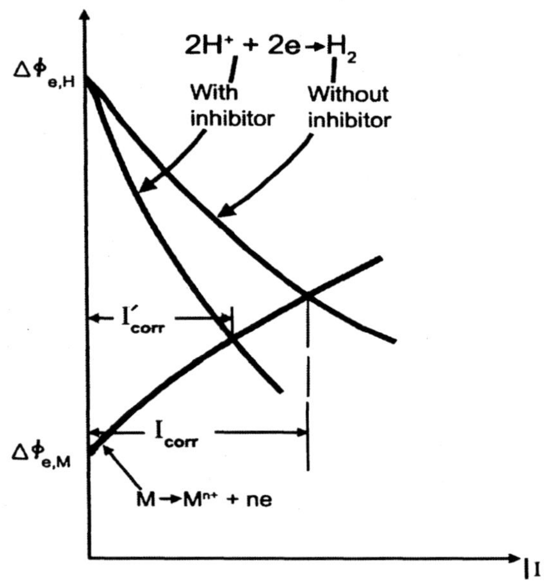

Figure 12-22. If hydrogen-evolution reaction is inhibited, the potential-current relation changes in such a way that the corrosion current is reduced.

Actually, corrosion inhibition by adding chemicals to the electrolytic environment of a corroding metal has some general principles. In terms of the Evans diagram, a reduction of the exchange current densities of the electronation reaction will reduce the corrosion current. Figure 12-22 illustrates that if hydrogen evolution occurs in the electronation reaction, the addition of phosphorus, arsenic, or antionary compounds (e.g., As_2O_3) will reduce the exchange current density for hydrogen production. On the other hand, if oxygen electronation is the reaction at the electron-source area, in other words, if the electronation is:

$$O_2 + 4H^+ + 4e \rightarrow 2H_2O$$

then by adding compounds that react with dissolved oxygen the O_2 concentration will be reduced and the exchange current density also reduced. Two substances that act in this manner are hydrazine and sodium sulfide:

$$2N_2H_4 + 5O_2 \rightarrow 4NO_2^- + 4H^+ + 2H_2O$$

$$2Na_2SO_3 + O_2 \rightarrow 2Na_2SO_4$$

The latter has been used as an additive in boiler feed water.

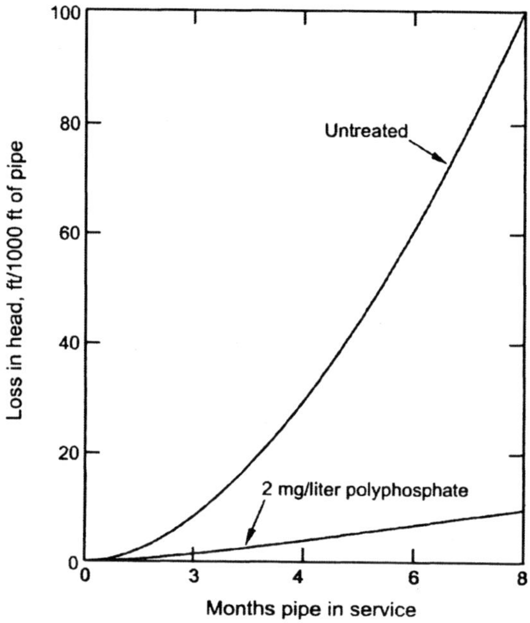

Figure 12-23. The effect of polyphosphate treatment of water on frictional heat loss. (From Betz Handbook of Industrial Water Conditioning, Betz Laboratories, Inc., Philadelphia.)

12.4.3 Inhibitors

Many polyphosphates including sodium tripolyphosphate, sodium hexametaphosphate, and sodium decaphosphate are used in municipal water treatment. As little as 2 mg/L may help in maintaining the flow capacity of the pipes (see Fig. 12-23). The active mechanism involves complex formation of iron-phosphate, which prevents the initiation of $Fe(OH)_3$ tuberculation for the "red water" problem. The dissolved ion may come from anodic areas.

Another type of corrosion inhibitor is the oleic imidazoline, whose main function is to stabilize metal surfaces by bonding to metal oxides and filling gaps in monolayers with their long hydrocarbon tails in such a manner as to restrict any access of surrounding electrolytes. The mechanism is illustrated in Figure 12-24. Its experimental effectiveness is shown in Table 12-7.

Imidazoline displaces water from the surface hence strong adsorption is achieved.

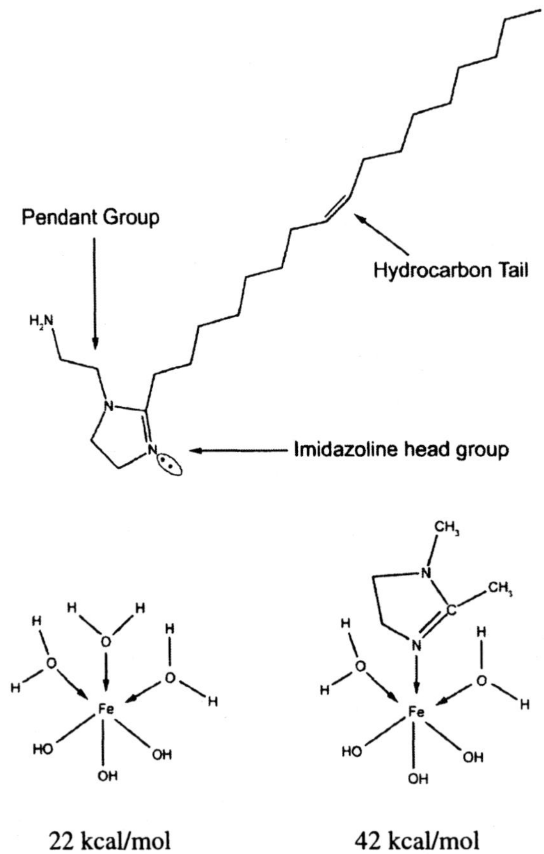

Figure 12-24. Adsorption mechanism of imidazoline corrosion inhibitors.

For an explanation of Evan's diagram, an inhibitor can reduce the exchange current density by absorption on the electron sink area of the corroding metal and slow down the metal dissolution reaction (see Fig. 12-25). Often used for this purpose are N-containing compounds (aliphatic and aromatic amines), S-containing compounds (thiourea and its derivatives), and O-containing compounds (aldehydes, quinones). Adsorption is complicated by the chemical nature of the adsorbate (free energy) and the electrode charge. The inhibitors must adsorb in the range of potential at which the corrosion reaction occurs. Differences in the degree of coverage arise when a metal is polarized cathodically or anodically with respect to the corrosion potential (see Table 12-8).

Table 12-8. Experimental Corrosion Rates and Formation of a Hydrophobic Barrier of Different Imidazolines

Pendant Group	Hydrocarbon Tail	Corrosion Rate (mpy)[a]	Coverage of Surface
$NH_2(CH_2)_2$	$CH_3(CH_2)_7$ $CHCH(CH_2)_7$	2.5	Full
H	$CH_3(CH_2)_7$ $CHCH(CH_2)_7$	7.0	Full
CH_3	$CH_3(CH_2)_7$ $CHCH(CH_2)_7$	3.0	Full
$HO(CH_2)_2$	$CH_3(CH_2)_7$ $CHCH(CH_2)_7$	3.0	Full
$NH_2(CH_2)_3$	$CH_3(CH_2)_7$ $CHCH(CH_2)_7$	3.0	Full
$HO(CH_2)_2$	$((CH_3)(CH_2)_8)$ $(CH_3(CH_2)_7)CH$	2.5	Full
$HO(CH_2)_2$	(CH_3CH_2)	26.0	Partial

[a] From Edwards et al., *Corr. Sci.*, 36, 315 (1994).

Table 12-9. Approximate Percent of Coverage of Certain Inhibitors (When Anodic and Cathodic) on Mild Steel in 5% Sulfuric Acid at 70°C

Solution Concentration of Inhibitor × 10^3 (moles liter^{-1})	Tolythiourea		β-Naphthoquinoline	
	Anodic	Cathodic	Anodic	Cathodic
0.003	3	3	4	
0.01	22	22	20	
0.03	70	70	40	
0.1	92	90	70	
0.3	99	90	85	< 5
1	> 99	90	95	20
3	> 99	90	98	35
10	—	—	99	60

Source: J. M. West, *Electrodeposition and Corrosion Processes*, D. Van Nostrand Co., Inc., New York, 1965.

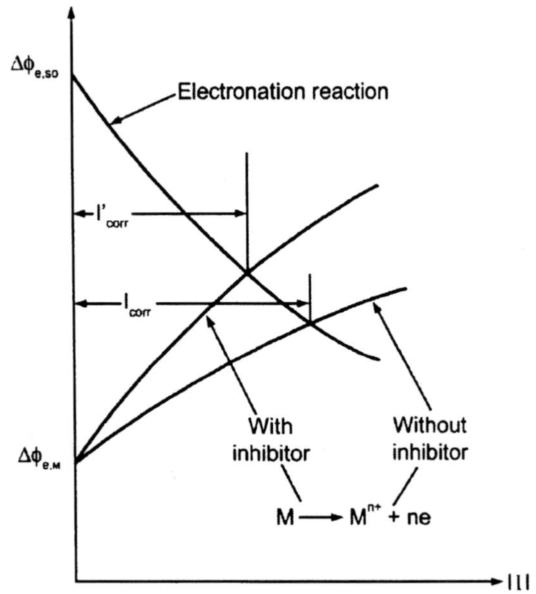

Figure 12-25. Metal-dissolution reaction can be inhibited by adsorption organics at electron-sink areas, in which case the corrosion current is reduced.

Table 12-10. Examples of Practical Organic and Inorganic Corrosion Inhibitors Used in Acidic, Neutral, and Alkaline Solutions

		Inhibitor	
Metal	Medium	Organic	Inorganic
Iron	Acidic	Aniline, ethyl- and diethylamine, pyridine, quinoline, β-naphthoquinoline, tolylthiourea, thiodiglycol, and formaldehyde	As_2O_3 $NaAsO_2$ $K_2Cr_2O_7$
	Neutral	Sodium benzoate and hydrazine	Na_2SO_3
Steel	Acidic	The same as for iron	
	Neutral	Sodium benzoate	
Copper	Acidic	Thiourea	$K_2Cr_2O_7$, $KMnO_4$
	Neutral	—	$Ca(HCO_3)_2$, $K[Fe(CN)_6]$, Chromates
Aluminum and Zinc	Neutral	—	$Ca(HCO_3)_2$, sodium and calcium hexametaphosphate
	alkaline	Glucose	S– and SiO_3–

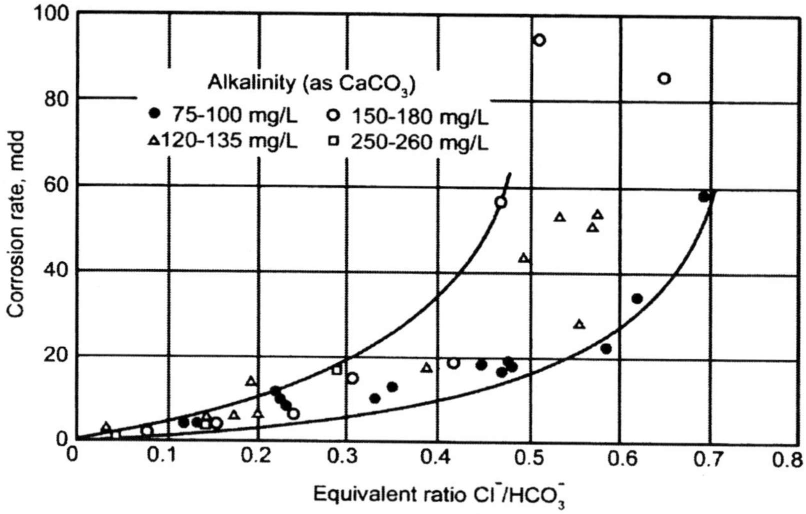

Figure 12-26. Effect of Larson's ratio on corrosion of mild steel.

Similarly, the exchange current density also can be shortened if absorption and blockage occurs on the electron-source area. Some common inhibitors used are listed in Table 12-9.

Zinc sulfate is used as a cathodic inhibitor because it reacts with OH^- as follows:

$$Zn^{2+} + 2OH^- = Zn(OH)_2 \downarrow$$

Automobile antifreeze usually consists of a mixture of 1.5% sodium benzoate and 0.1% sodium nitrate in ethylene glycol.

Carbonate, chloride, and sulfate ions in water-carrying conduits and pipes are related to corrosion. **Larson's ratio** has been used to indicate the corrosion rate.

$$LR = \frac{[Cl^-] + [SO_4^{2-}]}{[HCO_3^-]}$$

or

$$LR = \frac{[Cl^-]}{[HCO_3^-]}$$

The relationship is shown in Figure 12-26.

12.4.4 Cathodic Protection

Electrolytic methods also can be used for control of corrosion. Insulation with a dielectric is the most common method of connection to galvanic cells. The cathodic protection is usually furnished by a galvanic cell formed when the metal to be protected is connected to a sacrificial metal located higher in the electromotive series. The sacrificial metal dissolves (or corrodes), and the flow of electrons is toward protection of the cathodic surface. For steel tanks or pipes (and those made of zinc), aluminum has been utilized, but should be replaced periodically. Figure 12-27 demonstrates this concept. Also, an external electromotive force with a low voltage 6–10 V can be imposed to counteract the potential created by the natural corrosion cells. This power supply can be furnished by any source of direct current. Areas that ordinarily become anodic are made cathodic by the formation of a negative change. Currently Pt-clad silver-copper anodes are used for submarines as cathodic protection. External electromotive force is used for drilling platforms.

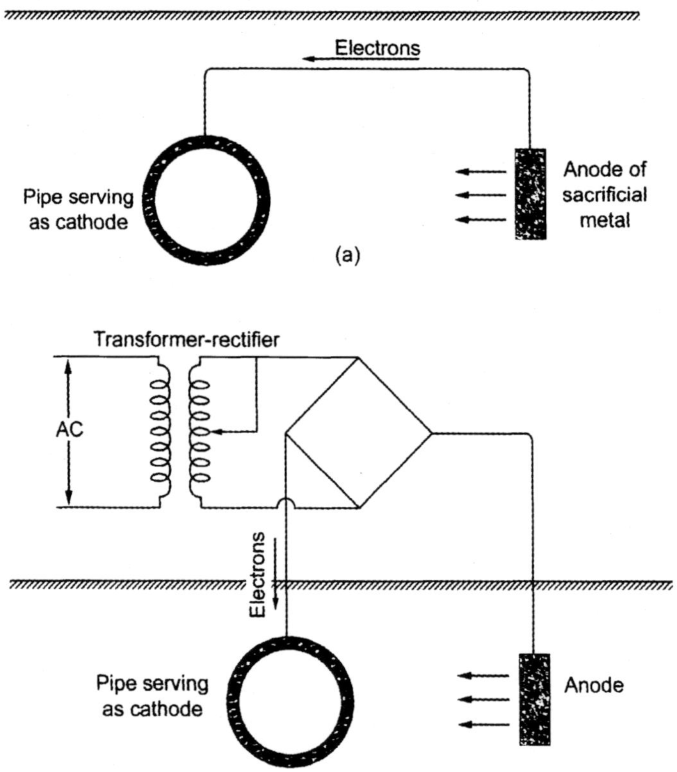

Figure 12-27. Cathodic protection of pipelines by (a) galvanic action and (b) external electromotive force.

REFERENCES

12-1. J. Baylis, Prevention of corrosion and 'red water', *JAWWA*, 15, 598 (1926).

12-2. J. Benson and H. Klein, Lead in drinking water. Investigation of a corrosive water supply, *J. Environ. Health*, 45, 179–181 (1983).

12-3. G. H. Booth and A. K. Tiller, Polarization studies of mild steel in cultures of sulfate-reducing bacteria, *Trans. Faraday Soc.*, 56, 1689 (1960).

12-4. G. Butler and H. Ison, *Corrosion and Its Prevention in Waters*, Reinhold, New York (1966).

12-5. F. M. Donahue, Corrosion and corrosion control, in *Physicochemical Processes for Water Quality Control* (W. J. Weber, Jr., ed.), Wiley, New York, 1972, pp. 457–501.

12-6. J. Flis, *Corrosion of Metals and Hydrogen-related Phenomena*, Elsevier, Amsterdam, 1991.

12-7. M. Fontana and N. Greene, *Corrosion Engineering*, McGraw Hill, New York, 1978.

12-8. R. Gramer, P. I. Moreland, and G. Pini, *A Literature Review of Polarization Resistance Constant (B) Values for the Measurement of Corrosion Rate*, National Association of Corrosion Engineers, Houston, 1982.

12-9. W. Kolle and H. Rosch, Untersuchungen an Rohrenetzinkrustierungen unter mineralogischen Gesichtspunkten, *Vom Wasser*, 55, 159 (1980).

12-10. C. S. Lin, E. D. Denton, H. S. Gaskill and G. L. Putnan, Diffusion controlled electrode reactions, *Ind. Eng. Chem.*, 43, 2136 (1951).

12-11. P. Moreland and J. Hines, Corrosion monitoring–select the right system, *Hydrocarbon Proc.* 57(11), 251–255 (1978).

12-12. J. S. Newman, *Electrochemical Systems*, 1973.

12-13. B. Pathak, Testing in fresh waters, in *Handbook on Corrosion Testing and Evaluation*, (Ailor, ed.), Wiley, New York, 1971.

12-14. M. Pourbaix, *Atlas of Electrochemical Equilibria*, Pergamn Press, Oxford, 1966.

12-15. M. Pourbaix, *Thermodynamics of Dilute Aqueous Solution, with Applications to Electrochemistry and Corrosion*, Edward Arnold, London 1949.

12-16. J. Goodisman, *Electrochemistry, Theoretical Foundations*, Wiley, New York, 1987.

12-17. A. J. Bard and L. R. Faukner, *Electrochemical Methods, Fundamentals and Applications*, Wiley, New York, 1980.

12-18. J. O'M. Bockris and A.K.N. Reddy, *Modern Electrochemistry*, Plenum Press, New York, Vol. 1 and Vol. 2, 1970.

12-19. Z. Zhong and G. Mei, *Application of Diagrams of Chemical Potential in Hydrometallurgy Energy and Purification Wastewater* (in Chinese), Central South University Press, Xiang Fang, China, 1986.

12-20. L. Antropov, *Theoretical Electrochemistry*, MIR, Moscow, 1972.

12-21. S. A. Bradford, *Corrosion Control*, Van Nostrand-Reinhold, New York, 1993.

12-22. C. P. Dillon, *Corrosion Control in the Chemical Process Industries*, McGraw-Hill, New York, 1986.

12-23. D. A. Jones, *Principles and Prevention of Corrosion*, MacMillan, New York, 1992.

12-24. L. S. Van Delinder, *Corrosion Basics — An Introduction*, National Association of Corrosion Engineers, Houston, 1984.

12-25. 12U. R. Evans, *An Introduction to Metallic Corrosion*, Edward Arnold, London, 1963.

PROBLEM SET

1. Compute the corrosion potential and corrosion current density of iron in acid solution.

2. The figure(Figure 12-19) represents a shell and tube heat exchanger that is used in a certain process. The process water has an average temperature of 105°F while in the exchanger and is assumed to be air saturated. Previous experience with this type of equipment has indicated that the corrosion rate of the tube (per unit area) is equal to the mass transfer rate of the oxygen corrosion rate (A/cm^2) for the tube.

   Fluid Properties: mean viscosity = 6.9 cm/sec

   Mean Kinematic Viscosity = 9.39×10^{-3} cm^2/sec

   Oxygen Property: mean diffusion coefficient = 1.2×10^{-5} cm^2/sec

   Find the corrosion rate of the shell and tube heat exchanger

APPENDIX A

Thermodynamic properties

A code sentence was used to describe the relationship between the *T, G, P, H, S, E, V* and *A* functions:

"**T**he **G**ibbs **P**otential **H**as **S**hown **E**ndless **V**aluable **A**pplications"

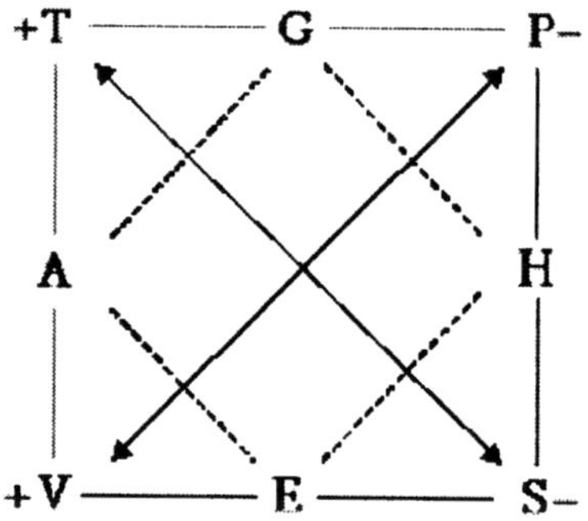

A number of relationships can be obtained. For example, the trapezoids, *EATS, AGPV*, and etc.

E-A=TS

A-G= -PV

G-H= -ST

H-E=VP

At the four corners of the diagram above:

$$\left(\partial V/\partial S\right)_E = T/P$$

$$\left(\partial T/\partial V\right)_A = -P/S$$

$$-\left(\partial P/\partial T\right)_G = -S/V$$

$$-\left(\partial S/\partial P\right)_H = V/T$$

The symmetrical property of this diagram originates from the group theory, i.e., Jacobean Theory

APPENDIX B

IUPAC Periodic Table of the Elements

Key:

atomic number **Symbol** name standard atomic weight

1	2	3	4	5	6	7	8	9	10	11	12	13	14	15	16	17	18
1 **H** hydrogen 1.007 94(7)																	2 **He** helium 4.002 602(2)
3 **Li** lithium 6.941(2)	4 **Be** beryllium 9.012 182(3)											5 **B** boron 10.811(7)	6 **C** carbon 12.0107(8)	7 **N** nitrogen 14.0067(2)	8 **O** oxygen 15.9994(3)	9 **F** fluorine 18.998 4032(5)	10 **Ne** neon 20.1797(6)
11 **Na** sodium 22.989 769 28(2)	12 **Mg** magnesium 24.3050(6)											13 **Al** aluminium 26.981 5386(8)	14 **Si** silicon 28.0855(3)	15 **P** phosphorus 30.973 762(2)	16 **S** sulfur 32.065(5)	17 **Cl** chlorine 35.453(2)	18 **Ar** argon 39.948(1)
19 **K** potassium 39.0983(1)	20 **Ca** calcium 40.078(4)	21 **Sc** scandium 44.955 912(6)	22 **Ti** titanium 47.867(1)	23 **V** vanadium 50.9415(1)	24 **Cr** chromium 51.9961(6)	25 **Mn** manganese 54.938 045(5)	26 **Fe** iron 55.845(2)	27 **Co** cobalt 58.933 195(5)	28 **Ni** nickel 58.6934(2)	29 **Cu** copper 63.546(3)	30 **Zn** zinc 65.409(4)	31 **Ga** gallium 69.723(1)	32 **Ge** germanium 72.64(1)	33 **As** arsenic 74.921 60(2)	34 **Se** selenium 78.96(3)	35 **Br** bromine 79.904(1)	36 **Kr** krypton 83.798(2)
37 **Rb** rubidium 85.4678(3)	38 **Sr** strontium 87.62(1)	39 **Y** yttrium 88.905 85(2)	40 **Zr** zirconium 91.224(2)	41 **Nb** niobium 92.906 38(2)	42 **Mo** molybdenum 95.94(2)	43 **Tc** technetium [98]	44 **Ru** ruthenium 101.07(2)	45 **Rh** rhodium 102.905 50(2)	46 **Pd** palladium 106.42(1)	47 **Ag** silver 107.8682(2)	48 **Cd** cadmium 112.411(8)	49 **In** indium 114.818(3)	50 **Sn** tin 118.710(7)	51 **Sb** antimony 121.760(1)	52 **Te** tellurium 127.60(3)	53 **I** iodine 126.904 47(3)	54 **Xe** xenon 131.293(6)
55 **Cs** caesium 132.905 4519(2)	56 **Ba** barium 137.327(7)	57-71 lanthanoids	72 **Hf** hafnium 178.49(2)	73 **Ta** tantalum 180.947 88(2)	74 **W** tungsten 183.84(1)	75 **Re** rhenium 186.207(1)	76 **Os** osmium 190.23(3)	77 **Ir** iridium 192.217(3)	78 **Pt** platinum 195.084(9)	79 **Au** gold 196.966 569(4)	80 **Hg** mercury 200.59(2)	81 **Tl** thallium 204.3833(2)	82 **Pb** lead 207.2(1)	83 **Bi** bismuth 208.980 40(1)	84 **Po** polonium [209]	85 **At** astatine [210]	86 **Rn** radon [222]
87 **Fr** francium [223]	88 **Ra** radium [226]	89-103 actinoids	104 **Rf** rutherfordium [261]	105 **Db** dubnium [262]	106 **Sg** seaborgium [266]	107 **Bh** bohrium [264]	108 **Hs** hassium [277]	109 **Mt** meitnerium [268]	110 **Ds** darmstadtium [271]	111 **Rg** roentgenium [272]							

57 **La** lanthanum 138.905 47(7)	58 **Ce** cerium 140.116(1)	59 **Pr** praseodymium 140.907 65(2)	60 **Nd** neodymium 144.242(3)	61 **Pm** promethium [145]	62 **Sm** samarium 150.36(2)	63 **Eu** europium 151.964(1)	64 **Gd** gadolinium 157.25(3)	65 **Tb** terbium 158.925 35(2)	66 **Dy** dysprosium 162.500(1)	67 **Ho** holmium 164.930 32(2)	68 **Er** erbium 167.259(3)	69 **Tm** thulium 168.934 21(2)	70 **Yb** ytterbium 173.04(3)	71 **Lu** lutetium 174.967(1)
89 **Ac** actinium [227]	90 **Th** thorium 232.038 06(2)	91 **Pa** protactinium 231.035 88(2)	92 **U** uranium 238.028 91(3)	93 **Np** neptunium [237]	94 **Pu** plutonium [244]	95 **Am** americium [243]	96 **Cm** curium [247]	97 **Bk** berkelium [247]	98 **Cf** californium [251]	99 **Es** einsteinium [252]	100 **Fm** fermium [257]	101 **Md** mendelevium [258]	102 **No** nobelium [259]	103 **Lr** lawrencium [262]

Notes

- "Aluminum" and "cesium" are commonly used alternative spellings for "aluminium" and "caesium."
- IUPAC 2005 standard atomic weights (mean relative atomic masses) as approved at the 43rd IUPAC General Assembly in Beijing, China in August 2005 are listed with uncertainties in the last figure in parentheses [M. E. Wieser, *Pure Appl. Chem.*, in press]. These values correspond to current best knowledge of the elements in natural terrestrial sources. For elements that have no stable or long-lived nuclides, the mass number of the nuclide with the longest confirmed half-life is listed between square brackets.
- Elements with atomic numbers 112 and above have been reported but not fully authenticated.

APPENDIX C

Some Useful Constants for Lithosphere, Hydrosphere, Atmosphere and Biosphere.

Quantity	Unit of Measurement	Symbol	Numerical Value
Area of continents	km^2	S_C	149×10^4
Area of world oceans	km^2	S_O	361×10^4
Mean height of continents above sea level	m	h_C	875
Mean depth of world oceans	m	h_O	3794
Mean position of earth's surface with respect to sea level	m	h_m	2430
Mean thickness of lithosphere within the limits of the continents	m	$h_{c.l.}$	35
Mean thickness of lithosphere within the limits of the ocean	km	$h_{o.l.}$	4.7
Mean rate of thickening of continental lithosphere	km	$\frac{\Delta h}{\Delta t}$	10 – 40
Mean rate of horizontal extension of continental lithosphere	m/ 10^6hr	$\frac{\Delta l}{\Delta t}$	0.75 – 20
Mass of lithosphere	km/ 10^6yr	m_t	2.267×10^{25}
Mass of water released from mantle and core in course of geological time	gr		3.400×10^{24}
Total reserve of water in the mantle	gr		2×10^{26}
Present day content of free and bound water in the earth's lithosphere	gr		$2.2 - 2.6 \times 10^{24}$ $1.8 - 2.7 \times 10^{24}$
Mass of hydrosphere	gr	m_h	1.664×10^{24}
Amount of oxygen bound in the earth's crust	gr		1.300×10^{24}
Amount of free oxygen	gr		1.5×10^{21}
Mass of atmosphere	gr	m_a	5.136×10^{21}
Mass of biosphere	gr	m_b	1.148×10^{19}
Mass of living matter in the biosphere	gr		3.6×10^{17}
Density of living matter on dry land	gr/ cm^3		0.1
Density of living matter in ocean	gr/ cm^3		15×10^{-8}

APPENDIX D

Miller indices and the use of square brackets [ ], the parenthesis ( ) and the braces { }, in crystals.

Mineralogists and crystallographers describe the points in unit cell as shown in Figure C-1 with the system of *uvw*. As with point locations, the directions in term of unit cell are given as [*uvw*] in square brackets. In Figure C-2, direction of OA has indices [132]. In terms of vector, OA is equivalent to 1***a*** + 3***b*** + 2***c***. [132] is the direction from origin to the point 132. Please notice that direction [1/3 1 2/3] is the same as [132] when the fractions are cleared and the common denominators are eliminated. Figure C-2c indicates several lines on a two-dimensional lattice. The direction [120] (OC) is the same as [240] (OD). Also, parallel lines have identical indices, so OC, O′C′ and O″C″ have the expression of [120]. Line OF has indices [230]. The bar over a number indicates a negative sign. All lines in Figure C-2c lie in the *a-b* plane, thus [*uv*0] is used. As shown in Figure C-3, if two indices are 0, the direction is parallel to an axis; e.g., [100] is parallel to *a*, [010] parallel to *b* and [001] parallel to *c*. Also, in Figure C-3c, the line [111] can also be designated as [111].

Miller indices (W.H. Miller, 1839 at Cambridge University) are used to indicate the planes in crystals. As shown in Figure C-4, Miller indices show the intercepts with three, two or one axis. In general, Miller indices are represented by (*hkl*) with parenthesis only. Miller indices are also correlable with various crystal forms. In the cubic form of crystal, the six identical faces are indicated by (001), (010), (100), (001), (010) and (100). In order to generalize these faces, a symbol {100} with braces indicating the form contains all faces with the numerals of 1, 0 and 0 in their Miller indices. Figure C-5 illustrates this designation of crystals with different designations. In this manner of designation {*h*00} will always be a cube, {*hhh*} an octahedron, and {*hh*0} a decahedron for all values of *h*.

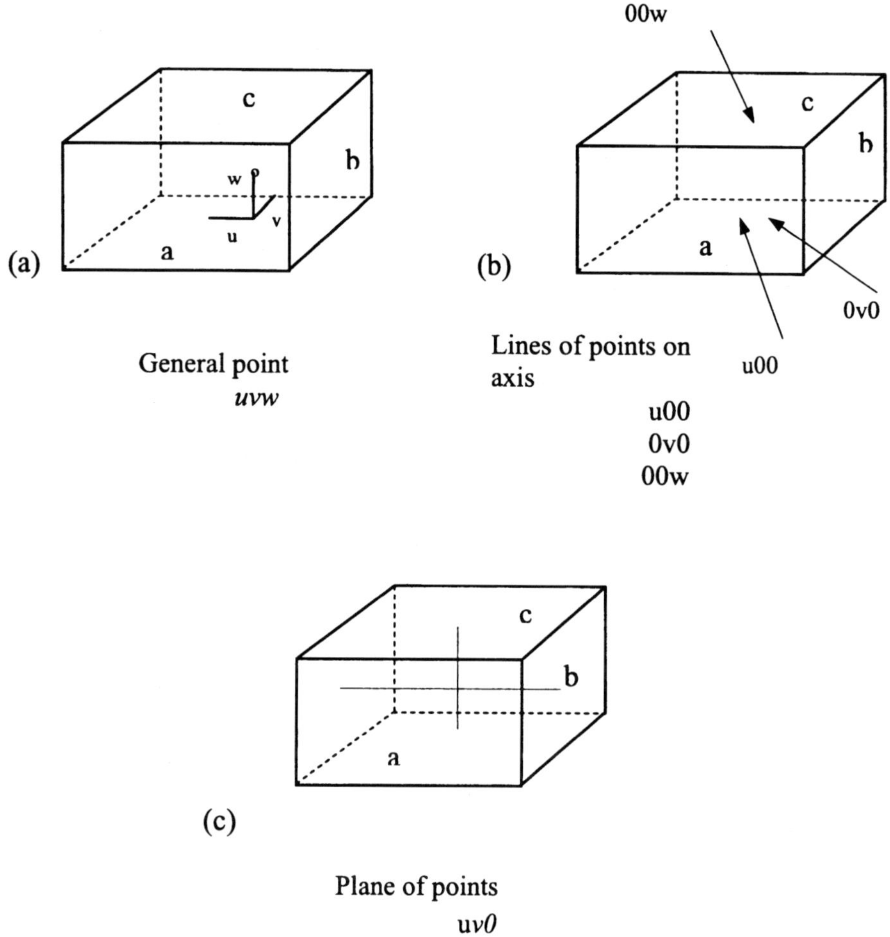

Figure C-1. General points and special points in a unit cell. (a) A general point in a unit cell has nonspecial values for its indices u, v, and w. (b) Special points lying on crystallographic axes must have two indices equal to 0. (c) Special points within a plane including two axes have one index equal to 0

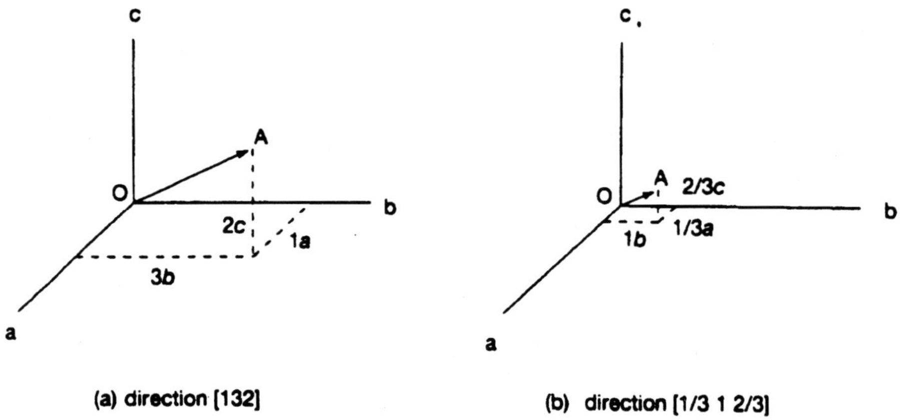

Figure C-2. Vector addition to describe a line. (a) Vector $\overrightarrow{OA}$ is described by giving its components in the a, b, and c directions: 1a, 3b, 2c, given in abbreviated form as 132. (b) Vector 1/3 1 2/3 parallels 132 but is short

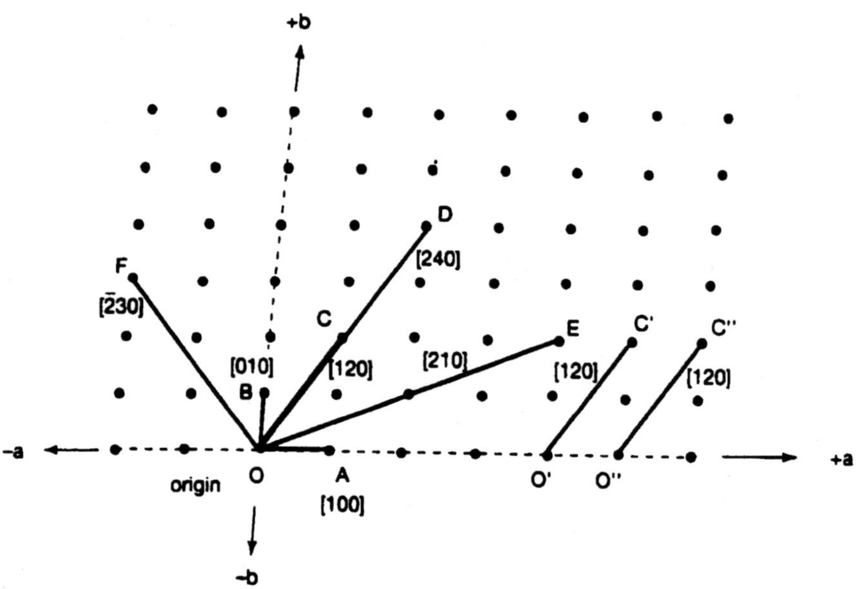

Figure C-2c. Lines and their indices on a two-dimensional lattice. All lines are assumed to lie in the a-b plane.

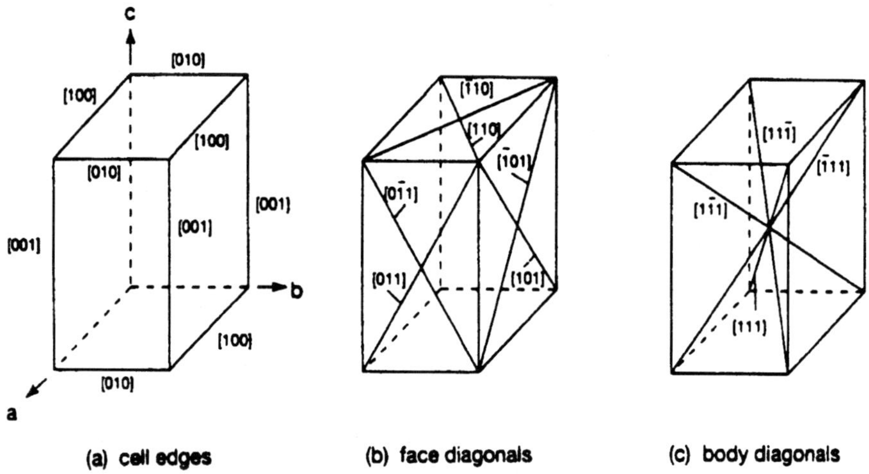

Figure C-3. Special directions and inches in a unit cell: (a) directions parallel to unit cell edges; (b) directions parallel to face diagonals; (c) directions parallel to body diagonals.

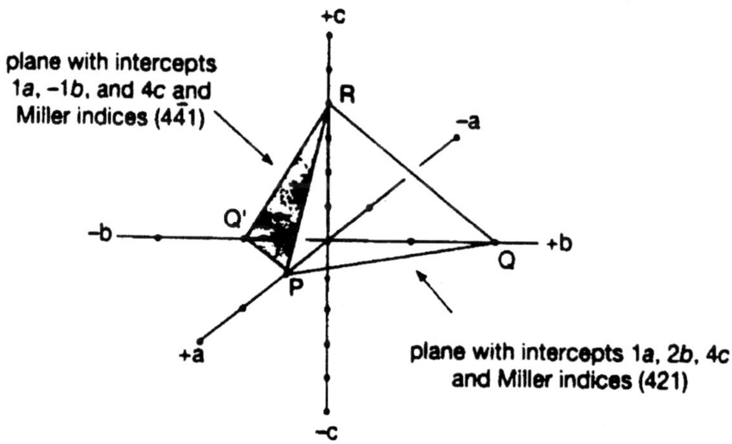

(a)

Figure C-4. (see next page)

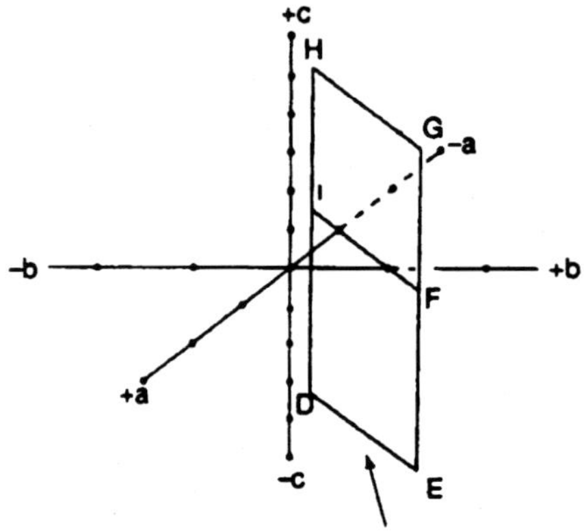

(b) plane parallel to c with intercepts -1a, 1b and •c and Mliller indices ($\bar{1}10$)

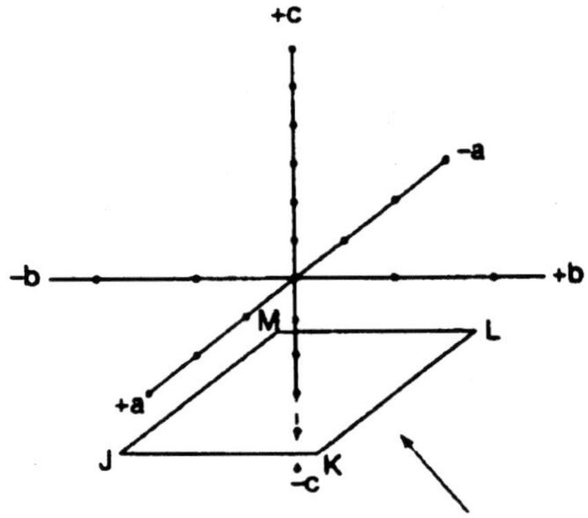

(c) plane parallel to a and b with intercepts •a, •b, and 3c and Mliller indices ($00\bar{1}$)

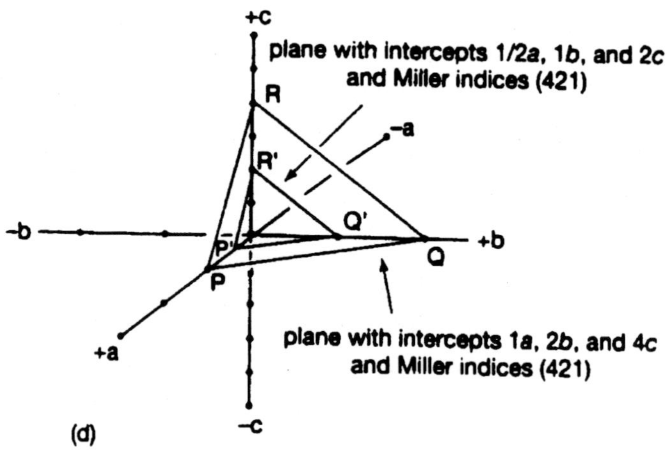

(d)

Figure C-4. Planes intersecting three, two, or one axis: (a) planes intersecting at all three axes; (b) a plane parallel to c axis; (c) a plane parallel to two axes; (d) two parallel planes intersecting all three axes.

Figure C-5a

Cubic forms and Miller indices: (a) cube; (b) tetrahedron; (c) octahedron; (d) dodecahedron; (e) a combination of cube, octahedron, and dodecahedron; (f) a different combination of cube, octahedron, and dodecahedron.

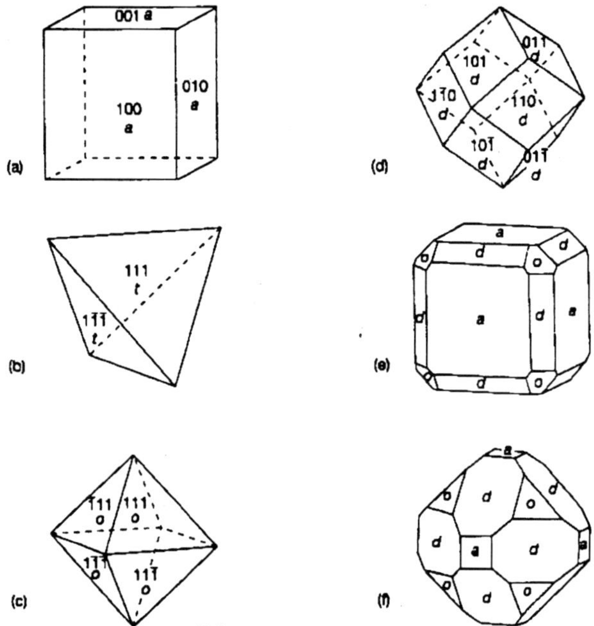

C-5a

Figure C-5b

The general form (hexoctahedron) and some special forms (trapezeohedron, cube) for point group 4/m2/m. Note that all faces of a form have similar Miller indices.

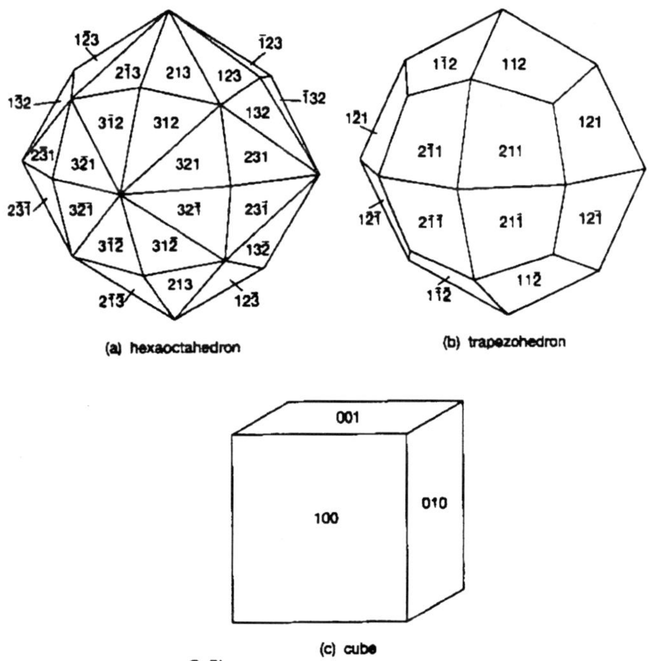

C-5b

APPENDIX E

Similarity and common dimensionless parameters

Some of the more frequently used dimensionless groups including numbers, coefficients and factors are listed below (Please refer to Section 5.3.1):

Numbers and Factors	Symbols	Definition	Summary of Ratio
Reynolds	Re	$\rho uL/\mu$	Inertial to viscous force
Fronde	Fr	$u/(gL)^{1/2}$	Inertial to gravitational forces
Mach	Ma	u/c	Inertial to elastistic forces
Weber	We	$u(\rho L/\sigma)^{1/2}$	Inertial to surface tension
Euler	Eu	$\Delta p/\rho^{1/2}u$	Friction to velocity heads
Nusselt	Nu	hL/κ	Total to conductive heat transfer
Nusselt (mass)	Nu, m	$k_L\delta/D_A$	Interfacial film mass flux to molecular
Grashof	Gr	$g\beta\Delta TL^3\rho^2/\mu^2$	Inertial to viscous times buoyant to viscous forces
Stanton	St; (St = $\mathrm{NuRe^{-1}Pr^{-1}}$)	h/C_pu	Heat transfer to fluid heat capacity
Péclet	Pe; (Pe = Pr · Re)	uL/D_A	Advective to dispersive mass transport (tubular reactor)
Prandtl	Pr	uC_p/κ	Momentum to thermal diffusivities
Schmidt	Sc	$u/\rho D_A$	Dynamic viscosity to molecular diffusivity

Sherwood (also as Taylor number)	Sh	$k_L L/D_A$	Overall mass to molecular diffusivities
Capillary	Ca	$u\mu/\sigma$	Viscous to surface tension force
Biot	Bi	$k_L\delta/D_A$	Interfacial to interior domain mass transfer rates
Damkohler	Da	rL/uC	Chemical reactions to bulk mass flow rates
Damkohler II	Da, ii	RL^2/D_AC	Chemical reaction to molecular diffusion rates
Bodenstein	Bo	$uL/D_{A(x)}$	Advective to axial dispersive transport in tubular reactors
Newton	Ne	$I/\rho u^2L^2$	Frictional impedance (resistance) to inertial force
Drag Coefficient	C_D	$(\rho_p-\rho_u)gL/\rho_p u^2$	Gravitated to inertial force
Colburn Mass Transfer Factor	J_D	$k_L\delta Sc^{2/3}/uL$	Viscous, inertial and diffusion impedance
	$(J_D = Sh\,(Sc)^{-1/3}\,(Re)^{-1})$		
Leverett Function	J_L	$Pc(\phi/\varepsilon)^{0.5}/uL$	Characteristic surface curvature to pore dimension
Archimedes (for particle or droplet)	Ar	$\frac{d_p^3\rho_p}{u}\left(\rho_p-\rho_l\right)$	Gravitational to viscous forces times Re
Bond (for particle or droplet)	Bd	$(\rho_p-\rho_m)L^2g/\sigma$	Gravitational to surface tension force
	$(Bd = We/Fr \text{ as } \rho_m \rightarrow 0)$		
Galileo (for particle or droplet)	Ga	$g\rho d^3/u^2$	Inertial to viscous times gravitational to viscous force

Fluidization	Fl	u_b/u_t	Bulk flow to initial fluidization velocity
Thiele Modulus	Tm	$(\text{Da, ii})^{1/2}$	Surface reaction to intraportile diffusion
Fanning Friction Factor	F_t	$d_p\Delta\rho/2\rho u^2 L$	Shear stress at pipe wall expressed as numbers Velocity heads
Power Factor	F_p	$P_w/\rho\varpi^3 L^5$	Inertial force on agitation (impeller)
Dispersion	d	D_A/uL	Dispersive to advective mass transport in tubular reactor

Symbols used for the above are as follows:

L	characteristic length
u	fluid velocity
μ	viscosity
ρ	density
c	sound velocity
C	concentration
σ	surface tension
$\Delta\rho$	pressure drop
k_L	mass transfer coefficient
ε	porosity
ϕ	permeability
ω	angular velocity
δ	pore depth
κ	thermal conductivity
I	impedance
P_w	power
C_p	specific heat at constant temperature
h	film coefficient
g	gravitational acceleration
β	thermal expansion coefficient
ΔT	temperature difference
r	chemical reaction rate
D_A	diffusion coefficient

INDEX

The French poem by Paul Verlaine depicts life after the loss of a loved one. Although other conditions remain unaffected, the presence of the departed lives on in the mind of the speaker. Similarly, the Chinese poem by Chou Bang Yan, written nearly one thousand years before, shows that the memory of a place endures even after the speaker is forced to move on, physically. In environmental engineering, sometimes what seems like the ideal solution may bring with it disastrous consequences. In order to maintain the best possible environment, we must never lose sight of the ultimate goal, an ideal environmental situation. Among all the chemical processes mentioned in this volume, scientists currently work to find a method that is sustainable, one which will have no negative impact upon the environment.

APRÈS TROIS ANS **Paul Verlaine**	**After Three Years**
Ayant poussé la porte étroite qui chancelle, Je me suis promené dans le petit jardin Qu'éclairait doucement le soleil du matin, Pailletant chaque fleur d'une humide étincelle.	Having pushed open the narrow, wobbly door, I I took a walk in the little garden Softly lit by the morning sun. Which spangled each flower with a humid spark
Rien n'a changé. J'ai tout revu: l'humble tonnelle De vigne folle avec les chaises de rotin ... Le jet d'eau fait toujours son murmure argentin Et le vieux tremble sa plainte sempiternelle.	Nothing had changed. Again: the humble bower Of wild vine, rattan chairs, The silvery murmur of the fountain, And the old aspen's eternal complaint.
Les roses comme avant palpitent; comme avant, Les grands lys orgueilleux se balancent au vent, Chaque alouette qui va et vient m'est connue.	The roses quivered as before; still The great proud lilies swayed in the wind; Each skylark that came and went I knew.
Même j'ai retrouvé debout la Velléda, Dont le plâtre s'écaille au bout de l'avenue, Grêle, parmi l'odeur fade du réséda.	I even found standing the Veleda statue, Its plaster flaking, at the end of the avenue, Blemished amidst the faded odor of the mignonette.

宋.周邦彥,一O九三.

Three Seasons of Plum Blossoms
By Chou Bang Yan – Circa 1093 AD (Song Dynasty)

粉牆低,
梅花照眼,
依然舊風味.

露痕輕綴,
疑淨洗鉛華,
無限佳麗.

去年勝賞
　曾孤倚,
冰盤同宴喜.

更可惜,
雪中高樹,
香篝薰素被.

今年對花
　最悤悤,
相逢似有恨,
依依愁悴.

吟望久,
青苔上,
旋看飛墜.

相將見,
脆丸薦酒,
人正在,
空江煙浪裏!

但夢想,
一枝瀟灑,
　黃昏斜照水。。。

The decorated walls appear lower,
As the brilliant color of your petals penetrate my sight;
You welcomed me as a familiar acquaintance.

Hanging with tiny dew droplets of crystals,
Your face has been cleansed of heavy rouge
To reveal your true beauty and splendid grace.

Last season I was irresistibly drawn
To resting on your bough,
So happy to join you in dining on those icy plates.

The vivid image of a tall solid trunk is etched in my mind,
Rising through thick layers of white snow
As the warm fragrance invaded my blanket

Now I hasten to have a glimpse of your grace,
Our eyes full of regret as we meet,
Each suspended with a sad fate.

During my patient gaze—long silence,
Yet above the moss pathway
Alas! Your leaves have fallen into a pattern of circles.

Soon I will journey south along the lonely river
　under the misty waves;
Still I want to see you then,
To taste your tender fruit with fermented nectar.

Yet the only thing I can dream about
Is a branch of your sparkling petals
Amidst the sunlight of the setting sun
　reflected by the water....